BUILDING CONSTRUCTION MATERIALS AND TECHNIQUES

AF661661

P. PURUSHOTHAMA RAJ

Former Dean
College Development Council
University of Madras
Tamil Nadu

Former Principal
Pondicherry Engineering College
Puducherry

Former Director
Adhiparasakthi Engineering College, Tamil Nadu
V.R.S. College of Engineering and Technology, Tamil Nadu
Sri Aravindar Engineering College, Tamil Nadu

Pearson

Associate Editor—Acquisitions: R. Dheepika
Senior Editor—Production: C. Purushothaman

Copyright © 2016 Pearson India Education Services Pvt. Ltd

This book is sold subject to the condition that it shall not, by way of trade or otherwise, be lent, resold, hired out, or otherwise circulated without the publisher's prior written consent in any form of binding or cover other than that in which it is published and without a similar condition including this condition being imposed on the subsequent purchaser and without limiting the rights under copyright reserved above, no part of this publication may be reproduced, stored in or introduced into a retrieval system, or transmitted in any form or by any means (electronic, mechanical, photocopying, recording or otherwise), without the prior written permission of both the copyright owner and the publisher of this book.

Although the author and publisher have made every effort to ensure that the information in this book was correct at the time of editing and printing, the author and publisher do not assume and hereby disclaim any liability to any party for any loss or damage arising out of the use of this book caused by errors or omissions, whether such errors or omissions result from negligence, accident or any other cause. Further, names, pictures, images, characters, businesses, places, events and incidents are either the products of the author's imagination or used in a fictitious manner. Any resemblance to actual persons, living or dead or actual events is purely coincidental and do not intend to hurt sentiments of any individual, community, sect or religion.

In case of binding mistake, misprints or missing pages etc., the publisher's entire liability and your exclusive remedy is replacement of this book within reasonable time of purchase by similar edition/reprint of the book.

ISBN 978-93-325-4479-6

First Impression, 2016
Seventh Impression, 2022
Eighth Impression, 2023

Published by Pearson India Education Services Pvt. Ltd, CIN: U72200TN2005PTC057128,

Head Office: 1st Floor, Berger Tower, Plot No. C-001A/2, Sector 16B,
Noida – 201 301, Uttar Pradesh, India.
Registered Office: Featherlite, 'The Address' 5th Floor, Survey No 203/10B, 200 Ft
MMRD Road, Zamin Pallavaram, Chennai - 600044
Website: in.pearson.com, Email: companysecretary.india@pearson.com

Compositor: SRS Technologies, Puducherry
Printed in India at Saurabh Printers Pvt. Ltd.

Brief Contents

Contents

Preface

One of the basic needs of human beings is shelter. Housing fulfils this requirement. It is an important indicator of the level of a country's social programme.

Housing involves the construction of buildings with suitable building materials and by following certain construction practices depending on the location. Buildings are constructed not only for housing purposes but also for different purposes such as schools, hospitals, industries, etc.

Thus, the knowledge of building construction materials and building construction techniques are essential for a civil engineer. A couple of courses are available, which are offered in B.E/B.Tech. programmes in universities – Building Materials and Building Construction.

Recently, several universities combined these courses into a single course to provide continuous and comprehensive material. This book is one which deals with materials and construction techniques needed for buildings.

This book comprises of 32 chapters, 13 of which are devoted to Building Materials and 19 to Construction Techniques and Maintenance. All conventional building materials and modern construction materials are explained in detail.

Building construction procedures are covered in detail starting from conventional methods to modern techniques. Basic and special service requirements are dealt in Chapters 27 and 28. Protection of buildings from termites, dampness and fire are explained in Chapter 29. Chapter 30 has been dedicated to preventive measures and maintenance of buildings, while Chapters 31 and 32 deals with construction planning, scheduling and equipment.

Salient points are provided at the end of each chapter to get a better grip on the subject matter. Review questions have also been included for practice.

Acknowledgements

A book of this magnitude would not have been possible if not for the treasure of references in the form of textbooks, manuals, special reports and Indian Standard Codes for which the author expresses his profound gratitude. References have been made to the authors and authorities wherever feasible. Any omissions were inadvertent.

A book of this form would not have been possible without the encouragement of my wife Indrasena.

A special thanks to Sojan Jose, R. Dheepika and C. Purushothaman of Pearson India Education Services Pvt. Ltd, Chennai, for bringing out this book successfully.

Last but not the least, I expresses my profound and sincere thanks to Pearson India Education Services Pvt. Ltd, for the opportunity given to me to write this book.

I welcome comments and suggestions for improving the text and contents of the book from students and teachers.

P. Purushothama Raj

1 PROPERTIES OF CONSTRUCTION MATERIALS

1.1 INTRODUCTION

Many naturally available materials have been in use for construction purposes since pre-historic times. Some of the materials are soil (earth), stone, rock, wood and water. These materials are available abundantly in nature. While some of these materials may be directly used without processing, others may need additional refinement to fit in the exact requirement.

Naturally available materials are used as raw materials to produce construction materials, e.g., clay is used to produce bricks and tiles. These materials may be used in different fields of engineering.

There is a need to know the various properties of these construction materials in order to ensure their proper application.

Properties of construction materials, in particular, building materials are discussed in detail.

1.2 CLASSIFICATION OF ENGINEERING MATERIALS

Engineering materials may be broadly classified into civil engineering materials, electrical engineering materials and mechanical engineering materials. However, some materials may be used in other disciplines also.

Soils, building stones, bricks and clay products, lime, cement, concrete, mortar, timber, metals, alloys, etc., may be mainly grouped under civil engineering materials.

Conductors, semi-conductors, insulators, magnetic materials come under the broad classification of electrical engineering materials.

Cast iron, steel, lubricating materials, etc., are the main mechanical engineering materials.

Metals, alloys, ceramic materials and organic materials are used in one or more of the disciplines.

1.3 QUALITIES OF CONSTRUCTION MATERIALS

Properties of construction materials, which are of use in engineering works, are classified as follows:

1. Physical properties
2. Mechanical properties
3. Electrical properties
4. Magnetic properties
5. Chemical properties

Properties of construction materials which are relevant to building construction are dealt with in the following sections.

1.3.1 Physical Properties

The important physical properties of building materials are as follows:

1. Bulk Density

Bulk density is the mass per unit volume of a material in its natural state. This is obtained by finding the ratio of the mass of the material to the volume of the material. Most of the technical properties such as strength, conductivity, etc., depend on the bulk density of the material. It is represented by the unit kN/m^3. Bulk densities of some of naturally available materials are granite: 25 to 27 kN/m^3, limestone: 18 to 24 kN/m^3, pinewood: 5 to 6 kN/m^3.

Mass per unit volume of a homogeneous material is termed as density. Density of steel is 78.5 kN/m^3. Density index is the ratio of bulk density to its density. Evidently the density index of most of the building materials is less than one.

2. Chemical Resistance

It is the ability of a material to withstand the action of chemicals like acids, alkalis, salt solutions and gaseous substances.

3. Fire Resistance

Fire resistance is the capacity of a material to sustain the action of high temperature without loosing strength and change in the original shape of the structure. Wood, one of the building materials, is highly affected by fire. Such materials should be treated so as to resist high temperature and fire.

4. Frost Resistance

When voids of a building material are filled with water and subjected to sub-zero temperature, it is affected by a process known as freezing. Thus, the ability of water-saturated material to resist freezing and thawing is known as frost resistance. It depends on the density of the material and its degree of water saturation. Generally, dense materials are frost resistant.

5. Spalling Resistance

Materials that could sustain repeated temperature changes is said to be of high spalling resistance. This mainly depends on the coefficient of linear expansion of the material.

6. Weathering Resistance

It is the capacity of a material to resist alternate wet and dry conditions without getting affected. Weathering causes a change in shape and decrease in mechanical strength.

7. Porosity

Porosity of a material is the ratio of volume of voids or pores to the total volume of the material. Denser the material, lesser will be its porosity. Bulk density, water absorption, thermal conductivity, strength, durability, etc., depend on the porosity of the material.

8. Water Absorption

The ability of a material to absorb and retain water is called water absorption. It depends on the porosity of the material, the size and shape of the pores.

9. Water Permeability

It is the capacity of a material to allow water to pass through it under pressure. For example, wood may have high water permeability compared to steel.

10. Hygroscopicity

It is the property of a material to absorb water vapour from the air. It is governed by the porosity of the material, air temperature and the nature of the substance involved.

11. Coefficient of Softening

It is the ratio of compressive strength of a fully saturated material to that of the same material when dry. Materials that have a coefficient of softening more than 0.8 are said to be water-resistant material.

12. Durability

The property of a material to resist the action of atmospheric and other factors continuously is called the durability of the material. For example, the durability of a building depends on the resistance offered by various materials used to construct it under different conditions.

1.3.2 Mechanical Properties

Various mechanical properties of materials are discussed below:

1. Abrasion

Abrasion is the property of the material which resists the force caused by a moving load.

2. Elasticity

Elasticity is the ability of a material, which under stress, recovers its original shape after the removal of the external load. If the material regains its original shape then it is said to be perfectly elastic. Steel, copper, aluminium, etc., may be grouped under perfectly elastic materials within certain limits of deformation. Thus, for each material, there is a critical value of load, generally known as the *elastic limit*, which makes for the partial breakdown of elasticity. Loading the material beyond this point leads to permanent deformation.

3. Plasticity

Plasticity is the property of a material by which a strained material retains the deformed position even after the removal of the external load which caused the deformation. Under large loads of forces, most materials become plastic.

4. Ductility

The ability of a material by which it can be drawn into a wire by external forces is known as ductility. Thus, a ductile material can withstand large deformation before failure. During the process of extension, a ductile material may show a certain degree of elasticity together with a considerable extent of plasticity. Some of the ductile materials include copper, aluminium, gold, etc.

5. Brittleness

Brittleness is the property of a material by which it is not capable of undergoing a significant amount of deformation due to the application of an external load but breaks or ruptures suddenly. This is the most undesirable property of a construction material. Some brittle materials include glass, porcelain, etc.

6. Malleability

Malleability is the property of a material by which it can be uniformly lengthened or widened by hammering or rolling without rupture. A malleable material possesses a high degree of plasticity. This property has a wide use in forging, hot rolling, drop stamping, etc. Some malleable materials are wrought iron, copper, mild steel, etc.

7. Strength

Strength is the property of a material determined by the maximum stress that the material can withstand prior to failure. Strength is defined based on the nature of loading and the nature of stress. There is no unique value which can define strength in all cases.

For example, an adequately designed structural member is not expected to fail under normal operating conditions. This is ensured when the material of the member is strong enough to withstand the force exerted on it.

8. Hardness

Hardness of a material is the ability of the material to resist penetration by a hard material or object. The hardest material is diamond, and the one with least hardness is talcum.

9. Toughness

Toughness is the property of a material which enables the material to absorb energy without fracture. This is a very useful property of a material, which is applicable in cyclic or instantaneous loading.

10. Stiffness

Stiffness is the property that enables a material to withstand high stress without large deformation. Stiffness of a material depends on its elastic property. For example, in a structural element, large deformations are undesirable, and the material should be stiff enough to withstand the load.

11. Fatigue

When a material is subjected to repetitive fluctuating stress, it will fail at a stress well below than that required to cause a fracture under steady loads. This property is known as fatigue.

12. Stability

Stability is the overall property of a member to maintain the overall equilibrium preventing complete collapse. For example, a component member made out of a particular material is just long enough to prevent buckling when subjected to a force acting along its axis.

1.3.3 Electrical Properties

Physical properties and mechanical properties are important for solid materials, which are used as building construction materials. However, knowledge of electrical, magnetic and chemical properties is also needed to deal with construction activities of buildings under different environments. Electrical properties of materials are discussed briefly below.

1. Resistivity

It is the property of the material to resist the flow of electricity through it. It is a useful property of a conductor.

2. Conductivity

Electrical conductivity is that property of a material due to which the electric current flows easily through the material. It is the reciprocal of electrical resistivity.

3. Dielectric Strength

Dielectric strength is the insulating capacity of a material against high voltage. Hence, a material having high dielectric strength can withstand sufficient high voltage before it will break down and conduct.

4. Superconductivity

It is the phenomenon of abrupt drop of resistivity of some metals at a temperature called superconductivity transition temperature.

1.3.4 Magnetic Properties

Magnetic materials are those in which a state of magnetization can be induced. Such materials create a magnetic field in the surrounding space. Some of the important magnetic properties are as follows:

(i) Permeability
(ii) Coercive force
(iii) Magnetic hysteresis

Permeability is the ratio of the flux density in a material to the magnetizing force producing that flux density.

Coercive force is the magnetizing force that is necessary to neutralize the magnetism completely in an electromagnetic field.

Hysteresis is that quality of a magnetic substance due to which energy is dissipated in it on reversal of its magnetism.

1.3.5 Chemical Properties

Engineering materials when they come in contact with other substances are likely to react and suffer from chemical deterioration. Some of the chemical properties of importance are corrosion resistance, chemical composition and acidity or alkalinity.

SALIENT POINTS

1. Engineering materials may be broadly classified as civil engineering materials, electrical engineering materials and mechanical engineering materials.
2. Soils, building stones, bricks and clay products, lime, cement, concrete, mortar, timber, metals, etc., may be mainly grouped under civil engineering materials.
3. Conductors, semi-conductors, insulators and magnetic materials are the broad classification of electrical engineering materials.
4. Cast iron, steel, lubricating materials, etc., are the main mechanical engineering materials.
5. Properties of construction materials are (i) Physical properties, (ii) Mechanical properties, (iii) Electrical properties, (iv) Magnetic properties and (v) Chemical properties.
6. Physical properties are bulk density, chemical resistance, fire resistance, frost resistance, spalling resistance, weathering resistance, porosity, water absorption, water permeability, hygroscopicity, coefficient of softening and durability.
7. Mechanical properties are abrasion, elasticity, plasticity, ductility, brittleness, malleability, strength, hardness, toughness, stiffness, fatigue and stability.
8. Electrical properties are resistivity, conductivity, dielectric strength and super conductivity.
9. Magnetic properties are permeability, coercive force and magnetic hysteresis.
10. Chemical properties are corrosion resistance, chemical composition and acidity or alkalinity.

REVIEW QUESTIONS

1. How are engineering materials classified?
2. List different properties to be considered for construction materials.
3. What are the two important properties to be considered for solid materials? Name the properties in each case.
4. Explain the physical properties of building materials briefly.
5. Briefly explain the resistance needed for building materials against chemicals, fire, frost, spalling and weathering.
6. What is porosity? Name the properties that depend on porosity.
7. Distinguish between elasticity and plasticity.
8. Distinguish between ductility and brittleness.
9. What do you understand by hardness and toughness?
10. Write short notes on the following:
    (i) Malleability
    (ii) Strength
    (iii) Fatigue

2

BUILDING STONES

2.1 INTRODUCTION

The earth's crust is composed of rocks. Rock is a mineral mass of a more or less uniform composition. It may consist of a single mineral (monomineralic) or of several minerals (polymineralic). A mineral is a natural body, homogeneous in chemical composition and physical properties. Further minerals are the products of physical and chemical processes occurring in the earth's crust.

Monomineralic rocks are quartz sand, pure gypsum and magnesite, and the polymineralic ones are granite, basalt and porphyries.

Natural rock materials are widely used for building purposes owing to their universal occurrence and physical and mechanical properties. They are used as mined materials, e.g., gravel, sand, or mechanically processed materials, e.g., crushed stone, sawn, dressed and polished stones, and for various architectural and decorative purposes.

Rocks are the main source of material for the manufacture of mineral binding materials, such as gypsum, lime, cement and of artificial rock materials, e.g., brick, glass, items for concrete and mortar.

2.2 CLASSIFICATION OF ROCKS

The origin and conditions of rock formation predetermine their chemical and mineralogical composition, crystalline structure and texture. These characteristics, underlying the strength and decorative properties of rocks, are the starting point for the engineering and economic evaluation of rocks for various structures of buildings and installations.

The three major classifications are based on the following aspects:

1. Geological or genetic classification
2. Physical classification and
3. Chemical classification

2.2.1 Geological or Genetic Classification

According to the geological classification, rocks are subdivided into three large groups, viz., igneous rocks, sedimentary rocks and metamorphic rocks.

(i) **Igneous rocks:** Cooling and hardening of molten magma result in the formation of igneous rocks. Depending on the rate of cooling the sizes of crystals are formed. Granites, syenites, diorites and gabbros have large crystals, while basalts, rhyolites and andesites have small crystals. Igneous rocks are generally very hard.

(ii) **Sedimentary rocks:** Transportation agents like wind, water and ice may move the loose weathered rock materials and deposit them in the form of layers called sediments. Such sediments when subjected to heavy pressure undergo compaction and cementation, resulting in sedimentary rocks. Limestone, dolomite, shale, sandstone, etc., are sedimentary rocks.

(iii) **Metamorphic rocks:** Sedimentary rocks and to some extent igneous rocks when subjected to changes brought about by the combination of heat, pressure and plastic flow (called metamorphism) undergo changes in the structure, texture and mineral composition, and this results in the formation of metamorphic rocks. Gneiss, schist, slate, marble, etc., are metamorphic rocks.

2.2.2 Physical Classification

The basis for such a classification is the physical properties of rocks, the manner and arrangement of different particles and mass forming a stone. They are classified as follows:

(i) Stratified rocks
(ii) Unstratified rocks
(iii) Foliated rocks

(i) Stratified Rocks

Stratified structure is formed due to the splitting of a series of parallel layers of sedimentary rocks. They exhibit distinct layers which can be separated. The plane of separation is called a cleavage plane. Limestone, slate and sandstone are examples of stratified rocks.

(ii) Unstratified Rocks

They are of granular or crystalline structure and become solid on cooling. They do not show any sign of strata, e.g., igneous rocks like granite, basalt, etc.

(iii) Foliated Rocks

These rocks show definite parallel arrangement of minerals showing a tendency to split in a specific direction, e.g., metamorphic rocks like gneiss, schist, etc.

2.2.3 Chemical Classification

The presence of chemical constituents in the rocks is the basis for their type of classification. They are as follows:

(i) Siliceous rocks
(ii) Argillaceous rocks
(iii) Calcareous rocks

(i) Siliceous Rocks

In these rocks silica is the main constituent. The presence of silica in the free state is called sand, and in the combined state is a silicate. Rocks containing silica in the free state are harder, and those found in the combined form are likely to be disintegrated. Examples of silicate rocks are sandstone, quartzite, etc.

(ii) Argillaceous Rocks

In these rocks, clay or alumina is the main constituent. Examples of argillaceous rocks are slate, laterite, etc.

(iii) Calcareous Rocks

Calcium carbonate or lime is the main constituent in these rocks. They are readily acted upon by dilute hydrochloric acid. Examples are limestone, marble, etc.

2.3 PROPERTIES OF GOOD BUILDING STONES

Different civil engineering constructions use stone. It is necessary to find their suitability under different conditions. The following properties need to be examined before their use:

2.3.1 Appearance and Colour

Stones must look good in appearance and be of uniform colour. Such a quality is essential for stones to be used for decoration work. Light-coloured stones are preferred as they resist weathering action in a better way. Stones with iron oxide should not be used as the presence of iron oxide disfigures the stones and brings about disintegration.

2.3.2 Strength

Stones are used as a compression member and should have sufficient compressive strength. In general, all stones possess a reasonable degree of strength. The crushing strength of stone should be greater than 100 N/mm^2. Igneous rocks have a strength of around 100 N/mm^2 and some of the metamorphic rocks also possess these strengths. But sedimentary rocks have a lower strength.

2.3.3 Weight

In general, good building stones should have a high weight to resist higher compressive forces. Good building stones must have a specific gravity greater than 2.70. A heavy stone possesses more compactness and less porosity.

2.3.4 Hardness and Toughness

Stones must be hard and tough so that they can resist wear and tear. Hardness is assessed by scratching and toughness by hammering. A good building stone should have a wearing resistance less than 3%, and if it is more than 3% it is not satisfactory. Stones used for road work should be hard to resist wear and tear.

2.3.5 Porosity and Absorption

All stones possess porosity. Stones that have a porosity over a certain percentage are unsuitable for building purposes. The presence of a higher percentage of pores may absorb rain, which may deteriorate the stone. A good stone should not absorb water more than 0.6% by weight. It must be capable of withstanding effects of atmosphere. If stones in a cold climate absorb water, they may freeze and even split.

2.3.6 Compactness

A good stone must have a compact, fine, crystalline structure and must be strong. A compact stone is capable of withstanding the effects of external agencies effectively.

2.3.7 Fineness of Grain

Stones that are fine grained are suitable for moulding purposes. Such stones are easily carved and dressed. Non-crystalline structured stones are likely to disintegrate under the action of natural agencies.

2.3.8 Resistance to Fire

Stones must be fire resistant, i.e., they must retain their shape when subjected to fire. Limestone resists fire up to about 800°C. Sandstone can resist fire in a better way. Although argillaceous stones are poor in strength, they are highly fire resistant.

2.3.9 Durability

Stones must be durable. Basically, stones that have their natural bed perpendicular to the direction of pressure are durable. Durable stones are those which are compact, homogeneous, acid resistant and have negligible water absorption.

2.3.10 Dressing

Dressing is the act of shaping a stone for a decorative purpose or for other purposes. For this, the stone should possess uniform texture and softness so that they may be easily dressed. Hard stones can not be dressed.

2.4 TYPES OF BUILDING STONES

Stones of geological origins have been identified and classified as building stones which are used for different purposes. They are discussed subsequently.

2.4.1 Granite

It is a deep-seated igneous rock, hard, durable and available in different colours. It has high crushing strength and is capable of sustaining high weathering.

2.4.2 Basalt and Trap

Basalt and trap are also of igneous origin. These are effective solid rocks which have been formed on the earth's surface in the absence of pressure by rapid cooling of the magma which also carries crystals of various minerals. These are hard, tough and durable and are available in different colours.

2.4.3 Limestone and Chalk

These are sedimentary rocks which have been formed of remnants of seaweeds and living organisms consolidated and cemented together. They are easy to work with and contain a high percentage of calcium carbonate.

2.4.4 Sandstone

This is another form of sedimentary rock formed by the action of mechanical sediments. It shows a sandy structure which is low in strength and easy to work with and dress. This is available in different colours.

2.4.5 Laterite

This is a metamorphic rock and is a sandy clayey stone. It is porous and soft. It can be cut easily into blocks and contains a high percentage of iron oxide.

2.4.6 Gneiss

It is a metamorphic rock which can be easily split into thin slabs and is easy to work on.

2.4.7 Marble

It is a metamorphic rock which can be easily cut with a saw and can be carved.

2.4.8 Slate

It is also a metamorphic rock which is black in colour and can be split easily.

2.4.9 Gravel

It is available in river beds in the form of pebbles and is of different shapes.

2.4.10 Quartzite

It is a metamorphic rock which is hard, brittle and crystalline. It is difficult to work with and is very durable.

2.5 USES OF COMMON BUILDING STONES

Common building stones are used for the following purposes:

1. For foundations, walls, columns, arches, lintels, roofs, floors, etc.
2. For facing work of masonry
3. For concrete and road construction as coarse aggregate
4. For railways as ballast
5. For bridges as floors, piers, abutments, retaining walls, etc.
6. For lighthouses and dams

2.6 USEFUL BUILDING STONES OF INDIA

2.6.1 Granite

Granite is used for stone columns, road metal, ballast for railways, bridge components, etc. It is found in Karnataka, Maharashtra, Rajasthan, Uttar Pradesh, Madhya Pradesh, Punjab, Assam, Tamil Nadu and Kerala.

2.6.2 Sandstone

Sandstone is used for building and ornamental purposes and also as road metal. It is available in Uttar Pradesh, Madhya Pradesh, Rajasthan, Himachal Pradesh and Tamil Nadu.

2.6.3 Limestone

Limestone is used for flooring, roofing, pavements and in general building construction. Also used for manufacturing cement and lime, it is found in Maharashtra, Andhra Pradesh, Punjab, Himachal Pradesh and Tamil Nadu.

2.6.4 Gneiss

Gneiss is used for flooring and pavement and not for major purposes because of its weakness. It is found in Andhra Pradesh, Karnataka, Gujarat and Tamil Nadu.

2.6.5 Marble

Marble is used for ornamental purposes, flooring, facing works, etc. It is found in Rajasthan, Gujarat and Andhra Pradesh.

2.7 QUARRYING OF BUILDING STONES

Quarrying is the process by which stones are obtained from rock beds. Quarry is the place from where the stone is obtained. Quarrying differs from mining in which various operations are carried out for exploring minerals such as coal, quartzite, etc. from a mine under the ground.

2.7.1 Quarry Location

Before starting to quarry, the quarry location has to be decided based on the following factors:

1. On the Earth's surface, a large quantity of stones of good quality should be available.
2. The location should be as close to highways and railways as possible.
3. Adequate space should be available for the installation of machinery, storing of stones and other materials.
4. If the blasting technique is to be used, the location should be away from any permanent structure and inhabitants.
5. Sufficient provision should be made for the drainage of rain water.

2.7.2 Methods of Quarrying

The following methods of quarrying are listed based on the outcrop of the rock and the purpose for which stones are intended to be used.

1. Digging

In soft rocks, stones are obtained merely by digging with the help of hand tools, viz., pick-axes, hammers, chisels, etc. Only those stones that occur in the form of detached nodules buried in the Earth can be easily quarried by this method.

2. Heating

This method is adopted by burning certain kinds of fuel (usually wood is used) and the exposed surface of the portion of the rock to be separated. On burning for several hours continuously, due to unequal expansion of the rock, the rock gets separated from the cold mass with a dull noise. The detached portion is then removed and made to the required sizes by using quarrying tools. This method is employed for obtaining road metals, ballast for railways and aggregates for concrete.

3. Wedging

Soft stratified rocks can be removed by wedging as these rocks are in the form of layers and are easy to split. Soft rocks are removed with the help of pick axes and crow bars. Limestone, marble, slate and laterite are removed by this simple method.

In hard rocks, holes are made and grooves are cut at shorter intervals. Blunt wedges are inserted into the grooves and are forced to get separated. Alternatively, conical pairs of steel are driven into the holes with a hammer and further force to get separated. In the case of very hard stones, the holes are originated by a pneumatic drill.

4. Blasting

This technique is employed for quarrying hard and compact rocks. Blasting is done in stages as discussed below.

Holes may be made with a steel bar with knife-edged ends called jumpers. In case large quantities of stones are required, holes may be drilled by a drilling machine. In order to facilitate the operation, water is used. The slurry produced as a result of drilling is removed by a scraper or a spoon or by a compressed air blast.

The drilled holes are charged with an explosive of suitable capacity. It is necessary for the holes to be made a day before charging with explosive.

Tamping of the explosives is done before firing in order to prevent the blasting within the hole itself. Tamping consists of filling the hole with stiff sandy clay and tamping with a brass rod. A fuse is inserted to fire the explosive. While tamping, a small quantity of gun powder is sprinkled to link the explosive to the bottom of the fuse.

The fuse is fired carefully. During the explosion, masses of stones around the hole are removed. A properly packed explosive shall produce a dull sound followed by displacing of the mass of rock around it.

2.7.3 Precautions in Handling Explosives

The explosives generally used are blasting-powder, dynamite, gun-cotton, blasting-gelatine, cordite and gelignite. From the time of getting an explosive for use, it has to be properly stored, cautiously handled and effectively utilised. The following precautious are to be meticulously adhered to:

1. Storing

Explosives are stored in a specially built buildings called magazines. A magazine should adhere to the following norms:

(i) It should be 100% leak proof and must be provided with an efficient lightening conductor.
(ii) It should be located away from petrol bunks, diesel storages, public places, residential and industrial areas.
(iii) Electrical wiring of magazines should be concealed.
(iv) It should be located at least 500 m away from any working establishment.

(v) Smoking/firing should be strictly prohibited within a radius of 50 m from the magazine.
(vi) The magazine should be properly protected by a high barbed wire fencing.
(vii) Outsiders should not be permitted to enter the magazines without prior permission.

2. Handling

(i) Explosive packages must be handled carefully.
(ii) Detonators should be kept entirely away from other explosives.
(iii) In case of power failure no open fire should be used, but only torches should be used for lightings.
(iv) During the time of charging and tamping, extraordinary precaution should be taken to prevent any fire hazard.
(v) It is necessary to give adequate time for the person to retire to a safe place before the explosions.

2.8 DRESSING OF STONES

Generally the stones obtained after quarrying have a rough surface and are irregular in shape. The process of bringing the stones to a regular shape and required finish is known as dressing. The purposes of dressing are as follows:

1. To set the stones in a regular shape and appearance.
2. To prepare the stones for a suitable site for easy handling and transport.
3. To make hammer-dressed surface, tooled surface, polished surface, rubbed surface or cut-stone surface to suit a particular stone masonry.
4. To secure proper bedding in stone masonry.

Two types of dressing are in use, viz., quarry dressing and site dressing. Quarry dressing is one where the stones are dressed at the quarry and sent to the work site. Site dressing comprises of dressing the stones at the site to the specific size and type. Proper choice depends on the quality of labour at the quarry, distance of the quarry and work site, etc.

2.9 DURABILITY OF STONES

Durability of stones depends on the environment at which they are constructed. For proper durability assurance, it is necessary to know the agencies which deteriorate stones and to take adequate steps to preserve the quality of stone during its lifespan.

2.9.1 Deteriorating Agencies of Stones

Various agencies which are responsible for the deterioration of stones are discussed below.

1. Rain

Rain is one among the main causes for wetness on stones. This wetness is dried by the sun. The process of wetting and drying go a long way in deteriorating the stone. Gases and acids contained in rain water may have damaging effect on stones.

2. Climate

Climate is another factor which causes deterioration. In hot climates there are frequent changes in temperature which result in deterioration. Similarly, stones exposed to cold climate causes freezing of water in the pores resulting in expansion, which causes splitting of the stone. Hence, porous stones should be avoided in cold regions.

3. Wind

Wind laden with grit and dust strikes the stones and causes wear of the stones. Also the impurities in the atmosphere along with moisture penetrate into the pores due to the wind and causes dampness resulting in deterioration.

4. Vegetation Growth

Vegetation growth along the cleavages of joints of stones secretes certain acids by the roots, and thereby cause deterioration. The presence of moisture in roots causes further decay.

5. Chemicals

Chemicals of different kinds move and react due to the usage of different stones such as limestone and sandstone. Such movements of chemicals from one to the other cause deterioration.

6. Living Organisms

Living organisms sometimes cause holes to form in the joints of stones and cause instability to the structure. Further, these holes provide room for the accumulation of moisture and thereby deteriorate the stones.

2.9.2 Preservation of Stones

Preservation of stones is intended to protect the structure against the action of weathering agents and also to protect the good appearance of the face work. The durability of stones increases remarkably if the stones are properly preserved. Preservation is effected in the following ways:

(i) Voids existing in the pores are filled properly.
(ii) Preservation of stones by coating with suitable oils like coal tar, linseed oil, etc.
(iii) Entry of moisture into the surface of the stones is to be avoided.
(iv) Stones with carbonate of lime should be discouraged from use in industrial towns.
(v) Brick masonry with a combination of limestone and sandstone or magnesium limestone and granular limestone should not be laid close to each other.
(vi) Building materials which are inactive with stones should be used.
(vii) Growth of plants on the joints of stones should be prevented.

2.10 SELECTION OF STONES FOR VARIOUS WORKS

Stones have to be selected for a specific work based on the strength and other properties including durability. Stones needed for building exposed to different environments and stones needed for other civil engineering works are discussed below.

2.10.1 General Building Works

Stones are used for foundations, walls and superstructure. Stones with a high mechanical strength and pleasant colours are widely used which are also highly durable. Sandstone is in general used for the above purpose.

2.10.2 Buildings Exposed to High Wind

Such buildings are those constructed near the sea or locations where more wind prevails. Here, apart from the general requirements the stones will be subjected to high weathering action. Hence, stones with a high crushing strength and the presence of silica in addition to hardness should be preferred. Granite and basalt may be used.

2.10.3 Buildings in Industrial Areas

Here the buildings may be subjected to polluted atmosphere which may contain harmful gases, acids and moisture. These elements adversely affect the stones leading to deterioration. Stones that could sustain the effect of acidity and smoke on them are preferred. Granite and compact sandstone may be used.

2.10.4 Buildings Exposed to Heat and Fire

Buildings constructed in regions where hot weather conditions are prevalent throughout the year and chemical factories which may have hot air which are susceptible for frequent fires. For such cases, the stones should have high fire-resisting properties. In general, stones are free from calcium carbonate can resist fire.

2.10.5 Heavy Engineering Works

Heavy engineering works include docks, harbours, bridges, etc. These are constructed in general in an open area and on the seashore. Structures constructed for the above purposes should be able to resist thrust, salt environment, must be strong and highly durable. Granite and quartzite are generally preferred for such purposes, and gneiss is also used for the inside purposes of such structures.

2.10.6 Road Metals and Railway Ballasts

Road metals and railway ballasts are subjected to high compression and abrasion. Further, railway ballasts are exposed to atmosphere throughout their lifespan. Stones selected for such purposes should be hard, tough and highly abrasive resistant. Generally, igneous rocks are preferred for this purpose, viz., granite and basalt.

2.10.7 Facing and Architectural Items

Facing slabs and stones, parts of stairs and landings, parapets and guard rails are made of slabs sawn or split from natural stone and worked on by mechanical means. Sandstone, slate and marble are used for these purposes.

2.11 TESTS ON BUILDING STONES

Building stones must be tested to assess their properties to use for various purposes. Some tests are simple which can be conducted in the field, and some can be tested only in a laboratory. Accordingly, they may be grouped as field tests and laboratory tests (IS: 1121, 1974 and IS: 1124, 1998).

2.11.1 Field Tests

Simple field tests are the absorption test, Smith's test, the toughness test, the hardness test, the acid test and the crystallization test.

1. The Absorption Test

This is usually called the water absorption test. This is a simple test conducted on all stones. It consists of keeping a cubical specimen of the stone of about 50 g weight immersed in water for 24 hours and finding the volume of the absorbed water. This is represented as a percentage of absorption by weight. This percentage should not be more than 0.6%. This test is also called the *electrical resistance test*.

2. Smith's Test

This test is conducted to find the presence of earthly matter or the deterioration of stones when immersed in water. A specimen of the stone is placed in a glass container with water and stirred vigorously. It is placed in water for at least 24 hours. If the water shows turbidity, then the stone is assumed to contain earthly substances. This test is to be conducted for all stones.

3. The Toughness Test

It is a rough test without any rigid specification or procedure. A stone is struck by a hammer to gauge the toughness of the stone. The force required to break the stone reflects the toughness of the stone.

4. The Field Hardness Test

This test is also called Mohs' *scale of hardness test*. This is a simple test in which the hardness of the test is based on the hardness of the surface. The surface of the stone is scratched with the help of a pen knife, and the hardness is gauged based on the relative abrasiveness of minerals (the soft being talc and the hardest being diamond). Such a classification is given by Mohs and the scale is divided into the ten following scales:

1. Talc
2. Gypsum
3. Calcite
4. Fluorspar
5. Apatite
6. Orthoclase feldspar
7. Quartz
8. Topaz
9. Corundum (saphire)
10. Diamond

Thus, if a mark is made by the pen knife on the surface of the stone, the hardness may be equated to that of a calcite and hence the hardness, $H = 3$. A siliceous rock surface can not be scratched which may be equated to that of quartz and in this instance, $H = 7$.

5. The Acid Test

This test is performed to determine the presence of alkaline or lime content. This test consists of placing a cube of 50 to 100 gm weight in 1% hydrochloric acid for 1 week. If the corners of the cube show roundish and loose particles deposited on the surface, the stone has some alkaline content. If the stones exhibit florescence when subjected to the action of acids, then it reflects the presence of lime content. Stones that do not react to sulphuric acid are highly fire resistant.

6. The Crystallization Test

Because of crystallization on reaction with some chemicals, the stones may show a loss of weight and some defects. This test consists of immersing a cubical specimen of 40 mm side in a sodium sulphate solution for 2 hours and then drying it in an oven at 100°C. This procedure is repeated for five times. The loss of weight and the presence of cracking on the surface are noted. The presence of visible defects and loss in weight should be minimal indicating high durability and good resistance to weathering.

2.11.2 Laboratory Tests

Laboratory tests on stones comprise of the attrition test, the crushing test, the freezing and thawing test, the hardness test, the impact test and the microscopic test.

1. The Attrition Test

This is also called the abrasion test. This test is performed to determine the percentage of the wear of stones that are used for road construction. This test is carried out in Deval's Attrition Testing Machine.

In this test, a certain quantity of stone is placed on Deval's Attrition cylinder. The cylinder is rotated about its horizontal axis at the rate of 30 rpm for 5 hours. After the end of rotation, the contents of the cylinder are sieved, and the quantity of material retained on the stone is weighed. The percentage weight is the ratio of loss in weight to the initial weight which is expressed as a percentage.

2. The Crushing Test

This test is performed on the stones which are to be subjected to compressive stress. In this test a cubical specimen of 40 mm side is tested in a compression testing machine. The maximum load at which the stone specimen fails by crushing is noted. Then the crushing strength of the stone is the ratio of maximum load at failure and area of bearing face.

3. The Impact Test

The toughness of a stone is determined by the impact test using an impact testing machine. Here, a stone specimen of 25 mm diameter and 25 mm height is placed on the machine. A steel hammer of 2 kg weight is allowed to fall down axially on the cylinder from 1 cm height for the first blow, 2 cm height for the second blow, 3 cm height for the third blow and so on. The blow at which the specimen breaks is noted, which is the Toughness Index of the stone.

4. The Laboratory Hardness Test

This test is different from the field hardness test and the attrition test. This test is carried out in Dorry's testing machine. A cylinder of 25 mm of the stone is rubbed against a steel disc spindled with coarse sand. The specimen is subjected to a pressure of 1.25 kg/mm^2. After 1000 revolutions in the machine, the loss in weight is found, and the coefficient of hardness is determined as

$$\text{Coefficient of hardness} = \frac{20 - (\text{Loss of weight in gm})}{3}$$

5. The Microscopic Test

This is a geologist's test. In this test, a thin section of a stone is taken and placed on a microscope to study the following properties:

1. Mineral constituents
2. Texture of stone
3. Average grain size
4. Nature of cementing material
5. Presence of pores, fissures and veins

6. The Freezing and Thawing Test

This test is conducted for the use of stones in regions where the temperature goes below the freezing point. The test consists of keeping a specimen of stone in water for 24 hours and then placing it in a freezing mixture at –12°C for 24 hours. The specimen is taken out and thawed (warmed) to atmospheric temperature. The procedure is repeated at least seven times after which the specimen is examined for any damage.

2.12 ARTIFICIAL STONES

Artificial stones are known as cast stones or reconstructed stones. In many localities natural stones are not available at a reasonable cost. In such cases, these stones are useful. However, because of their high cost, they can not be treated as a substitute for natural stones.

2.12.1 Types of Artificial Stones

Different types of materials are combined, and a variety of stones are manufactured. Some of the types of artificial stones are discussed below.

1. Cement Concrete Blocks

These are made using cement, fine aggregate, coarse aggregate and water similar to that of concrete. They may be cast-in-situ or cast-in-moulds in the required shapes. Sometimes they are precast with steel and are therefore called reinforced cement concrete blocks.

2. Artificial Marble

It is made using sand with Portland cement or one of the gypsum or magnesite cements. It is either precast or cast-in-situ and is used for external work.

Another form of artificial marble is made by fusing and moulding constituent mixture, which is composed of 60 parts of plaster in powder, 20 parts of pulverised marble, 15 parts of sulphate of potash with 5% solution of glue mixed with water. Such marble stones are mostly used in France.

3. Terrazo

It is prepared using mixtures of marble chips with white cement and some pigment. It is also either precast or cast-in-site. It is generally used in bathrooms, residential buildings, etc.

4. Mosaic Tiles

These are precast tiles where the upper surface is made of marble chips. They are available in different sizes and colours.

5. Reconstructed Stone

This is made from the debris of limestone quarries by crushing them into grit. This is mixed with lime made from dolomite and heated in a closed retort up to 980°C to drive out CO_2. This is slaked, mixed with water and consolidated into blocks under pressure. It is then dried, and CO_2 is admitted until the carbonization of the hydrate of lime blocks is complete.

6. Bituminous Stone

Diorite and other granite stones are impregnated with prepared or refined tar to form bituminous stones. These are preferred where noise-, wear- and dust-resistant stone surfaces are needed.

2.12.2 Advantages of Artificial Stones

The advantages of artificial stones are as follows:

1. In areas where natural stones are costly and are unavailable, artificial stones are used.
2. Any desired size and shape can be attained easily.
3. Can be made in-site, and thereby the cost of dressing the stone and transport can be saved.
4. They are practically defect free when compared with natural stones.
5. Cavities in artificial stones can be made for wiring or plumbing work easily.
6. Good in resisting deterioration and disintegration caused by various atmospheric agencies.
7. The strength of artificial stones can be assured as it is design based.
8. It is comparatively more durable than natural stones.

SALIENT POINTS

1. Rock is a mineral mass of a more or less uniform composition. It may consist of a single mineral (monomineralic) or of several minerals (polymineralic).
2. Monomineralic rocks are in quartz sand, pure gypsum and magnesite.
3. Polymineralic rocks are granite, basalt and porphyries.
4. According to geological classification, rocks are subdivided into three large groups, viz., igneous rocks, sedimentary rocks and metamorphic rocks.
5. The cooling and hardening of molten magma resulted in the formation of igneous rocks. Depending on the rate of cooling, various sizes of crystals are formed.
6. Transportation agencies like wind, water and ice may move the loose weathered rock materials and deposit them in the form of a layer called sediments. Such sediments when subjected to heavy pressure undergo compaction and concentration resulting in sedimentary rocks.
7. Sedimentary rocks and to some extent igneous rocks when subjected to changes brought about by the combination of heat, pressure and plastic flow (called metamorphism) undergo changes in structure, texture and mineral composition, and this results in the formation of metamorphic rocks.
8. Physical classification is based on the physical properties of rocks, the manner and arrangement of different particles and mass forming a stone.
9. The presence of chemical constituents in the rocks are the basis for chemical classification.
10. Granite is a deep-seated igneous rock, hard, durable and available in different colours.

11. Basalt and trap are of igneous origin, which are effective solid rocks.
12. Limestone and chalk are sedimentary rocks which have been formed by remnants of seaweeds and living organisms consolidated and cemented together.
13. Sandstone is another form of sedimentary rock formed by the action of mechanical sediments.
14. Laterite is a metaphoric rock and is a sandy clayey stone.
15. Gneiss is a metamorphic rock which can be easily split into thin slabs.
16. Marble is a metamorphic rock which can be easily cut with a saw and then be carved.
17. Slate is a metamorphic rock which is black in colour and can be split easily.
18. Gravel is available on river beds in the form of pebbles of different shapes.
19. Quartzite is a metamorphic rock which is hard, brittle and crystalline.
20. Quarrying is the process by which stones are obtained from rock beds.
21. Methods of quarrying are: digging, heating, wedging and blasting.
22. Explosives generally used for blasting are blasting-powder, dynamite, gun-cotton, blasting-gelatine, cordite and gelignite.
23. Dressing is the process of bringing the stones to a regular shape with the required finish.
24. Quarry dressing is one where the stones are dressed to the required size and type at the quarry and sent to the work site.
25. Site dressing comprises of dressing the stones at the site to the specific size and type.
26. Deteriorating agencies of stones are rain, climate, wind, vegetation growth, chemicals and living organisms.
27. Preservation of stones is intended to protect the structure against the action of weathering agents and also to protect the good appearance of the face work.
28. Field tests conducted on stone are: the absorption test, Smith's test, the toughness test, the hardness test, the acid test and the crystallization test.
29. Laboratory tests on stones comprise the attrition test, the crushing test, the freezing and thawing test, the hardness test, the impact test and the microscopic test.

REVIEW QUESTIONS

1. Name the geological, physical and chemical classifications of rocks.
2. Distinguish between sedimentary and metamorphic rocks.
3. Explain the physical classification of rocks.
4. How are igneous rocks formed?
5. Give examples of igneous, sedimentary and metamorphic rocks.
6. Enumerate the properties of building stones.
7. Explain the properties of good building stones.

8. Briefly describe some important varieties of building stones.
9. Classify and give the properties of the following stones:
   (i) Basalt
   (ii) Sandstone
   (iii) Laterite
10. What are the requirements of good building stones?
11. Briefly describe the building stones of India, and name the places where they are available.
12. Discuss the methods of quarrying.
13. What are the precautions to be adopted in handling explosives?
14. What are the purposes of dressing?
15. Discuss the factors that affect the durability of stones.
16. What is the necessity of preserving stones? How is it done effectively?
17. Explain the factors to be considered for the selection of stones for various civil engineering works.
18. What are the simple field tests you would suggest to determine the suitability of stones for various purposes?
19. Explain the various laboratory tests that determine the crushing strength and freezing and thawing qualities of stones.
20. Discuss the factors which are responsible for the deterioration of stones.
21. What are artificial stones? Give their advantages.

3 BRICKS

3.1 INTRODUCTION

Bricks have been in use since the dawn of civilization. In the initial stages they were used as sun-dried bricks. Burnt bricks have succeeded sun-dried bricks. It has been reported that bricks have been used to build monuments in different parts of the world. Refined brick making and burning techniques have remarkably improved the quality of buildings and other works.

Now, brick plays a prime role in construction at all levels starting from thatched roofs to multi-storeyed buildings. But, the process of brick making has not changed over many years in India except for minor refinements. It is the oldest construction material which has been extensively used at present because of its durability, easy availability and low cost.

3.2 CONSTITUENTS OF BRICK EARTH

The main constituents of good brick earth are alumina, silica, lime, oxide of iron and magnesia.

3.2.1 Functions of Constituents of Brick Earth

1. Alumina

This is the main constituent of every kind of brick earth. This imparts plasticity to the earth so that it can be moulded. If alumina is present in excess, it leads to shrinking and warping during drying and burning. This can be prevented by the addition of sand.

2. Silica

It exists in brick earth as free sand or in a combined form as silicate of alumina. The presence of silica prevents cracking, shrinking and warping of raw bricks. Thus, it imparts uniform shape to bricks. The addition of silica also increases hardness, durability and resistance to heat. Excess of silica removes the cohesion between particles and makes the brick brittle.

3. Lime

It enables the silica to melt during burning and bind the particles together. Lime should be in the form of powder, otherwise lumps of lime are converted into quick lime after burning.

Quicklime slakes and expands in the presence of moisture and results in the splitting of bricks into pieces. Excess of lime causes the brick to melt too readily, and the shape will be lost.

4. Iron Oxide

It acts as a flux and helps the grains of sand to melt and bind the particles of clay together. It is responsible for imparting a red colour to the brick. It improves the durability of the brick. Excess of iron oxide makes the bricks dark blue or blackish, and less quantity of iron oxide makes the brick yellowish in colour.

5. Magnesia

A small quantity of magnesia in brick earth imparts a yellow tint and decreases shrinkage. Excess of magnesia causes decay of bricks.

3.2.2 Desirable Chemical Composition of Brick Earth

The desirable chemical composition of brick earth for good quality bricks are given below (IS: 1077, 1976):

1. Alumina or clay – 20 to 30% by weight
2. Silica or sand – 35 to 50% by weight
3. Silt – 20 to 25% by weight
4. Other required ingredients are
   (i) Iron oxide
   (ii) Magnesia
   (iii) Lime
   (iv) Sodium potash, etc.
   } 1 to 2% by weight
5. In the case of alluvial soils, the total lime and magnesia should not be more than 1%; in other soils it should be less than 15%.
6. Water-soluble materials in total should be less than 1%.

3.2.3 Harmful Ingredients in Brick Earth

Naturally available soil may contain ingredients other than those required for making quality bricks, which are discussed below.

1. Alkalis

Alkalis in the form of soda and potash lower the fusion point of clay, and cause bricks to fuse, twist and warp during burning. Alkalis present in bricks absorb moisture from the atmosphere and create dampness. Such dampness dries and leaves behind grey or white deposits on the surface of the wall.

2. Pebbles

Although the presence of pebbles causes harm chemically, they are not desirable as they do not permit the clay to be thoroughly mixed uniformly resulting in weak and porous bricks. Regular breaking of bricks during masonry work is not possible because of the presence of pebbles.

3. Iron Pyrites

The presence of iron pyrites in brick earth causes the earth to decompose and oxidize resulting in the splitting of bricks.

4. Vegetation and Organic Matter

Vegetation and organic matter when present in brick earth assists in burning. Incomplete burning of these materials causes the bricks to become porous.

5. Limestone

Limestone and kankar nodules present in brick earth are harmful as that of lumps of lime. On heating limestone is converted into lime, which comes into contact with water, swells and finally results in splitting and crumbling.

3.2.4 Field Testing of Brick Earth

Brick earth in the field has to be tested for consistency, moulding properties and shrinkage.

In order to test for the first property, a handful of soil sample is taken and formed into a ball. The ball is dried in the sun and then examined. Crumbling of the dried ball shows the presence of excess sand. If the ball is hard and shows cracks on the surface, it indicates the presence of less sand. Thus, the soil has to be modified by mixing different proportions of sand or clay such that the above-mentioned defects do not occur.

The soil found or made suitable in the first test is taken and ground well. Then a required quantity of water is added and mixed. The mixed soil is kneaded well to be rolled to form threads of about 3 mm diameter without crumbling. Such a mixture of soil and water is used to make a brick using a mould. The moulded brick should be with clear edges and corners which ensures perfect requirement of water. Otherwise water is added or removed to get a satisfactory brick.

Sample of bricks moulded as above are burnt in a clamp until bricks become red hot. These bricks are examined for shrinkage and deformation after cooling. Bricks which have shrunk evenly and do not show any defects are separated out and others are rejected. The percentage of sorted out bricks should be high, otherwise the soil is modified to get a better percentage of good bricks.

3.3 SELECTION OF SITE FOR BRICK EARTH

The following factors are to be considered in the selection of a site for brick earth:

(i) The site to be selected should be capable of providing an adequate quantity of soil during the entire planned production period.
(ii) Additional materials, such as sand or silt, required to make good bricks should also be available near the site.
(iii) Water and fuel should also be available near the site; otherwise extra cost may be incurred for transport.
(iv) It should be connected by road and rail for easy conveyance of the produced bricks.
(v) The ground of the site must be situated as a plain ground.
(vi) The location of the water table should be at a deeper depth.

3.4 MANUFACTURING OF BRICKS

The manufacturing of bricks involves four stages, viz., preparation of soil, moulding, drying and burning.

3.4.1 Preparation of Soil

1. Removal of Top Soil

The preparation of soil involves first renewal of loose materials at the top of the ground for a depth of about 200 mm. As it contains a lot of impurities, this material is not used for the preparation of bricks.

2. Digging and Spreading

The earth is dug out 200 mm from the ground. The soil is spread on the level ground, and heaps of clay are about 600 to 1200 mm.

3. Cleaning

The soil after being spread on the ground should be cleaned of stones, pebbles, vegetable matter, etc. If non-clay materials are in excess, the clay should be washed and screened. Such a process will be expensive and cumbersome. All the lumps of soil should be broken into a powder form.

4. Weathering

The soil is then exposed to the atmosphere for softening or mellowing. The period of exposure may last for a few weeks depending on the nature of soil. This imparts plasticity and strength to the soil.

5. Blending

To increase the quality of soil, additional soil such as sandy or calcareous clays may be added in suitable proportions with a small quantity of coal, ash, etc. The whole mass is mixed uniformly and water is added.

6. Tempering

It is the process of kneading the soil under the feet of men or cattle after adding the necessary quantity of water in order to make the soil stiff and homogeneous. In general, a soft plastic clay could be prepared by using about 25 to 30% of water. This procedure is adopted for the majority of common handmade bricks. For making superior bricks on a large scale, the earth is tempered in a pug mill.

A pug mill comprises of a truncated conical tub of 120 cm diameter at the top and 75 cm diameter at the bottom with a height of about 180 cm. A long vertical shaft is placed at the centre (Fig. 3.1). The central shaft is provided with a number of horizontal arms attached with cutting knives for breaking clay lumps, if any. A long horizontal arm is fitted at the top of the vertical shaft. The shaft is rotated with the help of bullocks or sometimes by electric power. Clay and water in the ratio of 1:3/4 are fed into the vessel from the top. The tempered clay is collected from the hole provided at the bottom of the vessel. This type of pug mill, as shown in Fig. 3.1, can be used to get sufficient soil for a daily output of about 20,000 bricks.

3.4.2 Moulding of Bricks

Bricks are made in traditional size (in inch) and in metric size (cm) as prescribed by the Bureau of Indian Standards. Metric size bricks are called modular bricks. Nominal size of bricks is the size including thickness of the mortar. Hence, the actual size of modular bricks is 19 cm × 9 cm × 9 cm and the nominal size of modular bricks is 20 cm × 10 cm × 10 cm.

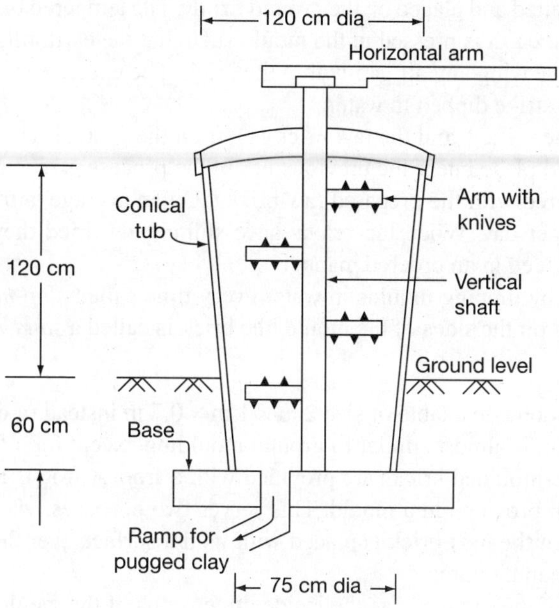

Figure 3.1 Pug mill

A brick mould is a rectangular box of steel or wood. Both ends, the top and bottom, of the box are open. The inside dimensions of the mould are 20 cm × 10 cm × 10 cm (Fig. 3.2).

Moulding of bricks is carried out either by hand or by machine.

1. Hand Moulding

Two types of hand moulding, viz., ground moulding and table moulding, are adopted in India. In hand moulding, bricks are moulded manually. This is preferred in areas where the manpower is cheap and available readily and where only a small quantity of bricks is needed.

(i) Ground Moulding

The process of moulding bricks on the ground by manual labour is called ground moulding. At first a level ground is prepared, and a thin layer of fine sand is spread over the ground.

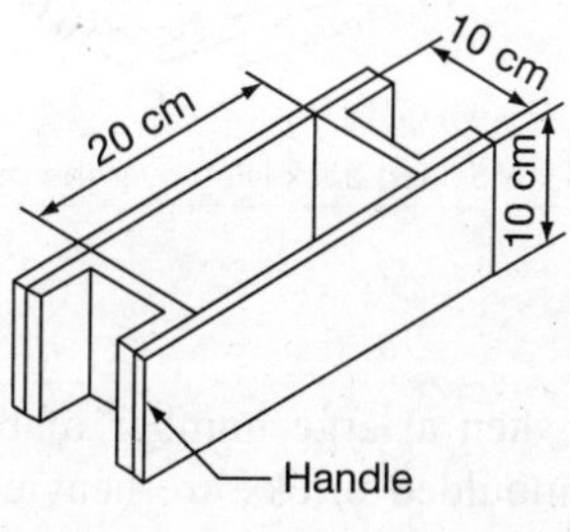

Figure 3.2 Typical steel brick mould

The mould is wetted and placed on the ground firmly. The tempered brick earth is dashed into the mould. The earth is pressed in the mould such that the earth fills all the corners of the mould without leaving any air gap within the brick. The excess earth is removed using a wooden or metal strike dipped in water.

The mould is then lifted, and the raw brick is left on the ground. The mould is cleaned, dipped in water and placed near the previous brick; the process is repeated till the ground is completely covered with the prepared raw bricks. On an average, a moulder can mould about 750 bricks per day. When the bricks have sufficiently dried they are taken to the drying shed and placed in an ordered manner.

Brick prepared by dipping moulds in water every time called *slop-moulded brick*, and if sand is sprinkled on the sides of the mould, the brick is called a *sand-moulded brick*.

(ii) Table Moulding

Table moulding is done on a table of size 2 m × 1 m × 0.7 m instead of on the ground. The process of moulding is almost similar to ground moulding except for a few changes.

Invariably, table-moulded bricks are provided with a frog. A frog is a mark of depth of about 10 to 20 mm provided in a mould. This serves two purposes, viz., it provides a key for the mortar when the next brick is placed with its flat surface over this and to place the trademark of the manufacturer.

A stock board of the same site as the inside dimensions of the mould with a projection for the frog with the trademark of the manufacturer is kept on the moulding table, and the moulder stands behind the table.

The mould is placed to fit the stock board and the tampered earth is dashed against the mould, carefully filled, and excess earth is removed. Then a thin board called the pallet board is placed on the mould; the mould and pallet board are lifted together followed by lifting the mould leaving the brick on the pallet board. Another pallet board is kept on the brick and carried to the drying yard where it is placed on its edge and the pallet boards are removed. This procedure is repeated. Figure 3.3 shows a moulded brick lying over the pallet board.

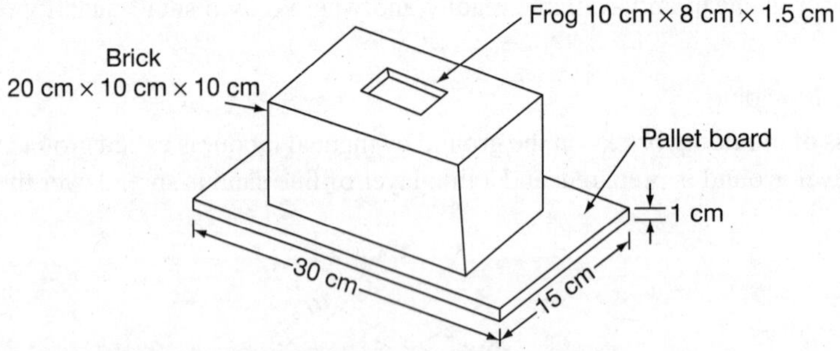

Figure 3.3 Moulded brick lying over the pallet board

2. Machine Moulding

Moulding machines are used when a large number of bricks are to be manufactured within a short time. Machine-moulded bricks are heavier and stronger than the hand-moulded ones. These bricks have a sharp regular shape and size, a smoother surface and sharp edges.

There are two types of machine moulding, viz., plastic method and dry method.

(i) Plastic Method

In the plastic method, pugged earth is used. The machine adopted for the plastic method contains a rectangular opening of size equal to the length and width of the brick. The pugged earth is placed in the machine and a beam of the moulded earth comes out. This is cut into strips of width equal to the depth of the brick by wires fixed in the frames. These bricks are also known as *wire-cut bricks*.

(ii) Dry Method

In the Dry Method, the machine first converts the hard earth into a powder form. A small quantity of water is added to the powder to form a stiff plastic paste. The plastic paste is placed in the mould and pressed by the machine to form hard and correct-shaped bricks. These bricks are called pressed bricks. Such bricks do not require drying and can be sent directly to the burning section.

3.4.3 Drying of Bricks

Moulded bricks can not be burnt directly as they may get cracked or distorted. Hence, before burning they are dried. Natural drying or artificial drying may be resorted to. The bricks are left to dry for about two weeks.

1. Natural Drying

It is also called *hack drying*. It comprises placing moulded bricks in rows on their edges on a slightly raised ground called a hack. A small space is given between bricks for the circulation of air. Direct exposure to sun is avoided by providing a cover and it is also protected from the rain. The air- and sun-dried bricks are strong enough and can be used for the construction of small structures.

2. Artificial Drying

When bricks are needed continuously and to a large scale, artificial drying is resorted to. The bricks are dried in special dryers which receive heat from special furnaces that are made especially for this purpose. Hot flue gases from the chambers of a kiln and waste steam from engines may also be used for the artificial drying of bricks.

3.4.4 Burning of Bricks

Bricks are burnt in kilns after moulding and drying so as to impart hardness and strength and to increase the density of the brick so that it absorbs less quantity of water.

Physical and chemical changes take place in burning of bricks. Heating brick earth up to about 640°C produces only physical changes. At this temperature moisture and water of crystallization are driven out, and the organic matter is burnt. Such a brick that can absorb moisture from the air can get back hydrated. Such bricks are said to be poorly burnt and disintegrate when subjected to moisture contact.

If brick earth is heated up to 700–1,000°C, it undergoes chemical changes. During chemical action alumina and silica in brick earth fuse together resulting in a compound which is strong and stable. After this chemical transformation, it does not turn back to break earth on cooling. Further, it does not crumble when immersed in water. Thus, the break earth burnt above 700°C is different from the original brick earth. Hence, to get a good quality brick it has to be heated to the required temperature.

On heating the brick earth beyond 1,300°C, the above materials get vitrified. The bricks begin to lose their shape.

1. Clamp or Open Kiln

Clamp or open kiln is a temporary structure where dried bricks are stacked in alternate layers of fuel, i.e., both bricks and fuels are placed in alternate layers. Locally available materials such as grass, rice, husk, wooden chippings, cheap quality woods and dried cow dung are used as fuel. About 20,000 to 1,00,000 bricks are available after burning and cooling. It takes around 3–6 months to complete the burning and cooling of the bricks (Fig. 3.4) in the clamp.

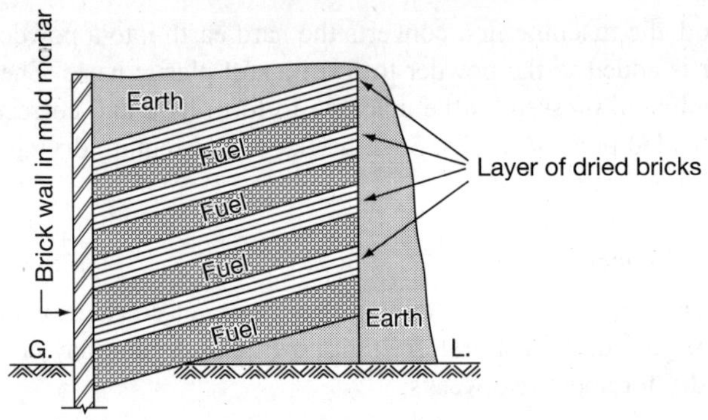

Figure 3.4 Typical arrangement of fuel and bricks in clamp or open kilns

The main advantages of this method are that its initial cost is low, fuel cost is low and there is no need of a permanent structure and skilled labourers. The regulation and circulation of heat are not possible, and hence only 60% of good-quality bricks can be expected. Further, only a small quantity of bricks can be manufactured at a time.

2. Intermittent Kiln

When a large quantity of good-quality bricks are needed, intermittent or continuous kilns have to be used. In intermittent kilns, the process of burning is discontinuous.

Figure 3.5 shows the plan of an intermittent kiln. The operations are in the order of loading the bricks, firing, cooling and unloading, which are performed one after the other. It is rectangular in shape with thick outside walls and is constructed over the ground.

Trenches are dug across the floor of the kiln. Small openings are made in between the trenches. Sun-dried bricks are laid above the trenches with the bricks being laid on edges with gaps in between. This gap enables the hot flue gases to flow to each brick.

Flue gases are let in through the longitudinal walls through the small openings. Dampers are provided in the flue openings to regulate the air supply. Circulation of flues is kept up for 3–4 days. The bricks are cooled for a week's time.

3. Continuous Kilns

In continuous kilns, the process of burning is continuous. There are three types of continuous kilns, viz.,

(i) Bull's Trench Kiln
(ii) Tunnel Kiln
(iii) Hoffman's Kiln

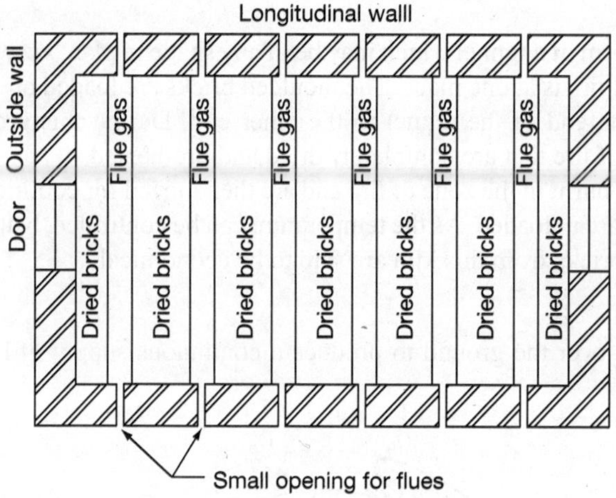

Figure 3.5 Plan of an intermittent kiln

(i) Bull's Trench Kiln

This type of kiln is usually oval in plan and is constructed in a trench excavated in the ground. It may be fully underground or partly projecting above ground. The depth of trench is about 2 metres. The outer and inner walls are constructed of bricks, and the flue holes are provided in the outer walls (Fig. 3.6). Dampers are provided to conveniently divide the kiln into sections.

Bricks are arranged in sections in such a way that flues are formed. Fuel is placed in the flues, and the top surface is covered with earth and ashes to prevent the escape of heat. The fuel is burnt through the flue holes. Additional flue holes are provided at the top to insert fuel when burning is in progress. Two movable chimneys are used to form a draught. These chimneys are placed before the section is fired. This arrangement makes the hot gases leaving the chimney warm up the bricks in the next section. When the burning is over in a section the flue holes are closed and the bricks are cool down gradually. The fire is advanced to the next section and the chimneys are moved forward.

As loading, burning, cooling and unloading are carried out simultaneously, a continuous supply of bricks is available. This is the mostly used kiln in India.

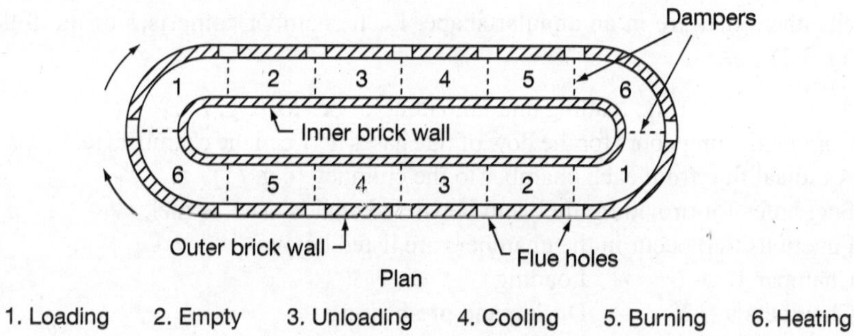

Figure 3.6 Bull's trench kiln

(ii) Tunnel Kiln

This kiln is in the form of a tunnel which may be of any shape in plan, viz., straight, circular or oval. The zone of fire is at one place. The moulded bricks are loaded on a trolleys, which are moved from one end of the tunnel to the other end. During this process when they approach the zone of fire that are completely dried and pre-heated.

These bricks are burnt in the zone of fire and are then moved for cooling. After adequate cooling, the bricks are unloaded. As the temperature can be controlled, better quality bricks are produced. The bricks from this kiln are said to be economical.

(iii) Hoffman's Kiln

This is constructed over the ground to produce a continuous supply of bricks on a large scale.

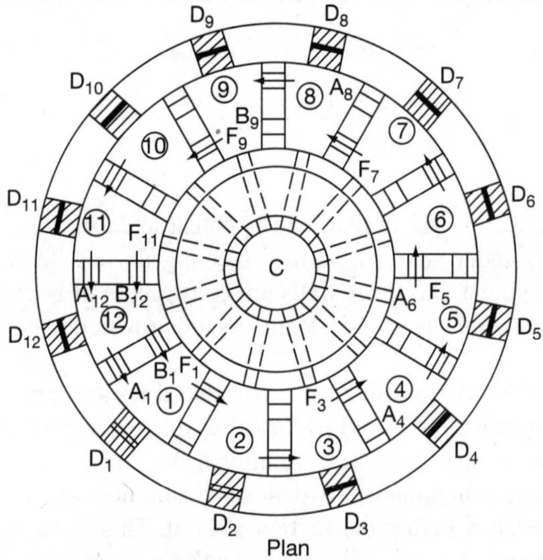

D_1 to D_{12} – Main doors; D_1 and D_2 – Opened doors, D_3 to D_{12} – Closed doors
1 to 12 – Chambers; C – Chimney, F_1 to F_{12} – Radial flues
A_1 to A_{12} and B_1 to B_{12} – Communicating doors

Figure 3.7 Hoffman's continuous kiln

This is circular in plan and is provided with a chimney at the centre. Around the chimney are 12 chambers that are in an annular shape. Each chamber comprises of the following parts (Fig. 3.7):

(i) A main door for the loading and unloading of bricks (e.g. D_1).
(ii) Communicating doors for the flow of flue gases between the chambers (e.g. A_1 and B_1).
(iii) A radical flue from each chamber to the chimney (e.g. F_1).
(iv) Fuel holes for providing fuel, and powdered coal is used as fuel.

Functions that occur in the chambers are listed below:

Chamber 1 – Loading
Chambers 9–12 – Drying and pre-heating
Chambers 7 and 8 – Burning
Chambers 3– 6 – Cooling
Chamber 2 – Unloading

The working of a kiln is as follows:

(i) Cool air enters through Chambers 1 and 2 as they are open doors.
(ii) It crosses the cooling Chambers 3–6 and enters the burning Sections 7 and 8 in a heated condition.
(iii) It moves to Chambers 9–12 to dry and pre-heat the raw bricks.
(iv) It escapes into the atmosphere through the damper of Chamber 12 and the chimney.

The flow of air and fuel gas are shown by arrows in each chamber. Although the initial cost is more, Hoffman's Kiln claims several advantages:

(i) High-quality bricks with uniform burning are obtained with regulation of heat.
(ii) Supply of bricks are continuous in all seasons because the top of the kiln is closed, and the working is not stopped.
(iii) Considerable saving in fuel due to pre-heating of raw bricks by fire gases.
(iv) No air pollution in the locality, as the exhaust gases do not contain black smoke or dust particles.

3.5 QUALITIES OF GOOD BRICKS

The qualities of good bricks are as follows:

1. Bricks should have perfect edges, must be adequately burnt, should be uniform red or copper in colour and should be free from cracks.
2. Bricks should have rectangular plane surfaces with parallel sides and sharp right-angled edges. The size of a standard brick is 190 mm × 90 mm × 90 mm.
3. It should be hard enough such that no impression is left when scratched with one's finger nails.
4. Bricks when struck with each other should produce a ringing sound.
5. Bricks should not break when dropped flat from a height of 1 m.
6. Bricks should be homogeneous and compact throughout, and the brick should not have any voids or grit.
7. Bricks should have a percentage of absorption of water by weight less than 20%.
8. Bricks should not show deposits of salts when immersed in water and dried.
9. Bricks should have less thermal conductivity and must be sound proof.
10. Brick should have a minimum crushing strength of 3.5 N/mm^2.

3.6 CLASSIFICATION OF BUILDING BRICKS AND USES

Bricks are broadly classified into two broad categories as follows:

1. Sun-dried bricks and
2. Burnt bricks

Sun-dried bricks also called un-burnt or *katcha bricks*, and these are dried directly under the sun after the process of moulding. These bricks are of inferior quality and are used for the construction of temporary and cheap structures. Such bricks should not be used in areas exposed to heavy rains.

Burnt bricks are of superior quality, which are generally used for civil engineering constructions. These are discussed in depth below.

3.6.1 Quality Classification of Burnt Bricks

The classification of bricks under the following four categories on the basis of constituents, preparation and burning is called quality classification:

1. First-class Bricks

First-class bricks are table-moulded bricks and are burnt in kilns. These bricks should not have any defects like cracks, stones or lumps of clay. They should be of standard size and have uniform colour, sharp edges, even surfaces, correctly burnt and hard. These bricks are used for superior quality works and works of permanent nature.

2. Second-class Bricks

Second-class bricks are ground moulded and are burnt in kilns. In general, they have to satisfy the requirement – quality of first-class bricks. However, they may have a slightly irregular shape, rough uneven surfaces or may have slight cracks. These bricks are hard and are correctly burnt and used where the brick masonry is to be plastered.

3. Third-class Bricks

Third-class bricks are ground moulded and burnt in clamps. These bricks have irregular edges with less sharpness, uneven surfaces and are not hard enough. They give a dull sound when struck against each other. They are used for unimportant and temporary constructions.

4. Fourth-class Bricks

Fourth-class bricks are either over burnt or under burnt with irregular shapes, edges and surfaces. These are used as aggregates for concrete in road, floor and foundation construction.

3.6.2 Indian Standard Classification of Burnt Bricks

As per the Indian Standard (IS: 3102 – 1971), bricks are classified according to their strength as given in Table 3.1.

Table 3.1 Classification of bricks

Class designation	Compressive strength requirement (not less than)	Additional requirements
10	10 N/mm^2	Dimensional tolerance ± 3%, surface must be smooth, corners should be sharp, should produce a ringing sound when struck against each other
7.5	7.5 N/mm^2	Dimensional tolerance ± 8%, permitted to have slight distortion but it should not cause difficulty while laying
5.0	5.0 N/mm^2	Dimensional tolerance ± 8%, permitted to have slight distortion but it should not cause difficulty while laying
3.5	3.5 N/mm^2	Dimensional tolerance ± 8%, permitted to have slight distortion but it should not cause difficulty while laying

Source: IS: 3102–1971.

3.7 TESTS ON BRICKS

Before recommending bricks for construction work, their suitability is to be assessed by conducting the following tests:

1. Compressive Strength Test
2. Water Absorption Test
3. Efflorescence Test
4. Dimensional Tolerance Test
5. Hardness Test
6. Soundness Test
7. Structure Test

The sampling and testing of bricks are carried out as per IS: 3495 – 1992 and are as shown in Table 3.2.

Table 3.2 Sampling and testing of bricks

Class designation	Sampling size	Lot size	Tests to be conducted
10	20 bricks	50,000 or more	1. Compressive Strength 2. Water Absorption 3. Efflorescence 4. Dimensional Tolerance 5. Hardness 6. Soundness 7. Structure
7.5–3.5	20 bricks	10,000 or more	Tests under the direction of the engineer-in-charge

Source: IS: 3495–1992.

3.7.1 Compressive Strength Test

As per norms, five bricks are taken at random and their dimensions are measured accurately to 1 mm. They are immersed in water at 25–29°C for a period of 24 hours. After that they are taken out, and excess moisture is allowed to drain. If the bricks have frogs they are filled with C.M 1:3. They are again kept under a jute bag for another 24 hours. They are again immersed in clean water for three days.

Just at the time of testing they are taken out, one at a time, from the water and wiped dry. The horizontal and mortar-filled surface is placed facing upwards with three thin plywood sheets on a brick-testing machine.

Load is applied on the brick at a rate of 140 kg/cm^2 per minute till the failure of the brick. An average of five test values of bricks is reported. While computing the average value, any single value obtained as compressive strength which is higher than the upper value of the class of the bricks tested should be taken only as the upper limit of the class. Test values less than 20% of the average should be rejected. Also the average value should not be less than the specified value of the class of the brick.

3.7.2 Water Absorption Test

For the test, five bricks are taken at random from the lot. They are dried in an oven at 110–115°C till they attain constant weight. Generally it takes 48 hours. Bricks are then cooled at room temperature, which generally takes 4–6 hours and are then weighed. Let the weight of a dry brick be W_1.

Bricks are then kept in clean water at 27 ± 2°C for 24 hours. They are taken out, wiped dry with a damp cloth, and the wet weight W_2 is noted.

$$\text{Percentage water absorption} = \frac{W_2 - W_1}{W_1} \times 100$$

The average percentage of water absorbed as a percentage of dry weight is reported. This value should not be more than the standard value of a particular class of brick.

3.7.3 Efflorescence Test

The presence of soluble salts causes efflorescence on the surface of the brick. Here also five test samples of bricks are taken at random. The brick is placed in a dish with 2.5 cm immersed in distilled water. The brick is allowed to absorb the water fully, and then the water evaporates through it. When the brick has dried, an additional and equal quantity of water is placed in the dish. As before the water is allowed to evaporate. After the second evaporation, the brick is examined and evaporated as under:

Observation	Effect of efflorescence
(i) No perceptible deposit of salt	Nil
(ii) Not more than 10% of the area of the brick is covered with salt	Slight
(iii) Salt deposit covering up to 50% of the area but not powder or flakes	Moderate
(iv) Salt deposit covering more than 50% with powdering or flaking of surface	Heavy
(v) Heavy deposit of salt extremely high accompanied by powdering and/or flaking of the surface and likely to increase more when subjected to further wetting	Serious

For general construction, bricks should not have more than slight to moderate efflorescence.

3.7.4 Dimensional Tolerance Test

Twenty bricks are taken at random and their dimensions, length, width and depth are measured. Variations in dimensions generally allowed up to ± 3% for class one and ± 8% for other classes.

3.7.5 Hardness Test

The hardness of a brick surface is determined by making a scratch on the surface of a brick with the help of a finger nail. If no impression is left on the surface of the brick, the brick is considered to be hard.

3.7.6 Soundness Test

Two bricks are struck against each other slightly. A good brick has a clear ringing sound, and the bricks should not get break.

3.7.7 Structure Test

In this test a brick is broken, and the broken surfaces are examined for the structure. It should be uniformly burnt, homogeneous in structure, compact and free from any defects such as holes, lumps, etc.

3.8 TYPES OF BRICKS

There are different kinds of bricks which differ from the conventional building bricks with respect to their shape, specifications and special purpose for which they are made. Those bricks are discussed in the subsequent sections.

3.8.1 Specially Shaped Bricks

1. Bull-nosed Bricks

A brick moulded with a rounded angle is termed as the bull nose. It is used for rounded quoin. A quoin is a connection which is formed when a wall takes a smooth circular turn. The centre of the curved position is located on the long centre line of the brick (Fig. 3.8(a)).

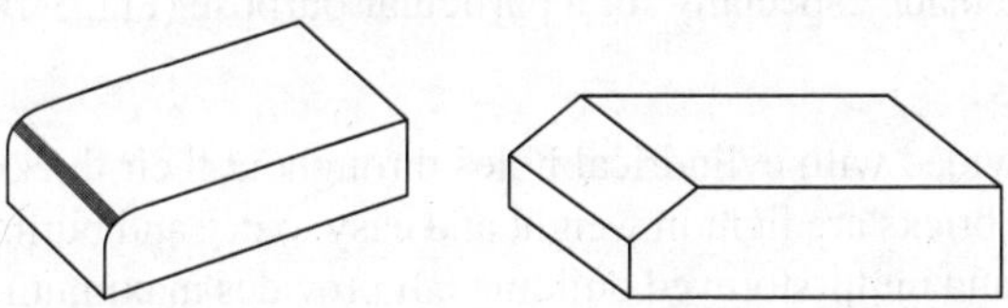

(a) Bull-nosed brick (b) Cant or plinth brick

Figure 3.8 Typical shapes of bull-nosed and cast bricks

2. Cant or Plinth Bricks

This has a slant-cum-straight edge at one end which is used in a plinth or in a door and window joints (Fig. 3.8(b)).

3. Circular Bricks

These bricks are provided with internal and external faces to be curved to meet the requirement of the particular curve and radius of the wall. These bricks are used for structures like towers, wells, etc. These bricks have to be specially made with a particular curvature (Fig. 3.9(a)).

4. Squint Bricks

These bricks have a special edge. These are used in the construction of active and obtuse squint quoins. Actual requirements should be provided for the manufacture of such bricks (Fig. 3.9(b)).

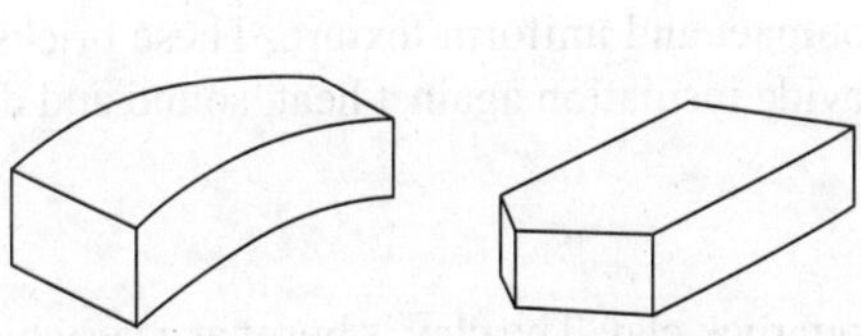

(a) Circular brick (b) Squint brick

Figure 3.9 Typical shapes of circular and squint bricks

5. Coping Bricks

These bricks are made to suit the thickness of walls on which coping is to be provided. Different forms such as chamfered, half-round or saddle-back can be made (Fig. 3.10(a)).

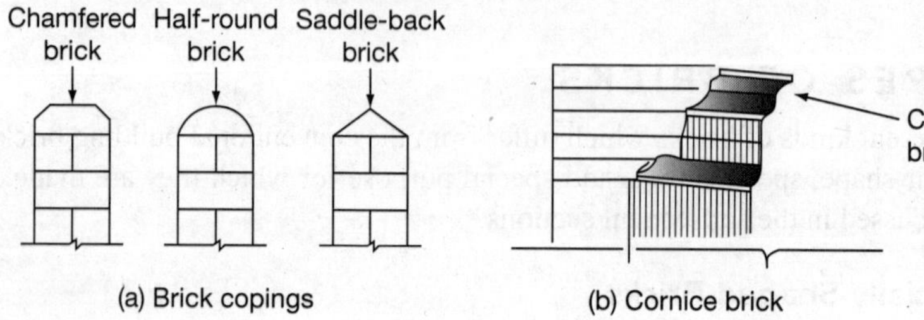

Figure 3.10 Typical shapes of brick copings and a cornice brick

6. Cornice Bricks

In the construction of a cornice, different shapes are made to give a beautiful appearance. Such bricks are to be made especially for a particular purpose (Fig. 3.10(b)).

7. Perforated Bricks

These bricks are provided with cylindrical holes throughout their thickness (Fig. 3.11(a)). Because of holes the bricks are light in weight and easy to dry and burn. These are used for panel walls in small and multi-storeyed buildings. It provides maximum amount of ventilation, and the perforations are placed such that they do not permit the entry of rats or mices.

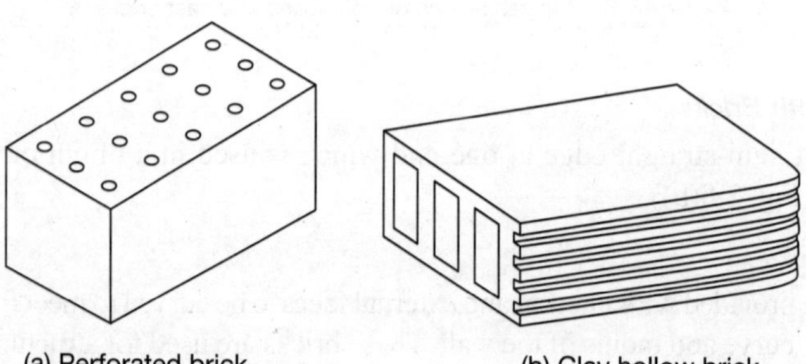

Figure 3.11 Typical shapes of perforated and clay hollow bricks

8. Hollow Clay Bricks

These bricks are made out of specially made homogeneous clay. They are uniform in colour and have a fine, compact and uniform texture. These bricks are of light weight. As bricks are hollow they provide insulation against heat, sound and dampness to the building (Fig. 3.11(b)).

9. Paving Bricks

Paving bricks are made from rock clay. The clay is burnt at a very high temperature than that of ordinary bricks. Paving bricks are used in roads and to resist the abrasive action of traffic.

The paving bricks may be plain or chequered. These bricks are non-slipping and are hence preferred for street pavements, garden walks, etc. (Fig. 3.12).

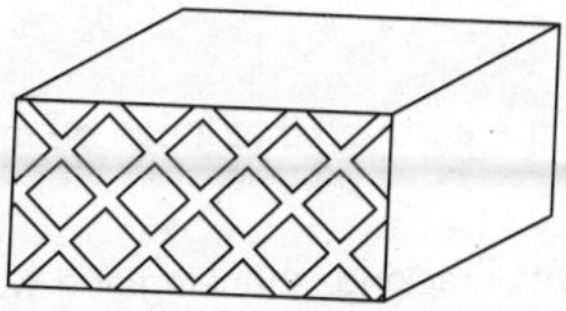

Figure 3.12 Typical chequered paving brick

3.8.2 Refractory Bricks

Refractory bricks, also called Refractory Fire Bricks, are prepared from fire clay in the same manner as ordinary bricks. Refractory bricks contain about 30% alumina and 70% silica. After drying, they are burnt in kilns at high temperature ranging from 1400–1900°C. These bricks are yellowish white in colour.

As the fire bricks can resist high temperatures without softening or melting, they are used for the linings of interior surfaces of furnaces, chimneys, kilns, ovens, fireplaces, etc. The compressive strength of these bricks varies from 200–220 N/mm^2, and the percentage absorption varies from 5–10.

3.8.3 Pressed Bricks

These bricks are made by pressing the clay to a high pressure of about 40 kg/sq.cm. They are directly burnt without drying. A special type of oil is used in the process which gives a glazed surface. The glazed surface may get peeled off when exposed to weather.

These bricks are of a regular shape, are compact and have high strength. These are used for decorative purposes of very high quality.

3.8.4 Sand–Lime Bricks

Autoclaved calcium silicate bricks are popularly known as sand–lime bricks. These bricks are made from a mixture of 95% of sand and 5% of lime (CaO) by weight. These bricks are hard, strong and are uniform in colour and texture. It presents a smooth and soft surface and hence may not be suitable for plastering. They are used for ornamental purposes.

3.8.5 Heavy-duty Bricks

These bricks are of very high quality, viz., high compressive strength, low water absorption, high durability and high bulk density. These bricks are free from any defects. They are used in heavy engineering works such as bridge structures, multi-storeyed buildings and industrial foundations.

3.8.6 Sewer Bricks

These bricks are manufactured using surface clay, fire clay or shale or a combination of these materials. They are used for the lining of walls, roofs and floors of sewers for ordinary domestic sewage. They are not suitable for industrial use as they are not acid resistant.

SALIENT POINTS

1. The main constituents of good brick earth are alumina, silica, lime, oxide of iron and magnesia.
2. Alumina imparts plasticity to the earth so that it can be moulded. Excess alumina leads to shrinking and warping.
3. The presence of silica prevents cracking, shrinking and warping of raw bricks. It imparts a uniform shape to the bricks.
4. Lime enables the silica to melt during burning and binds the particles together. Excess lime causes the brick to melt too readily, and the shape is thus lost.
5. Iron oxide acts as a flux and helps the grains of sand to melt and bind the particles of clay together.
6. A small quantity of magnesia in brick earth imparts a yellow tint and decreases shrinkage. Excess of magnesia causes decay of bricks.
7. The required chemical combination of earth is:

   Alumina or clay – 20–30% by weight

   Silica or sand – 35–50% by weight

   Silt – 20–25% by weight

   Others – 1–2% by weight
8. Harmful ingredients in brick earth are alkalis, pebbles, iron pyrites, vegetation and organic matter and limestone.
9. Brick earth in the field has to be tested for consistency, moulding properties and shrinkage.
10. The preparation of soil comprises of the removal of top soil, digging and spreading, cleaning, weathering, blending and tempering.
11. Moulding is the process by which wet bricks are made using a mould of size 200 mm × 10 mm.
12. Moulding of bricks is carried out either by hand or by a machine. Hand moulding consists of ground moulding and table moulding.
13. There are two types of machine moulding, viz., the plastic method and the dry method.
14. Bricks are dried by natural drying or artificial drying.
15. Bricks are burnt in kilns after moulding and drying so as to impart hardness and strength and to increase the density of the brick so that it absorbs less quality of water.
16. Kilns are of three types, viz., open kilns or clamps, intermittent kilns and continuous bricks. There are three types of continuous kilns, viz., Bull Trench Kiln, Tunnel Kiln and Hoffman's Kiln.
17. Bricks are broadly classified into two categories, viz., sun-dried bricks and burnt bricks.

18. The following tests are conducted on bricks to assess their suitability:
    a. Compressive Strength Test
    b. Water Absorption Test
    c. Efflorescence Test
    d. Dimensional Tolerance Test
    e. Soundness Test
    f. Hardness Test
    g. Structure Test
19. Specially shaped bricks are:
    (i) Bull-nosed bricks
    (ii) Cast or plinth bricks
    (iii) Circular bricks
    (iv) Squint bricks
    (v) Coping bricks
    (vi) Cornice bricks
    (vii) Perforated bricks
    (viii) Hollow clay blocks
    (ix) Paving bricks
20. Refractory bricks are prepared from fire clay in the same manner as ordinary bricks. Refractory bricks contain about 30% alumina and 70% silica and are burnt at a high temperature of 1400–1900°C.

REVIEW QUESTIONS

1. What are the constituents of good brick earth?
2. Explain the chemical composition of brick earth.
3. Explain briefly the harmful ingredients in brick earth.
4. How do you conduct field tests to assess the qualities of brick earth?
5. What factors are to be considered in the selection of a site for brick earth?
6. Discuss the operation of preparation of soil for the manufacture of bricks.
7. What is moulding of bricks? Explain.
8. Explain the advantages of machine moulding over hand moulding.
9. Why is drying of bricks needed?
10. Compare the merits and demerits of burning bricks in clamp and kilns.
11. Describe the process of burning bricks in intermittent kilns.

12. Explain the principles of continuous kilns.
13. Briefly explain the qualities of good bricks.
14. How do you classify bricks? Explain.
15. Distinguish between sun-dried bricks and burnt bricks.
16. What is the basis on which bricks are classified under the Indian Standard?
17. What are the field tests to determine the suitability of bricks for construction?
18. Enumerate specially shaped bricks and their uses.
19. What are refractory bricks? Where are they used?
20. Explain the advantages of sand–lime bricks.

4 TILES AND CERAMIC MATERIALS

4.1 INTRODUCTION

Naturally available clay is used directly or in the processed form to make tiles and ceramic materials apart from making bricks.

Tiles are thin slabs made out of brick earth by moulding and then burning them. The best clay for tile making is generally found a few metres below the brick earth. Tiles have been used for roofs, floors, walls, etc.

Ceramics is the term used for the act of making objects with a special type of clay. This term is derived from the Greek word *keramos* meaning potter's earth. Clay for ceramics is the product of chemical weathering of various types of rocks. Orthoclase feldspar is mainly responsible for the production of clays in nature. This mineral forms kaolinite during decomposition. Kaolin, a product having the composition of pure kaolinite, is fine white clay required to manufacture some ceramic articles.

Apart from bricks and tiles, other objects which are made using various types of clays and techniques are discussed below.

Earthenware is the term used for clay products burnt at low temperature. They have coarse surfaces and are opaque. General cooking vessels usually used in villages fall under this group.

Stoneware is a product of refractory clay, ground stone and pottery. These are burnt at high temperatures to vitrification. Stoneware jars are of great use. Earthenware and stoneware are together known as pottery.

Porcelain is a white translucent ceramic material of kaolin and feldspar. The objects are produced by firing at a very high temperature. Most crockery are made of porcelain and are glazed.

Terracotta is a hard brownish unglazed vitrified ceramic material used for ornamental purposes.

4.2 BUILDING TILES

As discussed earlier, building tiles are thinner than bricks and are susceptible to cracking and warping. Hence, special care has to be taken in the manufacture of tiles.

4.2.1 Characteristics of a Good Tile

A good-quality building tile should possess the following characteristics:

1. It should be uniformly burnt up to the requirement.
2. It should possess a uniform colour.
3. No cracks, flaws or bends must appear on the tiles.
4. The stipulated shape and size should be maintained.
5. It should produce a clear ringing sound when struck with another tile or by a light hammer.
6. It should have accurate measurement so as to fit well when placed in position.
7. When a tile is broken, the broken surface should exhibit even and compact surface.
8. It should be hard and durable.

4.2.2 Types of Tiles

Tiles may be grouped under two broad categories:

(i) Common clay tiles
(ii) Ceramic tiles

1. Common Clay Tiles

Common clay tiles are further classified as clay roof tiles, clay floor tiles, clay terracing tiles and clay ceiling tiles.

(i) Clay Roof Tiles

Clay roof tiles are made of plastic clay, which is different from brick clay. There are three types of roof tiles that are used for covering sloped roofs, viz., Mangalore tiles, country tiles or pot tiles and Allahabad tiles.

Mangalore tiles are available in various shapes and dimensions (Fig. 4.1(a)). They overlap the tile beneath them and also lock with the adjacent tiles. These tiles are entirely popular in South India, particularly in the states of Kerala, Tamil Nadu and Pondicherry.

The full specification is given in IS: 654–1972. The gap provided at the corners should not be more than 6 mm. Two batten lugs are to be provided for the tile. Further, there are two eave lugs. In order to prevent from lifting off by high winds, at least one hole is provided at the crossribs near the eave end for securing the tile to the reaper or batten.

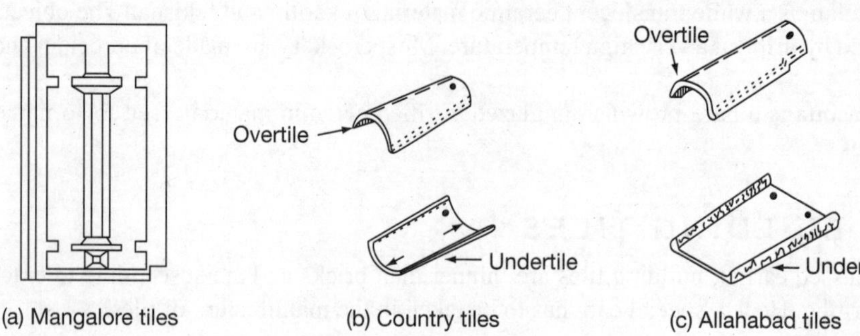

Figure 4.1 Common clay roof tiles

Country tiles are half round which are laid in pairs as undertiles and overtiles (Fig. 4.1(b)). This is achieved by placing the tiles on the sloped roof with the concave side up and with the longer end towards the ridge. Then another row of the same tiles with the convex side up and the small end towards the ridge is laid covering the adjoining edges of every pair of tiles below. These tiles are also sometimes used as ornamentation by placing the tiles over *AC* or *GI* sheets. These tiles are extensively used in villages.

Allahabad tiles consist of two shapes of tiles, one for bottom use and the other for top use (Fig. 4.1(c)). The overtiles are half round and are tapered, while the undertiles all are flat, tapered with upturned flanges at the sides. Specification are available in *CPWD* specification.

(ii) Clay Floor Tiles

These tiles are made from fine special blue clay which enables the tiles to be made thin and the surface smooth. In order to make the tiles hard and impervious, a mixture of ground glass and potteryware are added. In general, these materials are mixed together thoroughly and are extruded in the shape of a slab. These slabs are placed in moulds and are pressed to the required shape. They are burnt at about 1,300ºC of temperature. In order to give a good appearance they are salt glazed. These tiles are more suitable for walking on them barefooted.

(iii) Clay Terrace Tiles

These tiles are also flat tiles made in the same way as floor tiles. They may be hand made or machine made and burnt in a kiln. Large-size tiles are used nowadays. Detailed specifications are available in IS: 2690–1964.

2. Ceramic Tiles

(i) Glazed Ceramic Tiles

They are made from special ceramic clays in two operations. In the first operation the tiles are made and fired at 1,200–1,300ºC; then they are coated with glaze, decorations, etc. Again they are fired in ovens to get opaque glazing. Wear of the glazing is based on the temperature of firing and the thickness of glazing.

These tiles are used for walls or as floor tiles, where there is less usage. Large-size tiles are used for flooring in offices, airports, etc.

(ii) Vitrified Tiles

In this case special ceramic clay is mixed with oxides and burnt at a very high temperature. At a very high temperature, the clay becomes vitrified and we get coloured vitrified tiles. Compared to ordinary glazed tiles wherein the glazing is on the top surface, here the entire tile is made of vitrified clay.

The water absorption capacity is only 0.5%. Many types of fully vitrified tiles are now manufactured in India. These tiles are superior to ordinary ceramic tiles and inferior to marble tiles used for flooring.

These tiles after manufacturing are again ground to get smooth and straight edges so as to lay with very close joints. These tiles are preferred in kitchens compared to ceramic tiles.

(iii) Porcelain Tiles

Porcelain tiles are the latest addition to the ceramic tile family. These are prepared from clay, feldspar, quartz and minerals. They are available in different forms as plain, coloured and also with some decorative patterns, but are very brittle in behaviour. These tiles are very expensive.

4.2.3 Manufacture of Tiles

Manufacture of the common clay tiles involves the following processes:

(i) Preparation of clay
(ii) Moulding of tiles
(iii) Drying of tiles
(iv) Burning of tiles

1. Preparation of Clay

As stated earlier, the best blue clay for tiles is generally found beneath brick earth. Impurities such as grit, pebbles, etc. are removed from the selected clay. The clay is ground into a powder form in crushing roller mills. It is then pugged.

For the preparation of superior tiles, the powdered clay is mixed with water in a tank and allowed to settle down. The coarser particles settle down, and the solution containing the finer particles is run off into a lower tank. The collected material in the lower tank is allowed to dry leaving the fine clay ready for further processing.

2. Moulding of Tiles

Depending on the pattern of tile to be moulded, the moulding method is selected. Wooden patterns are used for tiles which are not of uniform thickness throughout the length and are moulded in wooden plattens. The pugged clay is pressed into the mould and the tiles are taken out for drying.

When the pattern of tile is perfectly circular in shape, Potter's wheel may be used. However, the diameter of the tile may vary along its length.

Mechanical method is resorted to for tiles having a uniform section throughout their length. The pugged clay is prepared through openings or orifices of the required shape and size. The moulded beam of clay coming out of the slot is received on a platform, and the tiles are cut to the required length by means of wire frames.

3. Drying of Tiles

After two days of moulding, they are slightly beaten with a flat wooden mallet to correct the irregularity in shape due to warping. They are then kept on edges for further drying under shade for 2 days. This is done to prevent warping and cracking.

4. Burning of Tiles

The tiles are burnt in a kiln which could accommodate 30,000–40,000 tiles. Such kilns are partly underground and are usually provided with a roof to protect the kiln from rain.

A layer of bricks is laid over the long narrow flues, and the flues are fed with wood. The bricks are so placed such that open spaces are left in between them so that flues can move.

Above the layer of bricks, the dried tiles are placed on edge layer by layer, until the kiln is fully filled with raw tiles. The doorways are closed.

The kiln is then fired initially with 800ºC, slaked for six hours and then increased to 1,300ºC for about 3 hours (Fig. 4.2). Once again it is slacked for 6 hours and then once more heated and kept steady for about 3–4 hours. Finally, the flues are filled with fuel and the doorways are closed by brickwork in mud. The kiln is cooled gradually over 72 hours. Then the tiles are removed.

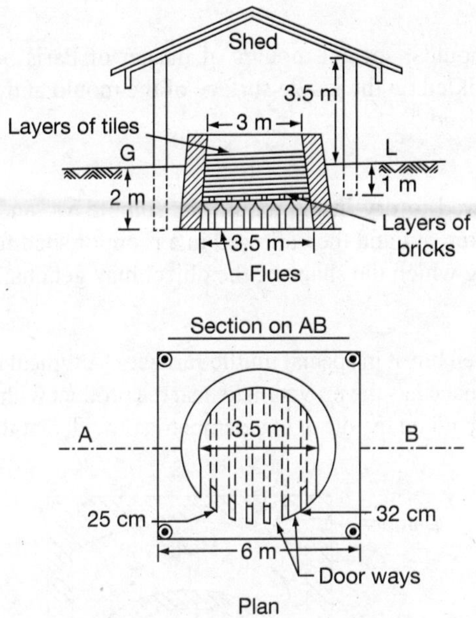

Figure 4.2 Circular kiln for burning tiles

4.3 EARTHENWARE

It has been discussed that earthenware is manufactured from ordinary clay burnt at low temperatures. Sometimes the clay is mixed with sand, crushed pottery, etc. Earthenware are susceptible to damage by atmospheric action since they are soft and porous. In general, they should not absorb more than about 20% of water by weight. In order to make them impervious, they may be glazed.

Glazed earthenware tiles are used for floors and walls of kitchens, bathrooms, water closets, hospitals, etc.

4.3.1 Terra-cotta

Terra-cotta is earthenware of vitrified material. It is high-quality earthenware. It is used as a substitute for stone in the ornamentation of buildings.

The manufacture of terra-cotta involves the same four stages as that of tiles.

1. Preparation of Clay

The clay selected for terra-cotta should contain about 5–8% of iron oxide and about 1% of lime. Sometimes clay with high alumina content is used. To these selected clays, sand, ground glass, old terra-cotta or pottery are added. The addition of these materials provides strength and rigidity to the terra-cotta and also prevents shrinking while drying.

Such a mixed clay should not have grit, pebbles, organic matter, etc. It is mixed with water, shovelled and kept for several days under damp conditions. It is pugged in a pug mill before being taken to the moulding section. At this stage, the required quantity of colouring substance is added to get the desired shade of colour after burning.

2. Moulding

Apart from wooden moulds, special moulds of plaster of Paris or templates of zinc are used. Fine sand is sprinkled on the inside surface of the mould and the clay is then pressed in the mould with the hand.

3. Drying

The terra-cotta is allowed to dry in the mould for some days, and then the object in the terra-cotta mould is taken out and then allowed in a room or shed for drying. Gradual drying is to be done failing which the shape of the object may get changed or cracked.

4. Burning

The dried objects are then burnt in special muffle furnaces. A typical muffle furnace is shown in Fig. 4.3. A muffle furnace has the provision to heat the product without coming into contact with fuels. In order to regulate the drought a damper is provided at the side of the furnace.

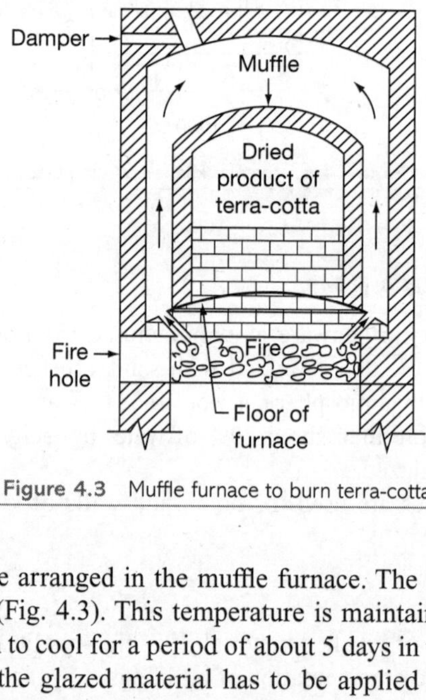

Figure 4.3 Muffle furnace to burn terra-cotta

The dried products are arranged in the muffle furnace. The temperature of the kiln is raised to about 1,200°C (Fig. 4.3). This temperature is maintained for 4 days. The burnt products are allowed then to cool for a period of about 5 days in the furnace itself. In order to get a glazed surface, the glazed material has to be applied by a brush on terra-cotta products before they are burnt.

4.3.2 Porcelainware

Porcelainware or merely porcelain indicates fine earthenware, which is white, thin and transparent. It is also sometimes called whiteware.

It is basically prepared from clay and mixed with feldspar, quartz and minerals. The ingredients are powdered and thoroughly mixed. The mixture is moulded to the required shape and then burnt. Porcelain has various uses. In buildings, they are used as sanitaryware, electrical insulators, etc.

4.4 STONEWARE

Stoneware are the products manufactured from refractory clay mixed with crushed pottery, powdered stone, etc.

The clay is prepared carefully and moulded to the required shape. The stoneware is then dried under the shade to prevent any cracking or deformation of the product. The dried material is burnt in a kiln, gradually heating to a high temperature. The high temperature is maintained for 1–3 days based on the size of the kiln and the material size. The stoneware are allowed to cool down gradually.

Stoneware are dense and compact when compared to earthenware. Stoneware are impervious to moisture, and can be kept clean easily. Glazed stoneware are highly resistant to acid or any atmospheric agencies.

SALIENT POINTS

1. Tiles are thin slabs made out of brick earth by moulding, drying and then burning.
2. Common clay tiles are: clay roof tiles, clay floor tiles and clay terrace tiles.
3. Clay roof tiles are: Mangalore tiles, country tiles and Allahabad tiles.
4. Ceramic is the term used for the act of making objects with a special type of clay.
5. Orthoclase feldspar is mainly responsible for the production of clay in nature. This mineral during decomposition forms kaolinite. Kaolin, a product composed of pure kaolinite, is a fine white clay required to manufacture some ceramic articles.
6. Ceramic tiles are classified as glazed ceramic tiles, vitrified tiles and porcelain tiles.
7. Manufacturing of common clay tiles involves the following operations: preparation of clay, moulding of tiles, drying of tiles and burning of tiles.
8. Earthenware is the term used for clay products burnt at low temperatures. They have a coarse surface and are opaque. Earthenware comprises terra-cotta and porcelainware.
9. Terra-cotta is a hard brownish unglazed vitrified ceramic material used for ornamental purposes.
10. Stoneware is a product of refractory clay, ground stone and pottery.

REVIEW QUESTIONS

1. What is meant by the term "tile"? What are the characteristics of a good tile?
2. What are common tiles?
3. Explain different kinds of roof tiles.
4. How are clay floor tiles manufactured?
5. Discuss the properties of clay tiles of flooring and roofing purposes.
6. Enumerate different types of ceramic tiles.

7. Discuss the steps involved in the manufacture of tiles.
8. Why do you consider that the cement tiles are superior to clay tiles? Explain.
9. What are earthenware? Explain.
10. Explain the manufacturing and uses of glazed earthenware tiles.
11. What is terra-cotta? How is it manufactured?
12. What is whiteware? How is it prepared?
13. Explain stoneware production and their uses.

5 LIME

5.1 INTRODUCTION

There are several naturally available civil engineering materials or materials directly produced from natural materials. Among them are binding materials, which play a major role. Lime, cement and bitumen are important for civil engineers.

Binding materials may be of air binding, hydraulic binding and autoclave binding.

Air binding materials, also known as air-setting binding substances, pass into a stone state in air, gain strength and then retain their mechanical properties. Lime, cement, gypsum, etc., are examples of air binding materials.

Hydraulic binding materials, also known as hydraulic-setting substances, pass into a stone stage in air and as well as in water, gain strength and then retain their mechanical properties. Hydraulic lime, cement, etc., are examples of hydraulic binding materials.

Autoclave binding materials, also known as autoclave-setting substances, set only when treated in autoclaves with saturated steam under pressure of 8–12 atmosphere and at a temperature between 170 and 200°C. Lime–silica, sand, Portland cement, etc., are examples of autoclave binding materials.

Lime is one of the oldest cementing materials as discussed in this chapter.

5.2 CONSTITUENTS OF LIMESTONE

Lime is not freely available in nature but is obtained by burning naturally available materials such as:

(i) Limestone found in limestone hills
(ii) Limestone boulders found in the bed of old rivers
(iii) Kankar found below ground and
(iv) Shells of sea animals.

Lime is more or less pure calcium carbonate. The main sources of obtaining lime are the stones which possess clay, soluble silica, magnesium carbonate, alkalis and metallic oxides, sulphates, iron compounds and carbonaceous matter.

5.2.1 Clay

It is the material which gives the property of producing hydraulicity in lime. In small quantity it retards slaking and is also capable of arresting slaking when it is in excess. It gives the property of insolubility for lime in water. In order to get the best quality lime it is recommended to have 8–30% of clay.

5.2.2 Soluble Silica

In order to develop hydraulicity in lime it is necessary to have silica and alumina during chemical combination with limestone. Hydraulicity in lime is caused due to silicates of calcium, magnesium and aluminium. These silicates are inert at low temperatures and they become active to combine with lime at a high temperature.

5.2.3 Magnesium Carbonate

Hydraulicity is rendered to lime when the magnesium carbonate content is about 30% and this can happen even with the absence of clay. But limes containing large proportions of magnesium carbonate are liable to crack. It has the important property of increasing the setting process but reducing slaking.

5.2.4 Alkalis and Metallic Oxides

These materials convert as soluble salts at a low temperature and thereby cause hydraulicity when present in small quantities (5%).

5.2.5 Sulphates

The presence of sulphates in small quantities tend to accelerate the setting and reduces the slaking action.

5.2.6 Iron

The presence of iron compounds lowers the temperature of calcination of limestone. But excess of iron is objectionable.

5.2.7 Pyrites

It is undesirable to have pyrite in limestone as it is harmful and liable to produce poor-quality lime.

5.3 PROPERTIES OF LIME

Lime is a binding material used in several ancient civil engineering works. A good quality lime should possess the following properties:

1. It is a flexible material and is easily workable.
2. When used as a mortar it provides strength to the masonry.
3. It possesses good plasticity.
4. It is highly resistant to moisture and is used for pointing works.
5. It solidifies with less time and hardens.
6. An excellent binding material which adheres to brick or stone masonry units perfectly.
7. As it is less shrinkable during drying it is highly durable when used in mortar.

5.4 CLASSIFICATION OF LIME AND THEIR USES

5.4.1 General Classification

In general, classification of lime is based on the calcination of limestone. They are

1. Fat lime,
2. Hydraulic lime and
3. Poor lime

1. Fat Lime

It is the lime which has high calcium oxide content. This can set and become hard only in the presence of carbon dioxide. This is white in colour.

This lime is also called high-calcium lime, pure lime, rich lime or white lime. It is mostly called fat lime as it slakes vigorously. By this process, the volume is increased two-fold.

It is obtained by calcining pure carbonate of lime, which contains about 95% of calcium oxide. The percentage of impurities will be less than 5%.

It has the following specific properties:

(i) Very slowly hardening
(ii) High degree of plasticity and
(iii) Soluble in water.

The uses of fat lime are as follows:

(i) It is used as lime mortar (lime and sand) for plastering and pointing works.
(ii) It is used for white washing of plastered surfaces.
(iii) Lime *surkhi* mortar is used for thick masonry walls, foundations, etc., where *surkhi* is the powder obtained by grinding burnt bricks.

2. Hydraulic Lime

It is the lime containing small quantities of silica, alumina and iron oxide, which are jointly in chemical composition with calcium oxide. Hydraulic lime can set and become hard even in the absence of CO_2 and can set even under water. This is also known as water lime.

Based on the percentage of clay, lime may be classified as (i) Feebly hydraulic lime, (ii) Moderately hydraulic lime and (iii) Eminently hydraulic lime. The increase in clay content makes slaking difficult and increases the hydraulic properties.

The uses of hydraulic lime are as follows:

(i) It can set under water and in thick walls where there is no free circulation of air.
(ii) As it contains about 30% of clay, it resembles that of cement and can be used for major civil engineering works.
(iii) It forms a thin paste with water.

3. Poor Lime

This lime is also known as impure lime or lean lime. More than 30% of clay present in poor lime makes the lime to slack slowly.

It sets and hardens very slowly and has very poor binding properties. It can form a thin paste with water. Because the lime is of inferior quality, it is used for inferior types of works.

5.4.2 Indian Standard Classification

Limes are classified into six categories as per Bureau of Indian Standards (IS:712–1984). They are explained below.

1. Class A: Eminently Hydraulic Lime

It is available in hydrated form. The compressive strength of lime sand mortar (1:3 by weight) should have 1.75 N/mm^2 and 2.8 N/mm^2 at 14 and 28 days, respectively. With such a strength they can be used for structural works such as arches, domes, etc.

2. Class B: Semi-hydraulic Lime

This is available as quick lime (lime obtained by calcining pure limestone) or as hydrated lime (dry power obtained by treating quick lime with water). The compressive strength of lime and mortar (1:3 by weight) should have 1.25 N/mm^2 and 1.75 N/mm^2 at 14 days and 28 days, respectively. This mortar is used for masonry work.

3. Class C: Fat Lime

This should be available in quick or hydrated form. It can be used as the finishing coat in plastering, white washing, etc. It is also used for masonry mortar with the addition of pozzolanic material (it is volcanic dust containing 80% clay with lime magnesia and iron oxide in varying proportions).

4. Class D: Magnesium or Dolomite Lime

This is also available in hydrated or quick form. It is used as the finishing coat in plastering and white washing.

5. Class E: Kankar Lime

The lime is produced by burning lime nodules (found in soils like black cotton soil containing silica) in the hydrated form. It can be used for masonry mortar.

6. Class F: Siliceous Dolomite Lime

It is available in quick or hydrated form. It is generally used as an undercoat and finishing coat of plaster.

5.5 MANUFACTURE OF LIME

Manufacturing of lime comprises three stages, viz.:

1. Collection of raw materials
2. Burning of limestone and
3. Slaking of burnt lime

5.5.1 Collection of Raw Materials

For manufacturing fat lime, the percentage of impurities in limestone should not exceed 5%. Kankar lime may be used for manufacturing hydraulic lime.

Limestones required for the manufacture of specific lime may be collected and slacked in sufficient quantity near the manufacturing site.

Fuels used for calcining limestone are wood, coal, coke, coal gas, charcoal, cinder, oil, etc.

5.5.2 Burning of Limestone

Limestones may be burnt in clamps or kilns. For kiln burning, an intermittent kiln or a continuous kiln may be used.

1. Clamp Burning

When the burnt lime required is small, this method may work out to be more economical. Further, this method can be adoptable in places where limestone and the fuel are available.

The clamp comprises of alternate layers of limestone and fuel. The fuel to be adopted in this case is wood. If the fuel is other than wood, viz., coal or charcoal, the fuel is mixed with the limestone and the clamp is formed. The clamp is of 3.6 m max height with sloped sides. The clamp is covered with mud plaster to prevent the escape of heat (Fig. 5.1).

Clamp burning has several disadvantages, viz., wastage of fuel, is uneconomical, is not a continuous process, can not be produced during the rainy season, etc.

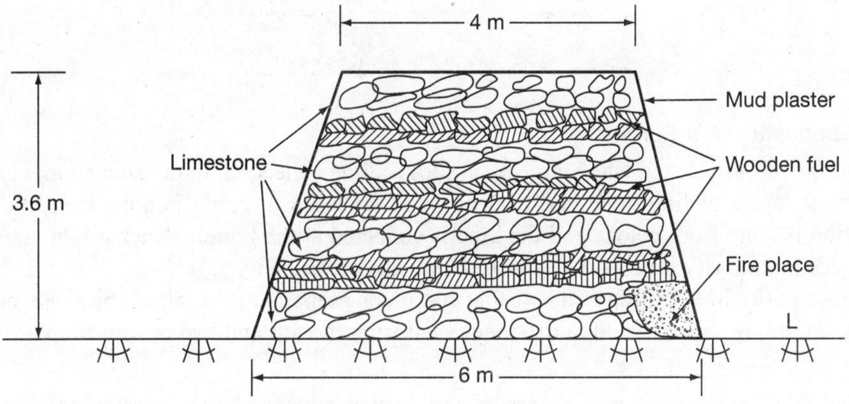

Figure 5.1 Clamp burning

2. Intermittent Kiln Burning

Various kinds of intermittent kilns are in practice, the most prevalent among them being the intermittent flame kiln and the intermittent flare kiln.

The intermittent flame kiln is similar to the clamp kiln. Here the limestone and fuel are arranged in alternate layers between two vertical walls. In some cases vertical flue is also provided. The top of the kiln is covered with unburnt material. The kiln is operated for three days. Then the kiln is cooled and unloaded and the process is repeated.

The intermittent flare kiln is one in which fuel is not allowed to come into contact with limestone. Big pieces of limestone are chosen, and a rough arch is formed over which small pieces of limestone are packed (Fig. 5.2).

The fuel is placed below the arch. When it is ignited, only the flame comes into contact with the limestone. After the limestone is adequately burnt, the kiln is cooled and unloaded. This process is then repeated.

This type of kiln is used only for the manufacture of lime on a moderate scale. The main drawback is the wastage of time for every operation.

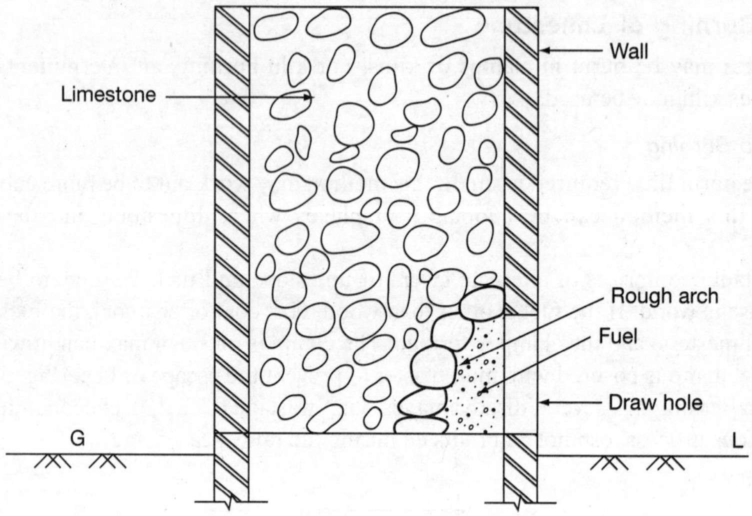

Figure 5.2 Intermittent flare kiln

3. Continuous Kiln Burning

This type of kiln is used when large quantities of lime is needed. In the conventional type, alternate layers of limestone and coal or wood are used. Feeding is done from the top, ignition is done from below, and the lime is collected at the bottom. Such a kiln does not produce high-purity lime as it gets mixed with ash of fuel (Fig. 5.3).

High-purity lime is prepared by using continuous separate feed kilns. The arrangement is shown in Fig. 5.3. Here the limestone is fed from the top and two or four fireplaces are

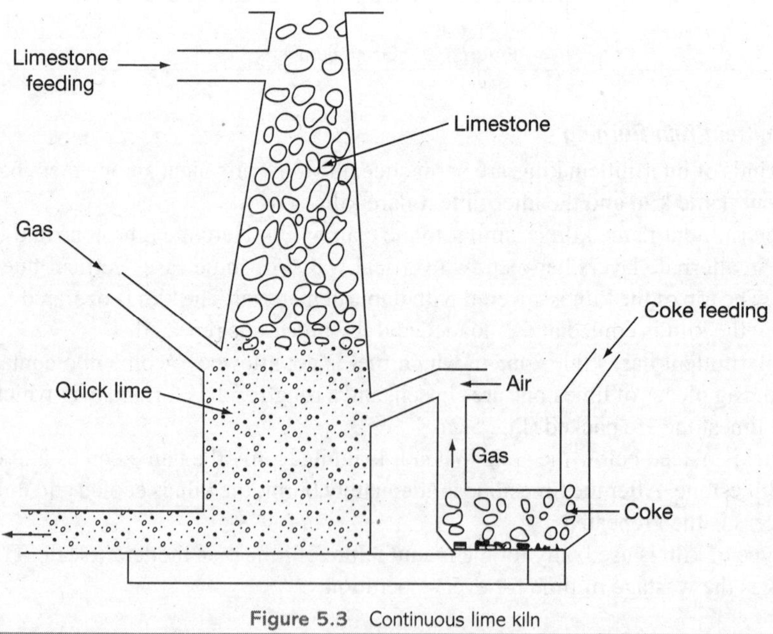

Figure 5.3 Continuous lime kiln

provided by the side of the kiln near the bottom. Coal or wood is used as the fuel, and hot gases of combustion move up the shaft of the kiln. The use of wood as a fuel has an added advantage in that long flames are produced, and the temperature of limestone is not raised too much.

5.5.3 Slaking of Burnt Lime

From quicklime, slaked lime and lime putty are prepared as discussed below:

1. Slaked Lime

Quicklime is heaped on a masonry or wooden platform. Then water is gradually sprinkled over it till the lime is slaked and reduced to a powder form. During the sprinkling of water, the heap is turned over and over again. This process is continued till no more water can be added than that required for the lime to convert into a powder form. The slaked lime is then sieved through an I.S sieve of mesh size 3.35 mm, and the residue if any is rejected. The end product is slaked lime.

2. Lime Putty

Sometimes before using lime in mortar, quicklime is converted into a putty form. For this purpose two large tanks are made, one 50 cm deep and the other 80 cm deep. The first tank is constructed as a pucca masonry at a higher level. The lower level tank is constructed at the ground level. The lower tank is made of dry brick masonry with joints filled with sand.

The upper tank is filled to half of its depth with water. Quicklime is added gradually till it fills half the depth of water. Only lime should be added to water and not otherwise. It is then stirred taking care to ensure that lime at no time is exposed to the air. For about 5 minutes the mixture is stirred till the boiling stops and it thickens. More water is added, if needed, and then the products are allowed to flow to the lower tank.

The mixture is then allowed to stand in the lower tank for 72 hours. As the lower tank is made of dry brick masonry with sand joints, water in the slurry is partly absorbed by the ground and partly through the brick joints. Thus, excess water is removed, and lime putty is obtained as a paste. It can be stored for about two weeks if adequately protected from drying.

5.6 PRECAUTIONS IN HANDLING LIME

Lime easily affects the skin and also is a health hazard if breathed in. The moisture on the body reacts with lime resulting in skin diseases. This may be avoided by coating the skin with oil.

Slaking of lime is an exothermic reaction, i.e., it produces a large amount of heat. Necessary precautions should be taken by workers who are involved in the use of lime. It is customary to use goggles, respirants, gloves and boots as protection.

5.7 STORAGE OF LIME

Quicklime should be kept in air-tight vessels or in large heaps under cover; otherwise, it should be slaked soon after delivery or run into putty to prevent deterioration.

Slaked lime can be stored safely for 3–4 months if packed in gunny bags lined with polythene, craft paper, etc. The storing should be done in weather-proof sheds, which are perfectly dry. As per specifications, the lime should be slaked not less than 1 week and not more than 4 weeks before its use.

Hydrated lime can be safely stored for considerable time without deterioration provided the bags are kept in a dry place.

5.8 LABORATORY TESTING OF LIMESTONES

Physical and chemical tests are conducted on limestone to determine their suitability.

5.8.1 Physical Tests

Physical tests are based on the texture, appearance, colour and odour of limestone. Limestone showing amorphous textures are suitable for making lime.

Chalk and fat lime deposits are white or whitish-brown or grey in colour. Magnesium limestone is distinguished from its heaviness and hardness.

Limestone with white, bluish-grey, yellow or brown colours are suitable for producing hydraulic lime. These limestones show a compact texture with an irregular dull fracture. A freshly fractured surface of such limestone has an earthy smell and is clayey in taste.

5.8.2 Chemical Tests

1. Heat Test

A piece of dry limestone is weighed and heated on an open fire for a few hours. The limestone disintegrates emitting CO_2. Due to the removal of CO_2 from the sample, it loses weight. The sample is weighed again. The loss of weight indicates the amount of CO_2. From this, the percentage of calcium carbonate present in the limestone is determined.

2. Acid Test

The purpose of this test to classify the limestone and to determine the percentage content of calcium carbonate.

A teaspoon of powdered lime is placed in a test tube, and dilute hydrochloric acid is added. The contents are stirred with a glass rod and set aside for 24 hours.

Abundant effervescence indicates a high percentage of calcium carbonate. On the other hand, the formation of residue signifies that the limestone is impure or hydraulic. The absence of formation of gel signifies fat lime.

The formation of a thick gel shows primarily the presence of hydraulic lime, and gel of moderate thickness signifies feebly hydraulic lime.

3. Slaking Test

Quick lime in combination with water hydrates, and the process is known as slaking. A lot of heat is generated during slaking, which indicates that the percentage of calcium content is high in the limestone. In the case of hydraulic lime, the slaking action is very slow.

The Bureau of Indian Standards (IS: 6932–1973) specifies 10 laboratory tests for lime (Methods of Test for Building Lime – Parts 1 to 11). The reader may refer to the codes for details.

5.9 FIELD TESTING OF BUILDING LIME

The Bureau of Indian Standards (IS: 1624–1974) provides a number of field tests for building lime, which can be easily conducted in the field. They are discussed below.

5.9.1 Visual Examination

Class C lime is identified by its pure white colour.

5.9.2 Hydrochloric Acid Test

Hydrochloric acid of 50% strength is added to powdered lime kept in a test tube till effervescence ceases. Generally, a teaspoonful of powdered lime of about 10 cc is required. It is kept for 24 hours. The bubbling reaction indicates the presence of lime. The volume of insoluble residue signifies unwanted inert material.

Class A lime shows the formation of good thick gel after 24 hours of letting the mixture stand. Class B lime shows the formation of gel, whereas no gel is formed in the case of Class C lime.

5.9.3 Ball Test

Sufficient water is added to lime and a ball of the size of an egg is made. It is stored for 6 hours and then placed in a basin of water.

If it shows expansion and disintegration in a few minutes, it signifies Class C lime. If the expansion is less with a number of cracks, it can be categorised as Class B lime. Class A lime will not show any adverse effects.

5.9.4 Impurity Test

A known weight of lime is mixed with water in a beaker and the solution is decanted. The residue is dried in the sun for 8 hours, cooled and weighed. Based on the percentage of residue, the quality of lime is decided. Residue of less than 10% indicates that the lime is good, 10–20% indicates fair and above 20% indicates poor.

5.9.5 Plasticity Test

A small quantity of lime is mixed with water and left overnight. The so-formed material is spread like butter on a blotting paper with a knife to test its plasticity. Good lime indicates high plasticity.

5.9.6 Workability Test

An 1:3 lime mortar is made with adequate water. The mortar is thrown on a brick wall by a trovel. The sticking quality is examined. If it sticks well, then its workability is good.

5.10 POZZOLANIC MATERIALS

Pozzolanic materials are siliceous and aluminous materials, which do not possess any cementitious properties but form cementitious compounds by reacting with lime in the presence of water. It is volcanic ash containing about 80% clay.

The use of pozzolanic materials provides the following improvements:

(i) Improves workability
(ii) Lowers the heat of hydration
(iii) Assists in early setting and hardening of the mortar
(iv) Reduces shrinkage
(v) Improves hydraulic properties and resistance to chemical attacks.

Naturally available pozzolanic materials are clay and shale, diatomaceous earth, volcanic tuffs and pumicites. Artificially available pozzolanic materials are fly ash, blast furnace slag, silica fume, rice husk ash, metakaoline and *surkhi*.

SALIENT POINTS

1. Binding materials may be of the following kinds: air binding, hydraulic binding and autoclave binding.
2. Lime is not freely available in nature but obtained by burning one of the naturally available materials such as (i) limestone found in limestone hills, (ii) limestone boulders found in the bed of old rivers, (iii) *kankar* found below ground and (iv) shells of sea animals. Lime is more or less pure calcium carbonate.
3. The main sources of obtaining lime are the stones which possess clay, soluble silica, magnesium carbonate. alkalis and metallic oxides, sulphates, iron compounds and carbonaceous matter.
4. The general classification of lime based on the calcination of limestone is (i) fat lime, (ii) hydraulic lime and (iii) poor lime.
5. Fat lime has high calcium oxide content. This can set and become hard only in the presence of carbon-di-oxide. This is white in colour.
6. Hydraulic lime is lime containing small quantities of silica, alumina and iron oxide, which are jointly in chemical composition with calcium oxide. Hydraulic lime can set and become hard even in the absence of CO_2 and can set even under water. This is also known as water lime.
7. Poor lime is also known as impure lime or lean lime. More than 30% of clay present in poor lime makes the lime to slake slowly. It sets and hardness very slowly and has very poor binding properties. It can form a thin paste with water.
8. The Bureau of Indian Standards classifies lime into six categories as (i) Eminently Hydraulic Lime (Class A), (ii) Semi-hydraulic lime (Class B), (iii) Fat lime (Class C), Magnesium or Dolomite lime (Class D), *Kankar* lime (Class F) and siliceous Dolomite lime (Class F).
9. The manufacture of lime comprises three stages, viz., collection of raw materials, burning of limestone, and slaking of burnt lime.
10. Burning of limestone comprises clamp burning, intermittent kiln burning and continuous kiln burning.
11. Laboratory tests of limestone are of two kinds: physical tests and chemical tests. Chemical tests are the Heat Test, Acid Test and Slaking Test.
12. Field tests of building lime are Visual Examination, Hydrochloric Acid Test, Ball Test, Impurity Test, Plasticity Test and Workability Test.

REVIEW QUESTIONS

1. Briefly explain the constituents of limestone.
2. List the properties of lime.
3. What is meant by hydraulic lime? How can it be obtained?
4. Compare fat lime with quick lime.
5. Give the classification and properties of different types of lime.
6. How is lime classified by Indian Standards?
7. Describe the process of lime burning in India.
8. Explain the types of intermittent kilns.
9. Provide a sketch of the continuous kiln, and explain its working.
10. How do you proceed to get (a) lime putty, (b) quick lime and (c) slaked lime?
11. Explain chemical tests conducted on limestone.
12. What are the field testing of building lime?

6 CEMENT

6.1 INTRODUCTION

Among the binding materials, cement is of recent origin compared to those used in the construction of different monuments throughout the world.

Undoubtedly the credit goes to John Smeaton, a British Engineer, in the year 1756, who was given the responsibility of rebuilding the Eddystone lighthouse. Based on an exhaustive collection of information about the binding materials used in earlier days and in experiments, he found that limestone which contained a considerable proportion of clay yielded superior hydraulic properties. His basic investigation attributed to the invention of cement by Joseph Aspdin in 1824. He patented the cement as Portland Cement. As the hardened cement showed the resemblance of the natural stone occurring at Portland in England, this fancy name has emerged.

Cement has taken the topmost position among building construction materials. This material is made from limestone and clay in a powder form. Manufacturing of cement was started in India in 1904. Portland Cement is generally referred to as Ordinary Portland Cement (OPC).

6.2 INGREDIENTS OF CEMENT AND THEIR FUNCTIONS

The main ingredients in the manufacturing of cement are lime, silica, alumina and other oxides. The ingredients present in Ordinary Portland Cement are given in Table 6.1. The contributions by different ingredients are discussed below.

6.2.1 Lime

It is the main constituent which contributes to the strength. Its proportion has to be properly maintained so as to make the cement sound and strong. Excess of lime makes the cement unsound and causes the cement to expand and disintegrate. On the other hand, if lime is deficient, the strength of the cement is reduced and the cement sets quickly.

Table 6.1 Ingredients of Portland cement

Sl. No.	Ingredients	Range of percentage
1.	Lime, CaO	60–65
2.	Silica, SiO_2	17–25
3.	Alumina, Al_2O_3	3.5–9.0
4.	Iron oxide, Fe_2O_3	0.5–6.0
5.	Magnesium oxide, MgO	0.5–4.0
6.	Sulphur trioxide, SO_3	1–2.0
7.	Alkalis (soda and or potash), K_2O and Na_2O	0.4–1.3

6.2.2 Silica

It contributes to the strength of cement due to the formation of dicalcium and tricalcium silicates. The excess presence of silica prolongs the setting time of cement but increases its strength.

6.2.3 Alumina

The quick-setting quality is imparted to the cement by alumina. Excess alumina reduces the strength of cement.

6.2.4 Iron Oxide

It provides colour, hardness and strength to the cement.

6.2.5 Magnesium Oxide

The correct quantity of magnesium oxide imparts hardness and colour to the cement. Soundness of cement is seriously affected, if it is present in excess quantity.

6.2.6 Sulphur Trioxide

It makes the cement sound when present in small quantities. Excess of sulphur trioxide causes the cement to become unsound.

6.2.7 Alkalis

Alkalis should be present only in small quantities. Excess quantities of alkalis cause efflorescence and straining when used in concrete or mortar.

6.3 REQUIREMENTS OF QUALITY CEMENT

1. Cement should have a uniform colour.
2. Cement should be free from lumps.
3. It should be smooth and uniform. As a test, if a small quantity of cement is thrown in water, it should sink completely.
4. The ratio of percentage of alumina to that of iron oxide should not be less than 0.66 in cement.
5. The total sulphur content in cement should not be greater than 2.75%.

6. Magnesia content in cement should not exceed 5% by weight.
7. Insoluble residue in cement should not be greater than 1.5% by weight.
8. Cement should not lose more than 4% of its weight when heated.
9. The specific surface of cement as found from the fineness test should not be less than 2,250 mm^2/gm.
10. The initial setting of cement should be about 30 minutes, and the final setting time should be about 10 hours.
11. Expansion of cement should not be greater than 10 mm in a soundness test.

6.4 PROPERTIES OF ORDINARY PORTLAND CEMENT

Ordinary Portland Cement should possess certain physical and chemical properties for the best performance when it is used in a structure. Further, it is possible to compare the quality of cement from different sources for proper selection. Frequent tests are to be carried out on the cement directly, on hardened cement paste or continuous on the concrete made from the cement. Such tests are needed to maintain the quality of work within the specified limits. Important physical and chemical properties of Portland Cement are discussed below.

6.4.1 Physical Properties of Portland Cement

1. Fineness

Fineness of cement is a measure of the size of particles of the cement. It is expressed as the specific surface of cement. It is an important factor to decide the rate of gain of strength and uniformity. In general, finer the grains higher the surface area.

The rate of hydration will be higher with the increase in fineness of cement. This is because of the larger surface area available for chemical reaction. Further, this leads to earlier development of strength. The effect of fineness increases with strength as shown in Fig. 6.1. As per specification, the residue of cement should not exceed 10% when sieved through a 90-micron sieve.

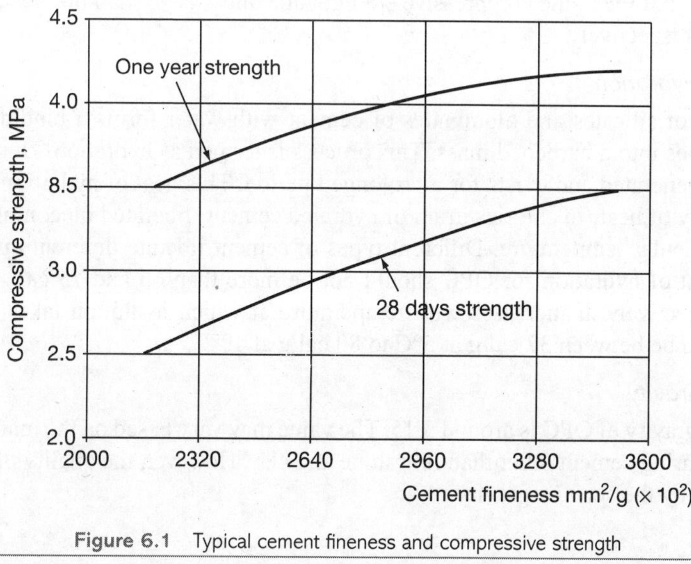

Figure 6.1 Typical cement fineness and compressive strength

2. Setting Time

When water is added to the cement and mixed, it forms a paste which gradually becomes less plastic and finally forms a hard material. The time taken for the entire process is called the setting time, which is divided into the initial setting time and the final setting time.

For convenience, the initial setting time is regarded as the time when the cement paste loses its plasticity, whereas the final setting time is the time elapsed between the time of water addition to the cement and the time when the paste has completely lost its plasticity, i.e., it has attained sufficient firmness to withstand certain pressure. For OPC the initial setting time should not be less than 30 min. and the final setting time not more than 600 min.

The initial setting time should be sufficiently long for finishing operations such as transporting and placing of concrete. The length of the setting time depends on the quantity of mixing water and the atmospheric temperature rather than on the cement.

3. Soundness

Cement which has less expansive qualities when mixed with water is termed as soundness. The soundness of cement is affected by the undesirable expansion of some of its constituents which could occur even after setting. This expansion causes disintegration and severe cracking. Mainly, the presence of free lime and magnesia in cement leads to unsoundness. The unsoundness may be controlled by:

(i) Limiting the MgO content to less than 0.5%
(ii) Fine grinding
(iii) Aerating the cement for aeration for several days
(iv) Thorough mixing

4. Compressive Strength

It is one of the important properties of cement. The strength of cement is generally assessed from compression tests on cement-mortar cubes (1:3) with an area of 5,000 mm^2. The graded standard sand to be used for the mortar cube preparation should conform to IS: 650–1991. For OPC, the compressive strength at 3 and 7 days shall not be less than 16 and 22 MPa, respectively.

5. Heat of Hydration

The reaction of silicates and aluminates of cement with water forms a binding medium which solidifies into a hardened mass. This process is termed as hydration. During hydration, heat is generated and exists for a prolonged period. This heat of hydration is defined as the quantity of heat (in calories/gram of hydrated cement) liberated on complete hydration at a particular temperature. Different types of cement liberate different quantities of heat. The heat of hydration for OPC should not be more than 66 and 75 cal/g for 7 and 28 days, respectively. It increases with temperature at which hydration takes place. For OPC it should be between 37 cal/g at 5°C to 80 cal/g at 40°C.

6. Specific Gravity

The specific gravity of OPC is around 3.15. The value may vary based on the materials used for manufacturing cement other than limestone and clay. However, the quality of cement is not governed by the specific gravity.

6.4.2 Chemical Properties of Portland Cement

Loss on ignition and insoluble residue are the two chemical properties of cement.

1. Loss of Ignition

Due to the ignition of cement at a high temperature, loss in weight occurs due to the evaporation of moisture and carbon dioxide, which are present in combination with free lime or magnesia. Loss in weight is a measure of the freshness of cement. Hydroxides and carbonates of magnesium and lime do not contain any cementing properties; thus, they are called inert substances. The quality of cement will be better if the inert materials are lesser, and hence lesser will be the loss in ignition. Normally, the loss is around 2% and should not be more than 4%.

2. Insoluble Residue

An inactive part of cement is called the insoluble residue. It is determined by a simple test.

One gram of cement with 40 ml of water and 10 ml of concentrated HCL is stirred. The stirred mix is boiled at constant temperature for 10 minutes. Lumps if any are broken and the solution is filtered. The residue on the filter is washed with Na_2Co_3 solution, water and HCL in this order and finally again with water. The filter paper is dried, ignited and weighed to give an insoluble residue. Lower the percent of residue better is the cement. The maximum allowable value is 0.85%.

6.5 MANUFACTURING OF ORDINARY PORTLAND CEMENT

Cement factories are to be set up in areas where raw materials, viz., limestone or chalk and shale or clay are abundantly available. It should have adequate transport facilities like road, rail, etc., for easy transport of the finished products. Further, adequate water and power should be available to run the industry without any interruption to production.

Production of cement is performed in three stages, viz.:

1. Mixing of raw materials in correct proportion
2. Burning of the mixture
3. Grinding

There are two processes in the manufacture of cement, viz., the wet process and the dry process, depending upon whether the mixing and grinding of raw materials are done in wet or dry conditions.

A semi-dry process is also in use wherein the raw materials are ground and then mixed with about 10–14% of water and further burnt to chinkering temperature. For many years, the wet process remained popular, which is replaced by the dry process as it consumes less fuel for burning.

6.5.1 Wet Process

Limestone is first crushed to small pieces. Then it is taken to a ball mill or a tube mill and mixed with clay or shale.

Ball mill is in the form of a steel cylinder of about 2 m diameter and length about 2 m. It is positioned in a horizontal position and rotates in a clockwise direction. The inside of the cylinder is fitted with perforated curved plates and steel balls. The material to be ground or mixed is fed from the top. During rotation, the material is ground and mixed by striking against the curved plates and steel balls. The ground and mixed material is collected at the bottom (Fig. 6.2).

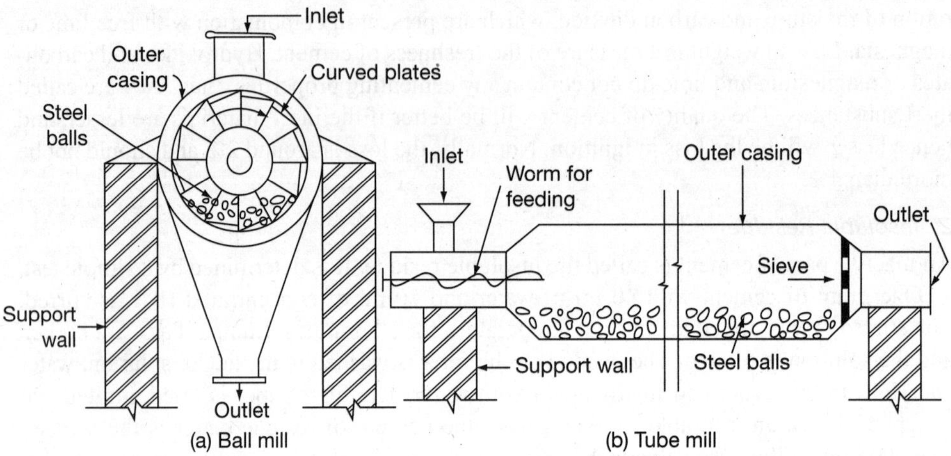

Figure 6.2 Ball and tube mills

The tube mill is in the form of a long horizontal cylinder of 1.5 m diameter and about 8 m long. The cylinder is partly filled with steel balls of varying sizes from 20 to 25 mm diameter. The action of the tube mill is similar to that of a ball mill.

Materials fed into the ball or tube mill are ground with the addition of water to a fine consistency of slurry. The slurry is pumped to slurry tanks or basins where it is kept in an agitated condition. The slurry is checked for the correct composition frequently. The slurry is stored in separate final storage tanks and agitated to keep the slurry in a homogeneous condition.

The correct slurry is sprayed on to the upper end of a rotary kiln. The sprayed slurry passes on the hot surface of a flexible chain (Fig. 6.3). Passing through the kiln of length 30–200 m, the slurry loses moisture and forms flakes. The flakes move from the top to the bottom of the kiln for easy collection of flakes.

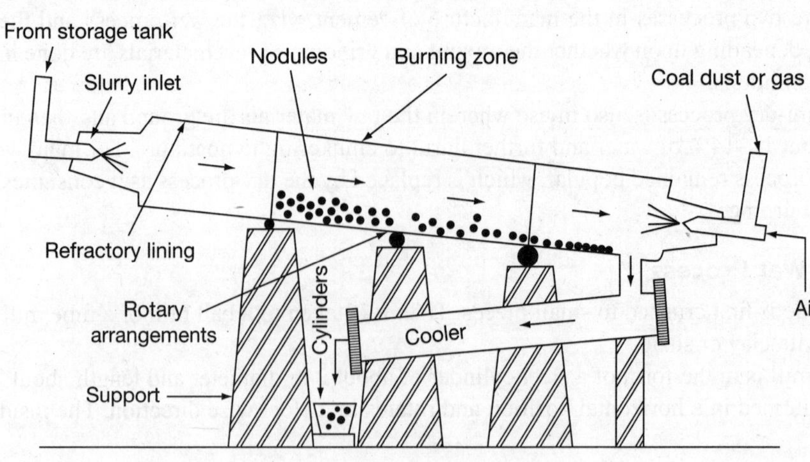

Figure 6.3 Rotary kiln

While passing through the kiln, the slurry undergoes a series of chemical reactions and gets fused and turns into a nodular form of size 3–20 mm known as clinker. The clinker is then cooled and ground in a ball mill with the addition of 3–5% of gypsum. Gypsum is added to prevent quick setting of the cement.

The manufacture of cement by the wet process is shown by a flowchart in Fig. 6.4.

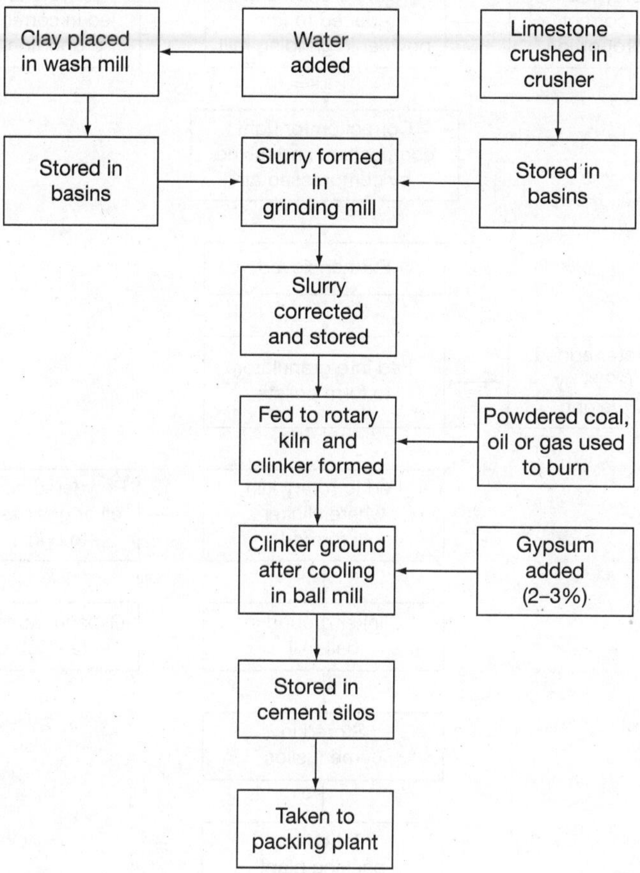

Figure 6.4 Flowchart for wet process

6.5.2 Dry Process

In the dry process, the raw materials are crushed dry, and the correct proportion of the materials is fed into a grinding mill. In the grinding mill, they are dried and finely powdered. The dry powder is then blended and corrected for its right composition and then mixed again by means of compressed air. The aerated powder behaves similar to that of a liquid. The blended mixture is sieved further and passed through a rotating disc called a granulator. In order to make the mixture as pellets, a small amount of water (about 12%) is added.

A small kiln is sufficient for the dry process. This process consumes less quantities of coal and is hence economical.

The pellets are passed through the kiln and form flakes after undergoing a series of chemical reactions and get fused. The fused material turns into nodular form of size 3–20 mm, known as clinker.

Then the clinker is cooled and ground in a ball mill with the addition of 2–3% of gypsum. The stages in the manufacture of cement by the dry process are shown in the flowchart (Fig. 6.5).

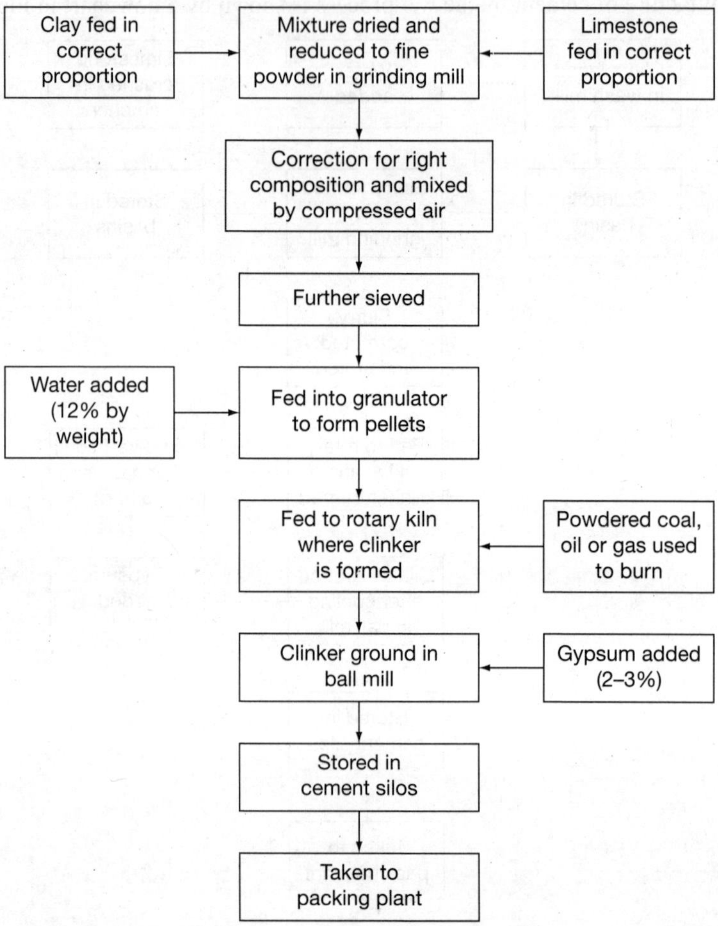

Figure 6.5 Flowchart for dry process

6.6 CHEMICAL COMPOSITION AND HARDENING OF CEMENT

The main constituents in the manufacture of cement, as discussed earlier, are lime, silica, alumina and iron oxide (Table 6.1).

These oxides interact with one another in the kiln at a high temperature, and more complex compounds are formed. The relative proportions of these oxides are fully responsible for the various properties of cement, rate of cooling and fineness of grinding.

Out of the complex compounds formed in the kiln at the high temperature, four compounds, viz., tricalcium silicate, dicalcium silicate, tricalcium aluminate and tetracalcium alumino-ferrite. The chemical formula and abbreviated formula are presented in Table 6.2.

Table 6.2 Major complex compounds

Name of the compound	Formula	Abbreviated formula
Tricalcium silicate	$3CaOSiO_2$	C_3S (54.1%)
Dicalcium silicate	$2CaOSiO_2$	C_2S (16.6%)
Tricalcium aluminate	$3CaO.Al_2O_3$	C_3A (10.8%)
Tetracalcium aluminoferrite	$4CaO.Al_2O_3.Fe_2O_3$	C_4AF (9.1%)

As the four compounds mentioned above were first identified by Mr R. H. Bogue, they are called Bogue's compounds.

The abbreviated notations stand for

C for CaO
S for SiO_2
A for Al_2O_3
F for Fe_2O_3

Of the four compounds tricalcium silicate and dicalcium silicate are the compounds which contribute towards strength. Both the compounds put together constitute about 70–80% of cement. In modern cement, the average C_3S content is 45% and that of C_2S is about 25%.

Apart from the four major compounds some minor compounds also form in the kiln, viz., K_2O and Na_2O.

The percentage of the four major compounds present, in general, in cement based on Bogue's concept is shown in brackets in Table 6.2.

When water is added to cement, these four compounds react in the decreasing order as C_3A, C_4AF, C_3S and C_2S.

During the initial period of hardening the strength gain of C_2S is small and is about 15% as that of C_3S. After 28 days, the hydration of C_3S comes practically to an end. The hydration of C_2S begins only after the full hydration of C_3S. The hardening of concrete is greatly speeded up by C_3A.

6.7 TYPES OF CEMENTS AND THEIR USES

6.7.1 Types of Portland Cement

Code (IS: 456:2000) permits the use of the following Portland cement for a specific work:

1. Ordinary Portland Cement (OPC)

Ordinary Portland Cement of grades 33, 43 and 53 are permitted. Only these cements are most commonly used in general concrete construction.

2. Rapid Hardening Portland Cement (RHPC)

This cement conforms to IS 8041:1990, which is similar to OPC. Here the percentage of C_3S is more and less C_2S and ground more finely. It has the quality of attaining high strength in a short period of time. Thus, it is used for construction purposes where speed is of importance.

3. Portland Slag Cement (PSC)

This cement conforms to IS 455:1990. Here the cement is made by inter-grinding Portland Cement clinker and granulated blast furnace slag. Slag is a waste product obtained from the manufacture of pig iron. Here the tricalcium content is restricted to a minimum. This has fairly high resistance to sulphates. Hence, it is suitable for environments exposed to sulphates.

4. Portland Pozzolana Cement (PPC)

Pozzolanic material is siliceous or aluminous material which by itself has no cementitous properties. However, in the presence of water it reacts with calcium hydroxide liberated in the hydration process of cement.

Fly ash and calcinated clay are pozzolanic materials used in the manufacture of PPC. Fly-ash-based and calcinated-clay-based Portland Pozzolana Cements conform to Parts 1 and 2 of IS 1489:1991. These cements hydrate and gain strength relatively slowly, and hence need more curing time. Because of the long curing time, this cement is preferred for construction purposes where time is not of prime importance, such as mass concreting.

5. Hydrophobic Portland Cement (HPC)

Hydrophobic is the water-repellent property of material. This cement conforms to IS 8043:1991. This type of cement is produced by inter-grinding Portland cement with 0.1–0.4% of oleic acid or stearic acid.

This water-repellent property is due to the formation of a water-repellent film around each particle of cement, but during mixing the film breaks and makes room for normal hydration to take place. This hydrophobic quality of cement facilitates its storage for longer periods in extremely wet climate conditions.

6. Low Heat Portland Cement (LHPC)

This contains low quantities of tricalcium aluminate and high quantities of dicalcium silicate. This cement conforms to IS 12600:1989. Compared to PPC the hydration process is slow because of which generation of heat is also low. This property is desirable for mass concreting purposes, such as the construction of gravity dams. Because of slow strength gain, care should be taken while removing the formwork early.

7. Sulphate-resisting Portland Cement (SRPC)

This cement contains a very low quantity (about 5%) of tricalcium aluminate and is ground finer than OPC. This cement conforms to IS 12330:1988. This cement is highly sulphate resistant caused by the reaction of C_3A in hardened cement. This is ideal for use in locations where sulphates are present in the atmosphere or soil.

Physical and chemical properties of different Ordinary Portland Cements are presented in Table 6.3.

6.7.2 Other Types of Cements

1. High Alumina Cement

It is the cement obtained by grinding high alumina clinker. This has a long initial setting time, high ultimate strength, high resistance to the action of acids and high temperature. This cement is used for furnace insulation, refractory concrete and for special structures which require imperviousness and corrosion resistance.

Table 6.3 Specifications for physical properties of cements (as per the Indian Standard)

Properties	Types of cement and relevant IS code requirements							
	OPC	RHPC	LHPC	HSPC	PPC	PSC	SSC	HAC
	(i)	(ii)	(iii)	(iv)	(v)	(vi)	(vii)	(viii)
	IS:269–1989	8041–1990	12600–1989	8112–1989	1489–1991 (Pt. I)	455–1989	6909–1990	6452–1989
1. Fineness								
(a) Residue (90 mic) not to exceed %	10	5	–	–	–	–	–	–
(b) Sp. surface (m^2/kg) not less than	225	325	320	350	300	225	400	225
2. Setting time (min)								
(a) Initial not less than	30	30	60	30	30	30	30	30
(b) Final not more than	600	600	600	600	600	600	600	600
3. Strength (N/mm^2)								
(a) At 1 day (24 ± 0.5 hour)	–	16	–	–	–	–	–	30
(b) At 3 days (72 ± 1 hour)	16	27	10	23	16	16	15	35
(c) At 7 days (168 ± 2 hours)	22	–	16	33	22	22	22	–
(d) At 28 days (672 ± 4 hours)	33	–	35	43	33	33	30	–
4. Soundness								
(a) Le Chatelier expansion not more than (mm)	10	10	10	10	10	10	5 modified	5
(b) Auto-clave expansion not more than (%)	0.8	–	0.8	0.8	0.8	–	–	–

(Continued)

Table 6.3 (Continued)

Properties	Types of cement and relevant IS code requirements							
	OPC	**RHPC**	**LHPC**	**HSPC**	**PPC**	**PSC**	**SSC**	**HAC**
	(i)	**(ii)**	**(iii)**	**(iv)**	**(v)**	**(vi)**	**(vii)**	**(viii)**
	IS:269–1989	**8041–1990**	**12600–1989**	**8112–1989**	**1489–1991 (Pt. I)**	**455–1989**	**6909–1990**	**6452–1989**
5. Max. heat of hydration (calories/g)								
(a) At 7 days not more than	–	–	65	–	–	–	–	–
(b) At 28 days not more than	–	–	75	–	–	–	–	–
6. Specific gravity (Bulk density kg/l)	–	–	–	–	–	–	2.8–2.9 (1.30)	–
7. Drying shrinkage (max percent)	–	–	–	–	0.15	–	–	–
8. Chemical requirements							–	
(a) Loss on ignition max. %	5.0	5.0	5.0	4.0	–	4	Insoluble residue max. (4%)	–
(b) Weight of MgO (max. %)	6.0	6.0	6.0	6.0	–	8	10.0	–
(c) Sulphuric anhydrate (max. %)	2.75	2.75	2.75	2.75	–	3	6.0 sulphide sulphur max.	–
(d) Insoluble material (max. %)	2	2	–	2	–	2.5	1.5	–
(e) Lime saturation factor %	0.66–1.02 max.	0.66–1.02 max.	–	–	–	–	–	–

OPC – Ordinary Portland Cement; PSC – Portland Slag Cement; HSPC – High Strength Ordinary Portland Cement; PPC – Portland Pozzolanic Cement; LHPC – Low Heat Portland Cement; HAC – High Alumina Cement; RHPC – Rapid Hardening Portland Cement; SSC – Super Sulphated Cement

2. Masonry Cement

It is a product obtained by inter-grinding a mixture of Portland Cement clinker with inert materials (non-pozzolanic) and gypsum and air entraining plasticiser. This type of cement is characterised by certain physical properties, such as slow hardening, high workability and high water retentivity, which makes it especially suitable for masonry work.

3. Oil-well Cement

This is hydraulic cement which contains retarders in addition to coarser grinding and/or reduced tricalcium aluminate content of clinker. This is suitable for use in high pressure and temperature.

4. Quick-setting Cement

This cement is produced by adding a certain quantity of aluminium sulphate and reducing the quantity of gypsum and a fine powder is made. This cement is used for under-water concreting.

5. Expanding Cement

This cement is obtained by adding an expanding medium like sulpho-aluminate and a stabilising agent to ordinary cement. Unlike conventional cement that shrinks, it expands during curing. This cement is used for repairing concrete surfaces.

6. Acid-resistant Cement

Materials like quartz, sodium silicate and sodium fluo silicate are added to the cement to attain an acid-resistant quality. It is used in the chemical industry.

7. White Cement

This type of cement does not contain colouring ingredients such as iron oxide, manganese oxide or chromium oxide, and the cement is burnt by oil. It is used for floor finish, plastering, pointing of masonry, manufacture of precast stones, tiles, runway markings and colour cement.

8. Coloured Cement

The desired colour for the cement can be obtained by initially mixing colouring materials with it. It is used for the external finishing of walls and floors, manufacturing of tiles and precast stones, and is also used for garden paths, swimming pools and tennis courts.

6.8 GRADES OF ORDINARY PORTLAND CEMENT

At present, Ordinary Portland Cement is available in three different grades in India as 33-grade (IS: 269–1989), 43-Grade (IS: 8112–1989) and 53-grade (IS: 12269–1987).

The numbers 33, 43 and 53 correspond to the 28-day compressive strength of cement-mortar as 33 MPa, 43 MPa and 53 MPa. The strengths are obtained from standard tests on cement–sand mortar specimens. These three grades are most commonly used in general concrete construction.

All the three grades of Ordinary Portland Cement are produced from the same basic materials. But higher strength is achieved by increasing the tricalcium silicate (C_3S) content and also by finer grinding of the clinker.

Physical properties of 33, 43 and 53 Grade Ordinary Portland Cement are shown in Table 6.4.

Table 6.4 Physical properties of OPC

Type of cement	Fineness m^2/kg	Soundness mm	Setting time		28 Days compressive strength of mortar (MPa)
			Initial min.	Final min.	
33 Grade OPC	225	10	30	600	33
43 Grade OPC	225	10	30	600	43
53 Grade OPC	225	10	30	600	53

6.9 LABORATORY TESTS ON CEMENT

The following tests are conducted on cement in the laboratory (IS: 4031 – PT 1 to 15, 1989):

1. Fineness Test
2. Consistency Test
3. Setting Time Test
4. Strength Test
5. Soundness Test
6. Tensile Strength Test
7. Heat of Hydration Test
8. Chemical Composition Test

6.9.1 Fineness Test

Fineness of cement is responsible for the rate of hydration and rate of evolution of heat and hence on the rate of gain of strength. Finer the grains, more is the surface area and faster the development of strength.

The fineness of cement can be determined by the following tests:

1. Sieve Test or
2. Air Permeability Test

1. Sieve Test

About 100 g of cement is weighed and sieved using a 90-micron IS sieve. Air-set lumps are broken and the cement is sieved continuously in a circular and vertical motion for a period of 15 minutes. A mechanical sieve shaker may also be used. The residue left on the sieve is weighed. This weight shall not exceed 10% for ordinary cement. This test is rarely used.

2. Air Permeability Test

Fineness of cement is represented by the specific surface, which is expressed as the total surface area in sq.cm/g. of cement. Blaine's Air Permeability Test is used to find the specific surface.

The principle is based on the relationship between the rate of flow of air through a cement bed. The cement bed is of a given porosity comprising the surface area of the particles. The surface area is more for finer particles.

6.9.2 Consistency Test

Consistency Test is conducted to find the setting times of cement, viz., initial setting time and final setting time. The test is conducted using a standard consistency test apparatus, viz., Vicat's apparatus (Fig. 6.6).

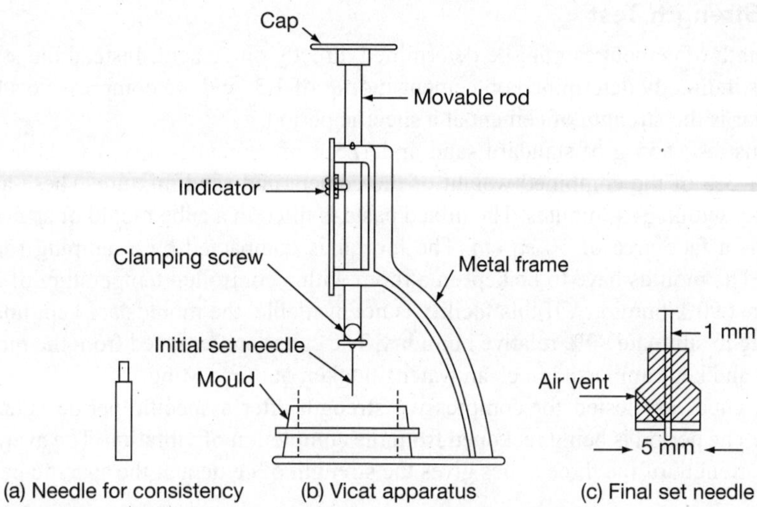

Figure 6.6 Vicat apparatus with various plungers

Standard or normal consistency of cement paste is defined as that water content which will permit a Vicat plunger of 10 mm diameter and 50 mm length to penetrate depths of 33–35 mm within 3–5 minutes of mixing. The test has to be repeated thrice. Each time the cement is mixed with water varying from 24 to 27% of the weight of cement. It is required that the test be conducted at a constant temperature of 27 ± 2°C and at a constant humidity of 20%.

The test procedure is simple. The paste prepared within 3–5 minutes is filled into the Vicat mould and shaken to expel air. The standard needle is brought down to touch the surface of the paste and quickly released allowing to sink into the paste by its own weight. The depth of the penetration is noted. Then that percentage of water which allows the plunger to penetrate only to a depth of 33–35 mm is known as the percentage of water required to produce a cement paste of standard consistency.

6.9.3 Setting Time Tests

An arbitrary division has been made for the setting time of cement as the initial setting time and final setting time. Vicat's apparatus is used for setting times also.

1. Initial Setting Time

About 500 g of the cement sample is taken and water content 0.85 P, where P is the percentage of water required as per the consistency test is added and made into a paste. The paste has to be made within a period of 3–5 minutes and filled into the mould.

For this test, a needle of 1 mm square size (needle C) is used. The needle is allowed to penetrate into the paste. The time taken to penetrate 33–35 mm depth is taken as the initial setting time.

2. Final Setting Time

The needle C is replaced by needle F. The attachment is lowered gently to cover the surface of the test block. The central needle makes an impression, whereas the circular cutting edge fails to do so. That is, the paste has attained hardness, and the centre needle does not penetrate the paste more than 0.5 mm. Thus, the time at which the needle does not penetrate more than 0.5 mm is considered to be the final setting time.

6.9.4 Strength Test

The strength of cement can not be determined directly on cement. Instead the strength of cement is indirectly determined on cement-mortar of 1:3, and the compressive strength of this mortar is the strength of cement at a specific period.

For this test, 555 g of standard sand and 185 g of cement is mixed with water quantity $P/4 + 3\%$ of the combined weight of the cement and sand mixture. These are mixed thoroughly within 3–4 minutes. The mixed paste is filled in a cube mould of size 7.06 mm, which has a face area of 50 sq cm. The mortar is compacted by a tamping rod or in a vibrator. The moulds have to be kept in a room with a controlled temperature of 27 ± 2°C and at least 90% humidity. If this facility is not available, the moulds are kept under a wet gunny bag to simulate 90% relative humidity. The cubes are removed from the mould after 24 hours and kept immersed in clean water till taken out for testing.

Three cubes are tested for compressive strength after a specific period (viz., 3, 7 or 28 days). The period is being reckoned from the completion of vibration. The average compressive strength of the three cubes gives the strength of cement at the specific period.

6.9.5 Soundness Test

The Soundness Test is performed to detect the presence of uncombined lime and magnesia in cement. This test is conducted in Le Chatelier's apparatus (Fig. 6.7). It consists of a small brass cylinder of 30 mm diameter, 30 mm height and 0.5 mm thickness, which is split along its generation with the split not exceeding 0.5 mm. On either sides of the split, two indicators with pointed ends are provided.

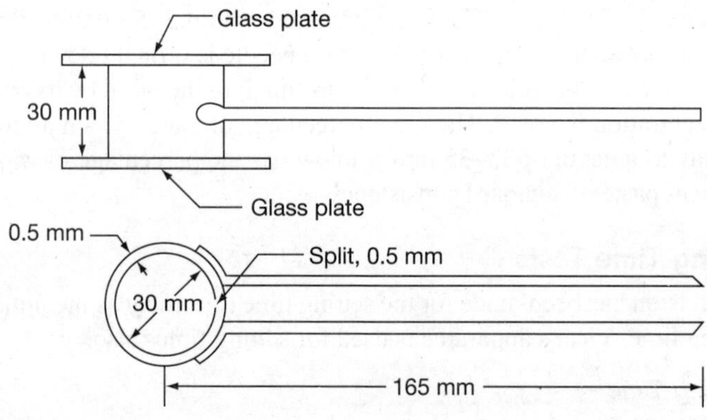

Figure 6.7 Le Chatelier's apparatus

The cement paste is prepared with water content at consistency. The cylinder is placed over a glass plate and is filled with the cement paste. It is covered with another glass plate on the top. The whole assembly is immersed in water at 24–35°C for 24 hours. After 24 hours, the gap between the indicators is measured.

The mould with the specimen is immersed in water again and boiled for 30 minutes. After boiling for 1 hour, the mould is taken out and cooled. Again the gap between the indicators is measured after the end of cooling. The increase in gap represents the expansion of cement.

6.9.6 Tensile Strength Test

This test is conducted using a cement-mortar briquette in a tensile testing machine. The dimensions of the briquette are shown in Fig. 6.8.

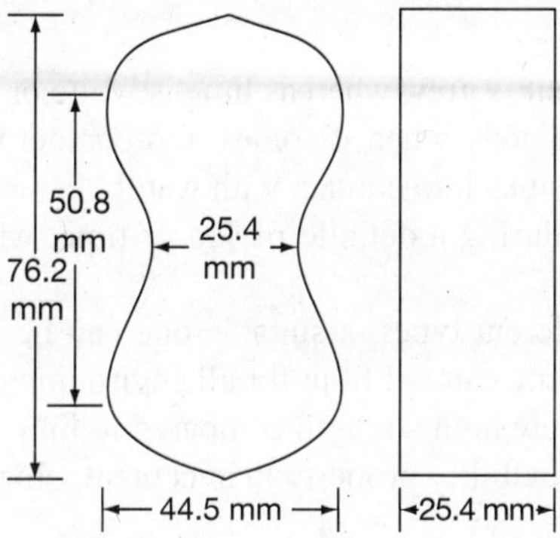

Figure 6.8 Standard cement-mortar briquette

A 1:3 cement–sand mortar with water content of 8% is mixed and moulded into a briquette in the mould. It is cured for 24 hours at a temperature of 27 ± 2°C and for 24 hours in an atmosphere at 90% relative humidity.

The briquette is tested under direct tension, the pull being applied through special jaws engaging the wide ends of the briquette. The area of cross-section at the neck is 1 sq inch (i.e., 25.4 mm × 25.4 mm). The average strength for six briquettes tested after 3 and 7 days is recorded.

6.9.7 Heat of Hydration Test

During hydration of cement, heat is produced due to chemical reactions. In a large-scale construction this heat may raise the temperature of concrete to a high degree to as high as 50°C. In such cases low-heat cement has to be used or cooling methods must be employed.

The test is conducted using a calorimeter adopting the principle of determining heat gain. Low-heat cement should not generate 65 calories per gram of cement in 7 days and 75 calories per gram of cement in 28 days.

6.9.8 Chemical Composition Test

Different tests are conducted to determine the quantity of various constituents of cement. The requirements as stipulated in IS: 269–1998 are as follows:

1.	Ratio of percentage of alumina to that of iron oxide	$\nless$ 0.66
2.	Lime Saturation Factor (LSF), i.e., ratio of percentage to that of alumina, iron oxide and silica	$\nless$ 0.66 and $\ngtr$ 1.02
3.	Total loss on ignition	$\ngtr$ 4%
4.	Total sulphur content	$\ngtr$ 2.75%
5.	Weight of insoluble residue	$\ngtr$ 1.50%
6.	Weight of magnesia	$\ngtr$ 5%

6.10 COMPARISON BETWEEN CEMENT AND LIME

Although cement and lime are cementing materials, they have different features and properties:

1. The colour of cement is grey, whereas lime is white or greyish white in colour.
2. Lime slakes immediately when it comes into contact with water, but cement does not slake when it comes into contact with water.
3. Cement paste sets during a definite period of time, whereas lime takes more time to set.
4. Cements are of different types, a suitable one can be selected to fit in for a given environment. But lime can not fit in for all environments.
5. Cement has extremely high strength compared to lime.
6. Lime has a better flexibility property than cement which acts as a rigid material.

SALIENT POINTS

1. Cement is made from limestone and clay in a powder form, which on addition of the requisite quantity of water is capable of hardening.
2. The main ingredients in the manufacture of cement are lime, silica, alumina and other oxides.
3. Lime is the main constituent which contributes to the strength. Its proportion has to be properly maintained so as to make the cement sound and strong.
4. Silica contributes to the strength of cement due to the formation of dicalcium and tricalcium silicates.
5. The quick-setting quality is imparted by alumina to the cement.
6. Iron oxide provides colour, hardness and strength to the cement.
7. Correct quantity of magnesium oxide imparts hardness and colour to the cement.
8. Sulphur trioxide makes the cement sound when present in small quantities.
9. The fineness of cement is a measure of the size of particles of cement. It is expressed as the specific surface of cement.
10. When water is added to cement and mixed, it forms a paste, which gradually becomes less plastic and finally forms a hard material. The time taken for the entire process is called the setting time, which is divided into the initial setting time and the final setting time.
11. Initial setting time is regarded as the time when the cement paste losses its plasticity.
12. Final setting time is the time elapsed between the time of water addition to the cement to the time when the paste has completely lost its plasticity.
13. Cement that has less expansive qualities when mixed with water is termed as the soundness.
14. Strength of cement is generally assessed from compression tests on cement-mortar cubes (1:3) with an area of 500 mm^2.

15. Reaction of silicates and aluminates of cement with water forms a binding medium which solidifies into a hardened mass which is termed as hydration.
16. Heat of hydration is defined as the quantity of heat (in calories/gram of hydrated cement) liberated on complete hydration at a particular temperature.
17. Specific gravity of Ordinary Portland Cement is around 3.15.
18. Loss on ignition is the loss in weight due to ignition of cement at high temperature. Loss in weight is a measure of the freshness of cement.
19. An inactive part of cement is called the insoluble residue.
20. Production of cement is performed in three stages, viz., (i) Mixing of raw materials in the correct proportion, (ii) Burning of the mixture and (iii) Grinding.
21. There are two processes in the manufacture of cement, viz., the wet process and the dry process, depending upon whether the mixing and grinding of raw materials are done in wet or dry conditions.
22. Major complex compounds formed due to chemical compositions are: tricalcium silicate, dicalcium silicate, tricalcium aluminate and tetracalcium aluminoferrite.
23. Grades of Ordinary Portland Cement are: 33-Grade, 43-Grade and 53-Grade.

REVIEW QUESTIONS

1. Name the various ingredients of Ordinary Portland Cement. Explain the functions played by each.
2. Explain the basic requirements of good quality cement.
3. What are the important physical properties of cement?
4. Discuss the chemical properties of cement.
5. Discuss in detail the manufacturing process of Ordinary Portland Cement.
6. Explain the different zones in a rotary kiln.
7. Why gypsum is added during the manufacture of cement? Explain.
8. Bring out the difference between the wet-process and dry-process methods of manufacturing cement.
9. How does hardening of cement occur? Explain.
10. What is meant by hydration of cement? Explain.
11. What is meant by rapid hardening cement? What are the conditions under which this can be used?
12. Discuss different types of Ordinary Portland Cement.
13. Name the important physical and chemical tests to be conducted on common cement.
14. Discuss types of artificial cements used for different purposes.

7 MORTAR

7.1 INTRODUCTION

The cementing materials discussed in the previous chapters can not be used directly as a construction material unless they are combined with some other inert material like aggregates. Thus, a combined material formed by mixing intimately a binding material (cement or lime) with a fine aggregate (sand or *surkhi*) in a certain proportion and with an adequate quantity of water is called mortar. Here lime or cement enables a proper bond to be developed among the particles, and sand adds mechanical strength to the mortar.

The properties of binding materials, lime and cement, have been discussed at length in the previous chapters. Hence, it is necessary to know more about the fine aggregate and water to be used to get the best mortar.

Properties of mortar and their kinds along with the quality of sand and water to be added are discussed in the following sections.

7.2 QUALITIES OF SAND FOR MORTAR

Sand particles are small soil grains of silica. As per the Indian Soil Classification, particles of diameter greater than 0.075 mm and less that 4.75 mm are categorized as sand.

Sand is naturally available and may be classified as (i) pit sand, (ii) river sand and (iii) sea sand. Sea sand is not suitable for mortar.

For the purposes of mortar, sand is classified as fine, coarse and gravelly sand. Fine sand is one that passes through a screen with clear openings of 1.59 mm, coarse sand passes through a screen with clear openings of 3.18 mm and gravelly sand passes through a screen with clear openings of 7.62 mm. Fine sand is used for plastering, and coarse sand is used for masonry work.

The properties of sand required for mortar are as follows:

(i) It should be free from any organic matter.
(ii) It should be chemically inert.
(iii) It should be well graded.
(iv) The fineness modulus of sand should be between 2 and 3.
(v) It should be sharp, angular, coarse and clean.

7.3 QUALITIES OF WATER FOR MORTAR

Water is used for construction purposes of Civil Engineering Works such as the construction of earth structures, highway construction, mortar and concrete works, etc. Although water is an important constituent of construction materials, it does not receive the attention due to it particularly in the construction of earth structures and highway construction.

However, due importance is given in case of preparation of mortar and concrete. The general belief is that if it is suitable for drinking, then it is considered suitable for construction including mortar and concrete. The quality of water in connection with the preparation of mortar and concrete and for curing is further discussed.

As water helps to form cement gel, which is the governing factor for the strength of mortar or concrete, the quantity and quality of water are to be carefully decided.

The quality of mixing water may be determined based on any one of the following approaches:

(i) If the water is obtained from an unknown source, the strength of the mortar or concrete made with questionable water is compared with similar mortar or concrete prepared with pure water.
(ii) If the *pH* value of water lies between 6 and 8, the water is assumed to be free from organic matter and declared as suitable.

The presence of the following impurities may affect the mortar if excess quantity is allowed:

(i) Presence of suspended particles $\ngtr$ 0.02%
(ii) Presence of salts: chlorides $\ngtr$ 10,000 ppm
sulphates $\ngtr$ 3,000 ppm
(iii) Presence of acids and alkalis: harmful
(iv) Presence of algae: harmful

The guidelines to assess the quality of water for concrete discussed in Section 8.4.3 and Table 8.1 are also applicable for mortar.

7.4 FUNCTIONS OF INGREDIENTS OF MORTAR

As discussed earlier, the ingredients of mortar are:

1. Binding or cementing materials, viz., cement or lime
2. Fine aggregates, viz., sand, *surkhi*, cinder, etc.
3. Water

7.4.1 Functions of Cementing Materials

The primary requirement of cementing materials is to provide strength and durability to the structure.

For cement mortar mostly Ordinary Portland Cement is used. However, for special types of works other cements may be used appropriate to the work. Cement performs the following functions when used in cement mortar:

(i) Primarily, it fills the voids formed by fine aggregates and thereby makes the mortar impermeable.
(ii) It binds the fine aggregates into a solid mass with time on adding water to the mix.
(iii) It imparts strength after setting and hardening.

For lime mortar, different types of lime are used, viz., eminently hydraulic, semi-hydraulic and fat limes. For preparing lime mortar, the lime should be slaked before mixing it with the fine aggregate. However, the strength of lime mortar is less than that of cement mortar. Lime performs the following functions when used in lime mortar:

(i) It binds the fine aggregates into a solid mass. However, it takes more time to set.
(ii) It imparts adequate strength to the lime mortar on setting and hardening.

7.4.2 Functions of Fine Aggregates

The fine aggregate mainly used is sand. The required properties of good sand were discussed earlier. Sand plays the following functions in mortar:

(i) It reduces the shrinkage of the binding material and thereby prevents the formation of cracks.
(ii) It provides high resistance to the mortar to withstand compressive forces.
(iii) It divides the paste of the binding material and spreads to provide more surface contact area.
(iv) It makes the fat lime porous to assist in setting and hardening.
(v) It provides any proportion of the mortar.

Surkhi is another form of fine aggregate, which is usually prepared by powdering slightly unburnt bricks. It should be free from admixtures of any foreign materials. *Surkhi* is generally used with sand. If used on its own, it causes the mortar to disintegrate. When used in combination with sand, it performs all the functions as that of sand. In addition, it provides strength and improves hydraulic properties of the mortar.

Sometimes ash or cinder is provided as fine aggregate in lime mortar. It provides strength and improves hydraulic properties. It acts as an adulterant and hence increases the volume.

7.4.3 Functions of Water

The quality of water required for mortar has been discussed earlier. Water plays a significant role in the formation of mortar. It helps in mixing, transporting and in proper laying of the mortar. As a matter of fact the quality and quantity of water contribute to the strength of the mortar. Water plays the following roles in mortar:

(i) It wets the surface of the aggregates to provide better bonding.
(ii) It helps in the spreading of the cementing material over the surface of the fine aggregate.
(iii) By acting as a lubricant for the aggregate, the mortar is made workable.
(iv) It causes hydration of cement and lime because of which setting and hardening of cement mortar and lime mortar take place.

7.5 FUNCTIONS OF MORTAR

Quality mortar should satisfy the following functions:

1. It forms a binding or cementing layer between two courses of masonry in the structure.
2. Because of this binding property, it provides strength and stability to the structure.

3. In brick or stone masonry, it forms a uniform layer by filling open joints and holes as a grout in a thin liquid form.
4. It makes the masonry structure a homogeneous mass of structure, which can withstand both dead and live loads and loads due to natural reasons.
5. When used for pointing purposes, it provides high rigidity and protects the structure.
6. Provides a uniform smooth surface when used as a plaster.
7. It also forms a good surface for taking finishing materials like white/colour washing or painting.
8. It holds the aggregates together in concrete to form a homogeneous solid mass.

7.6 PROPERTIES OF MORTAR

In general, the properties required for a good mortar are as follows:

(i) The workability of mortar should be easy.
(ii) Setting should be quick, particularly in constructions that have a faster turnaround time.
(iii) Should be capable of developing good adhesion with building units such as bricks, stones, etc.
(iv) Should be in a position to develop the designed stresses.
(v) Should not react with the material on contact and affect its durability.
(vi) Cracks should not be formed on the plastered surface.
(vii) Joints should be intact and not peel off or crack, particularly in pointing.
(viii) Should have high durability.
(ix) Should be comparatively cheap.

7.7 TYPES OF MORTARS AND THEIR USES

Mortars using different binding materials have been used for various purposes. These are discussed below.

7.7.1 Lime Mortar

It is the mortar wherein fat lime or hydraulic lime is used as the binding material. The lime should be slaked before use. If fat lime is used, 2–3 times the volume of sand has to be used as it shrinks to a great extent. This lime mortar is not suitable for water-logged areas or in damp situations.

It is easy to handle and is highly plastic in nature. It has high cohesiveness and adheres to the surface and undergoes less shrinking. Although it is sufficiently durable, it hardens slowly. It is suitable for light-loaded structures and is used for construction purposes above the ground.

7.7.2 *Surkhi* Mortar

It is prepared using lime and *surkhi* instead of sand. In case of fat lime mortar only 50% needs to be replaced by *surkhi*. *Surkhi* should be powdered well.

Surkhi mortar is used for ordinary masonry work. It can not be used for pointing works and for plastering as the *surkhi* is likely to disintegrate.

7.7.3 Cement Mortar

It is obtained by using cement as the binding material. Cement mortars are assigned certain grades based on their minimum compressive strength at 28 days curing and mix proportion by volume. Some of the grades that are frequently used are given in Table 7.1.

Table 7.1 Cement mortar mix and strength

Grade	Mortar mix (Cement : Sand)	Compressive strength at 28 days N/mm^2
MM 1.5	1:7	1.5 to 2.0
MM 3.0	1:6	3.0 to 5.0
MM 5.0	1:5	5.0 to 7.5
MM 7.5	1:4 to 1:3	7.5 and above

Note: MM denotes Masonry Cement Mortar.

The recommended cement mortar mix for different purposes are as follows:

(i) MM 2 – For external works with protection and foundation masonry in very wet soil.
(ii) >MM 2 – For external walls without protection.
(iii) MM 3 – For foundation masonry in soil with saturated water and for masonry subjected to vibrations.
(iv) MM 5 to MM 7.5 – For plastering.
(v) MM 7.5 – For floor finishes and damp-proof courses.

Cement mortar is used for the following purposes:

(i) In brick and stone masonry.
(ii) For plastering of surfaces of structures like walls, beams, columns, etc.
(iii) For floor finishes.
(iv) As binding material for decorative and aesthetic works.
(v) For pointing of stone or brick masonry works.
(vi) For irregular surfaces of stone masonry to distribute the load uniformly on the bed.
(vii) For underground construction purposes, water-saturated soils, etc.

7.7.4 Lime–Cement Mortar

This type of mortar is also called Gauged Mortar or Composite Mortar. It is the one wherein the quality of lime mortar is improved by adding cement to get early strength. This makes lime mortar economical, strong and dense. The general proportion of cement to lime by volume is about 1:6–1.8. This mortar is useful for purposes like bedding and for thick brick walls.

7.7.5 Mud Mortar

Cohesive clay is puddled or pugged with water till the required consistency is attained. In order to prevent shrinkage and cracks, certain fibrous materials are also added. All the ingredients are mixed thoroughly. The life of mud mortar varies from 5–15 years depending on its water-proofing treatment and weather conditions.

Mud mortar has been in use in ancient days and mostly in villages. This can be used for brickwork of ordinary buildings. It is also used for surface floors and plastering internal walls and to some extent external walls also. In order to protect the exposed wall to rain, the mud plaster surface may be sprayed or painted with bituminous material. Instead the bituminous material is mixed in the preparation stage itself so as to add weather-proof properties to the mud mortar.

7.7.6 Fire-resistant Mortar

It is obtained by adding aluminuous cement to the finely ground powder of fire bricks. This is fire-resistant mortar, which can be used with fire bricks for furnace lining, fire places, etc.

7.7.7 Light-weight Mortar

It is obtained by mixing materials like saw dust, wood powder, asbestos fibres, jute fibres, etc., to the lime mortar or cement mortar. This type of mortar is used for sound-proof and heat-proof construction.

7.7.8 Packing Mortar

It is a special type of mortar possessing properties of high homogeneity, water resistance, predetermined setting time and the ability to form solid and water-proof plugs. This property of the mortar is useful for plugging cracks or voids of rocks, resistance to subsoil water pressure, etc., in oil wells. Different kinds of packing mortars are cement–sand, cement–loam and cement–sand–loam.

7.7.9 Sound-absorbing Mortar

This type of mortar is formed by compositioning Portland Cement, lime, gypsum, slag, etc. The crushed aggregates are selected from light-weight porous materials such as pumice, cinders, etc.

7.7.10 X-Ray Shielding Mortar

It is a heavy type of mortar, which is obtained from crushed heavy rocks, and suitable admixtures are added to get the required property. A plastering coat of this mortar is applied to the walls and ceiling of X-ray cabinets.

7.8 PREPARATION OF MORTARS

Preparation of lime mortar and cement mortar have different procedures, which are discussed in subsequent paragraphs.

7.8.1 Preparation of Lime Mortar

Ordinary lime mortar and plaster are made by grinding lime with sand. There are three methods of preparing lime mortar.

In the first method, a dry mixture of lime and sand are placed in a trough or on a platform in the specified proportion. The required quantity of water is sprinkled over the mixture in stages and mixed each time after the water is sprinkled. Then the mixture is ground to a plastic mixture by pounding with heavy wooden hammers.

The second method is adopted when a large quantity of mortar is needed. A circular annular trough is made and the mixture is placed into it. The mixture is ground by rotating a wheel through the trough with the help of bullocks. This type of mortar production is called a bullock-driven mortar mill. The mixture is ground for about 200 revolutions to get a suitable mortar.

The third method is the modern method, which is used for preparing large quantities of lime mortar. Here, a machine-driven pug or mortar mill is used. For brickwork, the grinding is done for three minutes and longer for plaster for plasterwork, pointing and terrace works. While the construction work is in progress, the mortar is raked up continuously.

Ordinary lime mortar can be stored for a maximum period of 3 days after mixing. Hydraulic lime should be used within 2–4 hours after grinding. The lime mortar should be kept moist by covering it with a wet sack cloth and frequently moistened.

7.8.2 Preparation of Cement Mortar

Cement mortar is mixed by hand for small construction purposes, while large quantities of cement mortar may be mixed by a machine.

1. Hand Mixing

Sand is measured in boxes, and the specified quantity of sand is spread first. Cement is measured by the weight of bags or by considering the volume of 0.035 m^3. Cement is spread over the sand. Then they are dry mixed by turning over and over and backward and forward using a shovel or a spade till the mixture attains a uniform colour.

The quantity of mix that has to be used within 30 minutes is separated and mixed again and formed into a heap. After making a small depression at the top, water is added to the centre of the depression and mixed well. Adequate water must be added, as excess water may draw the cement out. The water required is roughly about 70% of the weight of cement for 1:3 mortar. The whole mass is mixed thoroughly for 10–15 minutes by hand mixing.

2. Machine Mixing

Machine mixing of cement mortar is done in a concrete mixer. The sand and then the cement are put into the mixer. Initially, it is dry mixed and then water is gradually added. Wet mixing is then continued for more than a minute till the mixture is brought to a plastic consistency.

Cement mortar has to be used before it begins to set. The entire cement mortar has to be used within a maximum time of 2 hours. While using the cement mortar, water may be added to the mortar when the mortar gets stiffened. This process may be continued to restore the necessary consistency. Such retempering may be allowed only up to a period of 2 hours, beyond which the mortar should not be used.

7.8.3 Preparation of Lime–Cement Mortar

In this case, it should be ensured that the lime is slaked well. Then the lime is mixed with sand and ground well to form an intimate mixture. Water is then added and mixed well to form a regular lime mortar. The lime mortar is transferred to a mechanical mixer to which the required cement is added and mixed with additional water. Mixing is continued for 3–5 minutes to get the required consistency. This mortar can be used similarly as that of cement mortar within 30 minutes.

7.9 SELECTION OF MORTAR FOR DIFFERENT PURPOSES

Selection of mortar for particular civil engineering purposes are tabulated in Table 7.2.

Table 7.2 Selection of mortar for different purposes

Sl. No.	Nature of civil engineering work	Type of mortar and proportion	
		Cement mortar	**Lime mortar**
	(a) Masonry work		
1.	General brickwork (using stock bricks)	1:6	–
2.	Brick work below ground level	1:3 to 1:4	–
3.	Brickwork with thin joints	–	1:3 (fat lime)
4.	Reinforced brick work	1:3	–
5.	Damp-proof course	1:2	–
6.	Internal walls and surfaces of less importance	–	1:3 (ashes or cinders instead of sand)
7.	Partition walls and parapet walls	1:3	1:1 (moderately hydraulic lime)
8.	Stone masonry	1:6	1:2 (eminently or mod. hyd. lime)
9.	Arch work	1:3	–
	(b) Plastering and pointing		
10.	Pointing work	1:1 to 1:2	–
11.	Brickwork plaster (external)	1:5 to 1:6	–
12.	Brickwork plaster (internal)	1:5 to 1:6	–
13.	Ceiling plaster	1:3 to 1:4	–
	(c) Special structures		
14.	RCC tank and retaining walls	1:1	–
15.	RCC works—beams, slabs and columns	1:2	–
16.	Highly stressed structures	1:1	–
17.	Laying fire bricks	1:2 (aluminous cement and crushed fire bricks)	–

SALIENT POINTS

1. Mortar is a combined material formed by intimately mixing a binding material (cement or lime) with a fine aggregate (sand or *surkhi*).
2. The primary requirement of cementing materials is to provide strength and durability to the structure.
3. Sand reduces the shrinkage of the binding material, provides high resistance, divides and spreads the paste of binding material over a larger surface area.
4. Water helps in mixing, transporting and in proper laying of the mortar.
5. If lime mortar, fat lime or hydraulic lime is used as the binding material. The fine aggregate used is sand. If *surkhhi* is used, it is called *surkhi* mortar.
6. Cement mortar is obtained by using the binding materials, cement and sand. There are four grades of cement mortar (CM), viz., MM 1.5, MM 3.0, MM 5.0 and MM 7.5.
7. Lime–cement mortar is one where the quality of lime mortar is improved by adding cement to obtain early strength. This is also called gauged mortar or composite mortar.
8. Fire-resistant mortar is obtained by adding aluminous cement to the finely ground powder of fire-bricks.
9. Light-weight mortar is obtained by mixing materials like saw dust, wood powder, asbestos fibres, jute fibres, etc., to lime mortar or cement mortar.
10. Packing mortar is a special type of mortar possessing properties of high homogeneity, water resistance, predetermined setting time and the ability to form solid and water-proof plugs.
11. Sound-absorbing mortar is formed by compositioning Portland Cement, lime, gypsum, slag, etc. The crushed aggregates are selected from light-weight porous material such as pumice, cinders, etc.
12. X-ray shielding mortar is a heavy type of mortar, which is obtained from crushed heavy rocks, and suitable admixtures are added to get the required property.

REVIEW QUESTIONS

1. What is mortar? What are the functions of its constituents?
2. What are the properties of sand that are used in mortar?
3. Explain the importance of the effect of the quality of water on the strength of mortar.
4. Discuss the function of mortar.
5. Explain the properties of good mortar.
6. What are the properties of an ideal mortar? Explain.
7. How are mortars classified?
8. How is lime mortar prepared?

9. What are the mixing methods of cement mortar?
10. Discuss different types of mortar selected for a variety of civil engineering works.
11. What type of cement mortar would you recommend to use for external and internal walls of brick? Why?
12. What would be your recommendation of mortar proportion for the following works.
    (i) Masonry work in super structure
    (ii) Pointing works of stone masonry
    (iii) Plastering surface for paint works.

8 CONCRETE

8.1 INTRODUCTION

Concrete can be easily prepared by mixing binding material (lime or cement), fine aggregate (sand) and coarse aggregate (crushed stone or brick) and water in the required proportion.

The fine aggregates fill the voids of coarse aggregate, and voids of fine aggregate are filled with lime/cement and also the lime/cement reacts with water and binds the aggregates together to form a hard material called concrete. The quality of a concrete depends on the grade of lime/cement, type of aggregates, quality of water, the mix proportion, method of mixing, placing and compacting, temperature and curing method and its duration.

Concrete produced by mixing cement, coarse and fine aggregates and water is called cement concrete. If lime is used in the place of cement it is called lime concrete. *Surkhi* concrete is one when *surkhi* (brick powder) is added along with lime.

Cement concrete has been a construction material used in large quantities for several decades. The reason for its popularity can be attributed to the excellent technical properties and the economy of the material. Thus, cement concrete and lime concrete are discussed further in this chapter.

8.2 PROPERTIES OF CEMENT CONCRETE

The properties of cement concrete are as follows:

1. It has high compressive strength and varies with the grade of concrete.
2. It is weak in tension.
3. It hardens with age, and the rate of hardening decreases after it has attained adequate strength.
4. During the process of hardening, it shrinks due to loss of water.
5. It is not impervious perfectly unless formed with special additives.
6. It is free from corrosion.
7. It can sustain all the normal atmospheric effects.
8. It forms a hard surface, which is capable of resisting abrasion.
9. It has fire resistance capacity.
10. It is more economical and highly durable.

8.3 TYPES OF CEMENT CONCRETE AND USES

Different types of cement concrete are discussed along with their uses.

8.3.1 Plain Cement Concrete

It is the simplest form of concrete made with cement, sand, pebbles or crushed rock and water. It is used for a variety of works where the structural members are subjected to compression. Simple foundations to mighty dams may be constructed with plain cement concrete.

8.3.2 Reinforced Cement Concrete

The cement concrete provided with steel reinforcement for taking on tension, bending, torsion, fatigue, etc., is referred to as Reinforced Cement Concrete (RCC). The steel used in RCC works are: mild steel bars, medium tensile steel bars, hot-rolled deformed bars, cold twisted bars, hard-drawn steel wire fabric or rolled steel. It is used in various structural members, viz., columns, beams, footings, slabs, staircases, shelves, basements, water tanks, retaining walls, folded plates, shells, domes, etc.

8.3.3 Pre-stressed Cement Concrete

Here high tensile steel wires are used as reinforcement instead of mild steel bars. High tensile steel wires are stretched initially to the desired level and concrete is placed. After setting and completion of curing, the concrete is put into use. As the steel wires are initially stretched, the concrete so casted is known as pre-stressed concrete. This concrete is used in the manufacturing of railway sleepers and electric poles and in the construction of large span beams, large span bridge girders, etc.

8.3.4 Light-weight Cement Concrete

This concrete has less density compared to the conventional concrete. Light-weight concrete may be prepared (i) by omitting the fine aggregate, (ii) by using light-weight concrete or (iii) by aerating concrete.

The concrete prepared by omitting the fine aggregate is called *no-fine concrete*. It has low drying shrinkage. This is used for cast-in-situ external load-bearing walls, small retaining walls, where repetitive loads are expected to occur, etc. This has better insulating properties.

The light-weight aggregates used are pumice, expanded slates, furnace clinker, etc., and the fine aggregates used are crushed light-weight aggregate or natural sand. This type of light-weight concrete has better insulating and fire-resistant properties, is highly water absorptive and economical. This can be used for all the works stated above.

The light-weight concrete obtained by aeration is known as *gas*, *porous* or *foamed concrete*. This is prepared by (i) mixing air-entraining agents with cement or cement mortar in high-speed machines, (ii) mixing a specified quantity of pre-formed foam with cement or cement mortar in an ordinary mixer or (iii) adding aluminium or zinc powder to cement mortar or hydrogen peroxide to concrete. It has a very high drying shrinkage and is used for the manufacture of building blocks.

8.3.5 Precast Cement Concrete

The concrete used for the casting of various structural components with or without reinforcement is called precast cement concrete. Fresh concrete is placed and compacted in moulds made for different structural units. After setting and curing they are removed from moulds and used. This is discussed in detail in the next chapter.

8.4 PLAIN CEMENT CONCRETE

Concrete is an important construction material consisting of ingredients that are inert materials, viz., coarse and fine aggregates and binding materials, viz., cement or lime. Cement concrete is widely used as plain concrete, reinforced concrete, pre-stressed concrete and precast concrete members.

Properties of coarse and fine aggregates and the quality of water to make quality cement concrete is discussed in the following sections.

8.4.1 Fresh and Hardened Concrete

Cement concrete is prepared to a plastic consistency by adding water and this hardens on curing. Plastic concrete is called *fresh concrete*, which is a freshly mixed material that can be moulded into any shape. The concrete formed to the required shape and cured is said to be *hardened concrete*. The properties of the concrete, in both the stages, depend on the relative quality and quantity of cement, aggregates and water.

8.4.2 Aggregates

Aggregates are the major and important constituent of concrete. They form the body of the concrete as it occupies 70–80% of the volume of concrete. Although aggregates were originally considered to be inert material, they have recently been found to be chemically active to some extent. Cement is the only factory-made component, whereas aggregates (both coarse and fine) and water are naturally available materials.

Concrete can be considered as a two-phase material, viz., paste phase and aggregate phase. Paste phase is the role played by cement and water in concrete whereas aggregate phase is the role played by aggregates (both fine and coarse) in concrete. The requirement of paste phase has been dealt in previous chapter. Now the aggregates phase is discussed.

1. Classification

Aggregates may be classified as (i) normal weight aggregates, (ii) light-weight aggregates and (iii) heavy-weight aggregates.

Normal weight aggregates can be further classified into natural and artificial aggregates. They are also classified as coarse or fine aggregates depending on the size. Natural and artificial aggregates are as follows:

Natural aggregates	Artificial aggregates
Sand, gravel, crushed rock (such as granite, quartzite, basalt, sandstone)	Broken brick, air-cooled slag, sintered fly ash, bloated clay

2. Coarse Aggregate

As stated above, coarse aggregates are produced by the disintegration and crushing of rock. Coarse aggregates are usually those particles which are retained on an IS 4.75 mm sieve.

Stones that are hard and durable such as granite, basalt, quartzite provide good coarse aggregate. Naturally available river gravels and shingle obtained from sea beaches after removing shale may be used.

Blast furnace slag is used for making light-weight concrete. Brick-bats are used for lime concrete or for temporary or cheap concrete work. For reinforced concrete only crushed rock aggregates of size 20 mm are generally considered as satisfactory.

The aggregates should be absolutely clean, free from organic matter and other impurities. The aggregate must be capable of resisting weather.

The grading of coarse aggregate is very important for getting good quality concrete. Good grading of coarse aggregate implies that the quantity of aggregate used should contain all standard fractions of aggregate in required proportions such that the sample contains minimum voids. Thus, a sample of well-graded aggregate (both coarse and fine) containing minimum voids requires a minimum paste phase. Minimum paste require less quantities of cement and water. This is further to mean increased economy, higher strength, lower shrinkage and greater durability.

Fineness modulus is obtained by adding the cumulative percentages of aggregates retained on each sieve (ranging from 80 mm to 150 micron) and dividing their sum by 100. Fineness modulus is an empirical factor, larger the figure, coarser is the material. For a good concrete, the value of fineness modulus of coarse aggregate should be between 6 and 8.5.

One of the methods of arriving at the practical grading is by the trial and error method. Different size fractions of aggregates can be mixed, and the one that gives the maximum weight or minimum void may be used.

3. Fine Aggregate

Fine aggregate is sand, which is usually obtained from rivers or lakes. Sometimes beach sand is also used. In places where sand is not available or a large quantity of sand is needed, crushed stone dust is used. The fineness modulus (F.M.) of sand should be around 2 to 3.2. The following limits may be used as guidance.

	F.M.
Fine sand	2.2–2.6
Medium sand	2.6–2.9
Coarse sand	2.9–3.2

8.4.3 Water

Water is an important ingredient of concrete. As a general guidance, if the water is fit for drinking, it is fit for making concrete. However, some water containing small traces of sugar is not suitable for concrete. Another yardstick adopted is that if the *pH* is between 6 and 8 the water is considered to be suitable.

The best method is to use the water from the particular source and sample concrete is prepared for testing. The concrete is tested for 7 days and 28 days compressive strengths; if 90% result is obtained it can be considered to be suitable.

The following guidelines may also be adopted:

(i) Neutralizing 100 ml sample of water using phenoplhaline as an indicator should not require more than 5 ml of 0.02 N NaOH.

(ii) Neutralizing 100 ml of sample of water using a mixed indicator should not require more than 25 ml of 0.02 N H_2SO_4.

BIS (IS: 456–2000) recommends the permissible limits for solids enlisted in Table 8.1.

Table 8.1 Materials and permissible solids

Material	Permissible max limit of solids
Organic	200 mg/l
Inorganic	3000 mg/l
Sulphates (as SO_3)	400 mg/l
Chlorides (as Cl)	2000 mg/l for plain concrete and 500 mg/l for reinforced cement concrete
Suspended	2000 mg/l

Source: IS: 456–2000.

8.5 CONCRETE CHEMICALS AND APPLICATIONS

Admixture and construction chemicals are chemicals added to the ingredients of concrete to obtain the desired strength and durability.

8.5.1 Admixtures

Of late concrete is used for varied purposes and is made suitable for different occasions and environments. Ordinary concrete is not as versatile and does not suit all purposes. Thus, admixtures and chemicals are added to the ingredients of concrete. On the other hand, additives are materials that are added at the time of grinding cement clinker at cement factories.

The effect of admixture depends on the brand of cement, grading of aggregate, mix proportion and richness of mix. Thus, it is with caution that admixtures should be selected in correctly predicting the behaviour of concrete. There are several admixtures available, some important ones are discussed below.

1. Plasticizers and Superplasticizers

High degree of workability is needed in different situations. The addition of excess water will only help the fluidity and not the workability of the concrete. But the addition of plasticizers will improve the desirable qualities requires for plastic concrete.

Plasticizers are based on the following constituents:

(i) Anionic surfactants such as lignosulphonates and their modifications.
(ii) Non-ionic surfactants, such as polyglycol acid of hydroxylated carboxylic acids and their modifications.
(iii) Others such as carbohydrates.

Among the plasticizers, calcium, sodium and ammonium lignosulphonates are mostly used. The quantities used are 0.1–0.4 % by weight of cement.

Superplasticizers constitute a relatively new and improved form of plasticizers. These are chemically different from conventional plasticizers. The special quality of superplasticizers is their powerful action as dispensing agents, and they are high-range water reducers. They are chemically different from plasticizers. Superplasticizers permit reduction of water up to 30% without reduction in workability. Superplasticizers are used for the production of flowing, self-levelling, self-compacting, high-strength and high-performance concrete.

2. Retarders

A retarder is an admixture which slows the process of hydration because of which the concrete remains plastic and workable. If concreting is prepared in hot weather, retarders overcome the accelerating effect of high temperature. Retarders are used in consolidating large number of pours without the formation of cold joints and in grouting oil wells.

Calcium sulphate is the commonly used retarder. Other admixtures used as retarders are lignosulphonic acids and their salts, hydroxylated carboxylic acid and their salts. These admixtures increase the compressive strength by 10 to 20%.

Retarding plasticizers are available in the market. These important admixtures are often used in the ready-mixed concrete industry for purposes of retaining the slump loss, during high temperatures long transportation distances and to avoid cold joints.

3. Accelerators

Accelerators are another very useful type of admixture which is added to get early strength. The various purposes are:

(i) Early removal of formwork
(ii) Reduction of period of curing
(iii) Putting the structure early to use
(iv) Accelerating the setting time during cold weather
(v) Energy repair work

The commonly used accelerator in the past was calcium chloride. However, it is not in use now, instead soluble carbonates, silicates, fluosilicates and some organic compounds are used. Fluosilicates and organic compounds like triethenolamine are comparatively expensive. Some of the accelerators available of late can set cement into a hard state in a matter of five minutes. The availability of such accelerators has made underwater concreting very easy. Further, waterfront structures which need repairs in short time may be done using accelerators. These materials could be used in a cold environment up to 10°C. Accelerating plasticizers are also available.

4. Air-entraining Admixture

Air-entrained concrete is made by using air-entraining cement or by adding an air-entraining agent. Air-entraining agents produce a large quantity of air bubbles which act as flexible ball bearings and modify the properties of concrete regarding workability, segregation, bleeding and the finishing quality of concrete. Further, the hardened concrete gains resistance to frost action and permeability.

Natural wood resins, animal and vegetable fats, various wetting agents (such as alkali salts), water-soluble soaps of resin acids, etc., are various air-entrained agents used.

Different air-entraining agents behave differently depending on the elasticity of the film of the bubble formed and the extent to which the surface tension is reduced.

5. Pozzolanic Admixtures

Pozzolanic or mineral admixtures have been in use since the advent of concrete. The application of pozzolanic admixtures modify certain properties of fresh and hardened concrete. The proper addition of pozzolanic admixtures to cement enhances many qualities of concrete, such as lowering the heat of hydration, increasing the water tightness, reducing the alkali–aggregation reaction, resisting sulphate attack, improving workability, etc.

Siliceous materials and aluminous materials do not possess any cementitious materials. But on reacting with cement and moisture, they chemically react with calcium hydroxide liberated on hydration and form compounds possessing cementitious properties. This reaction is called as pozzolanic reaction.

Naturally available pozzolanic materials are clay and shale, diatomaceous earth, volcanic tuffs and pumicites. Artificial pozzolanic materials are fly ash, blast furnace slag, silica fume, rice husk ash, metakaoline and *surkhi*. Other mineral admixtures are finely ground marble, quartz and granite powder.

6. Damp-proofing Admixtures

Two important properties that concrete should possess with reference to water are:

(i) To resist seepage when subjected to the pressure of water and
(ii) To protect the absorption of surface water by capillary action. In general, properly designed and constructed concrete should be impermeable. But it has been accepted that the addition of some damp-proofing admixture may prove to be of some advantage in reducing the permeability.

Damp-proofing admixtures are available in powder or liquid form. They have the property of filling pores or being water repellents. The prime materials in pore-filling admixtures are silicate of soda, aluminium and zinc sulphates, and aluminium and calcium chloride. These are also more active and render the concrete more impervious and accelerates the setting time. Hence, mineral oils free from fatty or vegetable oils are used.

The production of low-permeability concrete depends on the uniform spreading of the admixture.

8.5.2 Construction Chemicals

The discussion so far has been centered on the modification of the properties of concrete using admixtures. Other chemicals that are used to enhance the performance are discussed in this section. Such chemicals are referred to as construction chemicals or building chemicals.

1. Concrete-curing Compounds

The measure adopted to prevent the loss of water from the surface due to evaporation or any other means and to ensure retention is called curing. Surface loss of water from concrete depends upon air temperature, relative humidity, fresh concrete, temperature and wind velocity.

Liquid membrane-forming curing compounds are used. Curing compounds are made with bases such as synthetic resin, wax, acrylic and chlorinated rubber.

Resin and wax-based compounds effectively seal the concrete and prevent surface evaporation. After 28 days of curing, these compounds peel off.

Acrylic-based membrane compounds have the additional advantage of better adhesion of plaster. The membrane does not need to be removed, but the plastering can be done over it. Because of the acrylic emulsion the bonding of the plaster is better.

Chlorinated rubber-curing compounds form a thin film on the surface of the concrete, which prevents drying and at the same time fills the pores on the surface of the concrete. The surface film will wear out.

2. Polymer-bonding Agents

Many a times new concrete is required to be placed over an old concrete surface. In such cases a perfect bond is required. By providing a bond coat between the new and old surface of concrete a bond can be achieved. A mixing of a bonding agent with the new concrete helps to provide a better bond. Such mixtures also improve the workability and reduces shrinkage. Many types of commercial products such as Roof-Bond ERB, Nitobond PVA, etc., are available.

Polymer-modified repair materials are available for the repair of concrete work. Such repair works include ceiling of concrete roof, hydraulic structures, prefabricated members, pipes, poles, etc.

3. Waterproofing Chemicals

Many of the admixtures discussed in the previous section directly or indirectly reduce the permeability of concrete thereby making the material waterproof. However, waterproofing of roofs, walls, bathrooms, toilets, kitchens, basements, swimming pools, and water tanks, etc., still poses some problems.

Different materials are available to make the concrete perfectly waterproof. They are integral waterproofing compounds, acrylic-based polymers, mineral-based polymers, chemical DPC, waterproofing adhesives for tiles, silicon-based water-repellent materials, injection grouts, joint sealants and protective and decorative coatings.

8.6 GRADES OF CEMENT CONCRETE

During 1976 there was only one type of cement available. Later OPC had been graded and the gain in strength after 28 days had been based on the grade and type of cement. The design was based on 28 days' characteristic strength of concrete unless there was evidence to justify a higher strength for a particular structure due to age (IS: 456–2000). Different grades of cement are given in Table 8.2.

Table 8.2 Grades of cement concrete

Group	Grade designation	Specified characteristic compressive strength of 150 mm cube at 28 days (N/mm^2)
Ordinary concrete	M 10	10
	M 15	15
	M 20	20
Standard concrete	M 25	25
	M 30	30
	M 35	35
	M 40	40
	M 45	45
	M 50	50
	M 55	55
High strength concrete	M 60	60
	M 65	65
	M 70	70
	M 75	75
	M 80	80

Source: IS: 456–2000.

8.7 MIX DESIGN CONCEPT

Mix design of concrete is the process of selecting the required ingredients of concrete and finding their relative proportions with the aim of producing an economical concrete of certain strength and durability.

It has been discussed earlier that concrete is based on two phases, viz., aggregate phase and paste phase. Workability of the concrete depends on the lubricating effect of the paste phase. The strength of concrete is predominantly governed by the aggregate paste and the contribution by the paste phase is limited. But the permeability of concrete is based on the quality and continuity of the paste phase. Further, the paste phase fully contributes to the control of drying shrinkage of the concrete.

For a given set of materials, the four factors to be considered in the design of concrete mix are:

(i) Water–cement ratio or cement content
(ii) Cement–aggregate ratio
(iii) Gradation of aggregate
(iv) Consistency

In general, all the four factors are inter-related and can not be dealt with individually to get the best concrete. However, two or three factors are fixed and the others are adjusted to get the required workability and economy.

The water–cement ratio represents the dilution of the paste, and cement concrete depends on the amount of paste. The gradation of the aggregate is done by adjusting the quantity of given fine and coarse aggregates. The required consistency or workability is obtained on the site of placement of concrete.

The effort in proportioning is to use a minimum quantity of paste which will lubricate the mixture while fresh, harden afterwards, will bind the aggregate particles together and fill the space between them. Excess cement should be avoided as it leads to greater cost, high shrinkage, high permeability, and more weathering. All these defects can be nullified by proper gradation.

There are over twelve mix designs of which two are explained, viz., the American Concrete Institute Method and the Bureau of Indian Standards Method. These two methods are popular in India.

8.7.1 American Concrete Institute (ACI) Method of Mix Design

This method has been used since 1944 and has undergone several revisions continuously. It has developed an identical procedure for angular or rounded aggregates, regular or light-weight aggregates and air-entrained or non-air-entrained concretes. The ACI mix design is based on certain factors which are established by field experience on large works. The factors are as follows:

(i) The fresh concrete of a given slump with a well-graded aggregate (of a given maximum size) will have constant total water content irrespective of variations in water–cement ratio and cement content.
(ii) It considers the relationship that the optimum dry-rodded volume of coarse aggregate per unit volume of concrete is based on its maximum size and the fineness modulus of the fine aggregate (Table 8.3) irrespective of the shape of the particles.
(iii) Irrespective of the method of compaction, there is a definite percentage of air that exists which is inversely proportional to the maximum size of the aggregate.

Table 8.3 Dry bulk volume of concrete aggregate per unit volume of concrete

Maximum size of aggregate	Bulk volume of dry-rodded coarse aggregate per unit volume of concrete for fineness modulus of sand of			
Fineness modulus	2.40	2.60	2.80	3.00
10	0.50	0.48	0.46	0.44
12.5	0.59	0.57	0.55	0.53
20	0.66	0.64	0.62	0.60
25	0.71	0.69	0.67	0.65
40	0.75	0.73	0.71	0.69
50	0.78	0.76	0.74	0.72
70	0.82	0.80	0.78	0.76
150	0.87	0.85	0.83	0.81

Source: ACI 211.1–91.
Reproduced with permission from the American Concrete Institute, Farmington Hills, MI (www.concrete.org)

The following step-by-step procedure has to be adopted:

(a) All the required data, viz., (i) given modulus of the selected fine aggregate, (ii) unit weight of dry-rodded coarse aggregate, (iii) specific gravity of coarse and fine aggregates in SSD condition, (iv) absorption characteristics of both coarse and fine aggregates and (v) specific gravity of cement.
(b) The maximum size of the aggregate has to be decided. Generally 20 mm and 10 mm are recommended for RCC and pre-stressed concrete work, respectively.
(c) Workability has to be decided based on the slump depending on the work. General guidance may be taken from Table 8.4.

Table 8.4 Recommended value of slump for different works

Type of construction	Range of slump (mm)
Reinforced foundation walls and footings	20–80
Plain footings, caissons and substructure walls	20–80
Pavements and slabs	20–80
Mass concrete	20–80
Beams and reinforced walls	20–100
Building columns	20–100

Source: ACI 211.1–91.
Reproduced with permission from the American Concrete Institute, Farmington Hills, MI (www.concrete.org)
Note: Upper limit of the slump may be increased by 20 mm for compaction by hand.

(d) Based on the selected slump and maximum of aggregate, the total water in kg/m^3 of concrete can be read from Table 8.5. Also the amount of accidentally entrapped air in non-air-entrained concrete may be obtained from Table 8.5.

Table 8.5 Approximate water content requirement for mixing and air content for different workabilities and nominal maximum size of aggregates.

Workability/ Air content	Water content in kg/m³ of concrete for indicated maximum aggregate size							
	10 mm	12.5 mm	20 mm	25 mm	40 mm	50 mm	70 mm	150 mm
Non-air entrained concrete								
Slump								
30–50 mm	205	200	185	180	160	155	145	125
80–100 mm	225	215	200	195	175	170	160	140
150–180 mm	240	230	210	205	185	180	170	–
Approximate in entrapped air content percent	3	2.5	2	1.5	1	0.5	0.3	0.2
Air-entrained concrete								
Slump								
30–50 mm	180	175	165	160	145	140	135	120
80–100 mm	200	190	180	175	160	155	150	135
150–180 mm	215	205	190	185	170	165	160	–
Recommended average total air content percent								
Mild exposure	4.5	4.0	3.5	3.0	2.5	2.0	1.5	1.0
Moderate exposure	6.0	5.5	5.0	4.5	4.5	4.0	3.5	3.0
Extreme exposure	7.5	7.0	6.0	6.0	5.5	5.0	4.5	4.0

Source: ACI 211.1–1994.

Reproduced with permission from the American Concrete Institute, Farmington Hills, MI (www.concrete.org)

(e) The cement content is calculated by dividing the total water content by the water–cement ratio.

(f) The bulk volume of dry-rodded coarse aggregate per unit volume of concrete is selected from Table 8.5 for the particular maximum size of coarse aggregate and the fineness modulus of fine aggregate.

(g) On multiplying the bulk volume by bulk density, the weight of the coarse aggregate in one cubic meter of concrete can be calculated.

(h) With the knowledge of the specific gravity of coarse aggregate, the solid volume of the coarse aggregate in a cubic meter can be calculated.

(i) Similarly, the solid volume of cement, water, and volume of air is calculated in one cubic meter of concrete.

(j) The solid volume of sand is calculated by subtracting the volumes of cement, coarse aggregate, water and entrapped air from the total volume.

(k) The weight of the fine aggregate is calculated by multiplying the solid volume of the fine aggregate by the specific gravity of the fine aggregate.

8.7.2 Bureau of Indian Standards (BIS) Method of Mix Design

Bureau of Indian Standards (BIS) has recommended a Mix Design Concept mainly based on the research works performed in national laboratories. This method can be applied for both medium-strength and high-strength concretes. The step-by-step approach of the BIS mix design is given below.

(i) The target mean compressive $\left(\overline{f}_{ck}\right)$ strength at 28 days is given by

$$\overline{f}_{ck} = f_{ck} + tS$$

where f_{ck} = characteristic compressive strength
t = a statistical value
S = the standard deviation

(ii) It is desirable to establish a relationship between concrete strength and water–cement ratio at the site. If such a relationship is not available the water–cement ratio corresponding to the target strength may be determined from the relationship given in Table 8.6. It is possible to design in an effective way if the strength of the cement is incorporated in the concrete mix. This is done in the BIS design. Thus, if the 28-days strength is known, the water–cement ratio can be obtained from Fig. 8.1.

Table 8.6 Strength and water–cement ratio

Sl. No.	Exposure condition	Max. water–cement ratio for normal density	Minimum density strength (MPa)
1.	Concrete intended to be watertight		
	(a) Exposed to fresh concrete	0.50	25
	(b) Exposed to brackish or sea water	0.45	30
2.	Concrete exposed to freezing and thawing in a moist condition		
	(a) Kerbs, gutters, guard rails or thin sections	0.45	30
	(b) Other elements	0.50	25
	(c) In the presence of de-icing chemicals	0.45	30
3.	For corrosion protection of reinforced concrete exposed to de-icing salts, brackish water, sea water or spray from these sources	0.40	33

Source: ACI 211.1–91.
Reproduced with permission from the American Concrete Institute, Farmington Hills, MI (www.concrete.org)

(iii) The air content can be obtained from Table 8.7 for the normal size of the aggregate used.

Table 8.7 Approximate air content

Maximum size of aggregate (mm)	Entrapped air as % of volume of concrete
10	3.0
20	2.0
40	1.0

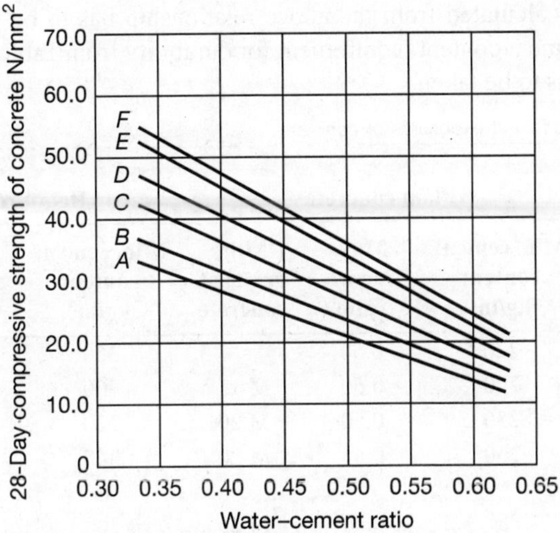

28-Day strength of cement tested according to IS: 4031

A = 31.9–36.8 N/mm^2
B = 36.8–41.7 N/mm^2
C = 41.7–46.6 N/mm^2
D = 46.5–51.5 N/mm^2
E = 51.5–56.4 N/mm^2
F = 56.4–61.3 N/mm^2

Figure 8.1 Relationship between free water–cement ratio and concrete strength for different cement strengths

Table 8.8 Approximate sand and water content per m^3

Maximum size of aggregate (mm)	Water content including surface water / cu.m. of concrete (kg) concrete	Sand as percent of total aggregate by absolute volume
	Applicable for concrete up to M 35 grade	
10	200	40
20	186	35
40	165	30
	Applicable for concrete above M 35 grade	
10	200	28
20	180	25

Source: IS: 10262–82.

(iv) The water content and percentage of sand are determined from Table 8.8 for medium-strength (less than M 35 grade) and high-strength (greater than M 35 grade) concrete.

(v) The cement content per unit volume of concrete is obtained as

$$\text{Mass of cement} = \frac{\text{Water content}}{\text{Water-cement ratio}}$$

The cement calculated from the above relationship has to be checked against the minimum cement content requirement for durability from Table 8.9, and the greater of the two has to be taken.

Table 8.9 Details for different exposures of concrete

Sl. No.	Exposure	Plain concrete: Min. cement content kg/m³	Plain concrete: Max. free w/c ratio	Plain concrete: Min. grade of concrete	Reinforced concrete: Min. cement content kg/m³	Reinforced concrete: Max. free w/c ratio	Reinforced concrete: Min. grade of concrete
1.	Mild	220	0.60	–	300	0.55	M 20
2.	Moderate	240	0.60	M 15	300	0.50	M 25
3.	Severe	250	0.50	M 20	320	0.45	M 30
4.	Very severe	260	0.45	M 20	340	0.45	M 35
5.	Extreme	280	0.40	M 25	360	0.40	M 40

Source: IS: 456–2000.

(vi) The quantity of aggregate can be calculated from the following relationship:

$$V = \left[W + \frac{C}{S_c} + \frac{1}{P}\frac{fa}{S_p}\right]\frac{1}{1000} \tag{8.1}$$

$$c_a = \frac{1-P}{P} \times f_a \times \frac{S_{ca}}{S_{fa}} \tag{8.2}$$

where V = absolute volume of fresh concrete
= (gross volume) – (volume of entrapped air) m³
W = mass of water per m³ of concrete, kg
C = mass of cement per m³ of concrete, kg
S_c = specific gravity of cement
P = ratio of fine aggregate to total aggregate by absolute volume
f_a, c_a = total masses of fine and coarse aggregates per m³ of concrete, respectively, kg
S_{fa}, S_{ca} = specific quantities of saturated, surface-dry fine aggregate and coarse aggregate, respectively.

(vii) The mix proportions have been calculated based on the assumption that the aggregates are saturated and surface dry. When there is any deviation, necessary correction has to be incorporated.

(viii) The above calculated mix properties have to be checked by means of trial batches.

(xi) (a) First the workability is checked, this forms Mix No. 1. If the measured workability is different from the assumed one, then the water content is changed (Table 8.10), and the whole mix design is modified keeping the new water–cement ratio constant.

(b) A minor adjustment in the aggregate quantity may be made to improve the finishing quality or freedom from bleeding and segregation. This forms Mix. No. 2.

(c) Now the water–cement ratio is changed by ±10% and the mix proportion is recalculated. This will form trial Mix Nos. 3 and 4.

(d) Testing for trial Mix Nos. 2, 3 and 4 are done simultaneously.

(f) These test results will provide adequate information including the relationship between compressive strength and water–cement ratio from which the correct mix proportion can be arrived at.

Table 8.10 Adjustment of value in water content and sand percentage for other conditions

Change in conditions stipulated	Adjustment required in	
	Water content	**% Sand in total aggregate**
For sand conforming to specific grade (Zone I to IV IS: 383–1997)	0	+1.5% for Zone I –1.5% for Zone III –3.0% for Zone IV
Increase or decrease in the value of compacting factor by 0.10	± 3%	0
Each 0.05 increase or decrease in water–cement ratio	0	± 1%

Source: IS: 10262–82.

8.8 MANUFACTURING OF CONCRETE

While manufacturing concrete, it should be ensured that every batch of concrete has the same proportions. This is a mandatory requirement so as to satisfy two aspects, viz., same workability and uniform strength. In the manufacturing of concrete the following steps are followed:

1. Proportioning of concrete
2. Batching of materials
3. Mixing of concrete
4. Conveyance of concrete
5. Placing of concrete
6. Compaction of concrete
7. Curing of concrete

8.8.1 Proportioning of Concrete

Selection of the proper quantity of cement, coarse aggregate, sand and water to obtain the desired quality is known as proportioning of concrete. Concrete is formed by successive filling of voids in aggregate by sand, the voids in sand by cement and voids in cement by water and undergoing a chemical reaction.

The concrete formed by proper proportioning of ingredients should satisfy the following properties:

(i) The fresh concrete should have adequate workability for uniform placement.
(ii) The hardened concrete after setting should have the desired strength and durability.
(iii) The concrete should be cheap considering the materials and labour.

There are two approaches in proportioning concrete. In the first method no preliminary tests are conducted. But based on experience, arbitrary ratios such as 1:2:4; 1:1½:3; 1:1:2, etc., are used. This method of proportioning by adopting an arbitrary ratio is called the *mix method*, and the concrete formed by this method is called *ordinary concrete*. This type of concrete is used for ordinary or common works such as columns and members subjected to medium loads, all general building RCC works, mass concrete work in culverts, retaining walls, compound walls, and ordinary machine bases. Ordinary concrete can also be used for long span arches with a mix of 1:1:2 and for heavy stressed members with a mix of 1:2:2.

In the *second method*, preliminary tests are conducted, the mix being designed by any one of the mix design methods to get the desired strength and durability. The concrete formed by this method is called *controlled concrete*. This type of concrete is used for all plain and reinforced concrete structures. The concrete mixes for controlled concrete are designed to have an average strength in preliminary strength test as shown in Table 8.11.

Table 8.11 Strength of concrete in preliminary strength test

Sl. No.	Desired strength (kg/cm^2)	Preliminary test value (kg/cm^2)
1.	100	135
2.	150	200
3.	200	260
4.	250	320
5.	300	380
6.	350	440
7.	400	500

8.8.2 Batching of Materials

After fixing, the desired proportion of quantity of required ingredients, viz., cement, coarse-aggregate, fine-aggregate, cement and sand, has to be measured out in batches for mixing. This process of measuring out ingredients is called batching. Batching may be done by weight or by volume. Volume batching is inferior to weight batching as using the former is liable to change the volume of sand in bulking or aggregate constant void feasibility.

1. Weight Batching

In this batching method all the ingredients of concrete are directly weighed in kilograms. As the weight of cement bag is 50 kg, 20 bags are needed for 1 tonne of cement. For all important works the batching method should be used. This is a slow process.

2. Volume Batching

In this batching method, two units of measurements are employed: liquids are measured in litres and solids in cubic metres. That is all ingredients, viz., water, cement, sand and coarse aggregates are measured in litres, while the end-product concrete is measured in cubic metres.

In volume batching, other quantities are measured keeping cement as the base. Considering that 1 litre of cement equals 1.44 kg, a bag of 50 kg cement has a volume of 3.5 litres. Hence, for measuring aggregates wooden boxes of an inner volume of 3.5 litres has to be used. A size of box of 40 cm × 35 cm × 25 cm satisfies this 3.5 litre volume requirement. Handles are provided on the sides for handling.

As the density of water is 1 g/ml, it can be measured by weight or by volume. The quantity of water required depends on the water–cement ratio. Thus, for a water–cement ratio of 0.50 the quantity of water required is 25 litres (0.50 × 50 = 25 litres).

For accurate batching, the moisture content and absorption of aggregates and bulking of sand have to be ascertained.

8.8.3 Mixing of Concrete

Mixing of concrete may be done by hand or by a machine. Mixing should be done thoroughly so that the ingredients are uniformly distributed, and this can be judged by uniform colour and the consistency of concrete.

On a clean, hard and water-tight platform cement and sand are mixed dry using shovels until the mixture shows a uniform colour. Then aggregates are added and the ingredients are thoroughly mixed. Based on the water–cement ratio, the quantity of water required is calculated and added to the dry mix. The mass is then turned to obtain a workable mass and placed in the required area within 30 minutes. Hand mixing can be used for small quantities of concrete, or due to the non-availability of a machine or where the noise of the machine should be avoided. In general, extra cement of 10% is used to compensate the possible inadequacy.

Mixing by machine is always preferred. Concrete mixers are used for mixing concrete and are of two types, viz., (i) continuous mixers and (ii) batch mixers.

Continuous mixers are used for purposes where large quantities of concrete are needed such as dams, bridges, etc. Batch mixers are also called drum mixers, which consist of drums with blades or baffles inside them, and they are rotated. In the batch mixer, all required materials are fed into the hopper of the revolving drum in correct quantity. When the mix has attained the desirable consistency, the mix is discharged from the drum and conveyed to the concreting yard.

8.8.4 Conveyance of Concrete

The mixed concrete should be conveyed to the concreting yard as early as possible but within the initial setting time of the cement. The choice of conveyance depends on several factors, viz., nature of work, distance from the mixing place to the construction site, height to be lifted, type of cement, etc.

During the transit from the point of mixing to the point of placement, the following factors have to be borne in mind:

1. Care should be taken not to allow segregation of aggregates.
2. The containers of the drums should be tight such that there is a minimum loss of water.
3. The mixed concrete should be brought to the site before the initial setting time of the cement.

For ordinary simple works, a temporary ladder is erected to convey the concrete using baskets, or it is passed from hand to hand, i.e., by manual labour. For larger and important works, various mechanical devices such as vertical hoists, lift wells for tall structures, wheel barrows, etc., are used.

8.8.5 Compaction of Concrete

Compaction of concrete has to be done as early as possible after placing the concrete in place. The purpose of compaction is to expel air and bring the particles closer so as to reduce the void and make the concrete denser. This increased density will give higher strength and make the concrete impermeable. Over-compaction leads to segregation while under-compaction makes the concrete lean. To check for correct compaction, the compaction should be stopped as soon as the cement paste starts appearing on the upper surface of the concrete.

Compaction by hand may be performed by rodding, tamping, ramming or hammering. Wooden or steel hammers are used for ramming massive concrete works. Tampering is adopted for compacting slabs or other horizontal surfaces. Rodding is followed for compacting vertical sections. In all cases of hand compaction high water–cement ratio should be adopted.

Compaction by machines is performed using mechanical vibrators. This method of compaction has several advantages as detailed below:

(i) The concrete produced is dense and impermeable.
(ii) A lesser water–cement ratio results in about 15% reduction in the use of cement.
(iii) A better bond exists between steel and concrete.
(iv) The surface of the concrete is uniform because of machine compaction.
(v) Because of a high aggregate–cement ratio, there is a possibility for the reduction in creep and shrinkage.
(vi) Filling small openings is feasible because of good consistency in concrete.
(vii) It is relatively fast in placing concrete.
(viii) It consumes comparatively less time, materials and labour and is hence economical.

8.8.6 Curing of Concrete

The following methods of curing are adopted depending on the type of work:

1. Direct Curing

In this method water is directly applied to the surface of curing. In this process, the surface is continuously cured by stagnating water, or using moist gunny bags, straws, etc. These methods are used for horizontal surfaces. Vertical surfaces can be cured by covering moist gunny bags or straws.

2. Membrane Curing

In this method, steps are taken to prevent water evaporation from finished concrete surfaces. This is done by covering the surfaces with water-proof papers, polythene papers or by spraying with patented compounds or covering with a bituminous layer to form an impervious film on the concrete surface.

3. Steam Curing

This approach is widely used in precast concrete units. Here the precast units are kept under the warm and damp atmosphere of a steam chamber.

4. Surface Application by Chemicals

Chemicals such as calcium chloride are spread as a layer on the finished concrete. The chemical absorbs moisture from the atmosphere and prevents evaporation of the moisture from the concrete surface.

8.9 FINISHING

Finishing is the last stage in concrete construction. After casting of a concrete, the concrete does not offer a pleasant architectural appearance. In some cases like beams, finishing may not be needed. For a residential building, airport or road pavement and culvert and bridges, finishing is a must. Now-a-days, the centering materials are so made such that the concrete exhibit a pleasant surface finish. Many of the prefabricated concrete units are made in such a way to give an attractive architectural effect.

Different types of finishes have been in use now-a-days. Surface finishes may be grouped as under.

1. Formwork finishes
2. Surface treatment
3. Applied finishes

8.9.1 Formwork Finishes

Concrete maintains the shape of formwork, i.e., centering work. Thus, keeping the required shape through formwork, viz., undulated fashionable shapes, V-shaped finishes, plain surfaces, etc., any pleasing surface can be obtained. The imaginative ideas of architects may be implemented by a careful formation of concrete surface.

A properly made out formwork can give a very smooth surface using the right proportioning of materials better than that made by the best mason. Because of increasing cost of labour, self-finishing concrete surfaces are preferred.

8.9.2 Surface Treatment

This is a commonly used method of surface finishing. It is important for a residential floor to be smooth. To obtain a smooth finish, first the proportioning of mix should be appropriate. The finishing of surface should be at the same rate as that of placing of concrete. Attention must be paid to the extent and time of transportation. Careful attention should be paid to the non-formation of laitance, that excess mortar does not remain and excess water is not accumulated on the surface. A poor and dissatisfactory surface is formed due to hurried completion.

Rough finishes are required on concrete pavement slabs, air-field pavements, on roads, etc. In such cases, the concrete is brought to the plane level surface, and then lightly raked, broomed, textured or scratched to create a rough surface. Other finishes are to give good look like exposed aggregate finish.

8.9.3 Applied Finish

Applied finish is the exterior application of rendering made on concrete structures. In this case the concrete surface is finished and kept wet after which mortar (1:3) is applied. The desired finish is then given to the mortar.

Sometimes the rendering applied on a wall is pressed with a sponge. On repeating this process, the sand is exposed and the surface obtains a finish which is known as *sand facing*. Another type of finish known as Rough Cast Finish is also done. In this type, a wet plastic mix (three parts of cement, one part of lime, six parts of sand and four parts of about 5-mm-size peagravel aggregate) is dashed on the wall surface using a scoop or plasterer's trowel.

Other finishes under this category are non-slip finish, coloured finish, wear-resistant floor finish, craziness finish, etc.

8.10 TESTING OF FRESH AND HARDENED CONCRETE

Testing of fresh and hardened concrete is important in concrete construction. Tests are conducted on fresh concrete to check the workability of concrete, and on hardened concrete to determine the strength, creep effects, durability, etc.

8.10.1 Testing of Fresh Concrete

The following tests are commonly employed to measure workability of fresh concrete:

1. Slump Test
2. Compaction factor Test
3. Flow Test
4. Kelly Ball Test
5. Vee Bee Consistometer Test

1. Slump Test

It is the most commonly used method of measuring consistency of concrete. This test can be conducted in the field or in a laboratory. However, this test is not suitable for very wet or very dry concrete.

The apparatus for conducting the slump test consists of a metallic mould in the form of a frustum of a cone with a 20 cm diameter at the bottom, 10 cm diameter at the top and 30 cm in height. A steep tamping rod of 16 mm diameter, 0.6 m long with a bullet end is used for tamping.

The internal surface of the mould is thoroughly cleaned and placed on a smooth non-absorbent horizontal surface. The mould is filled in four layers of equal height. Each layer is compacted by giving 25 blows with the tamping rod uniformly. After filling the mould and rodded, the excess concrete is shaken off and levelled.

The mould is lifted upwards from the concrete immediately by raising it slowly. This allows the concrete to subside. This subsidence is referred to as the *slump* of concrete. The difference in height of the mould and that of the subsided concrete is measured and reported in mm, which is considered to be the slump up of concrete.

The pattern of slump also represents the characteristics of concrete (Fig. 8.2). If the slump of the concrete is even, it is called a true slump. If one-half of the cone slides down, it is called a shear slump. In this case the average value of the slump is considered. The shear slump also indicates that the concrete is not cohesive and reflects segregation. Concrete mixes are classified based on the slump as given in Table 8.12.

Table 8.12 Slump and nature of concrete

Slump	Nature of concrete mix
No slump	Stiff and extra stiff mix
From 10–30 mm	Poorly mobile mix
From 40–150 mm	Mobile mix
Over 150 mm	Cast mix

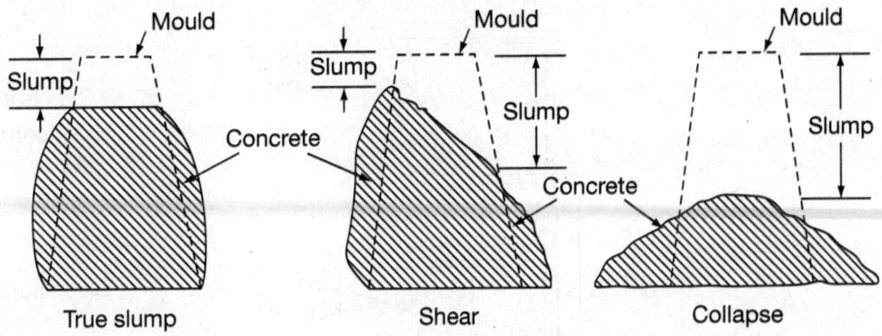

Figure 8.2 Types of slumps

Slumps recommended for various works of concrete construction are presented in Table 8.13.

Table 8.13 Recommended slump for concrete works

Sl. No.	Nature of concrete construction	Recommended slump
1.	Concrete to be vibrated	10–25 mm
2.	Concrete for road construction	20–40 mm
3.	Mass concrete	25–50 mm
4.	Concrete for horizontal tops of kerbs, parapets, piers, slabs and walls	40–50 mm
5.	Concrete for canal lining	70–80 mm
6.	Normal RCC work	80–150 mm
7.	Concrete for arch and side walls of tunnels	90–100 mm

The slump test can be conducted both in the laboratory and at the work site. The slump test results grant the facility to easily detect the difference in water content of successive batches of the identical mix.

2. Compacting Factor Test

This is a more refined test than the slump test. The test measures the degree of compaction obtained by using certain energy in overcoming the internal friction of the concrete. This property is a measure of workability.

The test apparatus consists of two conical hoppers with bottom doors and a separate cylinder kept at the bottom. The concrete is filled in the top hopper fully without compaction and released successively through the two hoppers and into the bottom cylinder (Fig. 8.3). After striking off the level in the cylinder the weight of the concrete (W_1) in the cylinder is determined. The same cylinder is filled with the same batch of concrete and compacted to get the maximum weight (W_2). The ratio of the observed weight, W_1, to the theoretical weight, W_2, i.e., W_1/W_2 is the *compacting factor*. The workability, compacting factor and the corresponding slump are given in Table 8.14.

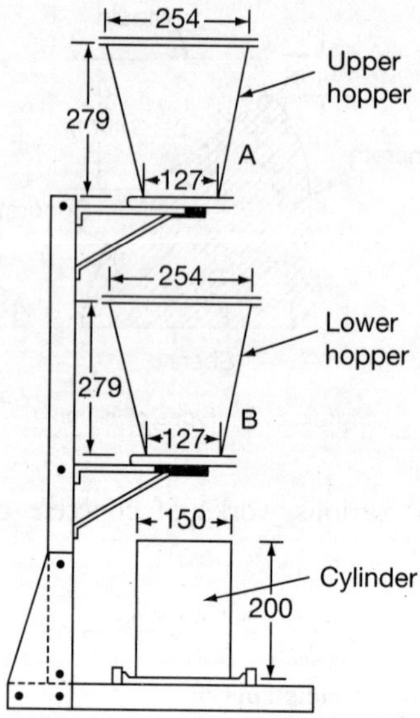

Figure 8.3 Compacting factor test apparatus (dimensions shown are in mm)

Table 8.14 Workability and compacting factor

Workability	Compacting factor	Corresponding slump
Very low	0.80	0–25 mm
Low	0.85	25– 50 mm
Medium	0.92	50–100 mm
High	0.95	100–180 mm

The compacting factor test measures the quality of concrete, which relates very close to the workability. This test clearly depicts the workability of concrete.

3. *Flow Test*

This test gives an indication of the quality of concrete with respect to consistency, cohesiveness and non-segregation. In this test a mass of concrete is subjected to floating, and the flow or spread of the concrete is measured. The flow is related to workability.

The test apparatus consists of a flow table of 76 mm diameter on which concentric circles are marked (Fig. 8.4). A mould similar to that used in the Slump Test with a base diameter of 25 cm, an upper diameter of 17 cm and a height of 12 cm is used. The mould is kept on a clean table, and concrete is filled in two layers with each layer being rodded 25 times with a tamping rod of 1.6 cm diameter and 61 cm long with a rounded end. The excess concrete at the top of the mould is levelled. The mould is lifted vertically upwards completely. The concrete stands on its own without support.

The table is raised and dropped 12.5 mm with the cam arrangement, 15 times in about 15 seconds. The diameter of the spread concrete is measured in 6 directions and the average value is taken. The flow of the concrete is defined as the percentage increase in the average diameter of the spread concrete to the base diameter of the spread concrete to the base diameter of the mould, i.e.,

$$\text{Flow (\%)} = \frac{\text{Spread in diameter in cm} - 25}{25} \times 100$$

The value varies from 0 to 150%.

The spread pattern of the concrete also reflects the tendency of the segregation. The flow test is a laboratory test.

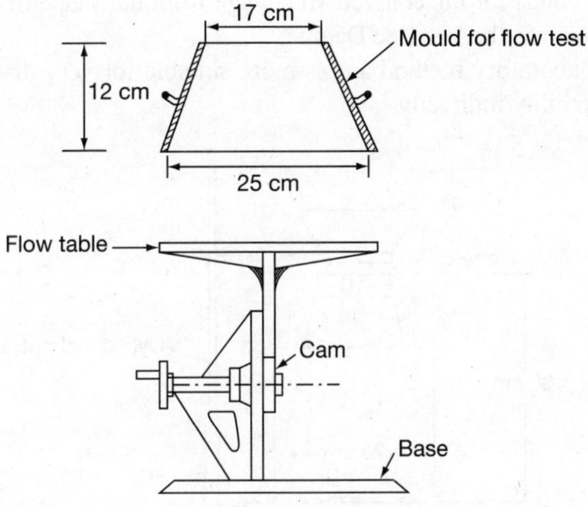

Figure 8.4 Flow table apparatus

4. Kelly Ball Test

It consists of a metal hemisphere of 15 cm diameter weighting 13.6 kg (Fig. 8.5). The concrete base should be 20 cm depth, and the minimum distance from the centre of the ball to the nearest edge of the concrete is 23 cm. The ball is gradually lowered to the surface of the concrete. The depth of the penetration is read immediately on the stem to the nearest 6 mm. The test can be done in a shorter periods of about 15 seconds. This test gives more consistent results than slump tests.

The test can be performed in the field and it can be performed on the concrete placed on the site.

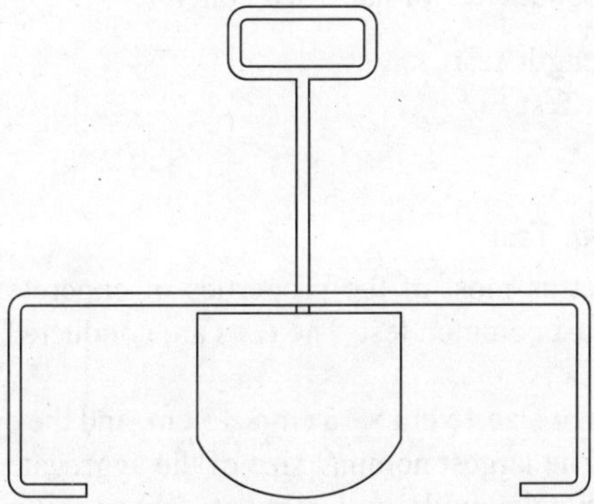

Figure 8.5 Kelly ball

5. Vee Bee Consistometer Test

This test consists of a vibrating table, a metal pot, a sheet metal cone and a standard iron rod (Fig. 8.6). A slump cone with concrete is placed inside the sheet metal cylindrical pot of the consistometer. The glass disc is turned and placed on the top of the concrete in the pot. The vibrator is switched on, and the stop watch is started simultaneously. The vibrator is kept on till the concrete in the cone assumes a cylindrical shape. The time is noted. The time required in seconds for the concrete to change from the shape of a cone to the shape of a cylinder is known as the Vee Bee Degree.

This is a good laboratory method and is more suitable for very dry concrete. This test measures the workability indirectly.

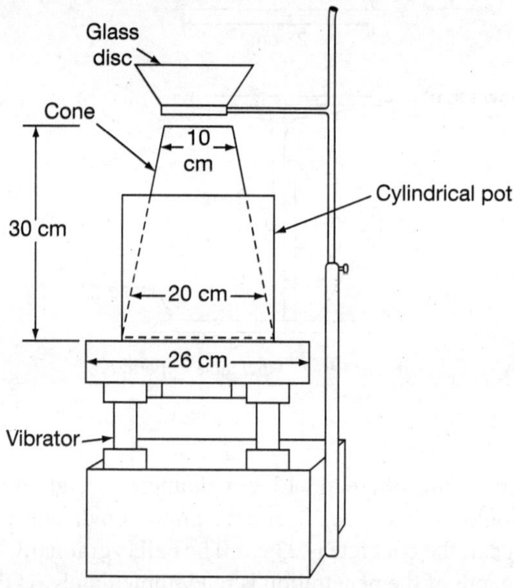

Figure 8.6 Vee Bee consistometer-type VBR

8.10.2 Testing of Hardened Concrete

The following tests are conducted for hardened concrete:

1. Compressive Strength Test
2. Flexural Strength Test
3. Split-tension test

1. Compressive Strength Test

This is an important test as most of the properties of concrete are qualitatively related to it. It is an easy and most common test. The tests are conducted on cubical or cylindrical specimens.

The cube specimen is of size 15 cm × 15 cm × 15 cm, and the cylinder is of 15 cm diameter and is 30 cm long. The largest nominal size of the aggregate does not exceed 20 mm. The moulds must be of metal moulds, preferably of steel or cast iron. The moulds are made in such a way that the specimens are taken out without damage. A tamping steel bar of 16 mm diameter and 0.6 m long with a bullet end is used for compacting.

The test cube specimens are made as soon as practicable. The concrete is filled into the mould up to approximately 5 cm. Each layer is compacted by the tamping rod (25–35 strokes depending on 10–15 cm depth) or by vibration. The top layer is compacted using a trowel. It is covered with a glass or metal plate to prevent evaporation. The specimens are demoulded after 24 hours and submerged in fresh clean water or saturated lime solution and kept there and taken out just prior to the test. The water should be maintained approximately at 27° ± 2 °C and on no account must the specimens be allowed to dry out.

The specimens are tested in a compression testing machine on completion of 7 and 28 days. Compression on the cube or cylinder undergoes lateral expansion owing to Poisson's ratio effect.

Cylindrical specimens are less affected by end restraints caused by plaster, and hence it is believed to give more uniform results than the cube. Further, the cylinder simulates the real condition in the field in respect of the direction of the load. Normally, the strength of the cylindrical specimen is taken as 0.8 times the strength of cubical specimens.

2. Flexural Strength Test

Concrete is relatively strong in compression and weak in tension. Tensile stresses can develop in concrete due to drying, shrinkage, rusting of steel reinforcement, temperature gradient and many other reasons. Hence, the tensile strength of concrete gains importance.

Direct measurement of tensile strength is not feasible. Hence, beam tests are found to be dependable to measure the flexural strength property of concrete. The Modulus of Rupture is taken to be the extreme fibre stress in bending.

The value of the Modulus of Rupture depends on the dimension of the beam and the type of loading. The loading adopted is central or two-third point loading. In the central point loading, the maximum fibre stress occurs below the point of loading where the bending moment is at the maximum. In the two-point loading, a critical crack may appear at any section, where the bending moment is at the maximum, or the resistance is weak. In general, the two-point loading yields a lower value of the Modulus of Rupture than the centre point loading.

The various types of loading are shown in Fig. 8.7. The size of the specimen is 15 cm × 15 cm × 70 cm. In case of concrete with an aggregate of size less than 20 mm, a beam size of 10 cm × 10 cm × 50 cm may be used. The mould may be of metal or steel or cast iron. The tamping bar may be of steel weighing 2 kg, 40 cm long and should have a ramming face of 25 mm^2.

The testing machine should have sufficient loading capacity with a specific rate of loading such that the permissible errors on the applied load should not be greater than ± 0.50%.

The flexural strength of the specimen is expressed as the Modulus of Rupture, f_b, as

$$f_b = \frac{3P \times a}{b \times d^2} \tag{8.3}$$

where P = maximum load in kg applied to the specimen

a = 17–20 cm for a 15.0-cm specimen or >13.3 cm for a 10.0-cm specimen

b = measured width in cm of the specimen

d = measured depth in cm of the specimen at the point of failure

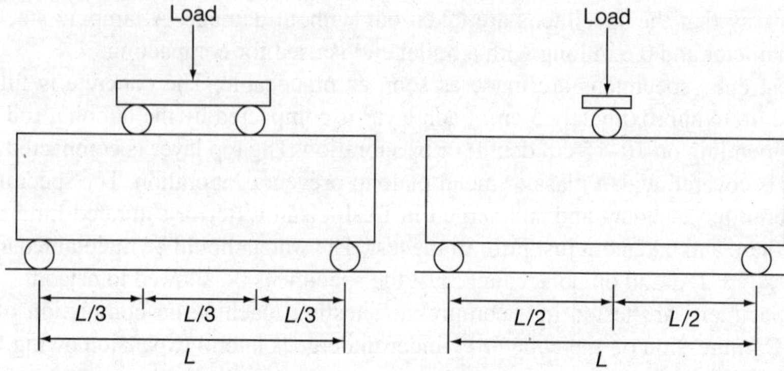

Figure 8.7 Loading arrangement in the flexural beam test

If *a* is less than 17.0 cm for a 15.0-cm specimen or less than 11.0 cm for a 10.0-cm specimen, the results of the test may be discarded.

3. Split-tension Test

This is an indirect tension test. This is also referred to as the Brazilian test. In this test a cylindrical specimen is placed horizontally between the loading surfaces of a compression testing machine. The load is applied until failure of the cylinder along the vertical diameter. The test specimen is shown in Fig. 8.8.

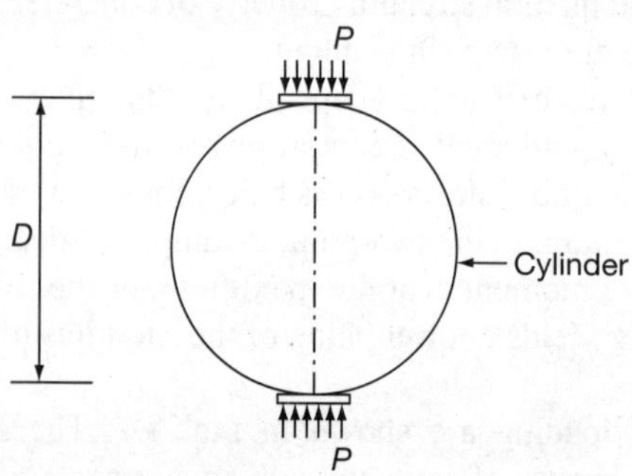

Figure 8.8 Split-tension test

When the load is applied along the diameter, compressive stresses develop immediately below the two generators to which the load is applied. But a larger portion about 5/6th of the depth is subjected to tensile stress.

The main advantage of the test is that the same compression testing machine and the same cylindrical specimen used for the compression test may be used. Narrow packing strips of suitable material such as plywood are used to reduce the high compressive stresses.

The Split-tension Test is simple to perform and generally gives more uniform results. The tensile strength from the Split-tension Test is almost near its true tensile strength than the Modulus of Rupture. The Split-tension Test gives 5–12% higher value than the direct tensile strength.

8.11 CRACKS IN CONCRETE

Factors that contribute to the formation of cracks in cement concrete are discussed below.

8.11.1 Plastic Shrinkage Cracks

Evaporation of water from fresh concrete is due to the absorption by formwork and during the hydration process. The surface of concrete dries up when the loss of water from the surface of the concrete is faster than the migration of water from the interior to the surface. This creates a moisture gradient which results in surface cracking while the concrete is stiff in a plastic condition. The magnitude of the crack depends upon the rate of evaporation of water from the surface of the concrete.

8.11.2 Settlement Cracks

Plastic concrete settles when vibrated. There will be no cracks if the concrete settles uniformly. Some cracks are bound to occur if there is any obstruction to uniform settlement due to reinforcement or larger pieces of aggregates. Such cracks are called settlement cracks. Such cracks are formed generally in deep beams.

8.11.3 Cracks Due to Bleeding

Water being light in weight moves upward in concrete, which leads to bleeding. This water evaporates to make the top surface porous having very little abrasion resistance. Masons often float the concrete when the bleeding water is still standing on the surface. Too much of working on the exposed surface leads to pressing the coarse aggregate down and bringing up fine particles of cement and water. Such a condition develops cracks on the surface.

8.11.4 Cracks Due to Delayed Concrete Curing

During the initial stages of curing, uninterrupted hydration is a must. Due to the hot sun, winds and lower relative humidity, young concrete dries faster leading to plastic shrinkage cracks.

8.11.5 Cracks Due to Construction Effects

Improper formwork with less rigidity may lead to sinking, bending, etc., and the wet concrete may cause cracks or deformation after compaction which may go unnoticed.

8.12 QUALITY CONTROL OF CONCRETE

Quality control implies that the assigned work is done according to the specifications agreed in the contract. Major civil engineering works such as multistoreyed buildings, dams, harbours, etc., have to be constructed with utmost care as they have to last and be used for decades. Specifications of work should be framed based on cost or standard processors so that they serve effectively as a guide to complete the work with high quality. The specifications are as important as the design of the project.

To make a quality concrete construction at a site, fieldwork has to be organized with the three divisions with mutual coordination, viz., the engineering division, the manufacturing division and the placing division. The engineering division looks after all forms, reinforcements details and installation of all embedded parts. The manufacturing division takes care of the control of materials, batching and mixing of concrete. The placing division takes care of placing, curing and other subsequent works.

The whole aim is to produce inexpensive high-quality concrete. The general requirements to produce high-quality concrete, i.e., densest, more workable and high strength, are as follows:

(i) Air bubble should be completely removed from the concrete.
(ii) Compaction of concrete should be such that a minimum void is present.
(iii) Adequate curing for 28 days has to be effected.

Resources such as supervisors, engineers, etc., who are involved in the production of high-quality concrete should be aware of all the factors affecting good-quality concrete. Accordingly, the execution of quality control of concrete should be carried out at every stage. Carefully constructed high-quality concrete work has the following advantages:

(i) The possibility of failure is minimized.
(ii) Lower cost of construction with long life.
(iii) Low-maintenance cost.
(iv) Possibility of using low-grade materials for some other purposes.

In summary, to obtain high-quality concrete all suitable precautions must be taken to ensure proper inspection of the ingredients, proper batching and mixing, proper transportation and careful placing, adequate curing and careful renewal of formwork and necessary finishing.

8.13 NON-DESTRUCTIVE TESTING

Non-destructive testing is done on hardened concrete. In non-destructive testing methods, some properties of concrete are used to estimate strength, durability, elastic parameters, crack depth, micro-cracks and progressive deterioration of concrete.

Such properties of concrete are hardness, resistance to penetration of projectiles, rebound number, resonant frequency, ability to allow ultrasonic pulse velocity, ability to scatter and transmit X-rays and gamma rays, its response to nuclear activation and acoustic emission. Various non-destructive methods have been developed using one or more of the above properties. Some of the important methods in use are explained below.

8.13.1 Schmidt's Rebound Hammer

This is a commonly adopted equipment for measuring surface hardness. It consists of a spring control hammer which slides on a plunger and is housed in a tube. Once the plunger is pressed against the surface of the concrete, the mass behind the spring rebounds. After impact, the spring control mass rebounds and takes the rider along the guide scale. Based on the position, the reading is taken. Considering the reading and the calibration, the actual strength can be determined.

8.13.2 Frequency Method

It is another important non-destructive method used to determine the compressive strength and other properties. The fundamental principle on which the method is based is the velocity of material through a material. A mathematical relation could be made between the resonant frequency of the material to the Modulus of Elasticity of the material. The property of homogeneous material can be made use of heterogeneous materials like concrete with judgement.

8.13.3 Nuclear Method

This is a new technique which is used to determine the moisture content and the cement content. This method employs the scattering of neutrons directed towards the concrete and the number of neutrons returned. With a standard relationship number of neutron and water content/cement content, the required water or cement content can be obtained.

8.13.4 Radioactive Method

Here X-rays and gamma rays are used. When X-rays and gamma rays are passed through concrete, the electromagnetic spectrum penetrates concrete but undergoes attenuation in the process. The degree of attenuation is a function of the kind of matter traversed, its thickness and the wavelength of the radiation. Further, the intensity of the incident gamma-rays and the emerging gamma-rays after passing through the specimens is measured. These two values are used to calculate the density of concrete.

The gamma-rays transmission method is particularly used to measure the thickness of concrete slabs of known density. This is achieved by passing gamma rays of known intensity to penetrate through the concrete. The thickness of the concrete is measured based on the intensity of gamma rays measured on the other end.

8.13.5 Pullout Test

Here a rod is embedded in concrete blocks. These are pulled out and the strength of the concrete is determined. The ideal way to use the Pullout test in the field is to incorporate assemblies for pull out in the structure itself. These could be pulled out and the strength determined.

8.13.6 Pulse Velocity Method

It consists of two parts, viz., the mechanical ionic pulse velocity method and the ultrasonic pulse velocity method.

The mechanical sonic pulse velocity method consists of measuring the time of travel of longitudinal or compressive waves generated by a single impact hammer blow or repeated blows. The ultrasonic pulse velocity method consists of measuring the time of travel of electronically generated mechanical pulses through the concrete. Of these two, the ultrasonic pulse velocity has gained popularity throughout the world.

The pull velocity methods have been used to evaluate the quality of concrete, concrete strength, durability, Modulus of Elasticity, detection of water, etc.

High pulse velocity readings in concrete are indicative of concrete of good quality. Table 8.15 gives the pulse velocity range of quality of concrete (Leslie and Chessman, 1949, reported by Shetty, 2006).

Table 8.15 Suggested pulse velocity for concrete

Pulse velocity m/s	General conditions
4575	Excellent
3660–4575	Good
3050–3660	Questionable
2135–3050	Poor
2135	Very poor

Pulse velocity techniques have been used successfully for the detection of cracks. This is possible only when the width of the crack is of considerable depth and of appreciable width. The basic principle in such a situation to detect the crack of the depth is that no signal will be received at the receiving transducers, the pulse will pass around the end of the crack and signal is received at the transducers. However, the pulse would have travelled a distance longer than the straight line path upon which pulse velocity computations are made. The difference in the velocity of pulse is used to estimate the path length and therefore the crack depth. Figure 8.9 illustrates the principle of crack detection.

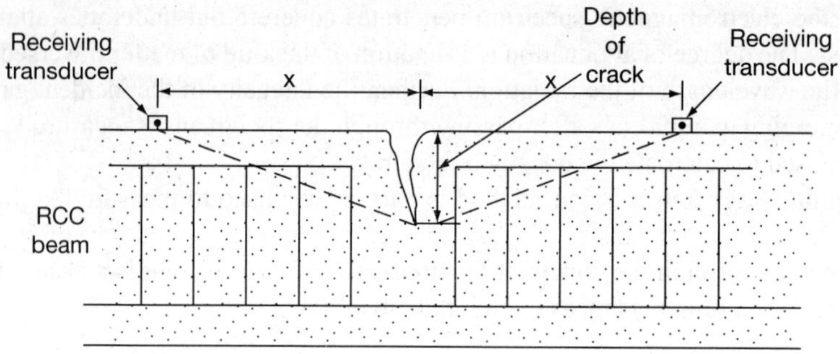

Figure 8.9 Pulse velocity technique

8.14 LIME CONCRETE

8.14.1 Ingredients of Lime Concrete

Lime concrete is made of lime, fine aggregate and coarse aggregate and mixed in suitable proportions in addition to water. Hydraulic lime is generally used as a binding material, sand, *surkhi* and cinder are used as fine aggregates, and broken bricks or broken stones are used as coarse aggregates.

8.14.2 Properties of Lime Concrete

1. It provides a good base and is capable of taking loads.
2. It has a certain degree of flexibility and adjusts very well with thé surface in contact.
3. It exhibits a certain degree of water proofing property.
4. It has adequate volumetric stability when matured lime is used.
5. It resists weathering effects and is quite durable.
6. It can be prepared easily with less cost.

8.14.3 Preparation of Lime Concrete

Good lime concrete is prepared using hydraulic lime. Fat lime is not used as it is not suitable to use in large masses. The sand to be used should be free from impurities. The coarse aggregates used for lime concrete are broken bricks or broken stones. The water used should be clean.

The mix proportion adopted for various works are as follows:

(i) Under floors – 1:2:3
(ii) Foundation work – 1:2:4

The following procedure is adopted while preparing lime concrete:

(i) Based on the mix, the quantity of coarse aggregate required is measured, slaked and soaked adequately with water, and a measured quantity of sand is added.
(ii) Dry mixing is done first, and then a sufficient quantity of water is added. The whole mass is again mixed thoroughly to obtain the correct consistency.
(iii) The wet concrete is laid in uniform layers in such a manner that after compaction, the thickness should not be more than 15 cm. The compaction is done with light hammers.
(iv) The concrete is allowed to set for 24 hours.
(v) The surface of the concrete is watered and rammed with heavy hammers till the concrete is thoroughly compacted.

8.14.4 Uses of Lime Concrete

Lime concrete is used for

1. Foundation bases of load-bearing walls, columns and floors.
2. Filling haunches over masonry arches.
3. Terrace finishing as it is the quality of volumetric stability.

8.14.5 Precautions

The following precautions have to be observed:

(i) The person working should wear rubber gloves and rubber gumboots, otherwise there is a possibility of formation of rashes on the skin due to lime.
(ii) Workers also should apply oil over their skins to protect themselves from rashes and skin cracking.

SALIENT POINTS

1. Concrete is an important construction material consisting of ingredients which are inert materials, viz., coarse and fine aggregates and binding materials, viz., cement or lime.
2. Plain cement concrete is the simplest form of concrete made with cement, sand, pebbles or crushed rock and water.
3. Cement concrete provided with steel reinforcement for bearing tension, bending, torsion, fatigue, etc., is referred to as reinforced cement concrete.
4. Pre-stressed concrete is one in which high tensile steel wires are stretched initially to the desired level and concrete is placed. After setting and completion of curing, concrete is used.
5. Light-weight concrete has less density compared to the conventional concrete. Light-weight concrete may be made (i) by omitting the fine aggregate, (ii) by using light-weight aggregate and (iii) by aerating concrete.
6. Concrete used for the casting of various structural components with or without reinforcement is called precast cement concrete.

7. Cement concrete is made to a plastic state by adding water and hardens on curing. Plastic concrete is called fresh concrete. The concrete formed to a required shape and cured is said to be hardened concrete.
8. Aggregates are the major and important constituents of concrete. Aggregates may be classified as (i) normal weight aggregates, (ii) light-weight aggregates and (iii) heavy-weight aggregates.
9. Concrete can be considered to be a two-phase material, viz., the paste phase and the aggregate phase. The paste phase is connected with cement, and the aggregate phase is connected with aggregates.
10. The Fineness Modulus is an empirical factor; the larger the figure, coarser is the material. For good concrete, the value of the Fineness Modulus of coarse aggregate should be between 6 and 8.5, and fine aggregate should be between 2 and 3.2.
11. As a general guidance, water fit for drinking is fit for preparing concrete. If the *pH* is between 6 and 8, the water is considered to be suitable.
12. Admixtures and construction chemicals are those added to the ingredients of the concrete or at a later stage to obtain the required mix for the desired strength and suitability.
13. Plasticizers are added to improve the desired qualities that are required for plastic concrete.
14. A retarder is an admixture which slows down the process of hydration because of which the concrete remains plastic and workable.
15. An accelerator is a type of admixture which is added to obtain early strength.
16. Air-entrained concrete is made using air-entraining cement or by the addition of an air-entraining agent.
17. Pozzolanic or mineral admixtures modify certain properties of fresh and hardened concrete, such as lowering the heat of hydration, increasing the water tightness, reducing alkali–aggregation reaction, resisting a sulphate attack and improving workability.
18. Damp-proofing admixtures have the property of filling pores or repelling water.
19. Mix design of concrete is the process of selecting the required ingredients of concrete and finding their relative proportions with the aim of producing an economical concrete of certain strength and durability.
20. For a given set of materials, the four factors to be considered in the design of a concrete mix are:
    (i) Water–cement ratio or cement content
    (ii) Cement–aggregate ratio
    (iii) Gradation of aggregates
    (iv) Consistency
21. Two popular methods of mix design are: the American Concrete Institute (ACI) Method of Mix Design and the Bureau of Indian Standards (BIS) Method of Mix Design.
22. The following steps are followed for the manufacture of cement:
    (i) Proportioning of concrete
    (ii) Batching of materials
    (iii) Mixing of concrete

(iv) Conveyance of concrete
(v) Placing of concrete
(vi) Compaction of concrete
(vii) Curing of concrete

23. The concrete formed based on the mix-design is called controlled concrete, whereas concrete formed by an adopting ratio (e.g., 1:2:4) is called ordinary concrete.
24. Batching of materials may be weight batching or volume batching.
25. The following tests are employed to measure the workability of fresh concrete:
    (i) Slump Test
    (ii) Compaction Factor Test
    (iii) Flow Test
    (iv) Kelly Ball Test
    (v) Vee Bee Consistometer
26. The tests conducted for hardened concrete are as follows:
    (i) Compressive Strength Test
    (ii) Flexural Strength Test
    (iii) Split-tension Test
27. The cracks formed in concrete are as follows:
    (i) Plastic Shrinkage Crack
    (ii) Settlement Cracks
    (iii) Cracks due to delayed concrete curing
    (iv) Cracks due to bleeding
    (v) Cracks due to construction effects
28. Quality control implies that assigned work is done according to the specifications agreed in the contract.
29. Non-destructive testing methods on hardened concrete use some properties of concrete to estimate the strength, durability, elastic parameters, crack depth, micro-cracks, and progressive deterioration of concrete.
30. Non-destructive testing methods are: the Schmidt's Rebound Hammer Method, the Frequency Method, the Nuclear Method, the Radioactive Method, the Pull out Test and the Pulse Velocity Method.
31. Lime concrete consists of lime, a fine aggregate and a coarse aggregate, mixed in proportion in addition to water. It is cheaper and has less strength than cement concrete.

REVIEW QUESTIONS

1. Distinguish between lime and cement concrete.
2. What is meant by proportioning of concrete?
3. Why proportioning of concrete is necessary?
4. Can sea water be used for making concrete? Explain.
5. What is meant by curing of concrete?

6. What are the quality requirements of water?
7. Discuss the importance of water–cement ratio.
8. How is compaction of concrete achieved?
9. What are the types of vibrators available for compaction of concrete and where they are used.
10. Explain the factors affecting evaporation of water from concrete.
11. Discuss in detail the various methods of curing.
12. What are admixtures? Explain any two.
13. What is mix design? Explain the ACL method of mix design.
14. How batching of materials is done?
15. Distinguish between fresh and hardened concrete.
16. What is meant by controlled concrete?
17. What do you understand by controlled concrete? How it is different from ordinary concrete.
18. Define and explain the workability of concrete.
19. Briefly explain the types of finishing.
20. List the tests conducted on fresh concrete. Explain any one test.
21. Explain the Compression Strength Test on hardened concrete.
22. How is quality control of concrete done?
23. Explain any one non-destructive method of testing concrete.
24. Explain crack detection in RCC structures with a sketch using ultrasonic non-destructive test equipment.

9 PRECAST CONCRETE UNITS

9.1 INTRODUCTION

The term cast-in-situ concrete is applied to concrete members which are cast or poured in forms that are set up at the final position of the structure. In contrast, the term precast concrete is applied to individual concrete members of various shapes, which are designed to fit in a part of a whole structure, which are cast in separate forms before they are placed in the proper position of the structure. Precast concrete units form a part of a new technique called pre-fabricated construction.

9.2 PREFABRICATION IN CONSTRUCTION

Prefabrication is most suitable for large-scale housing programmes. Prefabrication technique is used as a cost-effective measure. As the components of any structure can be manufactured under controlled conditions, there is vast improvement in quality. It helps to use the locally available materials with required characteristics. The speed of construction is generally fast compared to cast-in-situ constructions.

Materials that are used for prefabrication have to satisfy one or more of the following characteristics:

1. They should be available easily without any constraint.
2. Light-weight materials are preferred since they are easy to handle and transport.
3. They should preferably have a high thermal insulation property.
4. Workability of the material should be simple and easy.
5. Evidently the durability should be high and withstand all weather conditions.
6. Materials should preferably be non-combustible so as to prevent any problem during transit.
7. The overall economy should be given high importance.

As a matter of fact, materials to be used as prefabricated components are plenty. The latest trend is to use concrete, steel, treated wood, aluminium, cellular concrete, light-weight concrete, ceramic products, etc.

Further discussions are confined to prefabricated concrete elements.

9.3 ADVANTAGES AND DISADVANTAGES OF PRECAST CONCRETE

9.3.1 Advantages

Precast concrete has several advantages over cast-in-situ concrete:

1. When identical members are cast, it is advantageous for the same formwork to be used.
2. As better quality control can be exercised in the factory or yard site, concrete of superior quality can be produced.
3. Smoother exposed surfaces can be produced, which may not require plastering.
4. Precasting work can be carried out under all weather conditions if adequate protection is provided in the casting yard.
5. Precast units can be made to the required shape without much difficulty.
6. When required, they may be dismantled without any breakage.
7. Curing can be done more effectively and economically.
8. Constructions can be erected mostly under all climatic conditions.
9. Buildings may be constructed at a greater speed.

9.3.2 Disadvantages

There are certain aspects that discourage the use of precast concrete units under the following conditions:

1. Special equipment are needed for handling, transportation and erection. This is of particular concern for large-size units like beams, slabs, columns, etc.
2. Cost of construction may increase due to the employment of skilled labour and due to mechanized construction.
3. There is more chance of damage to the precast units while loading, transporting and erecting.
4. Combining different units and making them function as per the design is really difficult, particularly if different types of units are involved.
5. Future modification to the building constructed using precast units is difficult.

9.4 COMPONENTS OF BUILDING

9.4.1 Modular Co-ordination

For prefab construction and for the precast elements, modular co-ordination has to be followed. In this case, some basic module has to be adopted for general application to buildings and their components. The value of the basic module chosen is 100 mm for maximum flexibility and convenience. The symbol used for the basic module is M.

9.4.2 Dimensions of Precast Elements

The preferred dimensions for different elements of a building considering the basic module are as follows:

1. Flooring and Roofing Scheme

Precast slabs or other precast structural flooring units:

(a) Length – Normal length shall be in multiples of 3 M
(b) Width – Normal width shall be in multiples of 1 M and
(c) Overall thickness – Overall thickness shall be in multiples of M/4

2. Beams

(a) Length – Normal length shall be in multiple of 3 M
(b) Width – Normal width shall be in multiples of M/4 and
(c) Overall depth – Overall depth of the floor zone shall be in multiples of M/4

3. Columns

(a) Height – Overall depth (i.e., floor-to-floor or the clear height) shall be in multiples of 1 M for heights up to 2.8 M, and for heights above 2.8 M it shall be in multiples of 2 M
(b) Lateral dimension – Overall lateral dimension or diameter of columns shall be in multiples of M/4

4. Walls

Thickness – Nominal thickness of walls shall be multiples of M/4

5. Staircase

Width – Nominal width shall be in multiples of 1 M

6. Lintels

(a) Length – Nominal length shall be in multiples of 1 M
(b) Width – Nominal width shall be in multiples of M/4 and
(c) Depth – Nominal depth shall be in multiples of M/4

7. Sunshades/Chajja Projections

(a) Length – Nominal length shall be in multiples of 1 M
(b) Projection – Nominal length shall be in multiples of 1 M

Normally acceptable methods of finishes for these components are:

(i) Moulded concrete surface to design.
(ii) Laid-on finishing tiles fixed during casting.
(iii) Finishes obtained by washing, tooling, grinding and grooming of hardened concrete.
(iv) Exposed aggregates in-situ.
(v) Finishes added in-situ.

9.5 MANUFACTURE OF PRECAST CONCRETE ELEMENTS

Crucial factors that are to be considered for effective production and supply of precast concrete components are:

(i) Storage facilities
(ii) Suitable transport facilities
(iii) Erection equipment
(iv) Availability of raw materials

Manufacture of precast concrete units can be done in a centrally located factory, or in a site where a precasting yard is set up at or near the site of work.

9.5.1 Factory Prefabrication

Factory prefabrication is adopted in a centrally located plant for the manufacture of standardized components on a long-term basis. The production unit has to work throughout the year preferably under a closed shed to avoid the effects of seasonal variations. The feasibility of the introduction of the latest manufacturing technique should be considered. Further, there should be some room for the possibility of introducing enhanced technology. To ensure continuous production, work has to be organized in a factory-like manner with support from a team of workmen.

The disadvantages of factory prefabrication are:

(i) Extra cost incurred while transportation.
(ii) Limitation of the availability of transportation equipment for the size and shape of the prefabricated component and road contour.
(iii) Employment of organized labour and their service benefits cause a huge establishment.

9.5.2 Site Prefabrication

In this case the components are manufactured at the site or as near to the site as possible. This type of manufacturing is employed for specific jobs that last for a short span of time. Locally available labour force is as for as possible used and the equipment and moulds are rented. Generally, the work is carried out in open space.

Although this has some economical benefits, there are a few disadvantages:

(i) There are no elaborate arrangements for quality control.
(ii) Work may be disrupted due to weather conditions.
(iii) High degree of mechanization can not be adopted.

9.5.3 Process of Manufacture

Processes of manufacture comprise a main process supplemented by an auxiliary process and a subsidiary process (Sharma, 1988).

1. Main Process

The activities involved in the main process are given below:

(i) Providing and assembling the mould.
(ii) Placing a reinforcement cage in position on the reinforced concrete work and stressing the wires in the case of prestressed concrete elements.
(iii) Fixing tubes and inserts wherever necessary.
(iv) Placing the concrete in the moulds.
(v) Vibrating the concrete and finishing.
(vi) Demoulding the forms, removing the units and stacking the products for curing.
(vii) Curing is done. Preferably steam curing may be resorted to.

2. Auxiliary Process

The activities affecting the successful functioning of the main process are as follows:

(i) Mixing of proportionate ingredients and manufacture of fresh concrete which is done in a mixing station or by a batching plant.

(ii) Prefabrication of reinforcement cage is done in a steel yard or workshop.
(iii) Manufacturing of inserts and other finishing items needed for the precast units.
(iv) Taking on the job of finishing the precast products.
(v) Arranging for the testing of these products.

3. Subsidiary Process

The activities that help to keep the main production unit functioning properly are as follows:

(i) Storage of all the ingredients for concrete and other materials.
(ii) Arranging to transport cement, sand and aggregates.
(iii) Transporting the green concrete and reinforcement cages to the moulding yard.
(iv) Transporting the finished products to the stacking yard.
(v) Undertaking all repair works, maintenance of tools, machines, etc.
(vi) Arranging to produce steam for curing.

4. Overall Management

For the successful running of the entire production unit, the following aspects should be properly exercised:

(i) To ensure speed and economy while manufacturing concrete, a cyclic technological method has to be employed.
(ii) To improve quality and increase productivity, the mechanisation has to be improved.
(iii) Optimum production has to be achieved with quality elements produced.
(iv) To maintain speed to meet the supply.
(v) Working conditions should be improved for the people on the job.
(vi) To minimize the interruption of production due to weather or other causes.

5. Methods of Manufacturing

There are two methods of manufacturing precast concrete units, viz., the Stand Method and the Flow Method.

In the Stand Method, the moulds are stationary and remain at one place, where the various processes involved are carried out in a cyclic order at the same place.

In the Flow Method, the precast units under consideration are in movement from one place to another in a cyclic order according to the various processes involved in the work.

9.6 TYPES OF PRECAST UNITS

Different types of precast concrete units are discussed below.

9.6.1 Hollow Concrete Blocks

Hollow concrete blocks are manufactured in various shapes and sizes. Blocks of concrete are moulded in a machine. Details of standard size blocks, which are used for different works, are given below:

(i) Hollow concrete blocks – 39 cm × 19 cm × 30 cm
(ii) Hollow building tiles – 39 cm × 19 cm × 20 cm
(iii) Hollow concrete blocks for partition walls – 39 cm × 19 cm × 10 cm

The face thickness of blocks is maintained as 5 cm or more. Different forms of hollow concrete blocks are shown in Fig. 9.1.

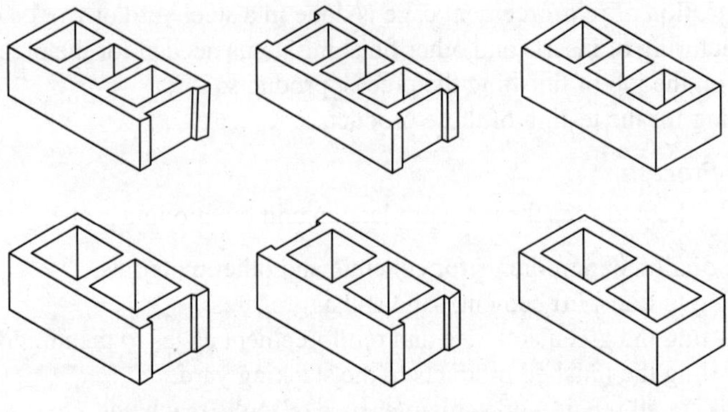

Figure 9.1 Different forms of hollow concrete blocks

Standard hollow blocks are used for the construction of walls and columns. To ensure maximum economy in construction, the overall length and height of walls should be so fixed so as to make maximum use of full and half-length blocks. The availability of the size of hollow blocks should be considered while determining the height of the door and window openings.

Specially shaped blocks are manufactured for corners, sills, lintels and jambs, wall closures, piers between doors and windows, etc., (Fig. 9.2).

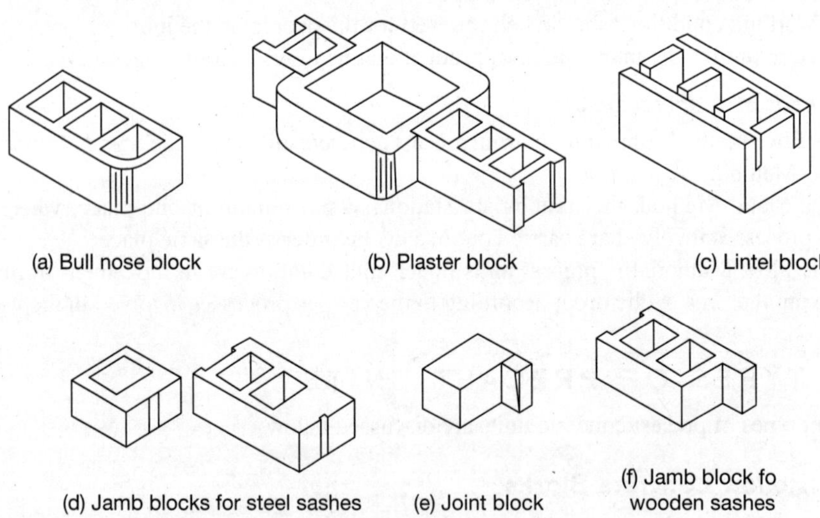

Figure 9.2 Special-shaped blocks of hollow concrete blocks

9.6.2 Lintel Units

Prefabricated RCC lintels can be used conveniently over windows, door openings and passage openings. Precast lintels indirectly accelerate the speed of construction as the curing period is eliminated.

Precast RCC lintels are preferred for small spaces up to 2 m (Fig. 9.3) and they are usually cast in wooden moulds and cured well in advance. While laying a precast lintel case care has to be taken to see that the marked face of the lintel is on the top or upper side. All pre-cast lintels are marked with tar or paint at the top so as to distinguish them from the bottom. This is essential, otherwise the position of reinforcement is reversed, which causes structural failure.

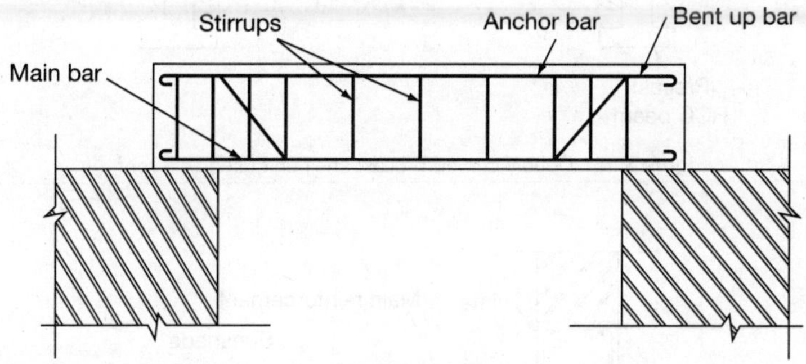

Figure 9.3 Cross-section of a precast concrete lintel

9.6.3 Door and Window Frames

Door and window frames can also be prefabricated. Steel bars of about 4–40 mm in diameter run through the frames of the doors or windows. Suitable hard wood blocks are used for fixing hinges, etc., to them. Concrete is vibrated on a table vibrator and then poured into the mould which ensures a durable mix.

9.6.4 Roofing and Flooring Elements

Prefabricated reinforced concrete battens are used instead of wooden battens. For sloping roof, precast reinforced and prestressed concrete trusses may be used. Plain precast concrete tiles can be used for roofing and flooring flat roofs.

9.6.5 Precast Shell Units

Plain or slightly reinforced concrete can be used in the form of precast shell roofing. A suitable frame is made to cast the shell unit. A hessian cloth is spread over it. The cloth is given a sort of sag to maintain the shape of the shell. To regulate the thickness a thin frame of 2.5 cm height is placed. Chicken mesh cut to fit this mould is placed on it. Cement concrete is poured and manually compacted. The frame is made to rest to permit the hessian to sag with wet concrete. Edge beams are also casted. The final positions of the edge beam and the steel are shown in Fig. 9.4. To obtain a level surface the depression is filled with concrete or earth. The shell units may be 1 m × 1 m to about 3 m × 3 m in size.

9.6.6 Sunshade Unit

Sunshades are made 75 mm at the cantilever end and tapered upwards to the depth of the lintel. They come monolithic with lintel. On an average the projection will be 60–90 cm. The main rods of the cantilever of the sunshade can be 8 mm rods at 10 cm centres with 8 mm distributors at 15 cm centres. At least a 20 mm cover is provided. A typical cross-section of a sunshade (*Chajja*) is shown in Fig. 9.5.

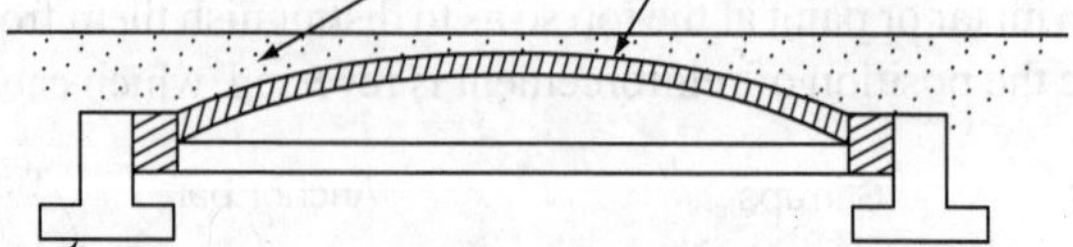

Figure 9.4 Doubly curved precast shell unit used as a roof

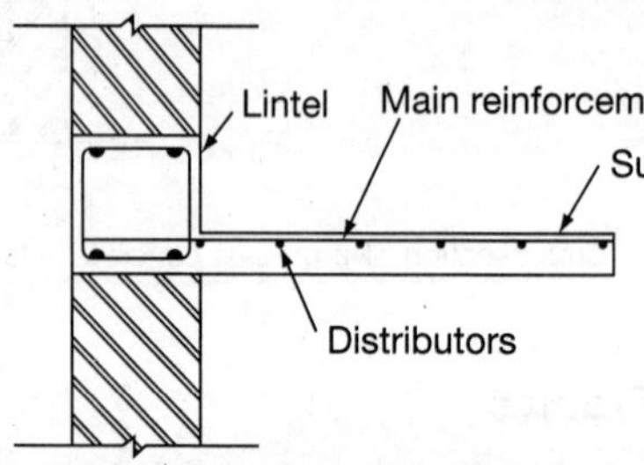

Figure 9.5 Sunshade precast concrete unit

SALIENT POINTS

1. The term precast concrete is applied to individual concrete members of various shapes, which are designed to fit in a part of a whole structure, which are cast in separate forms before they are placed in the proper position of the structure.
2. Precast concrete units form a part of a new technique called prefabricated construction.
3. Prefabrication techniques are used (i) for cost-effective measures, (ii) to produce units under controlled conditions, (iii) to get quality products, (iv) to use locally available materials and (v) to speed up construction.
4. Prefabrication materials are concrete, steel, treated wood, aluminium, cellular concrete, light-weight concrete, ceramic products, etc.
5. In prefabrication constructions some modular co-ordination has to be followed. In this case, some basic module has to be adopted for general application to the building and its components.
6. The value of the basic module is closer as 100 mm for maximum flexibility and convenience. The symbol used for the basic module is M.
7. Crucial factors that are to be considered for effective production and supply of precast concrete components are: (i) storage facilities, (ii) suitable transport facilities, (iii) erection equipment used and (iv) availability of raw materials.

8. Manufacture of precast concrete units can be done in a centrally located factory, or in a site where a precasting yard is set up at or near the site of work.
9. The process of manufacturing comprises a main process, which is supplemented by an auxiliary process and a subsidiary process.
10. Some of the types of precast units are: hollow concrete block units, lintel units, doors and window units, roofing and flooring elements, precast shell units, sunshade units, etc.

REVIEW QUESTIONS

1. What do you understand by prefabrication in construction? How it is different from cast-in-situ construction?
2. Explain the advantages and disadvantages of precast concrete.
3. What is modular coordination? Explain.
4. How are dimensions of prefabricated elements mixed?
5. Distinguish between factory prefabrication and site prefabrication.
6. Explain the process of manufacturing of precast concrete units.
7. Differentiate between the stand method and the flow method of manufacturing.
8. Name the precast concrete units which form the components of a building.
9. Discuss the uses of different shaped hollow blocks.
10. What type of curing method is adopted for precast concrete units?

10 TIMBER AND INDUSTRIAL TIMBER PRODUCTS

10.1 INTRODUCTION

Timber is a product of wood from felled trees suitable for construction purposes. Timber is one of the oldest and natural civil engineering construction materials. Three terms pertinent to timber one of interest. They are:

1. Converted timber – This is the timber that is sawn and cut into required sizes for commercial use.
2. Rough timber – This is the timber that is obtained after felling a tree.
3. Standing timber – This is the timber that is in a living stage.

Wood, as a building material, has a number of valuable properties, viz., relatively high strength, low bulk density, low heat conductivity and amenability to mechanical working.

Despite its remarkable uses, wood has several shortcomings, viz., different strengths because of anisotropy, susceptibility to decay and inflammability.

Properties of timber and timber-based products as a building material are discussed in the subsequent sections.

10.2 TREES

10.2.1 Basic Structure of a Tree

A tree consists of a trunk, a crown and roots. The functions of the roots are to implant the tree in the soil, absorb moisture and mineral substances it contains and in turn supply them to the trunk. The trunk supports the crown and serves to provide water and nutrients from the roots to the leaves through branches and from the leaves back to the roots.

10.2.2 Classification of Trees

According to the manner of growth, trees may be broadly classified as Endogenous trees and Exogenous trees.

1. Endogenous Trees

These trees grow by the formation of layers. In such trees new wood crosses and penetrates the fibres of the wood previously formed, e.g., bamboo, palmyrah, coconut, etc. Such trees are not useful for engineering works.

2. Exogenous Trees

These trees grow outwards by the addition of rings of young wood. A cross-section of these trees shows distinct approximate concentric rings called annular rings. Timber from such trees is mainly used in engineering works.

They are further classified into two types, viz., softwood and hardwood.

(i) Softwood

Softwoods have needle-like leaves and are also known as conifers as they bear cone-shaped fruits. Trees under this group are pines, deodar, spruces, etc.

The qualities of softwood are:

(a) Light in weight and colour
(b) Has straight fibres and is resinous
(c) Has distinct annular rings
(d) Is weak and can split easily

(ii) Hardwood

Hardwoods are mostly of broad-leafed trees. They are deciduous trees like teak, sap, etc.

Qualities of hardwood are:

(a) Heavy in weight and dark in colour
(b) Close-grained and non-resinous
(c) No distinct annular rings
(d) Strong and durable

10.2.3 Structure of Exogenous Trees

A cross-section of an exogenous tree is shown in Fig. 10.1. It shows almost distinct concentric rings called annual rings. *Pith* is the innermost central portion consisting of cellular tissues. The inner annular area surrounding the pith is called *heartwood*. Heartwood is the one which is used for all engineering works. The portion between the *cambium layer* and the heartwood is called the *sapwood*, which can not be of any use for engineering purposes since it contains large quantities of water. A very small amount of cells that grow in the horizontal direction occurs as groups or bundles of cells known as *medullary rays*.

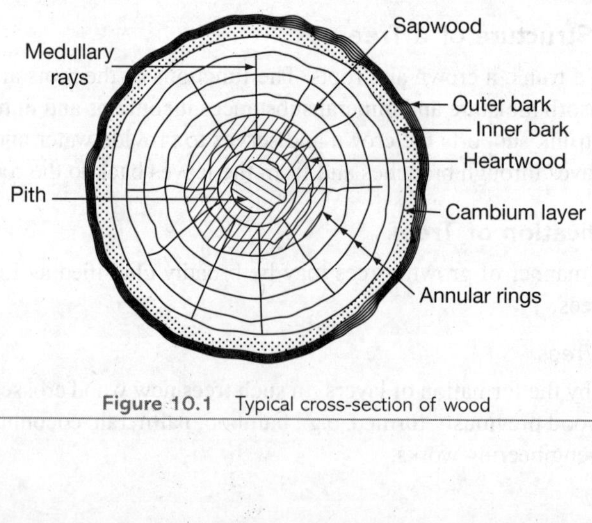

Figure 10.1 Typical cross-section of wood

Freshly cut trees contain large quantities of moisture. If the percentage of moisture is more than about 20%, then the cut trees are subjected to different types of fungus attacks and attacks by insects. Hence, the amount of moisture has to be reduced.

10.2.4 Growth of a Tree

A tree grows through the tops of branches. Just within the inner bark, the tree becomes thicker by the formation of cells. A young tree grows taller and thicker every year. The growth is rapid during spring and slower during late summer and early fall. In young trees, hollow cells are alive. During the growing season the cells are filled with sap, which circulates through each fibre by a series of openings and connecting channels. With time, the older channels become partly clogged and inactive and more mature tissues combine to form the heavier and stronger heartwood. Active growth takes place only in the cells of the inner bark. As long as the layer of cells just within the outer bark is healthy, growth continues. A few annual rings next to the inner bark are less mature, which is known as sapwood. Heartwood is more resistant to decay than sapwood.

10.2.5 Felling of Trees

The process of a cutting down or knocking down or causing trees to fall to the ground is called the felling of a tree. Trees have to be felled only when they are very close to maturity or have just matured. Generally, trees attain maturity between 50 and 100 years. Trees cut before maturity may not be durable. On the other hand, trees cut after maturity may cause the valuable heartwood to decay. The best time for felling trees is mid-summer or mid-winter when the sap quantity is minimum.

10.3 CLASSIFICATION OF TIMBER

As per Indian Standards (IS: 4021–1967), it is preferred that timber used for building construction purposes is of one of the following four types:

1. Hardwood like teakwood for permanent structures.
2. Softwood like deodar wood for permanent structures.
3. Hardwood like sal wood other than teakwood for permanent structures.
4. Softwood other than deodar wood used for any temporary structures.

10.3.1 Classification of Teak

Teakwood is classified into three grades, viz., *superior*, *first* and *second* depending on the defects. The classification of teak based on the permissible defects, viz., cross grain, knots, pith, pockets or streaks, sapwood, pin holes and worm holes is presented in Table 10.1.

10.3.2 Classification of Deodar

Only teak has been divided into three classes, whereas others are classified into two classes only, viz., first class and second class.

1. First-class Deodar

No individual 'hard and sound knot' shall be more than 25 mm in diameter. The total area of live knots shall not exceed 1% of the area of the piece. The wood should be free of sapwood.

Table 10.1 Permissible defects for the classification of teak

Sl. No.	Defects	Superior grade (Class)	First grade (Class)	Second grade (Class)
1.	Cross grain (not steeper than)	1/20	1/15	1/12
2.	Knots	≯ 12 mm	≯ 25 mm	≯ 40 mm
	(a) Percentage area of piece	≯ 0.5%	≯ 1%	≯ 1.5%
	(b) Maximum size	≯ 10 mm	≯ 25 mm	≯ 40 mm
3.	Pith, pockets or streaks	None	None	Permissible if filled with putty, etc.
4.	Sapwood	None	None	Traces allowed
5.	Pinholes	None	None	Filled up ones permitted
6.	Worm holes	None	None	One of ≯ 9 mm per member and treated

Source: IS: 4021–1967.

2. Second-class Deodar

No individual 'hard and sound knot' shall be more than 40 mm in diameter. The total area of all the knots in the piece should not exceed 1.5% of the area of the piece. The wood should be free from sapwood. However, traces of sapwood are allowed.

10.3.3 Classification of Other Woods

Sal, another hardwood, grown in the sub-Himalayan regions and in Madhya Pradesh, is heavier than teak, which has to meet the requirements of first-class deodar. Timbers of Mango, Benteak, etc., are categorized as second-class.

10.4 CHARACTERISTICS OF GOOD TIMBER

Characteristics of good timber are given below in the order of decreasing significance:

10.4.1 Strength

The timber should be strong enough to effectively act as structural members, such as joists, beams, rafters, etc. Further, it should be in a position to sustain loads applied gradually or suddenly to it. Preferably, the strength should be adequate in direct and transverse directions.

10.4.2 Hardness

The timber should be hard so that it is in a position to offer resistance when it is being penetrated by another body. As such hardness is imparted to the wood by chemicals present in the heartwood and the density of wood.

10.4.3 Fire-resistant

As timber is a bad conductor of heat, it should be dense enough so that it resists fire. A good timber should not contain resins and other inflammable oils since they accelerate the action of fire.

10.4.4 Structure

It should possess a uniform structure. This is ensured when there is firm adhesion of fibres and compact medullary rays. Further, the annual rings should be regular and closely located.

10.4.5 Elasticity

This is the property of a material which returns to its original shape when the load causing the deformation is removed. Timber should possess this property, as it is essential when timber is used for bows, carriage shafts, etc.

10.4.6 Durability

Another important quality required for timber is it should to be durable. It should have a long life. It should be capable of resisting the actions of fungi, insects, chemicals, physical agencies and mechanical agencies during its lifespan.

10.4.7 Defects

It should be free from defects such as shakes, flaws, dead knots or blemishes of any kind.

10.4.8 Weight

Quality timber should be of heavy weight and uniform along the length under consideration.

10.4.9 Fibres

Good timber should possess straight and close fibres, which signify uniformity in the timber.

10.4.10 Workability

The timber should be easily workable, i.e., it should not clog the teeth of the saw. It should be capable of being easily planed. A good timber when planed shows a firm and bright appearance with a silky lustre.

10.4.11 Colour

The colour of quality timber should be uniform and of dark colour. Light-coloured wood usually indicates timber of low strength.

10.4.12 Sound

Good timber should produce a clear ringing sound when struck. Decayed timber produces a dull and heavy sound.

10.5 DEFECTS IN TIMBER

Different kinds of defects are found in timber. Generally, they are caused by the nature of the soil on which the tree is grown, the fertility it gained during growth and changes occurred in the environment during its growth. The defects to be generally looked into while selecting timber for woodwork are discussed below.

10.5.1 Knots

These are the sections of the branches of the tree present on the surface of the wood. This appears in the form of hard and dark pieces. Knots are the spots of weaknesses in timber, particularly when the timber has to bear a compressive force (Fig. 10.2). When the diameter of the knot is less than 6 mm, it is called a *nail knot*; when it is 6–20 mm, it is called a *small knot*; when it is 20–40 mm, it is called a *medium knot*; and when it is more than 40 mm, it is called a *large knot*. *Tight knots* are those which are joined to the wood securely. A knot is said to be a *live knot* or a *sound knot* when the formation of the knot is free from decay and other defects, and is also intact with the surrounding wood. A knot that is not held firmly in the wood is called a *dead knot* or a *loose knot*. Timber containing a large number of knots should be rejected as they are weak in strength and poor in appearance.

Figure 10.2 A typical knot

10.5.2 Shakes

These are the cracks found in felled trees, which are formed due to some reason or the other. Different kinds of shakes are identified and are presented in Fig. 10.3.

Heart shakes are those which split or crack with more width at the centre and diminishes towards the circumference. This type of crack occurs in over-matured trees. *Star shakes* are radical splits or cracks wider at the circumference, which reduce towards the centre of the tree. These are caused mostly due to severe frost or more heat. These cracks are confined to sapwood only. *Cup shakes* are presumably formed by the rupture of the tissues in a circular direction across the cross-section of log, usually along the annular rings. *Radial shakes* are similar to star shakes, but they are thin, irregular and numerous. They are caused when there is a delay in sawing after felling. They usually radiate to a small distance from the bark towards the centre.

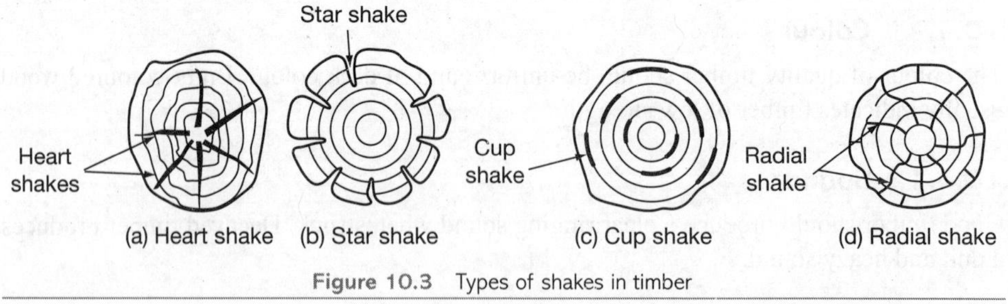

Figure 10.3 Types of shakes in timber

10.5.3 Twisted Fibres

These are also known as twisted grains or wandering hearts. They are caused by the twisting of young trees by fast-blowing winds. The fibres of the wood are twisted in one direction only which causes problems while sawing. However, they may be used as poles or posts without any planing (Fig. 10.4).

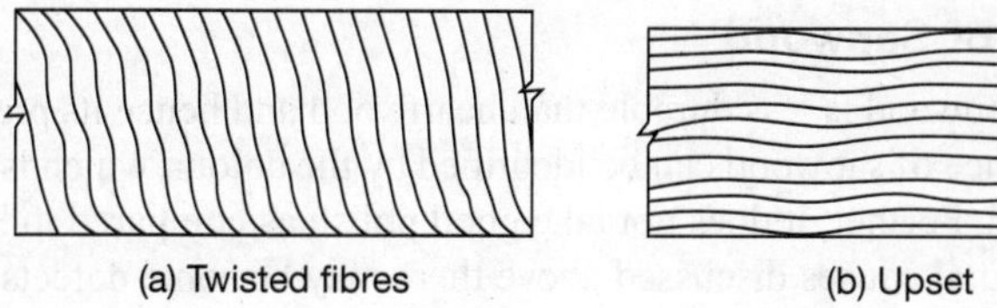

Figure 10.4 Typical twisted fibres and upset timbers

10.5.4 Upset

It is also called as *rupture*. This is caused due to some injury during the growth of the tree. It may be caused due to heavy winds and due to faulty felling of the tree.

10.5.5 Rind Galls

Rind means bark, and *gall* represents abnormal growth. Thus, rind galls are peculiar curved swellings formed on the body of a tree. These are caused due to the growth of layers over the injuries left over after improper felling of trees. The timber taken from this part is very weak and is not durable (Fig. 10.5).

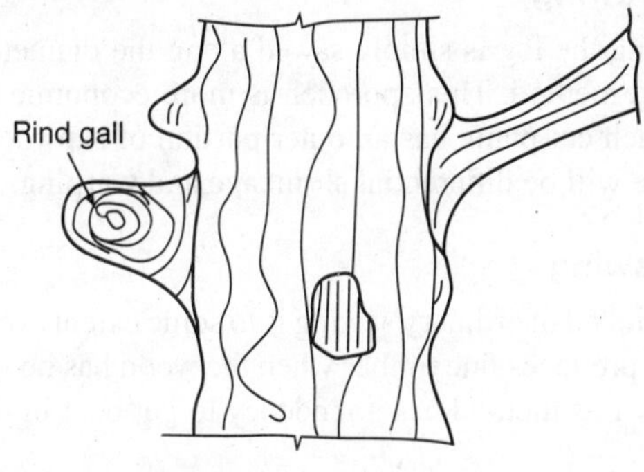

Figure 10.5 Rind galls

10.5.6 Wind Cracks

The exterior surface of a tree shrinks when it is exposed to atmospheric agencies. Such a shrinkage results in cracks known as wind cracks (Fig. 10.6).

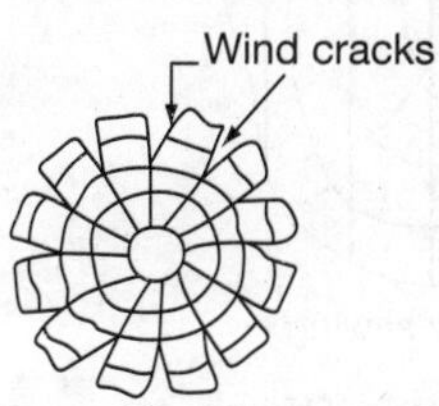

Figure 10.6 Typical wind cracks of timber

10.5.7 Sloping Grains

The cells in living trees do not always grow perfectly vertical or straight and parallel to the length of the truck but in a sloping manner. Such an action causes the grains to slope when the timber is sawn parallel to the pitch. Such sloping grains are considered to be a defect.

10.5.8 Presence of Sapwood

As discussed earlier sapwood is less durable than heartwood and hence its presence should be avoided. The presence of sapwood can be identified by the colour, which is much lighter than that of heartwood. Further, it does not take good polish as compared to heartwood.

Apart from the natural causes discussed above there may be some defects that occur in the process of seasoning, which is also to be considered while selecting timber for building construction works.

10.6 CONVERSION

After felling trees, the branches are removed and the trunk is cut into logs. Conversion is the process of cutting and sawing logs into suitable sections of timber. In earlier days, it was done manually using saws. Now-a-days, cutting is done by band and circular saws run by machines.

Sawing is done by four methods, viz., ordinary sawing, quarter sawing, tangential sawing, or radial sawing.

10.6.1 Ordinary Sawing

In this method of sawing the log is simply sawed along the diameter. This is the quickest and commonly adopted method. This approach is more economical as wastage of useful timber is minimum. Each cut plank has an outer portion of sapwood and an inner portion of heartwood. But there will be differential shrinkage and warping.

10.6.2 Quarter Sawing

The disadvantage mentioned in ordinary sawing is to some extent avoided in quarter sawing. This method of sawing produces fine timber when the wood has no distinct medullary rays. However, timber cut by this method has a tendency to get bent in the transverse direction (Fig. 10.7(a)).

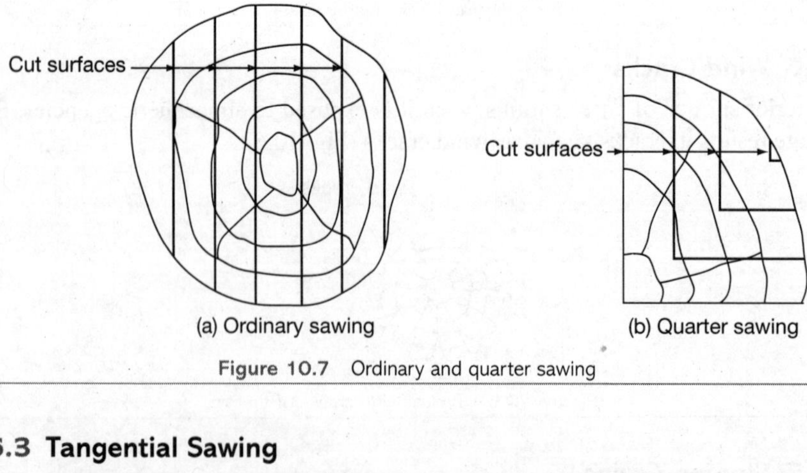

Figure 10.7 Ordinary and quarter sawing

10.6.3 Tangential Sawing

It is also called plain sawing or flat-grained sawing. This is done tangential to the annular rings. This method produces planks that are susceptible to warping as the medullary rays that give strength to the longitudinal fibres are cut. It is difficult to polish planks evenly (Fig. 10.8).

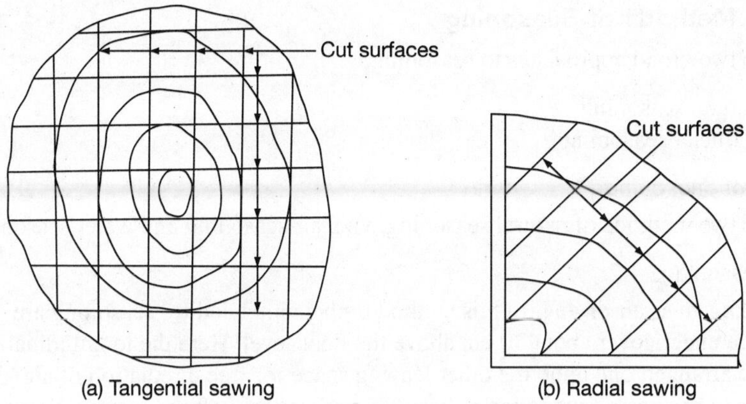

Figure 10.8 Tangential and radial sawing

10.6.4 Radial or Rift Sawing

This is the method adopted for sawing hardwood. This method of sawing produces planks, which do not shrink much. Because of the interplay of grains, it gives a decorative finish. However, maximum wastage occurs by this method of sawing.

10.7 SEASONING

Trees felled recently will contain large amounts of sap and moisture. Seasoning is the process of removing all the sap and moisture content under controlled conditions to avoid any splitting and distortion in the wood.

10.7.1 Objectives of Seasoning

Although the primary aim is to remove sap and moisture, the other objectives of seasoning are as follows:

1. As most causes of decay and other problems are more or less related to moisture, the main objective is to remove moisture thereby increasing the resisting power of timber.
2. To impart hardness, stiffness and strength, which improve the adoptability of timber for many building purposes.
3. To improve the resisting power of timber against electrical power and attack by fungi and insects.
4. To make the timber easily workable.
5. To totally remove the possibility of shrinking, warping and splitting.
6. To maintain the size and shape of the components of the timber structure of articles that are expected to be unchanged during the life span.
7. To make timber suitable for treating with paints, polishes, preservatives, etc.
8. To make the timber suitable for effectively joining with the use of glues.
9. To reduce the weight of the timber for easy handling and transportation to other places.
10. To give high order of durability for more years after construction.

10.7.2 Methods of Seasoning

There are two broad approaches to seasoning, viz.,

1. Natural seasoning
2. Artificial seasoning

1. Natural Seasoning

There are two methods of natural seasoning, viz., air seasoning and water seasoning.

(i) Air Seasoning

It is also known as air drying. In this method timbers intended for seasoning are stacked in a dry place under cover about 30 cm above the floor level. Here the longitudinal and cross pieces are arranged one upon the other leaving space for free circulation of air (Fig. 10.9). It is better to provide a foundation to keep the timber clean off the ground. Wood seasoned by such a process can be used for carpentry work after two years.

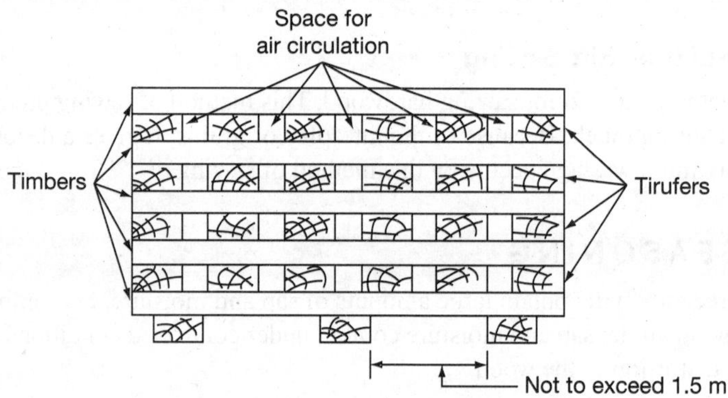

Figure 10.9 Typical air seasoning of timber

(ii) Water Seasoning

This method of seasoning timber is to completely immerse the logs of wood in water soon after cutting. This is better done in a running stream of water with the longer ends of the logs being kept pointing upstream. By this arrangement the sap, sugar, gum, etc. are leached out of the logs and in turn replaced by water. After adequate soaking the logs are dried in an open place. Although it is a quick process, it reduces the durability of the timber.

2. Artificial Seasoning

Artificial seasoning may be done by kiln seasoning, chemical seasoning and electrical seasoning. By artificial seasoning the moisture content can be brought under control from 4% to 12%. This is a quick process.

(i) Kiln Seasoning

In this system, complete control of temperature and humidity is maintained with proper air circulation and ventilation system. The timber is kept inside an airtight chamber. As a first step fully saturated air at 35–38°C is circulated. Then the humidity is reduced slowly followed by an increase of temperature till the moisture in the timber is reduced to the degree of moisture required. Then the kiln is slowly cooled. This process takes about a fortnight during which the timber is kept on a trolley. The quality of wood obtained by this method is inferior to those seasoned by natural methods.

(ii) Chemical Seasoning

This method of seasoning is also known as salt seasoning. In this method the timber is immersed in salt solution. After a specific period it is taken out and seasoned in the ordinary way. In this method of seasoning the interior surface of the timber dries before the exterior dries out. Here the chances of formation of external cracks are reduced.

(iii) Electrical Seasoning

This method is based on the principle that heat is produced when poor conductors are placed in the field of high frequency. Here, an induction coil producing field of high frequency is set up. The timber is made to pass through the coil. Due to electric induction the moisture is removed instantly and the wood is dried quickly. By this process the moisture gets evaporated uniformly and results in superior quality of timber. Because of high cost this method is not generally recommended.

10.8 DECAY OF WOOD

Once the woods get deteriorated they lose their engineering properties and the wood is said to be decayed. Several reasons contribute to the decay of wood, which are explained below:

1. Decay may be caused due to alternate dry and wet conditions.
2. Formation of fungi, which is responsible for the development of various diseases in wood, viz., dry rot, wet rot, blue stain, sap stain, etc.
3. Improper removal of sapwood from the wood.
4. Keeping timber in contact with a damp wall, damp earth, etc.
5. Improper and incomplete seasoning.
6. Non-application of preservatives on the surface of seasoned timber.
7. Using unseasoned wood after application of oily paint.
8. While younger, timber would have been subjected to shocks and impacts.
9. Improper storage while stacking the timber.
10. Permitting insects such as beetles, marine borers, termites, etc., to affect the timber.

10.9 PRESERVATION OF TIMBER

To increase the life span of wood, they should be treated with chemicals–this process is known as the preservation of timber.

10.9.1 Requirements for a Good Preservative

A good preservative should fulfil the following requirements:

(i) It should be safe to the user.
(ii) It should not have unpleasant smell.
(iii) It should be non-inflammable.
(iv) It should be capable of penetrating into the fibres.
(v) It should not corrode the metal pieces attached to the timber.
(vi) It should allow other materials such as paints or varnishes to be applied on it.
(vii) It should not be washed away by water.

(viii) It should not get deteriorated by heat, light, etc.
(ix) It should not reduce the strength of timber and warp the structure.
(x) It should be cheap and easily available in the market.

10.9.2 Types of Preservatives

There are three types of preservatives with different bases as prescribed in the Code (IS: 401–2001).

1. Oil Types

In this type coal tar *creosote* with or without admixtures of soluble oils is used.

2. Organic Solvent Types

These are toxic chemicals in non-aqueous solvents such as copper napthenate, zinc napthenate, benzene hexachloride, etc.

3. Water-base Types

These are toxic chemicals in water, such as zinc chloride, boric acid, copper chrome arsenic composition, etc.

10.9.3 Preservative Treatment of Timber

The method of preservative treatment to be used depends on the use of wood in construction. The general methods adopted are as follows:

1. Charring

In this method, the ends of the timber are burnt to about 1.5 cm and then quenched in water. This charred end act as a protective coat. This is particularly done to prevent dry rot and attacks by insects.

2. Tarring

This is applying a coat of tar or tar mixed with pitch. It is generally applied to rough types of works such as ends of door and window frames that are to be fixed to the walls, timber fences, etc.

3. Creosoting

Creosote oil is one which has wood preservative oil in it. The timber to be treated is placed in cylinders and closed tightly. Creosote oil under a pressure of 9 bars and at a temperature of 50°C is pumped into the cylinder until the desired absorption is reached. By this treatment the timber is preserved against rot and attack by white ants.

4. Treatment Using Preservatives

Some of the methods where preservatives are directly used are as follows:

(i) Surface Application Method

Surface Application Method is also known as the brush and spray method. Solvents or emulsions are used for this purpose. The timber to be treated is brushed or sprayed. At least two coats should be applied. This method is adopted only when naturally resistant timbers are used.

(ii) Soaking Treatment

Soaking Treatment is also known as steeping. In this treatment the timber is submerged in the preservative solution till the required absorption is obtained. Here, a water-soluble type preservative is used to avoid excessive evaporation loss. This treatment is recommended for timbers containing a lot of sapwood and for light and medium density timbers.

(iii) Hot and Cold Treatment

Hot and Cold Treatment is one in which the timber is placed in a steel tank immersed in a preservative (creosote) and heated to about 85–90°C. Then the tank is cooled slowly after some time with the timber immersed in the solution. This treatment is adopted for timbers containing sapwood and easily treatable heartwood.

(iv) Pressure and Vacuum Treatment

Pressure and Vacuum Treatment is of two types, viz., full cell process and empty cell process.

In the full cell process, the timber to be treated is placed in a closed vessel and is subjected to a low vacuum for about an hour. During this process, a preheated preservative is introduced till the vessel is full. A pressure up to 14 kg/sq.cm is gradually applied and kept constant for several hours. The pressure is then reduced and the solution is drained out. Timbers for marine works, railway sleepers, etc. are treated by this process.

In the empty cell process the quantity of preservative required is less and hence cheaper than the full cell process. The timber to be treated is stacked in a cylinder. After closing the door, the timber in the cylinder is then subjected to an air pressure of 1.75–5 bars from half an hour to one hour. The pressure is maintained while the cylinder is filled with preservatives. When the cylinder is completely filled, a pressure of 5–12 bars is applied and this pressure is maintained till the required absorption has taken place. The cylinder is drained off and vacuum is applied to remove the preservative dripping from the timber. This method is used for preserving timber of mixed species and timber containing sapwood and heartwood.

5. Treatment by Diffusion

This treatment is carried out for timber which has moisture content of over 50%, i.e., for green timber, which is felled recently. Extremely soluble boron compounds are applied to the surface of the timber. The timber pieces are stacked and covered with an impermeable cover to prevent evaporation. After a period of 1–2 months, the boron diffuses into the wet timber and acts as a preservative.

10.10 TESTING OF TIMBER

Generally, wood for building construction is not tested in the laboratory. It is a practice to classify them by their species and by visually examining them for defects. However, the following tests are conducted for important works:

1. Tensile Strength Test
2. Compressive Strength test
3. Shear Strength Test
4. Bending Strength Test
5. Moisture Content Test

10.10.1 Tensile Strength Test

The specimen used for tensile strength test is of size 5 cm × 5 cm × 20 cm long. The testing is done in a wood testing machine. The load may be applied either parallel to the grains or perpendicular to the grains. The tested result is compared with the standard value.

10.10.2 Compressive Strength Test

For this test also the specimen size is as that of tensile test. The load is applied parallel to the grains. The specimen shall not have a slope of grain more than 1 in 20 parallel to its longitudinal edges. The test result is compared with the standard value.

10.10.3 Shear Strength Test

The size of the specimen is 5 cm × 5 cm × 6.25 cm and notched in a corner so as to produce a failure on 5 cm × 5 cm surface in the radial or tangential surface as desired. Shear strength is determined only parallel to the grain. Shear strength is important when the timber is used as beams or slabs.

10.10.4 Bending Strength Test

The Bending Test is performed on specimens of size 5 cm × 5 cm × 7.5 cm. The slope of the grain should not be more than 1 in 20 parallel to the longitudinal edges. This test is performed to determine the Modulus of Rupture and Modulus of Elasticity of timber.

10.10.5 Moisture Content Test

The Moisture Content Test is an important test that has to be performed for all types of works. The presence of high moisture produces shrinkage of wood work, attack by organisms, causes warping, etc. Even blistering of paints occurs if the moisture content is high.

In this test, the weight of moist wood and the weight of oven dry wood are determined. Then

$$\text{Moisture content} = \frac{\text{Weight of moisture in sample}}{\text{Oven dry weight of sample}} \times 100$$

10.11 INDUSTRIAL TIMBER PRODUCTS

To meet the minor needs of wood works, natural wood can be made into many other industrial products by industrial processes for building construction. These industrial-made products are also called composite boards. The common industrial timber products are:

1. Plywood
2. Particle board or chip board
3. Hard board
4. Fibre board
5. Block board
6. Decorative laminates

10.11.1 Plywood

Plywoods are boards which are prepared from thin layers of wood or veneers. Veneers are thin sheets of slices of wood of superior quality. The thickness of veneers varies from 0.4 to 6 mm.

To make plywood, round logs of 1.5–2.5 m lengths are first cut. They are then steamed and veneers are cut from these logs with a wide sharp knife. About three or more veneers in multiples of odd numbers are glued one over the other. Care is taken to place the direction of the grains in one sheet at right angles to the other. This arrangement is needed to provide adequate longitudinal and transverse strengths. The arranged veneers with adhesives in between are pressed under a pressure of 0.7–1.40 N/mm^2 under hot or cold conditions (IS: 303–1989).

Three-ply plywood is made using only three plies. The thickness varies from 6 to 25 mm (Fig. 10.10). Multiply plywood is made using more than three plies but keeping the number of veneers used as odd in number.

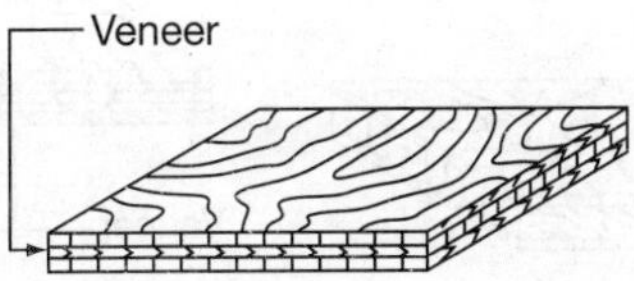

Figure 10.10 Typical three-ply plywood

10.11.2 Particle Board or Chip Board

These boards are made of particles of wood embedded in synthetic resins, which are then pressed under heat. They are made by extrusion pressing or by pressing between parallel plates. In the extrusion pressing process, the particles are oriented in the direction perpendicular to the plane of the board, whereas in the other case the particles are oriented parallel to the plane of the board.

Particle board is heavier than solid wood. Further, it provides broad and stable panels of reasonable strength. They can be sawn just like regular wood. Particle boards are preferred for furniture works than for construction works.

10.11.3 Hard Board

Hard Board is made as per the Indian Standard (IS: 1658–1966). It is made out of wood pulp that is compressed to make sheets usually of 3-mm thickness. Its top surface is smooth and hard and the rear side is rough with patterns or cross lines.

As per the CPWD specification, they are classified as given in Table 10.2.

Table 10.2 CPWD specification of hard board

Sl. No.	Type	Density (kg/m^3)	Thickness (mm)
1.	Medium	480–800	6, 8, 10 and 12
2.	Normal	800–1,200	3, 4, 5, 6, 9 and 12
3.	Tempered	More than 1,200	3, 4, 5, 6, 9 and 12

For door shutters only tempered hard board should be used.

10.11.4 Fibre Board

The technique adopted for the manufacture of fibre board is the combination of the process used is making particle board and hard board. In this case, wood chips are steamed to separate the fibres from each other. These fibres are blended with resin and wax. These are turned into sheets by pressing under controlled heat and pressure. The boards are available in thicknesses of 25–32 mm and as sheets of 2.44 m × 1.22 m size. They are suitable for flush doors, cabinets, etc.

10.11.5 Block Board

Block board is also known as batten-board or solid-core board. Batten board consists of a core solid block of thin wood sawn from natural wood. The central block is then glued between two or more outer veneers on either side (Fig. 10.11). Thicknesses are from about 19 mm upwards. They are suitable for door panels, partitions, table tops, etc.

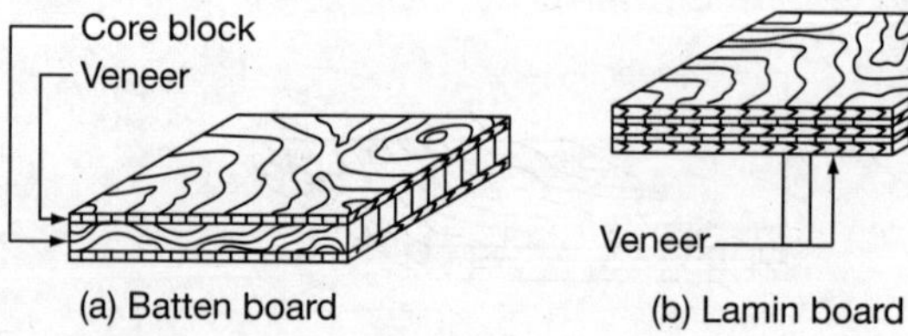

Figure 10.11 Typical batten board and lamin board

Lamin board is similar to that of batten board with the difference being the core is made up of multiply veneers. The thickness of veneers is about 6 mm, and the total thickness of the board is about 50 mm.

Another type of block board called the metal-faced plywood is the one where the core is sandwiched between thin sheets of aluminium, steel, copper, etc. This is a rigid type of plywood (Fig. 10.12).

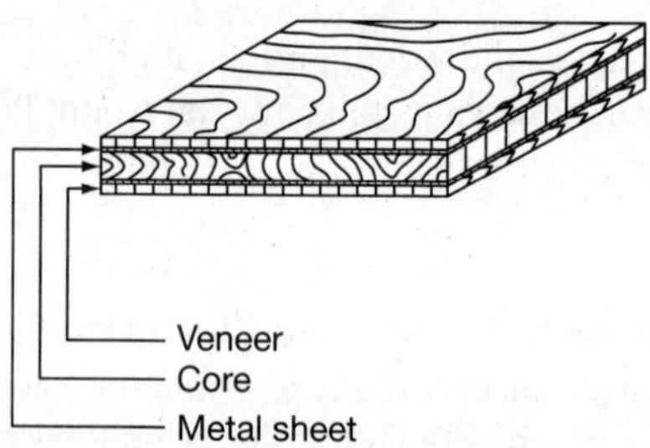

Figure 10.12 Metal-faced block board

10.11.6 Decorative Laminates

Laminates are the products made by bonding together two or more layers of materials. High-pressure decorative laminates are pasted on plywood sheets, which are used for large areas like cabinets, wide door shutters, tables, etc.

10.11.7 Advantages and Disadvantages of Processed Timber

1. Advantages

(i) Unusable trees such as branches of good wood and timber of fast-growing tress cannot be used for regular purposes. Such timber can be used as processed timber.

(ii) By conversion of expensive wood into thin veneers, large surface areas can be covered with less cost.

(iii) Processed timber is better dimensionally stable and does not absorb moisture from the air with the exception of particle board.

(iv) Properly made plywood and block board with better gluing properties are highly durable even in wet environment. Further, the durability depends on the grade of the adhesive used.

(v) Workability is better than that of regular wood.

(vi) Fasteners can be fixed easily as that of wood.

2. Disadvantages

Only processed timber can be used provided they are made out of reliable materials. However, doors and other wooden fittings should be restricted to wet areas like the bathroom.

SALIENT POINTS

1. Timber is a product of wood from felled trees, which is suitable for construction purposes.
2. Converted timber is that which is sawn and cut into required sizes for commercial use.
3. Rough timber is that which is obtained after felling a tree.
4. Standing timber is the timber which is in a living stage.
5. A tree consists of a trunk, a crown and roots.
6. Endogenous trees grow by the formation of layers.
7. Exogenous trees grow outwards by the addition of rings of young wood.
8. Softwoods have needle-like leaves and are also known as conifers as they bear cone-shaped fruits.
9. Hardwoods are mostly of board-leaved trees.
10. Pith is the innermost central portion consisting of cellular tissues.
11. The inner annular area surrounding the pith is called the heartwood.
12. The portion between the cambium layer and the heartwood is called the sapwood.
13. A very small amount of cells that grow in the horizontal direction occurs as groups or bundles of cells known as medullary rays.
14. Growth of a tree occurs through the growth at the tips of the branches.
15. The process of cutting down or knocking down or causing the trees to fall to the ground is called felling of a tree.
16. Defects in timber are knots, shakes, twisted fibres, upset, rind galls, wind cracks, sloping grains and the presence of sapwood.
17. Timber is sawed by four methods, viz., ordinary sawing, quarter sawing, tangential sawing or radial sawing.
18. Seasoning is the process of removing all the sap and moisture content under controlled conditions to avoid any splitting and distortion to the wood.
19. Methods of seasoning are natural seasoning and artificial seasoning.
20. Natural seasoning methods include air seasoning and water seasoning. Artificial seasoning methods comprise kiln seasoning, chemical seasoning and electrical seasoning.

21. Types of preservatives are: Oil types, organic solvent type and water-based type.
22. Preservative treatment of timber is done by charring, tarring, creosoting and by using preservatives. Treatment using preservatives are: surface application method, soaking treatment, hot and cold treatment, pressure and vacuum treatment and treatment by diffusion.
23. The following tests are conducted on timber. Tensile Strength Test, Compressive Strength Test, Shear Strength Test, Bending Strength Test and Moisture Content Test.
24. Industrial timber products are: plywood, particle board or chip board, hard board, fibre board, block board and decorative laminates.

REVIEW QUESTIONS

1. What is timber? Name three types of timbers.
2. Explain the basic structure of a tree.
3. How are trees classified?
4. With a sketch explain the different parts of an exogenous tree.
5. What are the factors that govern the growth of a tree?
6. What do you understand by the felling of a tree?
7. Name the types of timber used for building constructions as per the Indian Standards.
8. How is teakwood classified? Explain.
9. What are the characteristics of soft and hard timber?
10. What is the difference between first- and second-class deodar?
11. Enumerate the characteristics of good timber.
12. Explain the defects in timber.
13. Distinguish between dry rot and well rot in timber. What steps are taken in executing wood work to see that these defects do not occur.
14. What are the methods of conversion of timber? Discuss their merits.
15. Define seasoning. What are the objectives of seasoning?
16. Explain briefly the methods of seasoning.
17. Explain different causes for the decay of wood.
18. What are the requirements of a good preservative?
19. Name the types of preservatives.
20. How is the preservative treatment of timber done?
21. What are the different types of tests conducted to find the suitability of timber for building construction purposes?
22. Explain different types of industrial timber products.
23. Explain with neat sketches, the different forms of plywood.

11 METALS AND ALLOYS

11.1 INTRODUCTION

We use metals daily in one form or the other. Metals are also used for various engineering works such as structural members, roofing materials, reinforcing material, pipes, tanks, etc. Among the metals, iron is the most commonly used one. It is also available in abundance. It is stated that about 5% of iron is present in the crust of the earth.

Iron is extracted from solid naturally occurring mineral aggregates called ores. About five important iron ores are used to extract the metal iron.

Metals are grouped under two broad groups, viz.,

(i) Ferrous metals and
(ii) Non-ferrous metals.

Iron is the main constituent of ferrous metals. The important ferrous metals are, viz., pig iron, cast-iron, wrought iron and steel. The non-ferrous metals, viz., aluminium, copper, etc., do not contain iron as their main constituent. Alloy is an intimate compound of two or more metals.

11.2 FERROUS METALS

11.2.1 Iron Ores

Iron ores are compounds of iron with non-metallic elements. They also contain impurities such as carbon, manganese, phosphorous, silicon and sulphur. The iron ores are extracted from the earth by mining operations.

While selecting iron ores, the following aspects should be considered:

(i) The iron ores should be rich in metallic iron content and have less impurities.
(ii) The location of mines should not have constraints which may lead to complications during quarrying operations.
(iii) The location of mines should possibly be linked to the surrounding towns by suitable transport facilities.
(iv) The impurities should not contaminate the rich iron ore. However, there are some substances that increase the value of lean iron ore, which are to be considered.
(v) Any initial treatment made on the iron ore should be in a position to increase the output of iron.

Important varieties of iron ores are haematite, limonite, magnetite, pyrite and siderite.

11.2.2 Pig Iron

Pig iron is the basic material from which wrought iron and steel are manufactured. They are extracted from iron ores, viz., magnetite, haematite, limonite and siderite. The percentage of iron content, and their chemical formula are given in Table 11.1. Pig iron is extracted from these ores in a continuous working furnace called the blast furnace. The product obtained from the blast furnace is crude and is an impure form of iron.

Table 11.1 Iron ores and percentage of iron content

Name of the ore	Percentage of iron content	Chemical formula
Magnetite	70–73	Fe_3O_4
Haematite	65–70	Fe_2O_3
Limonite	55–65	$2Fe_2O_3H_2O$
Pyrite	45–47	FeS_2
Siderite	40–45	$FeCO_3$

1. Manufacture of Pig Iron

Three stages are involved in the manufacture of pig iron, viz.,

(i) Dressing
(ii) Calcination and roasting and
(iii) Smelting

Dressing is the process in which the iron ores obtained from mines are crushed into pieces of about 25 cm diameter sizes by rock crushers. If the ores contain impurities like clay, loam and other earthy materials they are removed by washing the ores.

After dressing, the ores are calcinated and roasted. Calcination is the process where the ores are heated in the presence of air to oxidise them. By this action carbon dioxide and water are removed. Roasting the ores enables the dissipation of the volatile parts such as sulphur.

Smelting is the process of melting so as to separate metal from the ore. It is carried out in a special furnace called the *blast furnace*. The blast furnaces of the old type had a capacity of producing 15,000 KN of pig iron per day, whereas the more modern types have a capacity of 25,000 KN.

2. Properties of Pig Iron

The properties of pig iron are as follows:

(i) Pig iron can be hardened, but tempering is not possible.
(ii) It is not possible to magnetise pig iron.
(iii) Riveting or welding is not possible.
(iv) Unlike steel it does not rust.
(v) Bending is not possible.
(vi) It is neither ductile nor malleable.
(vii) It has high compressive strength with less tensile strength.
(viii) It can be melted easily.
(ix) It is hard and brittle.

11.2.3 Wrought Iron

It is the oldest form of iron made by man. It was originally produced from iron ore in a huge fire. This gives a very impure iron which needs further refining by mechanical working. Wrought iron contains high-purity iron and iron silicate. It contains a very low percentage of carbon, and the iron silicate is distributed throughout the base metal. When fractured it shows a woody or fibrous appearance.

11.2.4 Cast Iron

Cast iron is an alloy of iron and carbon with the content of carbon being more than 2%. In addition, it contains impurities such as manganese, phosphorous, silicon and sulphur. The presence of manganese makes it brittle when it is more than 0.75%. The fluidity of cast iron is increased by the presence of phosphorous. Silicon decreases shrinkage and ensures softer and better casting when it is less than 2.5%.

1. Manufacture of Cast Iron

Cast iron is manufactured by re-melting pig iron with coke and limestone. This process of re-melting is done in a furnace called the *cupola furnace*. Except for the size, it is almost the same as that of blast furnace. To some extent, the impurities in the pig iron are removed by this process and comparatively pure iron is obtained in the molten stage from the bottom of the furnace. The slag is removed at regular intervals from the top of the cast iron. The molten cast iron is led into the moulds called cast-iron castings.

2. Properties of Cast Iron

As an engineering material, cast iron has the following properties:

(i) It has high compressive strength.
(ii) It has high machinable qualities based on the composition.
(iii) It has a fairly low melting point.
(iv) It has high fluidity which helps in making good casting impressions.
(v) It has fairly good corrosion resistance.
(vi) It lacks plasticity and is not suitable for forging.
(vii) It is hard and brittle but not ductile.
(viii) It can not be magnetised.

11.2.5 Steel

Steel is a very ductile alloy consisting of iron and carbon. Based on the percentage of carbon content, steel is classified as follows:

(i) Low-carbon or mild steel: carbon content up to 0.25%
(ii) Medium-carbon or medium hard steel: carbon content from 0.25 to 0.70%
(iii) High-carbon or hard steel: carbon content from 0.70 to 1.5%

1. Manufacturing of Steel

There are five processes by which steel is manufactured:

(i) Bessemer Process

A Bessemer convertor (a furnace) is used in this process, which is wide at the bottom and narrow at the top and rotates about the horizontal axis. Pig iron is filled in the convertor and hot blast air is forced into it. The impurities get oxidised by the blast air and a reddish-yellow flame is seen at the mouth of the convertor, and the flame diminishes gradually. Now the supply of air is stopped and adequate quantities of materials like ferromanganese is added to make steel of desired quality. The blast is again started for a few minutes and the molten metal is poured into moulds to form blocks called *ingots*.

(ii) Cementation Process

Here a dome-shaped furnace called a cementation furnace is used. In this process the pig iron is first converted into pure wrought iron. After adjusting the carbon content, steel is obtained.

(iii) Open-hearth Process

Here a mixture of pre-heated air and coal gas is used as a fuel. The steel produced by this process is homogeneous and is of high reliability.

(iv) Electric Process

Here the raw material is melted using electricity. The other procedure is the same as that of the Bessemer process and more control of temperature is feasible here. It is a quick process and special-grade steels are produced by this process.

(v) Duplex Process

It is a combination of the Bessemer and the basic open-hearth process.

Figure 11.1 shows a flow chart for the production of steel.

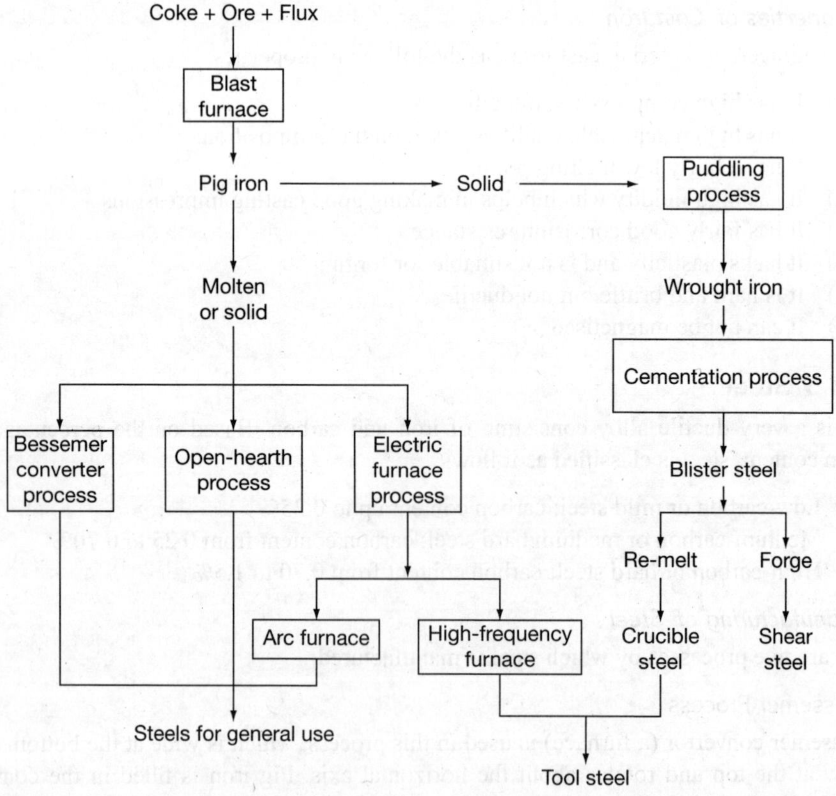

Figure 11.1 Flow diagram for the production of steel.

2. Uses of Steel

(i) Mild steel is used to manufacture tools, machine parts, tubes, sheet metal, tin plate, structural steel, etc.
(ii) Medium hard steel is used to manufacture machine/engine components, boiler plates, rails, hammers, pressing dies, structural steel, agricultural implements, aero-engine, cylinders, springs, etc.
(iii) Hard steel is used in the manufacture of plates, cutlery, springs, miner's drills, heavy tools, sledge hammers, axes, planning and slotting machines, lathes, etc.

3. Properties of Mild Steel

The properties of mild steel are as follows:

(i) It has a high tensile strength.
(ii) It has a high elastic property reflected by a high modulus of rigidity.
(iii) It develops an effective bond with concrete.
(iv) Because of a good bond, it is capable of transferring stresses.
(v) As it has an almost equal coefficient of thermal expansion and contraction, there is no room for the development of thermal stresses.
(vi) It is readily available and is comparatively cheap.

4. Properties of Medium Hard Steel

The properties of medium hard steel are as follows:

(i) It has higher strength than mild steel.
(ii) It can be hardened to a certain extent.
(iii) It is tougher and harder than mild steel.
(iv) It can not easily be forged or welded.
(v) It has a high resistance to shocks and vibrations.

5. Properties of Hard Steel

The properties of hard steel are as follows:

(i) It has a granular structure.
(ii) It is very hard and is tougher than other steels.
(iii) It is possible to magnetise permanently.
(iv) It has very high compressive and tensile strengths.
(v) It can be tempered and hardened.
(vi) It has a very high resistance to shocks and vibrations.
(vii) It is difficult to forge.

6. Suitability of Steel

The suitability of steel for different works as per Indian Standards is given in Table 11.2.

11.2.6 Alloy Steels

To obtain special properties, some elements such as nickel, chromium, manganese, etc., are added to the carbon steel. Steels thus formed are called alloy steels.

Table 11.2 Suitability of Indian Standard Steel as construction material

Steel type	Suitability
St 30	Structural steel sheets–plain drawn, tubes for oil well casing, steam, water and air passage
St 32, St 34, St 37 and St 44	General engineering purposes
St 39	Concrete reinforcement and other general engineering purposes
St 42	Structural steel bridges and building construction, tube piles and other engineering purposes
St 47 and St 50	Tube piles and steel piling
St 58	High tensile steel for bridges and general construction, and bars and wire for concrete reinforcement

The addition of alloying elements is effected to attain one or more of the following properties:

(i) To improve the fine-grain size content of steel.
(ii) To improve case-hardening properties.
(iii) To improve the elastic properties of steel.
(iv) To improve corrosion and fatigue resistance.
(v) To improve hardness and toughness.
(vi) To improve machinability.
(vii) To improve tensile strength.

Alloying materials have the following properties:

(i) **Nickel:** It increases toughness and improves responses to heat treatment. When added in large quantities, it provides special electrical and magnetic properties.
(ii) **Chromium:** Helps to provide stainless properties in steel. It is used in making tool steels and electrical plates.
(iii) **Manganese:** Improves heat treatment properties and is used in high-speed tool steels.
(iv) **Tungsten:** It retains the hardness property of steel and toughness at a high temperature. It is used exhaustively in the manufacture of tools, dies, valves, magnets, etc.
(v) **Silicon:** It provides high electrical resistance and magnetic permeability in electrical machinery.
(vi) **Copper:** It improves corrosive resistance even when added in minute quantities.

Some of the alloy steels are discussed below:

1. Silicon Steel

Silicon steel has high electrical properties and has very low magnetic hysteresis when present in about 3.5%.

Silicon steel is used in the manufacture of different materials in combination with other metals with a lesser percentage. The various materials made include springs, internal combustion engines, laminations of electrical machines, etc.

2. Nickel Steel

The percentage of nickel added varies from 2 to 40. Elastic limit and toughness are increased with about 4% of nickel. The tensile strength is highly increased with 20% nickel. To make the steel non-corrosive and non-magnetic, the nickel percentage should be about 27%. If the nickel content is 3.5% and about 0.3% of carbon, the nickel steel formed is highly ductile and such a material is used for long span bridges.

3. Manganese Steel

The steel attains very high tensile strength with toughness and non-magnetic properties when the presence of manganese is around 12.5%. Manganese steel is weldable and forgeable. Heat-treated cast manganese steel in bar form is extremely ductile that it can bent when cold without fracturing. Manganese steel has a high percentage of elongation. It is used in the jaws of crushers, rails, helmets, shields, etc.

4. Tungsten Steel

The hardening and magnetic properties are improved with the addition of tungsten to steel. The addition of a high percentage of tungsten makes the steel retain its hardness even at a high temperature. It is used for making permanent magnet and high-speed cutting tools.

5. Stainless Steel

Stainless steel is of chromium-based steel. They form two groups, viz., plain chromium and high-chromium low-nickel steel and chromium nickel steel. The first type is used for dies, valves and cutlery and can be heat-treated. These steels can be heat-treated chromium nickel steels, which are non-magnetic and can not be hardened. They have high resistance to corrosion. Further, they may be cold or hot worked, pressed, welded, barbed or soldered.

11.3 NON-FERROUS METALS

Metals which do not contain any percentage of iron is generally called non-ferrous metals. The important metals that fall under the non-ferrous metal category are discussed below.

11.3.1 Aluminium

Aluminium is abundantly available in the ore called bauxite. Bauxite is hydrated aluminium oxide. The manufacturing process involves the bauxite first being purified and then dissolving in molten cryolite. From this solution, aluminium is separated by electrolysis at about 900°C.

Aluminium has the following properties:

(i) It offers high resistance to corrosion.
(ii) It is ductile and malleable.
(iii) It is quite strong.
(iv) It has tensile strength varying from 95 to 160 MN/m^2.
(v) It easily forms alloys with iron, copper, zinc and other metals.
(vi) It is electron positive.
(vii) It is unaffected under normal atmospheric conditions.

11.3.2 Copper

Copper is one of the metals directly available in the metallic state. It is found in minerals like copper glance (Cu_2S), malachite ($CuCo_3.CuO_2H_2$) and azurite ($2CuCo_3.CuO_2H_2$). It is extracted by different methods from ores. Depending on the process by which copper is extracted, they are designated as:

(i) Tough-pitch copper
(ii) Oxygen-free copper
(iii) Deoxidised copper
(iv) Arsenic copper
(v) Free-cutting copper

Copper has the following properties:

1. It is one of the best conductors of heat and electricity.
2. It is highly resistant to corrosion by liquids.
3. It forms important alloys, viz., bronze and gun metal.
4. Its tensile strength varies from 300 to 500 MN/m^2.
5. It is strongly attacked by nitric acid.

11.3.3 Lead

The lead ore is galena or galenite consists of lead sulphide. It is widely distributed. In the manufacturing process, the roasted ore, silica, coke, metallic iron and lime are smelted together in a blast furnace. Lead oxide and sulphate react with iron to form ferrous oxide and sulphate, resulting in the formation of lead. Ferrous oxide in combination with silica forms slag.

Further, the heavy metal formed at the bottom of the furnace is further oxidised in Bessemer's converter to get pure lead after the removal of impurities.

The properties of lead are as follows:

1. It has high metallic lustre when freshly cut.
2. Among the common metals, it is the softest and heaviest metal.
3. It is highly malleable and can be formed into foils.
4. It is least tenacious.
5. In hot conditions it may be extruded by a hydraulic press into tubes, rods and wires.

11.3.4 Tin

The chief source of tin is tinstone (or cassiterite) SnO_2. Tin is a bright shining white metal. It is soft and malleable and can be made into the form of thin foil. As it has high resistance to corrosion, it is used for coating purposes on metals and alloys. It is particularly used in low melting point alloys as a substitute for bismuth. It is also widely used for moisture-proof packing.

11.3.5 Zinc

Common ores of zinc are zincite (ZnO), calamine ($ZnCO_3$) and zinc blende (ZnS). It is a bluish grey non-ferrous metal. It becomes brittle at a high temperature of 200°C and can also be powdered at this temperature. It is used in the form of a sheet as a corrosion-resistant surface. Further, it is also used as a protective coating on iron and steel in the form of a galvanized or sprayed surface. It is also used in making brass and other alloys and in electric cells.

11.3.6 Magnesium

It occurs in nature in several minerals such as magnesite ($MgCO_3$), dolomite ($CaCO_3$, $MgCO_3$), kieserite ($MgSO_4 \cdot H_2O$) and carnalite ($MgCl_2 \cdot KCl \cdot 6H_2O$). The metal is prepared by the chloride or oxide process. The oxide process is not in use anymore.

In the chloride process, a melted mixture of chlorides of sodium, potassium and magnesium is electrolysed. To prevent decomposition of magnesium chloride during the heating process, alkali chloride should be present. When the magnesium rises during liberation, it rises to the surface and has to be ladled out. If there is a need for pure magnesium, it is prepared by distillation.

The properties of magnesium are as follows:

1. It is the lightest material.
2. It can be readily machined.
3. It hardens very rapidly with cold working.
4. It is easily attacked by weak acids and even by saline solution.
5. It ignites easily.
6. It can be welded.

11.3.7 Nickel

Nickel is found combined with iron sulphides in deposits.

The properties of nickel are as follows:

1. It is like silver, which can take high polish.
2. Its hardness is equivalent to that of steel.
3. It is malleable with a small amount of carbon.
4. It is resistant against many acids except nitric acid.
5. It retains its surface upon exposure to the atmosphere.
6. It is somewhat less ductile than soft steel. But with a small amount of magnesium, ductility improves remarkably.

11.4 NON-FERROUS ALLOYS

Non-ferrous alloys may be defined as a coherent non-ferrous metallic mass produced by combining two or more non-ferrous metals. Some important non-ferrous alloys are discussed below.

11.4.1 Copper Alloys

Two important principal classes of copper alloys are brass and bronze. Brass is primarily an alloy of copper and zinc, whereas bronze is a combination of copper and tin.

Brasses are formed with 5–45% of zinc and copper. Brasses possess excellent mechanical properties. They are corrosion resistant and are readily machinable.

The mechanical properties of brass and the corrosion resistance of the brasses are further improved by the addition of one or more non-ferrous metals. Thus, manganese brass, iron brass, tin brass, lead brass and aluminium brass are modified brasses.

Bronze has varied physical properties. An increase in the percentage of tin increases the tensile strength of bronze. Bronze is most ductile when it contains about 5% of tin, but it decreases with increase in the percentage of tin. Like brass, bronze in modified forms is available as phosphor bronze, silicon bronze and aluminium bronze.

11.4.2 Aluminium Alloys

Principal elements that are alloyed with pure aluminium to improve its tensile strength and hardness are copper, silicon, manganese, zinc, magnesium and nickel.

Copper is the main hardening element for aluminium alloy by the addition of a small percentage of magnesium, which improves the hardness and strength after heat treatment. Similarly, the addition of a small percentage of manganese and nickel to an aluminium alloy further improves the qualities.

Silicon is the next alloying element which in combination with magnesium forms a hard compound that is largely responsible for the hardness.

The addition of manganese prevents the formation of coarse crystalline structure during heat treatment.

Popular aluminium alloys are duralumin, hindalium, magnalium and Y-alloy.

11.4.3 Magnesium Alloys

Magnesium alloys comprise 3–10% aluminium, 1–3.8% zinc and 0.4% manganese. The two important magnesium alloys are dow metal and electron metal. Dow metal comprises 9.1% magnesium and 9% aluminium. Electron metal, a trade name of magnesium, is a base alloy that contains 4% zinc and small percentages of copper, iron and silicon.

The tensile strength of magnesium alloy is low. Heat treatment increases the resistance to shock. On exposure to the atmosphere, it develops a dark oxide film which resists corrosion. Magnesium alloy, dow metal, can be forged, welded and drawn as wires.

11.4.4 Nickel Alloys

Important nickel alloys are iconel and monel metals. The composition of iconel is 75% nickel, 15% chromium and 9% iron. Monel metal comprises two-thirds nickel, one-thirds copper with a small percentage of elements like iron, silicon, manganese and carbon.

Iconel can be cast, forged, rolled and cold drawn. It has brittle behaviour when the temperature is between 650 and 950°C. Both at ordinary and high temperatures, it has high corrosion resistance. It can be soft soldered. It is used to manufacture springs that can withstand high temperature.

Monel metal is superior to brass or bronze in resisting corrosion. At ordinary temperature it is magnetic and becomes non-magnetic at temperatures between 100 and 150°C. It can be heat treated when aluminium and beryllium are added. Because of its peculiar flow properties, it can be welded only with special techniques. Because of its excellent corrosion resistance characteristic, it is widely used for parts of water pumps, propellers, domestic water storage tanks and parts subjected to high temperatures like intense combustion engines.

SALIENT POINTS

1. Iron is extracted from solid naturally occurring mineral aggregates called ores.
2. Metals are grouped under two broad groups, viz., ferrous metals and non-ferrous metals.
3. Ferrous metals contain iron as their main constituent. The important ferrous metals are pig iron, cast iron, wrought iron and steel.

4. Non-ferrous metals, viz., aluminium, copper, etc., do not contain iron as their main constituent.
5. Alloy is an intimate compound of two or more metals.
6. Iron ores are compounds of iron with non-metallic elements. Important varieties of iron ores are haematite, limonite, magnetite, pyrite and siderite.
7. Pig iron is the basic material from which wrought iron and steel are manufactured.
8. Three stages involved in the manufacture of pig iron are dressing, calcination and roasting and smelting.
9. Dressing is the process in which the iron ores as obtained from mines are crushed.
10. Calcination is the process by which the ores are heated in the presence of air and are oxidised.
11. Roasting the ores enables the dissipation of the volatile parts such as sulphur.
12. Smelting is the process of melting so as to separate metal from the ore. It is carried out in a special furnace called a blast furnace.
13. Wrought iron contains high pure iron and iron silicate.
14. Cast iron is an alloy of iron and carbon, with the carbon content being more than 2%. In addition, it contains impurities such as manganese, phosphorous, silicon and sulphur.
15. Cast iron is manufactured by re-melting pig iron with coke and limestone. The re-melting is done in a furnace called the cupola furnace.
16. Steel is a very ductile alloy consisting of iron and carbon.
17. Based on the percentage of carbon content, steel is classified into low carbon or mild steel, medium carbon or medium hard steel and high carbon or hard steel.
18. Manufacturing of steel may be done by five processors, viz., Bessemer process, Cementation process, Open-hearth process, Electric process and Duplex process.
19. To obtain special properties, elements such as nickel, chromium, manganese, etc., are added to the carbon steel. The steel thus formed is called alloy steel.
20. Metals that do not contain any percentage of iron are generally called non-ferrous metals. The important non-ferrous metals are aluminium, copper, lead, tin, zinc, magnesium and nickel.
21. Non-ferrous alloys may be defined as a coherent non-ferrous metallic mass produced by combining two or more non-ferrous metals. Non-ferrous alloys are brass, bronze, aluminium alloys (duralumin, hindalium, magnalium), magnesium alloys and nickel alloys.

REVIEW QUESTIONS

1. Define an ore. Discuss the important varieties of iron ore.
2. What is the composition of pig iron? How is pig iron produced?
3. What is cast iron? How is cast iron produced?
4. How is steel classified based on the carbon content?

5. Discuss various methods of manufacturing steel.
6. Draw a flow diagram explaining the production of steel.
7. Enumerate the uses of steel.
8. Differentiate between the properties of medium hard steel and hard steel.
9. What are the various properties of mild steel?
10. State the effects of adding nickel, chromium and manganese to steel. How are the products named?
11. What are non-ferrous metals?
12. Explain the properties of aluminium.
13. What is copper? Explain their properties.
14. What is the chief source of tin, and what are the uses of tin?
15. Differentiate between brass and bronze.
16. What are aluminium alloys?
17. Discuss the two forms of nickel alloys.
18. Compare the properties of Cast iron, wrought iron and steel and enumerate their uses.
19. Mention the properties and uses of
    (i) Aluminium, lead and copper
    (ii) Alloys of Copper.
20. Explain the properties of magnesium.

12 SURFACE FINISHING MATERIALS

12.1 INTRODUCTION

Apart from protecting various building units from the weathering action of the atmosphere, the main objective of surface finishing materials is to provide a decorative finish to obtain a clean, colourful and pleasing surface.

Surface finishing materials are paints, varnishes, distempers and white/colour washing. The surfaces may be plastered wall surfaces, ceilings, wooden surfaces, metallic surfaces, etc. These surface finishing materials have to provide a hygienic surface and present a healthy surrounding to live in.

Paints are primarily used for woods, walls and ceilings, and for metal works of windows. Varnishes are preferred for woodworks and sometimes for walls.

Distempers and white/colour washing are used for walls. Different varieties of distempers are available in the market.

Apart from the above surface finish materials, other materials used are wallpaper, wall tiles, coal tar, waxing, etc.

This chapter discusses the above aspects.

12.2 PAINTS

Paint is mainly composed of two ingredients, viz., a base which is a solid material and a liquid called a vehicle. The vehicle carries the solid matter, distributes it evenly and enables the solid matter to adhere to the surface.

12.2.1 Functions of Paints

Irrespective of the type of paint, in general, the functions of paints are as follows:

1. It has to protect the surface from the weathering effects of atmosphere such as sun, wind, other liquids, fumes, etc.
2. It has to prevent corrosion of metal bases and decay of wooden surfaces.
3. It has to provide a smooth surface for easy cleaning.
4. It should provide the surface good aesthetic appearance and a hygienically sound surface.

12.2.2 Constituents of an Oil Paint

Metals and wooden surfaces are generally painted by oil paints. Sometimes walls are also painted by oil paints. Thus, it is worth discussing the constituents of an oil paint. An oil paint is made up of six main constituents, viz., base, inert filler, colouring pigment, vehicle, thinner and drier.

1. Base

The base forms the chief ingredient of a paint. The most important purpose of adding a base to the paint is to make an opaque coating so as to hide the surface to be painted and to reduce shrinkage cracks. The bases that are commonly used for the paints are white lead, red lead, oxide of zinc, oxide of iron, titanium white, etc.

2. Inert Filler

It is a cheap pigment added to a paint to reduce its cost. Further, it modifies the weight of the paint and makes it more durable. Commonly used inert fillers, also called as extenders, are barium sulphate, lithophone, silicate of magnesia or alumina, gypsum, etc.

3. Colouring Pigments

It is a white or coloured pigment added to the paint to get the desired colour of the paint.

4. Vehicle

Vehicle acts as a binder for various pigments, viz., bases, extender and colouring pigment. Vehicles are required for two purposes, mainly:

(i) To make it possible to spread the paint as a thin layer on the surface evenly and uniformly.
(ii) To provide a binder for the ingredients of a paint such that the paint sticks to or adheres to the surface.

Vehicles that are commonly used are: linseed oil, tung oil, poppy oil, nut oil, etc.

5. Thinner

It is also known as solvent. It is a liquid which thins the consistency of the paint and evaporates after the paint film has been applied. It increases the workability of the paint. The commonly used solvents are: turpentine, pure oils, petroleum spirit and highly solvent naphtha.

6. Drier

It is a material used in small quantities for accelerating the drying of paint film. It is a metallic compound that acts as a catalyst for the oxidation and polymerization of the vehicle used in the paint. The commonly used driers are: litharge, lead acetate, manganese dioxide and cobalt. In an oil paint only 10% (by volume) of the drier should be used. The use of excess drier leads to flaking.

12.2.3 Characteristics of a Good Paint

An ideal paint should possess the following qualities:

1. The consistency of the paint should be adequately workable.
2. It should be capable of being spread across a wide area and cover the specific area with the minimum quantity of paint.
3. The thickness of the paint film should be enough for good protection and decoration of the surface.

4. It should dry within a reasonable time and not too rapidly.
5. The dried surface should withstand adverse weather conditions and should have a long life.
6. It should have adequate resistance to moisture, cracking and flaking.
7. It should be attractive as well as pleasing in appearance.
8. With time the colour should not fade.
9. During the application of paint it should not harm the health of the workers.
10. It should be fairly cheap and economical.

12.2.4 Types of Paints

Various types of paints are discussed below:

1. Oil Paint

The various constituents of an oil paint are discussed in Section 12.2.4. These paints are self-priming and in general do not need any special primer or undercoat. When it is used for general decoration purposes, it is a common practice to apply at least three coats of varying compositions, i.e., primers, undercoats and finishing coats on the unpainted surfaces.

2. Aluminium Paint

Finely ground aluminium is suspended in either quick-drying spirit varnish or slow-drying oil varnish as per the requirement. After painting, the spirit evaporates and a thin metal film of aluminium is formed on the surface. Aluminium paint has several advantages. It is widely used for painting gas tankers, hot water pipes, radiators, etc.

3. Cement Paint

The term cement paint is applied to paints which are based on white cement with a pigment, a filler and an accelerator. It is available in a dry powder form. A variety of shades are available and are water-proof and durable. It is mixed with water immediately before its application. Cement paints are ideal for rough surfaces because its adhesive power is poor on smooth surfaces.

Two coats have to be applied. Before applying the first coat the surface is wetted. The second coat is applied not less than 24 hours after the first coat. To obtain best results, the surface has to be kept wet for about two days after the final coat. However, the cement paint is not recommended for regions under sub-zero temperatures.

4. Anticorrosive Paint

It primarily consists of oil and a strong drier. Very fine sand is mixed with a pigment and added to the paint. The pigment may be chromium oxide, lead, red lead or zinc chromate. The appearance of the paint is black and has a long life. It is comparatively cheap.

5. Emulsion Paint

It contains binding materials such as polyvinyl acetate, synthetic resins, etc., and it is easy to apply. It dries quickly and the colour of the paint is retained for a long period. The surface of the paint is tough and can be cleaned by washing with water.

Emulsion painting can be carried out either by a brush or sprayed using a gun. Two coats are needed for a long life. In general, a smooth surface is needed to apply emulsion paint.

6. Enamel Paint

It contains white lead or zinc white, oil, petroleum spirit and resinous matter, which are available in different colours. It forms a hard and durable surface but dries slowly. The surface painted by enamel paint is not affected by acids, alkalies, fumes of gases, steam, water, etc. This is quite suitable both for internal and external walls. To make the appearance look better, it is recommended to apply a coat of titanium white in pale linseed oil before applying enamel paint.

7. Luminous Paint

This paint contains calcium sulphide with varnish. The surfaces painted by this paint shines like radium dials after the source of light has been cut off. This paint should be used only on surfaces that are free from corrosion or any other lead paint.

8. Cellulose Paint

This is a special type of paint prepared from nitro-cotton, celluloid sheets, photographic films, etc. It hardens on evaporation of a thinning agent and due to oxidation. Although it is costly, it presents a flexible, hard and smooth surface. The surface coated with this paint can withstand adverse weather conditions and can be washed and easily cleaned.

9. Asbestos Paint

This is a peculiar type of paint, and it is applied on the surfaces that are exposed to acidic gases and steam.

10. Plastic Paint

This type of paint is also called plastic emulsion paint and is available in the market under different trade names. The paint is composed of plastic compounds such as vinyl acetate and acrylate. This paint has a pleasing appearance, and it is attractive in colour. This is used in showrooms and auditoriums.

11. Bituminous Paint

This paint is obtained by dissolving asphalt or mineral pitches or vegetable bitumen in any type of oil or petroleum. This has a black appearance and is used to paint iron work under water.

12.3 VARNISHES

Varnish is a clear, pale solution of a resinous substance dissolved in either oil, turpentine or alcohol. The resinous substances are amber, common resin, copal, lac or shellac. The solution spread on the surface after drying forms a hard, transparent and glossy film on the varnished surface.

12.3.1 Functions of Varnishes

In general, varnish is applied on wooden surfaces of doors, windows, etc. to decorate the surface without hiding the beautiful grains of wood and at the same time to protect the surface from atmospheric effects. Sometimes painted surfaces are also varnished to enhance the appearance of the paint and also to increase the durability of the painted surface.

As the preparation of varnish is tough, readymade varnishes are used.

12.3.2 Constituents of Varnish

The three ingredients of a varnish are: resins or resinous substances, solvents and driers.

1. Resins

As stated earlier, the commonly used resins are copal, lac or shellac and resin. Copal is available from the earth at places where pine trees existed in the past. It is a hard substance in a variety of forms. Lac or shellac is obtained by the exudation of some insects. Resin is obtained from pine trees.

2. Solvents

Based on the type of resin, the type of solvent is decided. Table 12.1 shows the solvent for different resins.

Table 12.1 Solvents for different resins

Solvent-Boiled linseed oil	Solvent-Methylated spirits	Solvent-Turpentine	Solvent-Wood naphtha
• Amber • Copal	• Lac or Shellac	• Mastic • Gum dammar • Rosin	• Cheap resins

3. Driers

To accelerate the process of drying, driers are added to the varnish. Common driers used in varnishes are litharge, white copper and lead acetate.

12.3.3 Properties of an Ideal Varnish

The properties of an ideal varnish are as follows:

1. It should always provide a glossy surface and be pleasing in appearance.
2. It should dry fast and provide a uniform finished surface.
3. When exposed to the atmosphere, the colour of the varnish should not fade.
4. The developed thin film on the surface should be tough, hard and durable.
5. During or after drying it should not show cracks or shrink.

12.3.4 Types of Varnishes

Based on the solvent used, varnishes may be classified as given below:

1. Oil Varnishes

Oil varnishes are prepared from the hardest resin or gums, such as amba, copal, etc., by heating and dissolving them in linseed oil with turpentine. Heat is added to increase the workability. Although it dries slowly, it is the hardest and most durable of all varnishes. In general, it can be used both for interior and exterior works. It is preferred for joinery and fitting purposes and for all outside works.

2. Turpentine Varnishes

These varnishes are prepared from soft resins such as gum dammar, mastic, common resins, etc., by dissolving them in pure spirits or turpentine. These varnishes are flexible, lighter in colour and quick in drying. Compared to oil varnishes they are weak.

3. Spirit Varnishes

These varnishes are also called lacquers. These are prepared from soft resins such as lac or shellac dissolved in methylated spirit. They are harder, dries quickly and brilliant in appearance. These are easily affected by weathering actions.

4. Water Varnishes

These varnishes are prepared using shellac, which is dissolved in hot water and mixed with an adequate quantity of ammonia or borax, potash or soda such that the shellac gets dissolved. Water varnishes are used to varnish wallpapers, maps, pictures, etc.

5. French Polish

This is a refined spirit varnish. It is prepared by dissolving 0.15 kg of black or light brown shellac in one litre of methylated spirit without heat. The polish is obtained by adding a suitable colour pigment. This polish dries out quickly within a few minutes. It gives a fine glossy surface. It is mostly used for superior wood works like furniture, hand rails, etc.

12.4 DISTEMPERS

12.4.1 Function of Distempers

Distempers are water paints consisting of whitening (i.e., powdered chalk), glue or casein as a binder and suitable colouring pigments. Distempers are used on plastered, cement concrete and on various wallboard surfaces. They are cheap, durable and are pleasing in appearance. They also safeguard the surfaces from adverse weather conditions.

12.4.2 Constituents of Distempers

Distemper comprises a base, a binder, a thinner and a pigment. Whitening or chalk forms the base, glue forms the binder, water forms the carrier or thinner and a colouring pigment.

Distempers are available in a powder form or a paste form. They are to be mixed with hot water before use. This is more or less used as a water paint. Here, whitening or chalk serves as the base instead of white lead and the water is used as a carrier instead of linseed oil.

12.4.3 Properties of Distempers

General properties of distempers are given below:

1. On drying the distemper shrinks which may lead to cracking and flaking. This is predominant if the surface is weak.
2. In general distemper coating is thicker and hence they are more brittle than other types of water paints.
3. They are less durable than oil paints.
4. The distemper layer is porous in nature and allows water vapour to pass through it.
5. They provide less workability.
6. It is not suitable in places where there is dampness.
7. They are available in light shades.

12.4.4 Types of Distempers

Different forms of distempers are discussed below:

1. White Distempers

This type of distemper is prepared by mixing only whitening (i.e., powdered chalk) and glue.

2. Coloured Distempers

In this type of distemper, the required colour pigment is added to the whitening before mixing with glue.

3. Oil Bound Distempers

A variety of oil paints are available in which drying oil is so treated that it mixes with water. Whenever required, such distempers are diluted or thinned with water. These distempers after allowing drying form a washable and durable surface.

4. Casein Paints

Casein is extracted from milk curds and acts as a binder. These paints are prepared by mixing a finely ground casein with a white base, normally slaked lime. This has high opacity. Thus, it is preferred for use on plastered surfaces, stuccos, fibre boards, cement blocks, etc.

12.5 WHITE AND COLOUR WASHING

White washing is the oldest method of surface finishing in use. The white-washing material is prepared from pure fat lime, which may be slaked at site. Slaking is carried out in a tub until the mixture shows a creamy consistency. The slaked lime is allowed to rest for 24–48 hours. The mixture is then strained through a coarse cloth. Three kilograms of gum boiled with 10 kilogram of rice are added to each cubic metre of the slaked lime.

The completed white-washed surface shall form an opaque coat of white colour. It also presents a smooth regular surface free from any powdery matter.

Colour washing is prepared by the addition of the necessary pigment to the pure slaked fat lime and mixed thoroughly. Then it is screened through a clean and fine cloth. During the time of application it is continuously stirred with a stick. To get uniformity, an adequate quantity should be mixed. Colour should not crack or come off readily on the fingers when rubbed.

12.6 MISCELLANEOUS SURFACE FINISHES

Apart from the surface finishing materials, the following methods are also used in specific cases:

12.6.1 Wallpapering

Wallpapers of a certain design are used for developing the aesthetic values of interior walls and ceilings. They may be exclusively of paper or combined with other materials. The papers used for papering are: pulp paper, satin paper or flock paper.

Pulp paper is the most commonly used type. Satin paper is easily maintained by cleaning but it is affected by dampness. The third type, flock paper gives a good aesthetic appearance but is prone to dust.

Varnished wallpaper is also available, which can be polished. There are certain wallpapers, which are washable for which oil paper may be used. Metal-coated wallpapers and wallpapers with sound-absorbing properties are also available.

12.6.2 Wall Tiling

Wall tiling is the process of lining or finishing the walls with special tiles up to a certain height. In general, the height varies from 60 to 120 cm above the floor level or in some cases up to the ceiling.

Particularly, wall tiling is done in kitchen, bathrooms, W.Cs, passages, fire places, staircase walls, boiler rooms, etc. Sometimes, it is also used on outer surfaces of decorative causes or to protect them from atmospheric influences.

Wall tiles are either of terracotta, china clay or marble available in different colours and designs.

12.6.3 Coal Tarring

For preservative purposes, coal tar is applied to wood work or iron work. The tarred surfaces can further be treated by varnishing or painting. The mixture is prepared by adding about 200 gm of unslaked lime for every litre of tar and then heated till it starts boiling. The hot mixture is thinned by adding 1/16th litre of country spirit to each litre of tar.

12.6.4 Waxing or Wax Polishing

To improve the elegance and at the same time to protect undercoats, wax polishing is done on varnished surfaces. Wax polishing presents a highly pleasing lustre.

Wax polish is prepared by mixing bees-wax with linseed oil, turpentine and varnish in the ratio 2:1½:1:½, respectively, by weight. For preparing this, bees-wax is first melted or dissolved in linseed oil on low fire and then the mixture is cooled down slightly. Turpentine and varnish are then added to the mixture and then stirred well.

12.6.5 Wood Oiling

Woodwork not exposed to weather is generally provided with wood oiling as it is cheaper. Further, oiling of woodwork is done to increase the durability and also to improve the appearance of the surface. Linseed oil or sweet oil is used.

Linseed oiling consists of a mixture obtained by heating three parts of double-boiled linseed oil and one part of bees-wax to which one part of turpentine is added by weight. Sweet oiling consists of a mixture of equal parts of common vinegar, country sweet oil and spirits of turpentine. Sweet oiling mixture gives a darker effect than linseed oiling.

SALIENT POINTS

1. Paint is mainly composed of two ingredients, viz., a base which is a solid material and a liquid called a vehicle.
2. The vehicle carrying the solid matter distributes it evenly and enables the solid matter to adhere to the surface.
3. Constituents of an oil paint are base, inert filler, colouring pigment, vehicle, thinner and a drier.

4. The most important purpose of adding a base in a paint is to make an opaque coating so as to hide the surface to be painted and to reduce shrinkage cracks.
5. Inert filler is a cheap pigment added to a paint to reduce its cost.
6. Coloured pigment is added to the paint to obtain the desired colour of the paint.
7. Thinner, also known as a solvent, is a liquid which thins the consistency of the paint and evaporates after the paint film has been applied.
8. Drier is a material used in small quantities for accelerating the drying of the paint film.
9. In aluminium paint, finely ground aluminium is suspended in either quick-drying spirit varnish or slow-drying oil varnish as per the requirement.
10. The term cement paint is applied to paint which is based on white cement with a pigment, filler and an accelerator.
11. Anticorrosive paint primarily consists of oil and a strong drier. Very fine sand is mixed with a pigment and added to the paint.
12. Emulsion paint contains binding materials such as polyvinyl acetate, synthetic resins, etc., and it is easy to apply.
13. Enamel paint contains white lead or zinc white, oil, petroleum spirit and resinous matter.
14. Luminous paint contains calcium sulphide with varnish. The surface painted by this paint shines like a radium dial after the source of light has been cut off.
15. Cellulose paint is a special type of paint prepared from nitro-cotton, celluloid sheets, photographic films, etc.
16. Asbestos paint is applied to surfaces that are exposed to acidic gases and steam.
17. Plastic paint is composed of plastic compounds such as vinyl acetate and acrylate.
18. Bituminous paint is obtained by dissolving asphalt or mineral pitches or vegetable bitumen in any type of oil or petroleum.
19. Varnish is a clear, pale solution of a resinous substance dissolved in oil, turpentine or alcohol.
20. The three ingredients of varnish are resins or resinous substance, solvents and driers.
21. Commonly used resins are copal, lac or shellac and resin.
22. To accelerate the process of drying, driers are added to the varnish. Common driers used in varnishes are: litharge, white copper and lead acetate.
23. Oil varnishes are prepared from the hardest resin or gums such as amber, copal, etc., by heating and dissolving them in linseed oil with turpentine.
24. Turpentine varnishes are prepared from soft resins such as gums dammar, mastic, common resins, etc. by dissolving them in pure spirits or turpentine.
25. Spirit varnishes are called lacquers. These are prepared from soft resins such as lac or shellac dissolved in methylated spirits.
26. Water varnishes are prepared using shellac, which is dissolved in hot water and mixed with an adequate quantity of ammonia or borax, potash or soda such that shellac gets dissolved.

27. French polish is prepared by dissolving 0.15 kg black or light brown shellac in one litre of methylated spirit without heat.
28. Distempers are water paints consisting of whitening (i.e., powdered chalk), glue or casein as a binder and suitable colouring pigments.
29. A distemper comprises a base, a binder, a thinner and a pigment. Whitening or chalk forms the base, glue forms the binder, water forms the carrier or thinner and a colouring pigment.
30. White-washing material is prepared from pure fat lime which may be slaked at site. Slaking is carried out in a tub until the mixture shows a creamy consistency.
31. Colour washing is prepared by the addition of necessary pigments to the pure slaked fat lime and mixed thoroughly.

REVIEW QUESTIONS

1. Explain the functions of a paint.
2. What are the ingredients of an oil paint? Explain.
3. List the characteristics of a good paint.
4. Enumerate the different types of paints.
5. What paints would you recommend for the following job, reason out your choice.
    (i) Steel bridge girder
    (ii) Steel Roof trunes
    (iii) Interior of a drawing room
    (iv) Walls of a cinema theatre
6. How do you select a solvent based on the resin to be used?
7. Explain the properties of an ideal varnish.
8. What are the different types of varnish?
9. What are the functions and essential ingredients of oil varnishes?
10. Why should distemper be preferred to white washing or colour washing?
11. Explain the object of distempering? Mention the ingredients and properties of distemper.
12. Where are wallpapers preferred?
13. What is waxing or wax polishing?
14. Enumerate the advantages of wall tiling.

13 OTHER BUILDING MATERIALS

13.1 INTRODUCTION

Apart from the main building materials such as stone, brick, lime, cement, mortar, concrete, wood, etc., there are different other materials which are used fully or partly as a building construction material.

The other materials which may be of concern for building construction are asbestos, fly ash, gypsum plaster, bituminous materials, glass, plastics and other miscellaneous materials.

Apart from the above, rubber and some adhesives are also used. New materials such as Geotextiles and Geomembranes also come into use in building construction industry.

A brief material constituents, applications and use in different aspects in building construction are dealt in this chapter. The varied uses of these materials have also been highlighted.

13.2 ASBESTOS

Asbestos is a naturally occurring fibrous material. It is a silicate of calcium and magnesium and contains a small amount of iron oxide and alumina. The original source of asbestos was the mineral fibrous mineral actinolite.

13.2.1 Types of Asbestos

Natural fibre of asbestos belongs to two groups, viz., those which are readily active with acid (i.e., non-acid-resistant) and others that are acid-resistant.

The first group primarily consists of chrysolite asbestos which is used for industrial purposes. The second group comprises of amonite and crocidolite asbestos. The second group is not recommended for commercial use as they are said to cause diseases related to lungs. The Indian industry uses the first group asbestos. However, it is advisable to use asbestos cement products with caution and sufficient care. As a matter of fact, in some countries the use of asbestos is not allowed in buildings.

Asbestos is mixed with cement and used mostly in building industry as asbestos cement.

13.2.2 Properties of Asbestos

Following are the properties of asbestos:

(i) It has high tensile strength along the fibre.
(ii) It is an excellent insulator for electricity (i.e., it is a non-conductor).
(iii) It is incombustible and hence acts as a good insulator for heat.
(iv) It is acid proof.
(v) It is flexible, soft and non-porous.
(vi) It is fire-resistant and the melting point is 1200–1500°C.
(vii) It possesses good adhesion with cement and hence the fibre can be used as a fibre reinforcement.
(viii) It is rust-free.
(ix) Its specific gravity is 3.10.
(x) It can be easily cut into pieces.
(xi) Its quality is critically affected by the length of fibres.
(xii) It is smooth like glass and silk.

13.2.3 Uses of Asbestos

Following are some of the uses of asbestos:

(i) Asbestos with cement as asbestos cement have plenty of uses as sheets and pipes. Sheets are used as roofing material and pipes are used for drain rain water, seepage water, etc.
(ii) Asbestos being a good heat insulator, it is used for insulating furnaces, steam and exhaust pipes, boiler, etc.
(iii) Asbestos being a high temperature resistant, it is often used for making moulded shapes to resist high temperature.
(iv) Fire fighting persons use fire-proof suits made of asbestos.
(v) Asbestos being a good sound insulator, it is used as small length fibres in different conditions.
(vi) In the form of fibres, asbestos is used for cable insulation and in sheet form, it is used in lining and partition in switches and fuse boxes.
(vii) In electric motors and coils exposed to high temperature, asbestos paper is used as a cover for electric wires.
(viii) Asbestos felt is used as a damp proof layer in combination with bitumen.
(ix) Asbestos is used for internal combustion engine exhaust pipes and silencer gaskets and sparkling plug washers in combination with copper shell.
(x) Asbestos, when powdered or in the form of fibres, is sometimes kneaded into a dough form with water and used for blocking holes and cracks in metal objects.
(xi) In combination with magnesium carbonate, or wool felt, it is used for lagging steam pipes.
(xii) It is used to form asbestos paints.

13.2.4 Products of Asbestos

Some of the products of asbestos which are manufactured are discussed below.

1. Asbestos Boards

Asbestos fibres and Portland cement are mixed and moulded under hydraulic pressures. These are available in board and sheet forms. Some of the commercial forms are claddings, roofing and decking units.

Asbestos boards are available in natural mottled grey colour and pigmented boards are also marketed. Boards are dense and rigid but workable easily using carpenter's tools. These boards can withstand temperatures up to 540°C.

Asbestos boards are marketed in two sizes, viz., Class A: thickness = 6.5 mm and width = 1.2 mm; Class B: thickness = 5.0 mm and width = 1.2 mm.

Asbestos boards are economical and completely incombustible. They can take up paints of any shade economically. These are easily workable and durable.

2. Asbestos Sheets

Asbestos sheets are flat and are classified as compressed and uncompressed. Compressed sheets are of minimum thickness unit which can withstand bending stress of 22 N/mm^2 with a density of 1600 kg/m^3, whereas uncompressed units can withstand a bending stress of 16 N/mm^2 with a density of 1200 kg/m^3. Corrugated types of sheets are with thickness varying from 4.76 to 6.35 mm and the corrugation has a pitch of 10 cm and the weight being 145 N/m^2.

Asbestos sheets are used for wall-lining, wall-panelling, partitions, side-cladding and false-ceilings. Also used for wardrobes, cupboards and kitchen-pantry shelves. Black-boards, hoardings, signboards, window panes and door panels are also of asbestos sheets.

3. Asbestos Cement Pipes

These pipes are made using 15–20% of asbestos fibres, 80–85% of cement and water. Pipes of different sizes, bands, functions, taper pipes, shoes, etc., are also manufactured to the required shape and size. They can be easily cut and need no paint to maintain.

These pipes claim several advantages because of its adaptability, easy installation, and are light, strong and can withstand normal handling.

Asbestos cement pipes are used as rain water pipes, valley gutters, soil waste and ventilating pipes, half-round gutters, etc.

4. Asbestos Paint

It is prepared by the addition of finely ground asbestos to an oil paint or distemper. It is highly fire resistant, heat and sound insulation. It is vermin-proof and prevents corrosion of metals.

5. Asbestos Fibres

Asbestos is also available in fibre form in the market. These fibres have adequate strength and flexibility and are used in making conveyor belts, boiler insulation, brake lining, etc.

13.3 FLY ASH

Fly ash is a fine grained residue resulting from the combustion of pulverised coal or lignite in boilers. As it is in ash form it is easily transported by the flue gases and collected by mechanical or electrostatic separators. Mainly it consists of spherical glassy particles ranging from 1 to 150 μm in diameter and major portion passes through a 4.5 μm sieve.

13.3.1 Constituents of Fly Ash

Main constituents of fly ash are:

Silicon dioxide (SiO_2) = 30–60%
Aluminium oxide (Al_2O_3) = 15–30%
Unburnt fuel (carbon) = up to 30%
Calcium oxide (CaO) = 1–7%
Magnesium oxide (MgO) = small amount
Sulphur trioxide (SO_3) = small amount

The chemical composition of fly ash varies with the type of fuel burnt, load on boiler and type of separator, etc. High fineness, low carbon content, good reactivity are the necessary qualities of a good fly ash.

13.3.2 Uses of Fly Ash

Fly ash is most widely used in many profitable applications such as the following:

(i) Used in concrete as an admixture or as a part of replacement of cement.
(ii) Used in making cellular concrete blocks.
(iii) Used in making building blocks.

1. Application in Concrete as Admixture

Fly ash is the widely used pozzolanic material all over the world. Addition of fly ash, apart from playing the role of an admixture, imparts the following properties to concrete:

(i) Cement aggregate reaction is reduced by the addition of fly ash.
(ii) Evolution of low heat when fly ash is added to the concrete.
(iii) Water tightness of concrete is greatly improved by the addition of fly ash in concrete.
(iv) Addition of fly ash improves the plasticity and cohesiveness of concrete mixture and thereby permits easy placing of concrete and finishing.
(v) Addition of fly ash improves the strength of concrete also.

The importance and use of fly ash in concrete has grown so much that it has become a common ingredient in concrete, particularly for making high-strength and high-performance concrete. The quality of fly ash for use in concrete should satisfy the chemical composition recommended by Indian Standards (IS: 3812–1981), as given in Table 13.1.

Table 13.1 Chemical composition requirement

Sl. No.	Characteristics	Requirement Mass Percentage
1.	(Fe_2O_3)	≥ 70.0
2.	(SiO_2)	≥ 35.0
3.	(MgO)	≤ 5.0
4.	(SO_3)	≤ 2.75
5.	(Na_2O)	≤ 1.50
6.	Loss on Ignition	≤ 12.0

Source: IS: 3812–1981.

As per the Indian standards (IS: 3812–1981), the physical requirements are as follows:

(i) Minimum specific surface should be 320 m^2/kg for use as pozzolana.
(ii) Maximum average compressive strength of lime-fly ash mortar cube should not be less than 4.5 N/mm^2 for use as pozzolana.
(iii) On testing a mixture of four parts of OPC and fly ash by weight, the increase in volume should not be more than 0.8%.
(iv) A mixture of OPC and fly ash, the compressive strength should not be less than 80% of the strength of corresponding plain cement mortar cubes.

2. Application in Cellular Concrete Blocks

Cellular concrete blocks are produced by autoclaving a set mixture of fine silicous material such as fly ash and binder in the form of lime.

The cellular concrete blocks have many technical advantages, viz., strength-to-weight ratio, good sound insulation, stability against variations in temperature and humidity, resistance to fire and water seepage, low thermal conductivity, consumes less quantity of mortar in masonry and plaster can be completely avoided as the blocks are machine finished.

3. Fly Ash Building Blocks

Fly ash building blocks are manufactured using fly ash, sand and lime which are mixed in the ratio 80:13:7. Bricks are made in hydraulic press and the semi-dried bricks are cured in a steam chamber at an appropriate temperature and pressure.

Compared to the conventional burnt bricks, the fly ash building blocks are better in shape, technical specifications, compressive strength and impermeability. Furthermore, the fly ash building blocks are about 20% light in weight and about 12% cheap compared to the conventional bricks.

13.4 GYPSUM PLASTER

Gypsum is hydrated sulphate of calcium ($CaSO_4 \cdot 2H_2O$). It is not available in nature in pure form but contains impurities such as alumina, calcium carbonate, magnesium carbonate and silica. It is a white crystalline substance less soluble in water but soluble in hydrochloric acid. As a binding material, it sets and hardens quickly.

Gypsum heated between 120 to 180°C loses about 14.7% of the water of crystallisation in the form steam and forms calcium sulphate hemihydrate, known as *plaster of Paris*. Further heating leads to removal of all water and forms calcium sulphate anhydrate. This plaster is mixed with a small of accelerator (i.e., alum potassium sulphate, raw gypsum) to make anhydrous or Keene's gypsum plaster.

13.4.1 Properties of Gypsum Plaster

Following are the properties of gypsum plaster:

(i) It is light in weight and unaffected by bacteria.
(ii) It shows good adhesion to fibrous materials but only slightly soluble in water.
(iii) It sets with less change in volume and shows negligible shrinkage on drying.
(iv) Because of its quick drying property, finishing coats can be applied immediately.
(v) A porous gypsum is an important sound-absorbing material.

13.4.2 Uses of Gypsum Plaster

(i) It is used for ornamental plaster work.
(ii) It is used as a heat-insulating material.
(iii) Gypsum plaster boards are light in weight, fire-proof and easy to work which are used for ceiling for internal lining of walls and for partition walls.
(iv) Gypsum blocks are very useful in the construction of non-load bearing partition walls.
(v) Scatted gypsum piles (acoustic tiles) are used for large offices and auditorium to reduce echo.
(vi) Plaster of Paris is used in dentistry, artwork, pottery, etc.

13.5 BITUMINOUS MATERIALS

Bituminous materials may be broadly classified as bitumen and tar.

13.5.1 Bitumen

Bitumen is a hydrocarbon material of either natural or pyrogenous origin, found in gaseous, liquid, semi-solid or solid form. It is a complex organic material which is completely soluble in carbon disulphide. It occurs either naturally or may be obtained during the distillation of petroleum. Bitumen consists of colloidal hydrocarbon materials of asphalte nets, resins and oils. Bitumen may be further divided as native asphalt or petroleum asphalt (generally called as bitumen).

1. Native Asphalt

Native asphalts are those which occur in a pure or nearly pure state in nature. These are further designated as (i) Lake asphalt, (ii) Asphalites and (iii) Rock asphalt based on their origin of occurrence.

(i) Lake Asphalt

These are found in depressions in earth's surface which have accumulated in lakes. Most of the lake asphalts possess moisture in some form. The refined asphalt has a very low penetration. This material can be softened by fluxing with a petroleum flux. This may be blended with asphalts produced from petroleum to produce blended asphalts.

(ii) Asphalites

This is asphalt-like substance presumably derived from metamorphosis of petroleum. It is hard, brittle and practically a pure bitumen. These types of asphalts are further divided into three classes, viz., gilsonite, glance pitch and grahamite. It is obtained from crevices and seams in rock formation.

(iii) Rock Asphalts

These are deposits of limestone or sandstone naturally impregnated with asphalt. This is extensively available in many parts of the world. The asphalt content in the rocks is about a maximum of 20% and varies from soft to hard.

2. Petroleum Asphalts

Petroleum asphalts are derived from petroleum crude oil which is taken from oil wells. Crude oil is transported to refineries where it is separated into various components. The character of asphalt depends on the nature of crude oil. Crude petroleum is divided into three classes, viz., asphaltic base, semi-asphaltic base and paraffin base.

(i) Cutback Bitumen

When the viscosity of a bitumen is reduced by a volatile dilutant, the bitumen is called as cutback bitumen. The viscosity of the cutback and the rate at which it hardens depends on the characteristics and quantity of both the bitumen and volatile oil.

(ii) Bituminous Emulsion

When the bitumen is suspended in a finely divided condition in an aqueous medium and an emulsifier is used to stabilise, the end product is known as bituminous emulsion.

Bituminous materials are widely used in highway construction because of their dual qualities, viz., binding property and water proofing property.

Use of bitumen for road construction depends on the type of mix and the type of construction. The following are the desirable properties bitumen should have in paving mixes:

(i) Easy to get mixed,
(ii) Attainment of desired stability,
(iii) Should maintain stability even under adverse weather condition,
(iv) Adequate flexibility to avoid cracking,
(v) Sufficient adhesion with aggregates.

13.5.2 Tar

Tar is the viscous liquid produced when natural organic materials such as coal, petroleum, wood, etc., are carbonised. Depending on the material from which tar is obtained, it is called as wood tar or coal tar. Coal tar is used mostly for road work because of its superior quality. Road tar is produced by undergoing three stages, viz., carbonisation of coal to produce coal tar, refining or distillation of crude tar and blending of distillation residue with distillate oil fraction.

There are five grades of road tar, viz., RT-1 to RT-5 based on their viscosity and other properties. Uses of different grades of road tar are given below.

RT-1: Used for surface painting under adverse cold conditions.
RT-2: Used for standard surface painting under normal weather conditions.
RT-3: Used for surface painting, renewal coats, pre-mixing chips for top course and light carpets.
RT-4: Used for pre-mixing tar macadam in base course.
RT-5: Used for grouting.

13.6 GLASS

Glass may be defined as a hard, brittle and transparent or translucent material. Technically glass is any substance or combination of substances which has solidified from the liquid state without crystallisation.

13.6.1 Constituents of Glass

The main constituents of glass are silica, sodium or potassium carbonate, lime, manganese dioxide, cullet and colouring materials.

1. Silica

Silica is the principal constituent of glass. Silica alone when fused at a very high temperature would give a good glass on cooling. However, it is essential to add some quantity of alkaline materials along with lime to make the molten silica glass sufficiently viscous so as to get good workability and resistance against weathering agencies.

2. Sodium or Potassium Carbonate

It is an alkaline material which is also an essential component of glass. This is being added in certain proportion so as to reduce the melting point of silica and to impart viscosity to the molten glass.

3. Lime

This is added in the form of chalk so as to impart durability to glass. In order to make the glass bright and shining, lead oxide is added in place of lime.

4. Manganese Oxide

This is added in certain proportion so as to correct the colour of the glass due to the presence of iron in raw materials of glass.

5. Cullet

Cullet is the old broken glass of the same type as that is intended to be manufactured.

6. Colouring Substances

At the time of manufacturing a coloured glass, a suitable colouring substance is added at the fusion stage to provide the desired colour to the glass.

13.6.2 Properties of Glass

Properties of glass are mainly governed by the factors such as composition of constituents, state of the surface, thermal treatment, dimensions of the object, etc. Glass has been popular and useful due to the following properties:

(i) It has no sharp melting point and no definite crystalline structure.
(ii) It is capable to absorb, refract or transmit light.
(iii) At elevated temperatures it acts as an excellent electrical insulator.
(iv) It can take up high polish and can be used as a substitute for any costly gem.
(v) It is not affected by air or water.
(vi) It can not be attacked by ordinary chemical agents.
(vii) It can be altered to meet certain requirements by changing fusibility, hardness, refractive power, etc.
(viii) It is transparent and translucent.
(ix) It is possible to weld pieces of glass by fusion.
(x) It is affected by alkalis.
(xi) Based on the advanced technology, it is possible to make glass lighter than cork or softer than cotton.
(xii) The only drawback is that it is brittle.

13.6.3 Types and Uses of Glass

1. Conventional Types

Based on composition and properties, the conventional glass may be classified as soda lime or crown glass, flint glass, and Pyrex or heat-resistant glass.

(i) Soda Lime or Crown Glass

The composition by weight, in general, is sand 75 parts, lime 12.5 parts, soda 12.5 parts, alumina 1 part and waste glass 50 parts. It can be easily fused at comparatively low temperatures. It is quite cheap and available in clean and clear state.

It is principally used for window glass, plate glass and container glass (such as bottles, tumblers, etc.).

(ii) Flint Glass

The composition of flint glass is sand 100 parts, lead 70 parts, potash 32 parts and waste glass 10 parts. It liquefies at a lower temperature compared to soda lime glass and has better lustre. As the lead compounds are reduced easily, the glass must be melted in oxidising atmosphere, etc.

It is used for tablewares and for optical works. It has also wide range of use as electric lamps, thermometers, electron tubes, laboratory apparatuses, container for foods, etc.

(iii) Pyrex or Heat-resistant Glass

Both the above two glasses when subjected to sudden temperatures are not able to sustain because of large coefficients of thermal expansion. Only the basic oxides make them susceptible to chemical attack by water and acids. Elimination of the basic oxides and inclusion of boron oxide produce a glass that is very resistant to thermal shock and to attack by water and acids. The composition of material for such a glass by weight are silica 80 parts, boron oxide 14 parts, sodium oxide 4 parts, alumina 2 parts, with traces of potassium oxide, calcium oxide and magnesium oxide. In order to melt such as mixture, a very high temperature is required.

These glasses are called borosilicate glasses which are extensively used for cooking utensils and laboratory wares.

2. *Special Types of Glasses*

Special types of glasses are sheet glass, plate glass, float glass, wired glass, translucent glass, glass blocks, safety glass, bullet-proof glass, tinted glass, structural glass, etc.

(i) Sheet Glass

This is mainly used for small panels of doors and windows. These glasses are to be free from blisters, scratches, waves, bubbles, etc. It is manufactured in thickness varying from 1.5–5 mm and sizes up to 1.5 m × 1 m.

(ii) Plate Glass

It is available in thickness varying from 3–32 mm. This type of glass is stronger and more transparent with very less waviness. It is generally used for large size panels such as waiting halls, shopping complexes, etc. There are three varieties of this type of glass, namely rough cast, rolled (or patterned) and polished glass. Generally the polished variety is denoted as plate glass.

(iii) Flutted Glass

These are the glasses which have corrugations on one side of the plate glass. The other side is wavy but smooth. In this type, the light is admitted without the glare of the sun. It is used in situations where privacy without obstruction of light is needed. Ribs may be in the horizontal or vertical directions. Horizontal ribs give more light in the middle and less at the sides. The upright ribs give more light at the sides and less in the middle. It is thus more ideal to use for skylight roofs and for windows of industries and railway stations.

(iv) Float Glass

This type of glass is prepared by passing the molten glass through a molten tin bath (float bath). Because of the free-standing surface of the tin bath is very smooth, the float glass is of uniform thickness. It possesses excellent optical clarity and aesthetic appearance. Finally it is annealed to relieve all the stress. This is mainly used for large-sized shop windows and facade of tall buildings.

(v) Wired Glass

In this type of glass, wire-netting material is embedded during the time of manufacture. It has high resisting power compared to ordinary plate glass. Because of wire-netting, it does not fall into pieces when broken. It is used for fire resisting doors and windows and for skylights and roofs.

(vi) Translucent Glass

This glass is also called as obscured glass or frosted glass. In such glasses, a pattern or texture is imprinted on one or both faces. This prevents the image being seen through. The textured side should face the inside of a place (to be made obscure) and the plane glass side should face the other side. This type of glasses are used for doors and windows of bedroom, bath room, lavatories, etc.

(vii) Bullet-proof Glass

This type of glass is prepared by following special techniques. Here, the glass is made of several layers of plate glass with alternate layers of vinyl resin plastics. Further the inner layers are thicker than outer layers. At the time of manufacturing, special care is taken for heating and cooling of layers. Thickness of this type of glasses vary from 15–75 mm. Special quality of the glass is that it will not allow bullet to pierce through it. It is used in special type of cars, bank counters, jewellery stores, etc.

(viii) Glass Blocks

These are glass units that consist of two halves fused together so as to form a hallow inside. These blocks provide insulator against heat, cold and noise. These blocks are used for constructing wall panels.

13.7 PLASTICS

Plastics are basically organic substances which may be natural or synthetic resins in origin with or without moulding compounds. Synthetic resins are phenol, formaldehyde, cellulose, vinyl, alkyd, etc. The moulding compounds are catalysts, filters, hardeners, lubricants, pigments, plasticisers, solvents, etc.

13.7.1 Types of Plastics and their Properties

Plastics may be classified into two broad groups based on their behaviour with respect to heat, viz., thermo plastics and thermoset plastics.

1. Thermo Plastics

These are the plastics which become soft when heated and become hard when cooled. These plastics can sustain repeated softening and hardening for indefinite time. Old scrap and worn-out materials of this type can be effectively used again.

Some of the important thermo plastics are: acrylic, cellulose acetate, polythene, perspex and poly vinyl chlorides.

Acrylic contains methyl metha crylate. Acrylics are transparent, rough and strong and withstand impacts. It is used for doors, windows and as safety glass in automobile and aircrafts.

Cellulose acetate is made from cotton seed. It is used for electric cables, hard rails, etc.

Polythene is transparent and chemically unaffected by temperature and moisture. This is mainly used for preparing pipes, covers, waterproof packing, etc.

Perspex is strong enough and are made in the form of sheets. This is used for electrical fittings for decorative purposes.

Poly vinyl chloride (PVC) is a product obtained from vinyl chloride and acetate. It is acid proof, light and withstands wear and tear. It is used for drainage pipes, floor finishing, etc.

2. Thermoset Plastics

These plastics become rigid due to chemical changes when moulded under pressure and temperature. These plastics once cooled and take a form, can not be changed further due to the application of heat.

Some of the commonly used thermoset plastics are phenol formaldehyde, phenol furfuraldehyde, urea formaldehyde and bakelite.

Phenol formaldehyde is formed by heating phenol and benzene with a suitable catalyst. Phenol furfuraldehyde is formed by digesting husks of rice, oat, groundnut with sulphuric acid, distilling the mixture to separate the furaldehyde vapour and then permitting the vapours to chemically react with phenol with a suitable catalyst.

Urea formaldehyde is formed by reacting urea with formaldehyde in the presence of a catalyst.

Bakelite is made by heating equal amount of phenol and formaldehyde with a base. This is orange in colour which is used with fillers to form different materials.

Thermoset plastics are used for various purposes such as dishes, plates, cabinets for electrical appliances, etc.

13.7.2 Uses of Plastics

1. They are clean, light in weight and have shining and attractive surface.
2. They have resistance to corrosion, vermin attack and other atmospheric effects.
3. They act as good heat and electrical insulators.
4. They have easy workability.
5. They need no painting or polishing.
6. They are available in different colours and different degrees of transparency.
7. Some plastics have high compressive strength.
8. They have good adhesiveness capacity.
9. They are good in absorbing shocks and vibrations.
10. They are cheap and freely available.

13.8 COMPOSITE MATERIALS

Composite materials are formed by combining two or more different materials to utilise their virtues and minimising their deficiencies. These materials are also called as composites.

Depending on the requirement, tougher and lighter materials may be combined. Materials utilised retain their physical and chemical properties independently in the finished product.

The composites comprise of the following two main constituent materials:

(i) Strong load-carrying material called as reinforcement or reinforcing fibres.
(ii) Weaker material know as matrix.

13.8.1 Reinforcing Fibres

The following are the functions of reinforcing fibres:

(i) It provides strength and rigidity.
(ii) It supports the structural loads.

Three most commonly used reinforcing fibres are: (a) glass fibres, (b) carbon and (c) aramid.

Glass fibres are the heaviest which provide high flexibility. Carbon is the strongest and heavier. Aramid has moderate stiffness. When more than one reinforcing material is used, that material is called hybrid composites.

13.8.2 Matrix

Functions of a matrix are as follows:

(i) It has to work as a binder.
(ii) It is to keep the position and orientation of the reinforcement.
(iii) It is to balance the loads between the reinforcement.
(iv) It has to protect the reinforcement from degradation.
(v) It has to retain the shape and form of the finished structure.

Thermosetting resins are the most common type of matrix. These resins have the property of reacting chemically under certain time and temperature conditions. Epoxy resins are the most widely used thermosetting resins. Polyester, vinyl ester, phenolic, bismaleimide, epoxy novolac, polymide, etc., are the other matrixes.

Composites may be natural composite or man-made. For example, wood, bones, teeth, etc., are natural composites. Wood comprises of cellulose fibres with good strength and stiffness in a resinous matrix of polysaccharide.

Man-made composites are many and they are as follows:

(i) Bricks made out of mud and straw.
(ii) Plywood is a composite laminate.
(iii) Concrete is an excellent example of a ceramic composite.
(iv) Reinforced plastics.
(v) Wire reinforced tiles.
(vi) Fibre-reinforced glass.

Keeping in view the basic concepts of reinforcement and matrix, any combination of materials can be tried to suit a specific need.

13.9 MISCELLANEOUS MATERIALS

Some of the materials which are used to a lesser extent in building activities are discussed in this section.

13.9.1 Rubber

Rubber is a natural material extracted from rubber trees and are also manufactured by chemical processes. The former is known as natural rubber and the latter as synthetic rubber.

Natural rubber has high strength, good resistance to wear and tear as well as in flexure. But it has low hysteresis and easily gets affected by solvents.

Synthetic rubbers can be produced to serve different purposes. In making different articles of rubber natural rubber is added to synthetic rubber. Rubber is also produced from used worn out articles like automobile tyres which are called reclaimed rubber. These are mainly used for making hose pipes and minor articles.

1. Use of Rubber in Mortar and Concrete

Latex or polymer latex used in cement mortar forms a polymer-modified system. It exhibits the following properties:

(i) Excellent adhesion,
(ii) Improved tensile, compressive and flexural strength,
(iii) Excellent resistance to penetrations of water, and
(iv) Improved resistance to chemicals.

However, addition of natural rubber does not yield these benefits.

Synthetic rubber is used for making polymer concrete for waterproofing. At the time of hydration of cement along with polymerisation of latex-based polymers, the pores in the cement mortar or concrete get coated with the latex film. Formation of this latex film in the pores gives it good waterproofing qualities by preventing capillary rise of water.

2. Use in Floors

In certain environment, there is a need for *resilient floors* such as computer room, libraries, etc. Rubber-based floors are noise proof and highly elastic. Rubber tiles laid on any rigid floor can be used for these resilient floors. These tiles are non-slip, quiet flooring of low thermal and electrical conductivity.

3. Other Uses

A synthetic rubber, called neoprene, is extensively used for bearing of bridges. Natural and synthetic rubber is used indirectly to make many products used in building industry.

Rubber is vulcanised with sulphur to improve the qualities of rubber with reference to friction, solvents, durability, etc. Very soft variety is obtained with the use of small quantity of sulphur (1–5%) and very hard variety is got with higher percentage of sulphur (about 30%).

13.9.2 Geosynthetics

Geosynthetics are synthetic fabrics which are sufficiently durable to last a long time in any environment. Geotextiles and Geomembrane are the two major products of geosynthetics.

1. Geotextiles

Geotextiles are porous fabrics manufactured from synthetic materials that are primarily petroleum products and others such as polyester, polyethylene, polypropylene and polyvinyl chloride, nylon, fibre glass, and various mixtures of these. They are manufactured in thicknesses ranging from 10–300 mils (1 mil = 0.0254 mm), widths up to 10 m and roll lengths up to about 600 m. Geotextiles are termed filter fabrics and have permeabilities comparable in range from coarse gravel to fine sand.

Geotextiles have been used in a variety of civil engineering works. The four major functions for which geotextiles are used are soil separation, filtration, drainage and reinforcement.

2. Geomembranes

Geomembranes are thin materials with very low permeability. They are flexible and are manufactured from synthetic or bituminous products. They may be strengthened, if necessary, with a fabric or film. Geomembranes differ from geotextiles with reference to the rate of permeability. The permeability is high (as in sand) in geotextiles and very low in geomembranes (as in colloidal clay). For all practical purposes, geomembranes may be considered to be impermeable to both gases and fluids. This makes them ideal for constructing of water-proof or gas-proof barriers between adjacent bodies of soil or soil and fluid. Geomembranes are used for sealing against fluid percolation and buffers against pollutants.

13.9.3 Adhesives

Adhesives are used to stick two or more parts together so that they behave as a single unit. Adhesives are of three types viz., organic adhesives, synthetic adhesives and rubber-based adhesives.

1. Types of Adhesives

(i) Organic Adhesives

Organic adhesives are made from animal protein, blood albumen, milk casein or starch, etc.

(ii) Synthetic Adhesives

Synthetic adhesives are made from natural resins or from manufactured plastic resins such as melamine resins, urea resins and phenolic resins.

(iii) Rubber-based Adhesives

Rubber-based adhesives are materials in which rubber is dissolved in benzene or ground with other resins of plastics.

2. Uses of Adhesives

Adhesives are used for the following building construction works.

(i) Use in Timber Construction

In woodwork, formaldehyde adhesives of phenol, urea and resorcinol are the accepted products. They are also used in timber joints to strengthen the joints. Four types of adhesives used for timber construction are: Weatherproof and Boil proof (WBP), Boil resistant (BR), Moisture resistant (MR) and Interior (INT).

(ii) Use in Fixing of Ceramic Tiles

Adhesives used for fixing ceramic tiles provide greater bond strength, reduction in weight of fixing material and make the wood cleaner and quicker. Further use of adhesives eliminates the need to soak the tiles in water and errors due to variation of the composition of the mortar. Latex-based adhesives are very popular.

(iii) Use for Joining Concrete

In practice, there may be instances where concrete has to be joined with other concrete work or other materials. Epoxide adhesives are the best adhesives to be used for bonding broken concrete face, setting metal noses in stairs, etc. The required bond strength is attained in short duration of time.

SALIENT POINTS

1. Asbestos is a naturally occurring fibrous material. It is a silicate of calcium and magnesium and contains a small amount of iron oxide and alumina.
2. Natural fibre of asbestos belongs to two groups, viz., those which are readily active with acid (i.e., non-acid-resistant) and other group is of acid-resistant.
3. The non-acid-resistant group primarily consist of chrysolides asbestos which is used for industrial purposes. The acid-resistance group comprises of amorite asbestos and crocidolite.
4. The acid-resistant group is not recommended for commercial use as they are said to cause diseases related to lungs.
5. Asbestos is mixed with cement and used mostly in building industry as asbestos cement.
6. Products of asbestos are asbestos boards, asbestos sheets, asbestos cement pipes, asbestos paint and asbestos fibres.
7. Fly ash is a fine grained residue resulting from the combustion of pulverised coal or lignite in boilers. Mainly it consists of spherical glossy particles ranging from 1 to 150 μm in diameter and major portion passes through a 45 μm sieve.
8. Main constituents of fly ash are as follows:

   Silicon dioxide (SiO_2) = 30–60%
   Aluminium oxide (Al_2O_3) = 15–30%
   Unburnt fuel (carbon) = upto to 30%
   Calcium oxide (CaO) = 1–7%
   Magnesium oxide (MgO) = small amount
   Sulphur trioxide (SO_3) = small amount
9. Gypsum is hydrated sulphate of calcium ($CaSO_4 \cdot 2H_2O$). It is not available in nature in pure form but contains impurities such as alumina, calcium carbonate, magnesium carbonate and silica.
10. Gypsum heated between 120 and 180°C loses about 14.7% of the water of crystallisation in the form of steam and forms calcium sulphate hemihydrate, known as plaster of Paris. Further heating leads to removal of all water and forms calcium sulphate anhydrate.
11. Bitumen is a hydrocarbon material of either natural or pyrogenous origin, found in gaseous, liquid, semi-solid or solid form. It is a complex organic material which is completely soluble in carbon disulphide.

12. Natural asphalts are those which occur in a pure or nearly pure state in nature. These are further designated as
    (i) Lake asphalt
    (ii) Asphalites
    (iii) Rock asphalt based on their origin of occurrence.
13. Lake asphalts are found in depression in earth's surfaces which have accumulated in lakes. Most of lake asphalts possess moisture in some form.
14. Asphalites is asphalt like substance presumably derived from metamorphosis of petroleum. It is hard, brittle and practically a pure bitumen.
15. Rock asphalts are deposits of limestone or sandstone naturally impregnated with asphalt.
16. Petroleum asphalts are derived from petroleum crude oil which are taken from oil wells. Crude petroleum is divided into three classes, viz., asphaltic base, semi-asphaltic base and paraffin base.
17. When viscosity of a bitumen is reduced by a volatile dilutant, the bitumen is called as cutback bitumen.
18. When the bitumen is suspended in a finely divided condition in an aqueous medium and an emulsifier is used to stabilise, the end product is known as bituminous emulsion.
19. Tar is a viscous liquid produced when natural organic materials such as coal, petroleum, or wood, etc., are carbonised. Depending on the material from which the tar is obtained, it is called as wood tar or coal tar.
20. Class may be defined as a hard, brittle, and transparent or translucent material. Technically glass is any substance or combination of substances which has solidified from the liquid state without crystallisation.
21. Main constituents of glass are silica, sodium or potassium carbonate, lime, manganese dioxide, cullet, and colouring materials.
22. Plastics are basically organic substances which may be natural or synthetic resins in origin with or without moulding compounds.
23. Plastics may be classified into two broad groups based on their behaviour with respect to heat, viz., thermo plastics and thermoset plastics.
24. Thermo plastics are the plastics which become soft when heated and become hard when cooled. These plastics can sustain repeated softening and hardening for indefinite time.
25. Thermoset plastics become rigid due to chemical changes when moulded under pressure and temperature.
26. Composite materials are formed by combining two or more different materials to utilise their virtues and minimising their deficiencies. These materials are also called as composites.
27. Composites comprise of two main constituent materials, viz.,
    (i) Strong load-carrying materials called reinforcement or reinforcing fibres.
    (ii) Weaker material known as matrix.

28. Rubber is a natural material extracted from rubber trees and also manufactured by chemical processes. The former is known as natural rubber and the latter as synthetic rubber.
29. Geotextiles are porous fabrics manufactured from synthetic materials that are primarily petroleum products.
30. Geomembranes are the materials with very low permeability. They are flexible and are manufactured from synthetic or bituminous products.
31. Adhesives are used to stick two or more parts together so that they behave as a single unit.

REVIEW QUESTIONS

1. What is asbestos? Explain the properties of asbestos.
2. Name the commercial products of asbestos? Explain each one.
3. What are the constituents of fly ash?
4. Explain briefly the applications of fly ash.
5. What is gypsum? How gypsum plaster is used?
6. Distinguish between asphalt and bitumen.
7. What are the uses of tar in building construction?
8. Describe the principal constituents of glass, giving the functions of each.
9. Explain briefly the following: (i) soda-lime glass, (ii) flint glass and (iii) heat-resistant glass.
10. Bring out the major properties of glass.
11. What is a plastic? What are the types?
12. List the uses of plastics.
13. What is a composite material? Explain the two main constituents.
14. How rubber is used in building industry?
15. What are Geosynthetics?
16. Explain different types of adhesives.

14 PLANNING OF BUILDINGS

14.1 INTRODUCTION

One of the primary needs of human beings is shelter (i.e., to live in a comfortable place called house). A building constructed using natural or man-made material forms the house. There are different national level organizations which deal with housing.

Apart from just living, buildings are needed for different purposes. Accordingly the components are different. For a residential building, there are a set of components that are to be judiciously positioned to attain the maximum benefit.

Of these things, orientation of a building also plays a major role as every building is exposed to natural elements such as wind, sun, rain, etc. Hence, depending on the geographical location, the buildings have to be so positioned to attain the maximum comfort throughout the year.

Building regulations and specifications are to be taken into account while planning a building to satisfy the requirements.

14.2 IMPORTANCE OF HOUSING

Housing is an important indicator of the level of country's social progress. Among the basic requirements for mankind to live, shelter takes the place next to food and clothing. Shelter is needed to protect oneself from natural calamities, wild, animals and from fellow beings. As a matter of fact, human beings started from caves and now live today in sophisticated air-conditioned houses.

14.2.1 Housing Programme

United Nations has recommended certain norms to improve the quality of life of human beings. The norms stipulated are: (i) settlement policies and strategies, (ii) settlement planning, (iii) institutions and management, (iv) infrastructure and services, (v) land and (vi) public participation.

Based on the above norms, housing programmes have to take the following steps:

(i) Policy framework has to be formulated emphasizing human settlements with employment opportunities and economical growth.

(ii) Keeping community participation as a basic principal, development of human settlement should be approached.

(iii) On the national frame work, development of human resources should be attempted.
(iv) Financial resources have to be generated by mobilizing domestic household savings.
(v) Provide shelter and infrastructure by developing new partnership.
(vi) Construction sector has to be promoted preferably utilizing the locally available materials.

14.2.2 Priorities in Housing

Major priority areas for housing and construction are briefed below:

1. Policy planning including programmes and projection of housing methodology with proper standards.
2. Development of land and infrastructure with related economies.
3. Development towards use of indigenous building materials and available equipment.
4. Labour organization and connected infrastructure for education and training.
5. Management of housing finance.
6. Necessary documentation, designation and application.

14.2.3 National Level Housing Organizations

There are different levels of housing organizations which involve in policy planning, stipulating standards, research on housing and building, etc.

1. National Building Organization (NBO)

National Building Organization established in 1954, works in close collaboration with the Planning Commission, Research Laboratories, Construction Departments, Housing Boards, etc. It is engaged in coordination and promotion of housing in practice.

2. Central Building Research Institute (CBRI)

The Central Building Research Institute (CBRI) is essentially a research body set up by the Council of Scientific and Industrial Research, established in 1951. The institute involved research in seven research divisions, viz., building materials, soil engineering, efficiency of buildings, building process, plant and productivity, architecture and physical planning, fire research, and rural building. The results of research are used in practice.

3. National Buildings Construction Corporation (NBCC)

National Building Construction Corporation (NBCC) is a Government of India enterprise. The mission of the company is to be an organization in construction engineering and consultancy services. It provides efficient and quality execution of projects, extends services conforming to international standards, and development of human resources. NBCC now occupies its rightful place as a pioneering construction organization in the country.

4. HUDCO

Housing and Urban Development Corporation Ltd., (HUDCO) was established in 1970. It is developed as a specialized organization which collects and disseminates information and ideas on improved designs, construction techniques, etc. It purely works with research organizations of national repute.

5. Hindustan Housing Factory

Hindustan Housing Factory is a Government of India organization involved in manufacture of pre-cast components. Because of cost and need for speedy construction, pre-fabrication techniques have developed. The factory has made remarkable progress in the construction of industrial structures by using pre-cast and pre-stressed concrete components. It works on partially pre-fabricated system and fully pre-fabricated system.

14.3 CLASSIFICATION OF BUILDINGS

According to National Building Code, buildings are classified based on occupancy and type of construction.

14.3.1 Occupancy Classification

Under occupancy classification, buildings are grouped under the categories mentioned below.

1. Residential Buildings

These include any building in which sleeping accommodation is provided for normal residential purposes, with or without cooling or dining or both facilities (e.g., houses, lodges, dormitories, apartment houses (flats), hotels, etc.).

2. Educational Buildings

These include any building used for school, college or day-care purposes involving assembly for instruction, education or recreation.

3. Institutional Buildings

These include any building which is used for purposes such as medical or other treatment, care of persons suffering from physical or mental illness, etc. (e.g., hospitals and sanitorium, custodial institutions, penal and mental institutions, etc.).

4. Assembly Buildings

These include any building where group of people congregate or gather for amusement, recreation, social, religious, patriotic, civil, travel, etc. (e.g., theatres, motion picture houses, auditoria, museums, etc.).

5. Business Buildings

These include any building which is used for transaction of business (other than mercantile buildings discussed below) for keeping of accounts and records, professional establishments, service facilities, etc. (e.g., town halls, courts, libraries, etc.).

6. Mercantile Buildings

These include any building which is used as shops, stores or market, for display and sale of merchandise (either wholesale or retail). Storage and service facilities incidental to the sale of merchandise and located in the same buildings shall be included under this group.

7. Industrial Buildings

These include any building in which products of materials of all kinds and properties are fabricated, assembled, manufactured or processed. (e.g., assembly plants, power and gas plants, refineries, diaries, sawmills, etc.).

8. Storage Buildings

These include any building used primarily for the storage or sheltering of goods, wares or merchandise, vehicles or animals, etc. (e.g., warehouses, cold storages, garages, hangers, grain elevators, etc.).

9. Hazardous Buildings

These include any building which is used for the storage, handling, manufacturing or processing of highly combustible or explosive materials or products which are liable to burn with extreme rapidity; manufacturing or processing toxic alkalis, acids or chemicals producing flame, irritant or corrosive gases, etc.

14.3.2 Types of Construction Based on Fire-Resistance

Under this classification, buildings are classified into four categories based on the fire resistance offered by a building as Type 1, Type 2, Type 3, and Type 4.

Type 1: That provides 4 hours fire resistance.
Type 2: That provides 3 hours fire resistance.
Type 3: That provides 2 hours fire resistance.
Type 4: That provides 1 hour fire resistance.

14.4 PRINCIPLES OF PLANNING OF BUILDINGS

The basic objective of planning of buildings is to arrange all the units according to their functional requirements. This has to be achieved making best use of the space available. The plan has to be prepared based on the factors such as climatic conditions, location of site, accommodation requirements, surrounding environment, local bylaws, etc.

In general, the principles detailed below are the factors which are to be considered in planning of buildings:

(i) Aspect
(ii) Prospect
(iii) Privacy
(iv) Grouping
(v) Roominess
(vi) Furniture requirements
(vii) Sanitation
(viii) Flexibility
(ix) Circulation
(x) Elegance
(xi) Economy
(xii) Practical considerations.

14.4.1 Aspect

Aspect means location of doors and windows in a building so as to allow the occupants of the building to enjoy the natural gifts such as sunshine, breeze, scenery, etc. The aspect is also important from the hygienic point of view.

14.4.2 Prospect

Prospect is the impression the building gains by an outside observer. This includes projections wherever needed, disposition of doors and windows, concealment of some parts which may distract the view, etc. Both prospect and aspect concerned with disposition of doors and windows properly. As a matter of fact, for the sake of either seeing or hiding certain views, window sites play a vital role.

14.4.3 Privacy

Privacy plays a major role in planning of all the buildings and residential buildings in particular. In privacy, two factors have to be considered, viz., (i) privacy of one room from another and (ii) privacy of building from outside.

Privacy of the first factor is taken care of by carefully planning the building with respect to grouping, disposition of doors, provision of corridor or lobby, etc. Temporarily privacy can also be attained by providing screens and curtains.

Privacy from the second factor is easily attained by carefully planning the entrance and steering it with greeneries.

Privacy is particularly needed in bed rooms, bathrooms, kitchen, etc.

14.4.4 Grouping

Grouping is concerned with the positioning of the rooms keeping in view their functions and connected sequence of operations. For example, in a residential building, dining room must be closer to the kitchen. Similarly, in a hospital building, the administrative department should be centrally located.

14.4.5 Roominess

Roominess is concerned with the deriving of maximum benefit from the minimum dimensions of a room. It is the economical way of using a space with maximum benefit without any cramping of the place. In general, in a residential building a large storage space is required.

14.4.6 Furniture Requirement

The utilization of a room or a space is solely based on the furniture requirements. This is a crucial factor that has to be considered mainly in non-residential buildings. In residential building also adequate care has to be taken before selecting proper furniture for hall, bed room, etc.

14.4.7 Sanitation

Sanitation comprises of providing ample light, ventilation and facilities for cleaning and sanitary conveniences. Light plays a dual role. Firstly it illuminates and secondly, it is needed for maintaining hygiene. For all buildings, good lighting is necessary. Light is provided through windows and ventilators.

Ventilation is the supply of outside air, either by intention or by infiltration into the building. Good ventilation is a must for a conductive living inside a building. Ventilation may be provided naturally or mechanically. However, poor ventilation should be avoided as it may produce headache, sleepiness, inability to fix attention, etc.

Apart from light and ventilation, general cleaning and upkeep of the building is the responsibility of the occupants. Particularly prevention of dust accumulation is essential. Sanitary conveniences in a building include provision of bathrooms, lavatories, etc., and their cleanliness.

14.4.8 Flexibility

Flexibility means that a planned and designed room or space should be in a position to be used for other purposes at the time of need. This is particularly important for designing the houses for middle class families. For example, a well-planned house within a small space must provide various activities such as a hall, a sit-in, child's study, etc.

14.4.9 Circulation

Circulation means easy movement horizontally without ups and downs. However, stairs may be provided for vertical circulation. Passages, Corridors, halls and lobbies serve the purpose of horizontal circulation.

As a general rule, the circulation area should be straight, sufficiently lighted and well ventilated to achieve efficiency, comfort and convenience.

14.4.10 Elegance

Elegance is the effect produced by the elevation and general layout of the plan. Elevation should be elegant and signify the type of building and the interior facilities which could be expected from the building.

14.4.11 Economy

Although economy does not come under the purview of planning, it is evidently a factor to be seriously considered. The economy may restrict the liberties of the owner and may require some alterations and omissions in the original place. However, economy should not have any adverse effect on the utilities and safety of the structure.

14.4.12 Practical Considerations

The factors, which need practical considerations, include strength and stability of structure, simplicity of plan and design, a long life, and provisions for expansion with less cost.

14.5 BASIC REQUIREMENTS OF BUILDING

Classification of buildings as per National Buildings Organization is dealt in Section 14.3. Irrespective of the type of building, some of the terminologies and basic requirements for certain components of building (discussed in the next section) are necessary to know which are discussed below.

14.5.1 Building Line

This is the line corresponding to the plinth (discussed in next section) of a building which adjoins the street or extension of the street. As per the municipal regulations, a minimum distance has to be maintained between the building line and the nearest edge line of the street.

14.5.2 Set Back Line

This is a line usually parallel to the plot boundaries, marked by the municipal regulations beyond which nothing can be constructed.

14.5.3 Covered Area

This is the ground area covered immediately above the plinth level by the building.

14.5.4 Plinth Area

This is the built-up covered area measured at the floor level by taking the external dimensions of the building, excluding the plinth offset.

14.5.5 Floor Area and Floor Area Ratio

It is the plinth area minus the area occupied by wall, door, openings, etc. Floor area ratio (FAR) is also known as floor space index (FSI) which is given as

$$\text{FSI of a flat} = \frac{\text{Total covered area of all floors}}{\text{Total plot area of the building}}$$

An FSI of 1.5 is now-a-days allowed for flats in most cities.

14.5.6 Carpet Area

This is the usable area of a building. In a residential building, it will exclude verandah, bathrooms, staircases, etc., and is about 50–65% of the total area.

14.5.7 Building Height

The height of a building is its height above the ground level. For a flat roof construction adjoining a street, the height is defined as the vertical distance from the highest point of the roof of the building to the average level of the centre line of the adjoining street. For a sloped roof without gables the building height is from the point where the external surface of the outer wall intersects the finished surface of the sloping roof (for the gabled roof, the reference point for height on the building is the mid-point between the eves level and the ridge) to the average level of the centre line of adjoining street.

14.5.8 Room Height

It is the clear vertical distance from the finished floor surface to the finished ceiling surface. In case there is no finished ceiling surface, the underside of the joints or beams or tie beams shall determine the upper point of measurement.

14.5.9 Habitable Rooms

Any room which is occupied or designed for occupancy by one person or more persons for the conventional usage is called as habitable room. The minimum height of these rooms should not be less than 2.74 m. Area of habitable room shall not be less than 9.5 m^2 with a minimum width of 2.4 m, if there is of only one room. If there are two rooms, one of these shall not be less than 9.5 m^2 and the other of 7.5 m^2 with a width of 2.4 m.

14.5.10 Bathroom

The size shall not be less than 1.5 m × 1.2 m or 1.8 m^2 and height shall not be less than 2.2 m. If it is combined with a water closet, the area shall not be less than 2.8 m^2 with a minimum width of 1.2 m.

14.5.11 Water Closet

The minimum floor area shall not be less than 1.1 m^2 and height 2.2 m. This does not include a bathroom. Privacy is needed with arrangement for flushing the pan with water.

14.5.12 Balcony

This is a horizontal projection facing the open area including a handrail or balustrade to serve as passage or sitting out place.

14.6 BUILDING COMPONENTS AND THEIR FUNCTIONS

Parts of a building fall under two broad categories, viz., substructure and superstructure and all components of a building fall in any one of these categories.

14.6.1 Substructure

Substructure is the lowest portion of a building which is usually considered to be located below the ground level. The function of the substructure is to transmit the load of superstructure to the soil. The substructure is nothing but the foundation which is provided to give stability to the building and to distribute the load coming from superstructure (Fig. 14.1). If the sub-soil near the ground surface is weak, the load of superstructure is transferred to deep-strong strata by means of long vertical members of either timber, concrete or steel called piles.

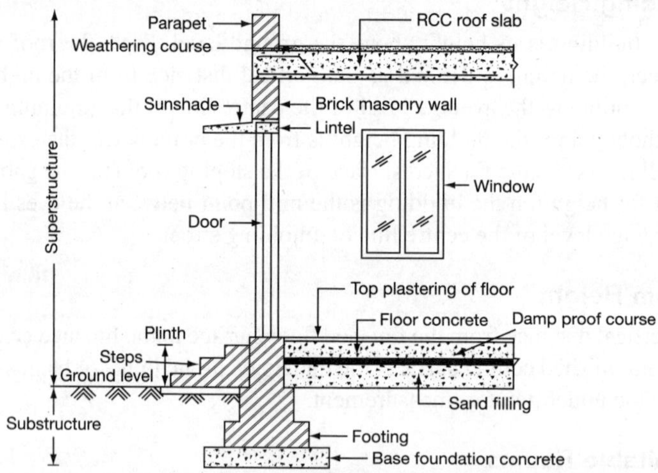

Figure 14.1 Some components of a building

14.6.2 Superstructure

Superstructure is that portion of a building which is constructed above the substructure. It is the major portion of a building which is fully used by human beings. It houses various facilities like number of storeys, floors, roofs, stairs, ventilation, etc. It is the one which provides adequate stability and withstands the adverse weather conditions. Apart from ensuring stability, it adds on to attractive aesthetic appearance. Various components constituting the superstructure are briefly explained below (Fig. 14.1).

1. Basement

Basement or cellar is the lower storage of a building constructed below or partly below the ground surface. It is not used for residential purposes. Basement is used for

(i) Storage of household or other goods,
(ii) As a strongroom for materials, documents and locker for banks,
(iii) Air-conditioning equipment,
(iv) Machines used for services and utilities of the building, and
(v) Vehicle parking, etc.

2. Plinth

This is the portion of a structure immediately above the ground and between the surface of the floor and the surrounding ground and the road surface so as to provide adequate drainage of the site. The minimum height of the plinth is 45 cm from the surrounding ground level. The top surface of the plinth is the surface over which the ground floor walls are raised and door frames rest.

3. Damp-proof Course

Dampness cause corrosion of materials used in the construction, crumbling and softening of plasters, loosening of floor coverages, blistering of painted surfaces, rotting of wooden parts, etc. Damp-proof materials should be highly impervious.

In order to prevent dampness of wall, a continuous impervious layer is generally provided at the top of plinth level and also in vertical faces of the basement floors.

The materials used for the damp-proof courses are flexible materials like bituminous sheet, plastic sheet, metal sheet, etc.; semi-rigid materials like mastic asphalt; and rigid materials like slates, bricks, stones, dense cement concrete, etc.

4. Wall

It is that part of superstructure which transfers the load of the roof and its self-weight to the foundation through the plinth surface. Materials used for the construction of wall are mud, brick, stone, structural units of clay or concrete. Walls which are responsible to transfer the loads are called load-bearing walls. Partition walls are generally of non-load-bearing and used to partition the space for different purposes. External or outer walls of a building protect the inmates of the building from adverse weathers.

5. Slab

It is a continuous flat portion of a building used as a roof or floor. This is provided to span small opening and distance between beams. It is generally of reinforced cement concrete with reinforcement at the bottom.

6. Beams

It is a horizontal structural member which carries a portion of the wall, floor slab and roof. Beams may be of wood, stone, steel or reinforced cement concrete. Now-a-days reinforced cement concrete beams are used extensively. Reinforced cement concrete beam in its simplest form has a rectangular shape with reinforcement at the bottom and at the top near the supports. In special cases, the beam can be of any shape with reinforcement at the appropriate places.

7. Arch

An arch is a structure comprising of a mechanical arrangement of wedge-shaped blocks. These arches are designed to support the superincumbent load of a wall and provided over doors, windows and passages. The materials used for the construction of arches are bricks, stones and concrete blocks. Steel and reinforced cement concrete are also provided which are built of a single unit and are of rigid type.

8. Lintel

It is a beam that supports the masonry over openings of door, window and passage. Lintels may be wood, stone, steel or reinforced cement concrete. Unlike arches, it does not use a strong supporting wall. Reinforced cement concrete lintels have reinforcements as that in a beam. For small openings, a reinforced brick lintel may be enough.

9. Roof

It is a cover to the inside of a building. Its function is to give a protective covering to the building so that rain, snow or wind may not damage the building. It is constructed at the highest part of the building. Roofs are constructed based on the climatic conditions, material used and type of lighting and ventilation needed. Roofs may be sloping, flat or many have special shapes like shell, folded plates or dome.

10. Floor

It is a level, firm and smooth surface provided at the plinth level in the ground floor for the comfortable use of the inmates of a building. In multi-storeyed buildings, floors divide the building into different levels one above the other for the purpose of providing accommodation within the restricted space. Top floors should also be level, firm, smooth and should be sufficiently strong so as to withstand the loads which come over the floor. It also forms a base surface to take floor coverings. The materials used for floor may be brick, wood, stone, or plain or reinforced cement concrete.

11. Stairs

It is provided in a building to afford a means of communication between floors. These are steps arranged in series and generally placed in the centre or to one side of a building. The stairs should be designed properly so as to provide maximum comfort, ease and safety with adequate ventilation and light. Stairs may be made of timber, brick, stone, mild steel, wrought iron or concrete.

12. Door

It is an opening provided to the outside of a building and for connecting inside rooms of a building. It also gives security and safety. Location of a door should be made so that there is free movement, adequate usable area, privacy and good ventilation. Movements of the door leafs may be swinging, sliding, rolling, revolving, folding or telescopic. Doors may be of wood, aluminium, steel, glass or any combination.

13. Window

It is an opening to the outside to provide adequate natural ventilation and light. While locating a window, the following aspects have to be considered:

(i) Distribution and control of day light.
(ii) Desired vision outside.
(iii) Privacy.
(iv) Control of ventilation.
(v) Pattern of air movement.

Windows may have one or more shutters. Windows may be of wood, aluminium, steel, glass or any combination.

14. Ventilator

It is provided for ventilation purposes. This can be combined with a door or window frame or provided independently. When provided independently, it is located near the ceiling of the room if the height of the room permits. It may be pivoted or louvered type. Only glass panels or fixed glass louvers are used.

15. Sunshade

It is a horizontal projection provided outside a building at the top of doors and windows to protect the inside of the building from sunlight and rain.

16. Column

It is a structural member intended to support compressive stress. This is provided to support beams both inside and outside a building. Columns may be made of timber, steel, stone, brick or concrete. Columns can be of any shape and provided with some decorative finishes at the top and at the bottom of the columns.

17. Weathering Course

It is a continuous layer provided over the flat roof for protecting the roof from wind, sunlight, rain and snow. The layer may be a combination of concrete with one or more layers of the flat tiles. The thickness of weathering course should be less such that the load on the roof is less, but at the same time should be sound enough to withstand the adverse effects of weather.

18. Parapet

It is a short wall provided on the top of a flat roof of a building to protect people from slipping from top.

14.7 ORIENTATION OF BUILDING

Orientation of a building refers to the direction of the normal to the long axis. For example, if the length of the building is north-south, its orientation is east-west. The chief purpose of orientation of a building is to derive physically and psychologically comfortable living conditions inside the building. The orientation will provide the maximum benefits from the elements of nature such as sun, wind and rain.

In other words, good orientation means proper organization of plan units of the building in relation to the natural causes like rain, sun, wind, etc. Poor orientation of the buildings results in discomfort conditions inside the building. Although needed comforts can be created by mechanical means, they will lead to extra cost.

14.7.1 Factors Affecting Orientation

Factors which may affect orientation are as follows:

1. Solar radiation
2. Clouds
3. Prevailing winds
4. Rainfall
5. Surrounding site conditions

1. Solar Radiation

Solar radiation is directly related to temperature. From the solar radiation point of view, the orientation should be such that the building as a whole should receive the maximum solar radiation in winter and the minimum in the summer. Hence, it is necessary to know the path of sun's movement throughout the year and its relative position with respect to the locality.

2. Clouds

Clouds reduce the direct radiation from sun. It is thus desirable to take note of the cloudy periods in a year. Sun protection may be provided accordingly.

3. Prevailing Winds

Prevailing winds minimize the adverse effects of high humidity and high temperature. Thus for the purpose of orientation it is necessary to study the velocity and direction of the wind periodically. It is generally accepted that variation up to 30° with respect to the prevalent wind does not materially affect the indoor ventilation inside the building.

4. Rainfall

The intensity of rainfall and the entire monsoon period should be known so as to decide the orientation of the building and also decide about the structure.

5. Surrounding Site Conditions

Location of the site, either rural, urban or sub-urban, also plays a role in the orientation of a building. Further, the surrounding greeneries or some factories, etc., are to be taken into account while fixing the orientation of the building.

14.7.2 Orientation Criteria under Indian Conditions

Major factors which control orientation of residential building are solar heat and humidity. Climate-wise, Indian climatic conditions may be divided into hot-arid zone and hot-humid zone.

1. Hot-Arid Zones

Such zones are also called as dry arid zones which are found mostly in the interior of the country away from the coast. Hence, to get the comfort, removal of hot air through the roof, doors, windows, etc., are to be thought of by proper orientation. North India and Central India fall under this zone.

2. Hot-Humid Zones

Such zones are also called as wet zones which are found generally along the coastal belts. Here, apart from providing opening for circulation, rain-protection measures also should be considered during orientation. West coast regions, east costs regions and Bengal fall under this zone.

14.8 CONSTRUCTION SEQUENCE OF RESIDENTIAL BUILDING

Before starting construction of a residential building, a site has to be owned at a desired place. A plan to be prepared as per the need of the person and necessary approval has to be obtained. The site clearance has to be done and then the actual work has to be started. The sequence of operations is discussed below.

Step 1: Excavation of foundation
Step 2: Base concrete
Step 3: Masonry construction
Step 4: Plinth beam construction
Step 5: Sand filling up to plinth level
Step 6: Preparation of doors and windows

Step 7: Floor concrete
Step 8: Starting of superstructure construction
Step 9: Fixing doors and windows
Step 10: Finishing superstructure upto lintel level
Step 11: Construction of lintels and sunshades
Step 12: Brickwork up to ceiling and roof
Step 13: Electrical pipe and board fixing
Step 14: Roof casting
Step 15: Removal of formwork
Step 16: Finishing all plumbing work
Step 17: Plastering
Step 18: Floor finishing
Step 19: Construction of compound or boundary walls
Step 20: Fixing doors and window shutters
Step 21: Electrical fittings
Step 22: Painting and polishing.

Above discussed sequence of operations is shown in a flow diagram, Fig. 14.2.

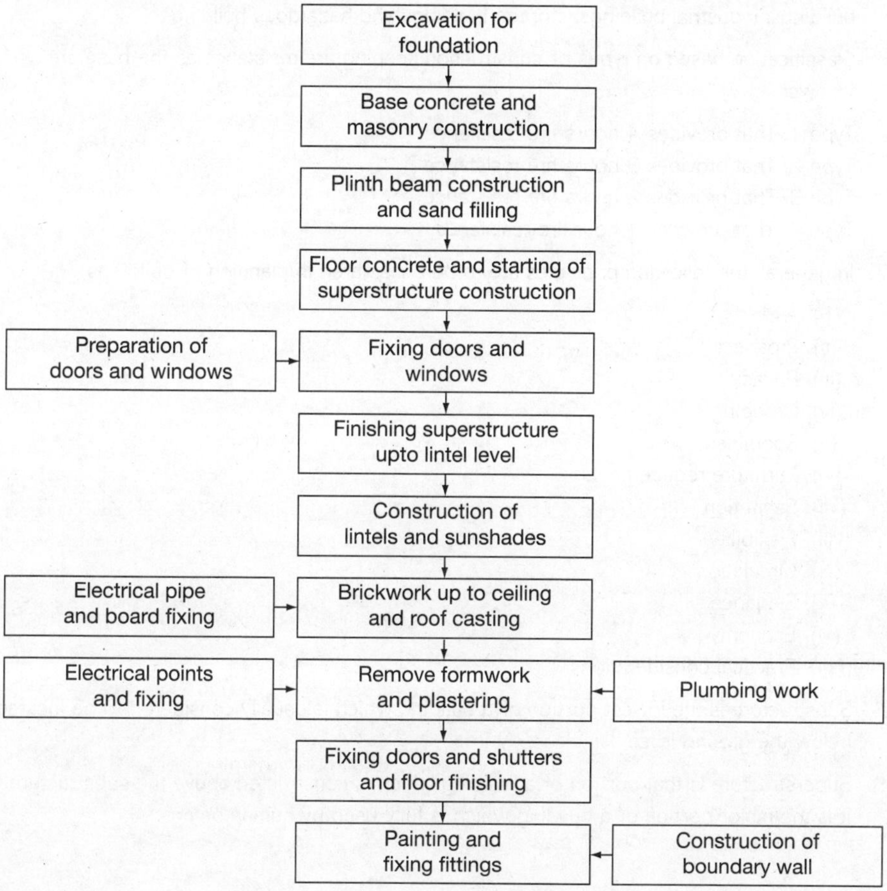

Figure 14.2 Construction sequence of residential building

SALIENT POINTS

1. Housing is an indicator of the level of country's social progress. Among the basic requirements for mankind to live, shelter takes the place next to food and clothing.
2. The norms stipulated by United Nations to improve the quality of life of human beings are: (i) settlement policies and strategies, (ii) settlement planning, (iii) institutions and management, (iv) infrastructure and services, (v) land and (vi) public participation.
3. There are different levels of housing organizations which involve in policy planning, stipulating standards, research on housing and buildings, etc. They are National Building Organization (NBO), Central Building Research Institute (CBRI), National Building Construction Corporation (NBCC), Housing and Urban Development Corporation Ltd. (HUDCO) and Hindustan Housing Factory.
4. According to National Building Code, buildings are classified based on occupancy and type of construction.
5. Under occupancy classification, they are classified as residential buildings, educational buildings, institutional buildings, assembly buildings, business buildings, mercantile buildings, industrial buildings, storage buildings and hazardous buildings.
6. Classification based on types of construction keeping fire resistance as the base are as follows:

   Type 1: That provides 4 hours fire resistance
   Type 2: That provides 3 hours fire resistance
   Type 3: That provides 2 hours fire resistance
   Type 4: That provides 1 hour fire resistance
7. In general, the following principles are to be considered in planning of buildings:
   (i) Aspect
   (ii) Prospect
   (iii) Privacy
   (iv) Grouping
   (v) Roominess
   (vi) Furniture requirements
   (vii) Sanitation
   (viii) Flexibility
   (ix) Circulation
   (x) Elegance
   (xi) Economy
   (xii) Practical considerations.
8. Substructure is the lowest portion of a building which is usually considered to be located below the ground level.
9. Superstructure is that portion of a building which is constructed above the substructure. It is the major portion of a building which is fully used by human beings.

10. Superstructure of a building comprises of basement, plinth, damp-proof course, wall, slab, beam, arch, lintel, roof, floor, stairs, doors, windows, ventilators, sunshades, columns, weathering course and parapet.
11. Orientation of a building refers to the direction of the normal to the long axis.
12. Factors affecting location are: solar radiation, clouds, prevailing, winds, rainfall and surrounding site conditions.

REVIEW QUESTIONS

1. What is the importance of housing for mankind?
2. How buildings are classified as per National Building Code?
3. Explain the details of occupancy classification.
4. Distinguish between occupancy classification and classification based on types of construction keeping fire-resistance as the norm.
5. What do you understand by principles of planning of buildings?
6. Explain the significance of aspect and prospect for residential buildings.
7. Distinguish between covered area, plinth area, floor area and carpet area.
8. What is called a floor area ratio? Mention the floor area ratio stipulated for flats.
9. How building height is determined for a pitched roof and a flat roof?
10. Discuss the criteria for determining the maximum permissible height of a buildings.
11. Distinguish between a substructure and superstructure.
12. Draw a neat sketch of a building and show various parts of the building.
13. Write short notes on
    (i) Basement
    (ii) Wall
    (iii) Arch
    (iv) Stairs
    (v) Weathering course
14. What are the three basic zones for the purpose of orientation of buildings? Explain.
15. What is orientation of building?
16. Discuss the factors affecting the orientation of buildings.
17. With a flow diagram explain the construction sequence of a residential building.

15

FOUNDATIONS

15.1 INTRODUCTION

A foundation is that part of the structure which is in direct contact with the ground and transmits the load of the structure to the ground. It includes the soil or rock of the earth's crust or any special part of the structure which serves to transmit the loads into the soil or rock. The main purpose of the transmissions of load can be satisfied by a particular type of foundation that takes into account the properties of the supporting soil.

Thus the supporting soil plays a major role in the performance of foundation. Hence it is of prime importance to know the soil which is done by a proper soil investigation. It is necessary to know about the types of soils and their distribution to decide a particular type of foundation.

The structural support is actually being provided by a soil-foundation system. This combination of soil and foundation (now referred to as soil-structure interaction) can not be separated. Although engineers are aware of this relationship, it is common practice to consider the structure to be sound and to attribute the failure of the foundations to the failure of the supporting soil.

15.2 SOIL INVESTIGATION

Ground investigation refers to the methodology of determining surface and sub-surface features in the proposed construction area. Information of surface conditions is necessary for planning the accessibility of the site, for deciding the disposal of removed material, for removal of surface water in waterlogged areas, for movement of construction material and equipment and other factors that could affect construction procedures.

Information on surface and sub-surface conditions is a more critical requirement in planning and designing the foundations of structures, dewatering systems, shoring or bracing of excavation, the materials to be used in construction, and site improvement methods.

Thus the purpose of ground investigation is to:

(i) Determine the geological condition of rock and soil formation.
(ii) Establish ground water level.
(iii) Select the type and depths of foundation.
(iv) Determine the bearing capacity of the site.

(v) Evaluate the anticipated settlement of the structure.
(vi) Locate and select the materials of construction.
(vii) Locate suitable transportation routes.

15.2.1 Planning the Ground Investigation

The ground investigation, irrespective of the magnitude of the project, consists of four phases, which are discussed below.

1. Available Information

This is the first phase in which the collection of published geological and topographical information of the area, hydrological data, details of load regulations for construction activity, etc., are made.

2. Reconnaissance

At this stage a thorough study of the existing structures for the type of construction and defects such as cracks and settlement, availability of water, etc., for the building project are collected.

3. Preliminary Investigation

This is an important phase of the entire programme. As the sub-soils were formed under different geological environments, the first step towards a ground investigation is a thorough understanding of the geology of the site, which enables an efficient working out of the investigation programme. The second step is to obtain more details about the sub-soil strata (e.g., thickness of each stratum) from one or two exploratory drill holes. All further steps depend on the magnitude of the job and the character of the soil profile.

4. Detailed Investigation

Additional borings are planned from the data obtained from the preliminary borings. If the sub-soil is uniform in stratification, an orderly spacing may be planned. Often, additional borings are made to locate weak soil or rock zones, outcrops, etc., which may influence the design and construction of the project. Sufficient samples are procured to obtain relevant parameters for design and construction.

15.2.2 Methods of Exploration

Methods of exploration are indirect methods, semi-direct methods and direct methods of exploration.

1. Indirect Methods

Indirect methods consist of geophysical and sounding methods. In these methods, depths to the principal strata are established based on some physical properties of the material, and the measurements are made on the ground surface. No samples are obtained in the geophysical methods, but in some sounding methods, representative samples are obtained.

2. Semi-direct Methods

Semi-direct methods are common boring and drilling methods combined with intermittent sampling. The depths of different layers are ascertained by the rate of advertisement of boring tools or by means of non-representative samples obtained in the course of boring operations. Borings provide access to a particular layer of sampling. In general, only major changes in the character of the sub-surface materials can be detected by this method. The different boring methods are wash boring, rotary drilling and auger boring.

3. Direct Methods

Direct methods are boring and sampling methods which continuously provide representative or undisturbed samples. All accessible exploration, such as test pits, trenches, large diameter boring, shafts, drifts, etc., are grouped under direct methods. These accessible explorations allow direct examination.of strata in-situ.

15.3 TYPES OF SOILS

Based on the method of formation, soil may be categorised as residual and transported soils. Residual soils have formed from the weathering of rocks and practically remain at the location of origin with a little or no movement of individual soil particles. Transported soils are those that have formed at one location (like residual soils), but transported and deposited at another location.

15.3.1 Residual Soils

Weathering (due to climatic effects) and leaching of water soluble materials in the rocks are the geological process in the formation of these soils. The rate of rock decomposition is more than the rate of erosion or transportation of weathered material, and results in the accumulation of residual soils.

15.3.2 Transported Soils

Weathered materials have been moved from their original location to new locations by one or more of the transportation agencies, viz., water, wind, gravity and glacier and deposited to form transported soil. Such deposits are further classified depending on the mode of transportation causing the deposit. For example, soils that are carried and deposited by river is called *alluvial deposits*, soils carried by wind and subsequently deposited are designated as *aeolian deposits*, gravity deposits are termed *talus* and the land formed after a glacier has receded is called a *till plain*.

15.3.3 Regional Soil Deposits of India

Among different types of soils spread over the Indian Peninsula, only six major deposits have been identified, viz., marine deposits, black cotton soils, lateritic soils, alluvial deposits, desert soils and boulder deposits.

1. Marine Deposits

They are generally formed in seawater areas. These deposits cover a narrow belt of tidal flats all along the coast from Porbandar in west to Puri in east. Marine deposits need a pre-treatment before application of any external load.

2. Black Cotton Soils

They are spread over a wide area of 300,000 sq km around Central India and in some parts of Tamil Nadu, Karnataka and Andhra Pradesh. The soil surface is hard in summer and becomes slushy during rainy season. Because of swelling and shrinking nature of the soils, there is a necessity for treatment of the soil. Special foundations are to be adopted in these soils to prevent failure of structures.

3. Lateritic Soils

In tropical regions of high moisture and temperature the lateritic soils are formed. These soils spread over an area of 100,000 sq km around Kerala, Tamil Nadu and West Bengal. Coarse grained soils of this type are called lateritic. The characteristic property of this type of soil is high strength when it is cut and dried in the sun.

4. Alluvial Deposits

The well-known alluvial deposits of India are in the Indo-Gangetic and Brahmaputra flood plains. North of Vindhya Satpura range is also covered with alluvial deposits. Alluvial deposits exhibit alternate layers of coarse and fine soils. The Bengal basin is another important alluvial deposit.

5. Desert Soils

Thar desert in Rajasthan covers most of the area which forms the desert soils of India. These are wind-blown deposits generally present in the form of sand dunes. Scarcity of water is a serious problem in these areas for any construction activity.

6. Boulder Deposits

Rivers flowing in hilly terrains and near foot hills carry large quantities of boulders. Such deposits are often found in the sub-Himalayan regions of Himachal Pradesh and Uttar Pradesh. The properties of these deposits depend on the relative proportions of the boulders and the soil matrix.

15.3.4 Re-claimed Soils

The term re-claimed soils comprises of all materials deposited on a site using various methods for different purposes. It is justifiable to name the re-claimed materials as soil, when it comes to the purposes of construction of structures on them considering them as foundation material. Industrial and commercial development of urban areas, development of navigation channels for ports and other waterfront structures require a large usable land which could be possible only by reclamation. Reclamations are also needed, though may be less important, for the disposal of garbage, industrial wastes, paper sludge, mine tailings, etc. Reclamations may be on unusable low level land areas or on large bodies of water. These reclamation especially near larger bodies of water lead to unsavoury odours, greater turbidity and toxicity of shore waters and affect in great extent the ecology of all marine life.

The geotechnical problems to be answered in these areas are control of settlement, increase of bearing capacity and biological stability of fill. Generally reclamation followed by ground treatment is preferable and economical than designing deep foundations.

The materials which are used in practice for reclamation purposes fall into five groups, viz., hydraulic fills of dredged soil, sanitary fill, paper sludge, flash including slag and rubbish sand debris.

15.4 BEARING CAPACITY

It is the earth that provides the ultimate support for most of the structures including buildings, bridges, dams, highways, etc. The behaviour of the supporting ground may therefore affect the stability of structures. The supporting ground is invariably the soil, particularly for buildings, which is weaker than any construction material like wood, concrete, steel or masonry. Soil is a particulate material, although weak, involved in carrying large loads.

15.4.1 Bearing Capacity Criteria

The design of foundation is primarily based on the concept of bearing capacity of the soil. Thus the bearing capacity is defined as the load or pressure developed under the foundation without introducing damaging movements in the foundation and in the super-structure supported on the foundation. Since damaging movements may result from foundation failure (collapse) as well as from excessive settlement, the following criteria should always be used in evaluating the bearing capacity:

(i) Adequate factor of safety against failure by collapse.
(ii) Adequate margin against excessive settlement.

The bearing capacity after allowing a certain factor of safety over the ultimate bearing capacity against failure (collapse) is termed as *safe bearing capacity.*

The bearing pressure which is safe against failure and at the same time does not cause settlement more than the permissible settlement is called *allowable bearing capacity* or *allowable soil pressure.*

Safe bearing capacity or allowable soil pressure is not a fixed value for a given soil but depends on density, strength, depth of foundation, cohesion, etc. In general, for cohesionless soils it varies from 100 to 450 kN/m^2 and for cohesive soils from 150 to 450 kN/m^2. The problematic soils are loose dry fine sand and expansive clays (black cotton soils). However, a suitable foundation can be provided based on the field condition.

15.4.2 Factors Affecting Bearing Capacity

The following factors directly or indirectly affect the bearing capacity of soil:

(i) Type of soil (i.e., homogeneous, layered, expansive, etc.) and its physical and engineering properties.
(ii) Initial stress condition of the soil due to pre-history and due to the existing structure in and around the proposed foundation.
(iii) Location of ground water in the soil and its fluctuations with time.
(iv) Type of foundation (i.e., shallow or deep), and other factors such as shape, size, and rigidity condition of the foundation.
(v) Depth and location of foundation.
(vi) Allowable settlement of the foundation which shall not be detrimental to the functioning of the foundation.
(vii) Natural calamities such as earthquake, flood, heavy wind, etc., of the region where the structure has to be located.

15.4.3 Methods of Determining Bearing Capacity

Based on the theoretical approaches, bearing capacity of a soil can be found knowing the strength parameters, density, depth of foundation and water table possible. Terzaghi's bearing capacity theory has been used mostly which suits most of the field conditions.

Field plate-load test can be conducted at the stipulated depth of foundation. Indirect methods such as penetration tests may be used.

Bureau of Indian Standards has given presumptive bearing capacity values, which is presented in Table 15.1.

Table 15.1 Safe bearing capacity

Types of rock or soil	Safe bearing capacity (kN/m^2)
Rocks	
Rocks without lamination and defects, e.g., granite, trap, diorite	3240
Laminated rocks, e.g., sandstone and limestone in sound condition	1620
Residual deposits of shattered and broken bed rock and hard shale	880
Soft rock	440
Cohesionless soils	
Gravel, sand and gravel, compact and offering high resistance	440
Coarse sand, compact and dry	440
Medium sand, compact and dry	245
Fine sand, silt (dry lumps easily pulverised by hand)	150
Loose gravel or sand-gravel mixtures; loose coarse to medium sand dry	245
Fine sand, loose and dry consolidation settlement	100
Cohesive soils	
Soft shale, hard or stiff clay, dry	440
Medium clay, readily indented with a thumb nail	245
Moist clay, and sand-clay mixtures which can be indented with strong thumb pressure	150
Soft clay indented with moderate thumb pressure	100
Very soft clay which can be penetrated easily with the thumb	50

Source: IS: 1904–1986.

15.4.4 Improvement of Bearing Capacity of Soils

If unsuitable soil conditions are encountered at the foundation site of a proposed structure, one of the following three procedures may be adopted:

(i) The unsuitable soil is bypassed by means of deep foundations extending to a suitable bearing material.
(ii) The poor material is replaced and either treated to improve and replaced or substituted by a suitable material.
(iii) The soil is treated in place to improve the properties.

Now-a-days various methods are available by which the characteristics of the construction site can be improved to facilitate construction operation, to allow increased bearing pressures or to reduce settlements. Soil improvement in its broadest sense is the alteration of any property of a soil to improve its engineering performance. The various techniques discussed are surface compaction, drainage methods, grouting and injection, chemical stabilisation, thermal stabilisation, soil reinforcement, and application of geotextiles and geomembrane.

1. Surface Compaction

One of the most widely used and the oldest technique of soil densification is compaction. Construction of a building on a loose foundation site needs a compacted base for laying the structures. If the depth to be densified is less then surface compaction may alone solve the problem. Surface compaction needs less skilled labour and is usually the most economical method.

2. Drainage Methods

Drainage method of densification of soil is lowering the water table temporarily or permanently by pumping using well-point systems.

3. Vibration Methods

Vibration methods comprise of vibro-compaction and vibro-displacement compaction. Another method of vibration method is heavy tamping.

The most basic and simplest way of compacting loose soil is by repeated dropping of a weight on the ground. This method, also known as deep dynamic compaction or deep dynamic consolidation, consists of allowing a very heavy weight (up to 400 kN) to fall freely on the ground surface from a height of 15–40 m. This leaves an impression on the ground. The tamping is then repeated either at the same location or over other parts of the area to be stabilised. In the case of non-cohesive soils, the impact energy causes liquefaction, followed by settlement as water drains. Fissures formed around the impact points sometimes facilitate drainage in some soils. This method can be adopted for densifying soils both above and below the water table. This method has been successfully used to treat various types of soils and fill deposits up to 20 m thick.

4. Pre-loading and Surcharge Fills

In this process, an earth fill or some other material is placed over the required site. The amount of fill is sufficient enough to produce a stress in the soil equal to the one anticipated from the final structure. The soft soil is allowed to consolidate prior to construction. Since the consolidation takes a very long time, the method is suitable only for stabilisation of thin layers.

The rate of pre-load and surcharge fill placement has to be controlled depending on the bearing capacity of the soil. If the bearing capacity of the soil is inadequate layers of fill can be placed only after a sufficient gain in shear strength is obtained. The two main requirements for preloading are enough space and availability of fill material. Heaping of fill is the most common method of pre-loading although pre-loading can be successfully effected by the weight of water or by lowering the water table. Among the fill materials, granular soil is the most desirable because it does not turn into mud during rains. Ores and industrial products are generally satisfactory, but clayey soils are less desirable.

5. Vertical Drains

For deep clay deposits, pre-loading alone will take more time because of the long drainage path available for consolidation. An efficient way to do this is by providing vertical drains. Vertical drains are continuous vertical columns of pervious materials installed in clayey soil for the purpose of collecting and discharging the water expelled during consolidation. Vertical drains in combination with pre loading will rapidly accelerate consolidation.

6. Grouting and Injection

Grouting is used for the following in connection with foundation:

(i) Void filling to prevent excessive settlement.
(ii) Stabilising loose sands against liquefaction.
(iii) Strengthening existing foundation.
(iv) Reduction of machine foundation vibrations.

As discussed earlier suspension or solution grouts are used in the above cases depending on the field condition.

7. Chemical Stabilisation

Chemical stabilisation in the form of lime, cement, fly ash and a combination of the above is widely used in soil stabilisation to:

(i) Increase bearing capacity
(ii) Decrease settlement
(iii) Expedite construction
(iv) Reduce permeability
(v) Improve shear strength

Chemical stabilisation may be used for surface soils more successfully. Such a stabilisation technique is sparingly used for building foundation.

8. Soil Reinforcement

Soil reinforcement is the process of strengthening weak soil by providing high-strength thin horizontal membranes. The modern form of soil reinforcement was first applied by According to Vidal's concept, the interaction between the soil and the reinforcing horizontal membrane is solely due to friction generated by gravity.

Reinforced soil is somewhat analogous to reinforced concrete. A wide variety of materials such as steel, concrete, glass fibre, rubber, aluminium and thermoplastic have been used successfully. High alloy steel, aluminium, glass-fibre reinforced plastics (GRP) and geosynthetics are non-corrosive and have long life.

15.5 FUNCTIONS OF FOUNDATIONS

A foundation by definition is that part of the structure which is in direct contact with the ground and transmits the load of the structure to the ground.

15.5.1 Load and Load Distribution

Foundations are subjected to three types of loading, viz., dead load, live load and wind load. Dead load is the self-weight of the various components of a building which include the proposed future expansion. Live load is not a constant load but a varying load, viz., weight of persons using the building, weight of material stored temporarily on the floor, weight of snow, etc. Wind load will be significant in tall buildings wherein the sides and roofs are exposed to wind pressure. Because of this the pressure on the wind-ward side is reduced and in lee-ward side is increased.

Foundation distributes the above loads to a large area (in shallow foundation) or through end-bearing and skin friction (in deep foundation) so that the intensity of stress and the settlement are within limits. It also provides a level surface for the super-structure to be raised.

15.5.2 Stability Requirements

Foundation imparts lateral stability to the super-structure by anchoring it to the ground. It also provides additional stability against sliding and overturning due to horizontal forces like wind, earthquake, etc.

15.5.3 Settlement Control

Settlement of a foundation may be classified as uniform (or total), tilt and non-uniform (or differential) settlement. Structures on rigid foundations undergo uniform settlement. When the entire structure rotates, the structure is said to be under uniform tilt. If foundations of different elements of a structure undergo varied settlements, the foundation is said to be under non-uniform or differential settlement. Foundations are capable of distributing the load evenly under non-uniform loading conditions and non-uniform soil conditions and thereby prevent differential settlement. This can be achieved by adopting suitable foundations such as combined footings, rafts, mats, etc.

15.5.4 Safety Against Natural Events

Foundations sustain large wind forces and earthquake forces and also provide safety against scouring or undermining by flood water or burrowing animals. Distress or failure due to seasonal variations causing volume changes in soils are minimised by providing special type of foundations.

15.5.5 Requirements of Good Foundation

Thus the foundation should satisfy the following requirements.

1. Depth of Foundation

(a) Foundations should be carried well below the top soil, miscellaneous fill, abandoned foundation, debris or muck.
(b) Foundation should be carried below the depth of weathering.
(c) Foundation on sloping ground should have sufficient edge distance as protection against erosion.
(d) Difference in elevation of foundation should not be so great as to introduce undesirable overlapping of stresses on soil.

2. Shear Failure of Foundation

Foundation should be safe against breaking into the ground (i.e., against shear failure). In order to satisfy this requirement an adequate factor of safety on the bearing capacity of the soil is provided.

3. Settlement of Foundation

Foundation should not undergo exercise total and differential settlements. The limiting total and differential settlement should satisfy the requirement specified by building codes for different structures and different soils.

15.6 SHALLOW AND DEEP FOUNDATIONS

Structural foundations may be grouped under two broad categories – shallow foundations and deep foundations. This classification indicates the depth of foundation installation.

A shallow foundation is one which is placed on a firm soil near the ground, and beneath the lowest part of the super-structure. A deep foundation is one which is placed on a soil that is not firm, and which is considerably below the lowest part of the super-structure.

15.6.1 Types and Suitability of Shallow Foundations

Shallow foundations are all suitable for building and are sub-divided into a number of types according to their size, shape and general configuration. They are discussed below.

1. Spread Footings

These footings are the most common of all types of footings with minimum cost and complexity of construction (Fig. 15.1(a)). It necessarily provides the function of distributing the column load over a wide area taking care of the strength and deformation characteristics of the soil. These types of footings are also known as pad footings, isolated footings and square or rectangular footings (for length of footing, L, and width of footing, B, ratio less than 5).

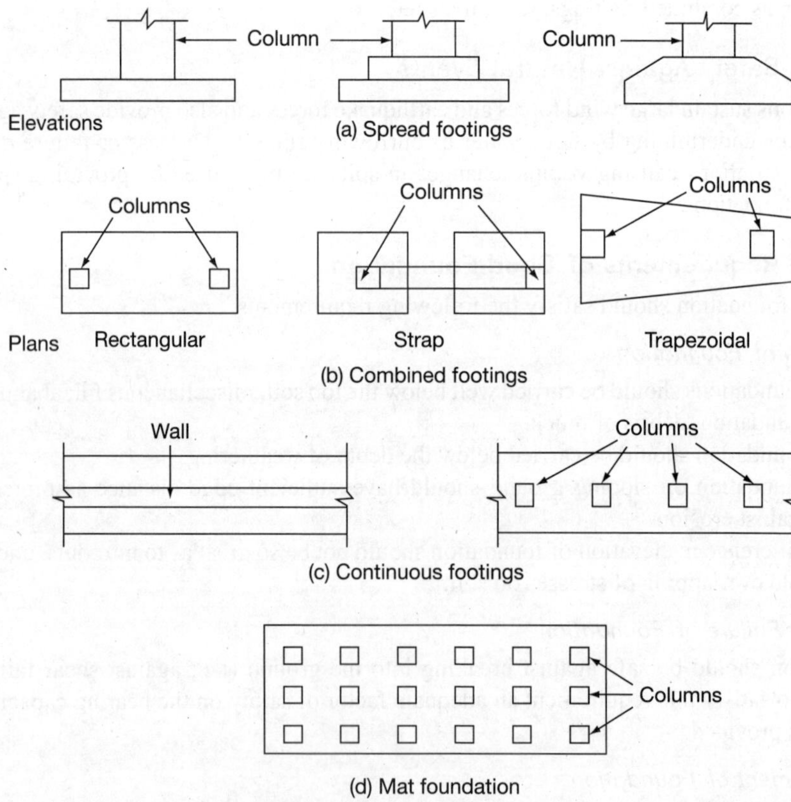

Figure 15.1 Types of shallow foundations

2. Combined Footing

These footings are formed by combing two or more equally or unequally loaded columns into one footing. This arrangement averages out and provides a more or less uniform load distribution in the supporting soil. Further distribution prevents variation of settlement along the footing. These footings are usually rectangular in shape. It may be modified to a trapezoidal shape so as to accommodate unequal column loadings or column close to property line. It may be provided with a strap to accommodate wide column spacing or columns close to property line (Fig. 15.1(b)).

3. Continuous Footing

These footings carry closely spaced columns or a continuous wall such that the load distribution is uniform and load intensity is low on the supporting soil (Fig. 15.1(c)). These footings are also named as strip footings or wall footings (for L/B ratio greater than 5).

4. Mat or Raft Foundation

These are characterised by the feature that columns frame into the footing in two directions. Any number of columns can be accommodated with as low as four columns (Fig. 15.1(d)). In the majority of the cases, mat foundations are used where the soil has low bearing capacity. By combining all individual footings into one large mat, the unit pressure in the sub-soil is reduced.

Since the bearing capacity increases with increasing depth and width of the foundation and the settlement decreases with the increasing depth of foundation, the advantage of mat foundation is two-fold. Mat foundation is also preferred when the total area of the footings exceeds 50% of the total plinth area.

15.6.2 Types and Suitability of Deep Foundations

The design and construction of deep foundations for transferring the weight of the superstructure through soft or weak soils, to deep load bearing strata is a challenging job for a civil engineer. Piles, piers and caissons are the most common types of deep foundations. For any system the mechanism of deriving support from the soil or rock below and adjacent to the foundation is similar. However, each system differs in its method of construction.

1. Pile Foundations

Piles are slender structural members normally installed by driving by hummer or by any other suitable means. The piles are usually placed in groups to provide foundations for structures. Piles may be classified according to their material composition, installation method, group effect and their function as a foundation.

(i) Classification Based on Materials

Under this classification, piles may be further classified as timber, steel, concrete or composite piles. *Timber piles* are the oldest types of piles made from tree trunk. The maximum length of pile is 20 m. The life of timber piles may be increased by treating them with preservatives (Fig. 15.2(a)).

Steel piles consist generally of either pipe piles or rolled steel H-section piles. Because of high strength there can be no restriction on length but steel piles are affected by corrosive agents such as salt, acid, moisture and oxygen. In order to prevent steel piles from corrosion, the thickness is increased, encased in concrete or chemical coating is applied (Fig. 15.2(b)). *Concrete piles* are precast to specified lengths and shapes with reinforcement.

The reinforcement is provided to enable the pile to resist the bending moment developed during lifting and transportation. Concrete piles are also cast in-situ (Fig. 15.2(c)).

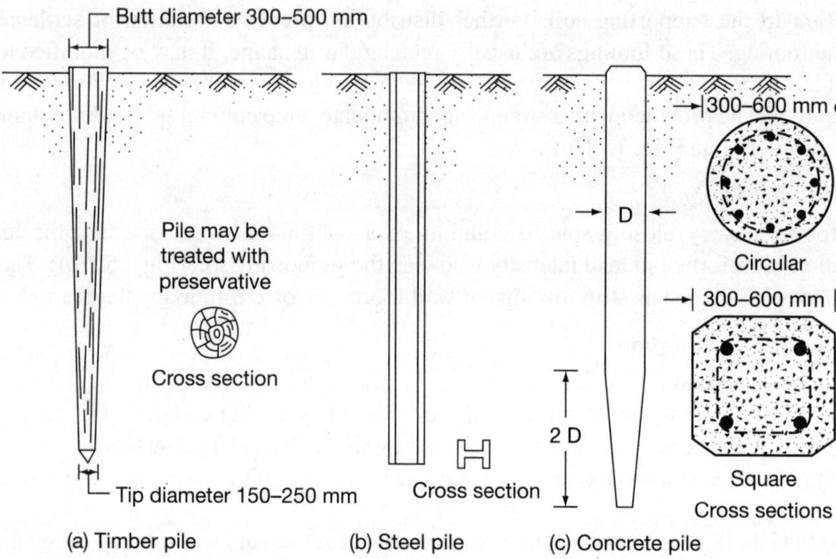

Figure 15.2 Classification based on materials

(ii) Classification Based on Installation Methods

Based on installation techniques piles are classified as *driven piles* and *cast-in-situ piles*. Driven piles may be concrete, steel or timber. Concrete piles are classified as driven pre-cast concrete piles, and bored cast-in-situ concrete piles. Driven precast concrete pile is the one casted in a casting yard subsequently driven to the required location.

Driven cast-in-situ pile is formed within the ground by driving a closed bottom casing and subsequently filling with concrete in the hole so formed with adequate reinforcement. Bored cast in-situ pile is formed within the ground by excavation or boring with or without the use of a temporary casing and subsequently filling it with plain or reinforced concrete.

(iii) Classification Based on Ground Effects

Piles are also used to compact soils and such piles are referred to as displacement or *compaction piles*. These piles displace a substantial volume of soil during installation. In granular soils, there is a tendency for compaction, whereas in clays heaving of the ground surface often results. Driven piles installed in pre-drilled holes are also called as non-displacement piles. Piles are also used to prevent the movement of earth slopes and to safeguard the foundation from damage due to shock.

(iv) Classification Based on Functions

Where the top soil is soft or too weak to support the super-structure, piles are used to transmit the load to the underlying bed rock, such piles are called *end-bearing piles* or *point bearing piles*. If the bed rock does not exist at a reasonable depth below the ground surface, the load is transferred through friction along the pile shaft such piles are called *friction piles*. Transmission towers, off-shore platforms, and basement mats are subjected to uplift forces and piles are used to resist the uplift forces, which are called *uplift piles* or *tension piles*. In order to resist horizontal and inclined forces in water and earth retaining structures *batter piles* are used. Application of piles for providing anchorage to sheet piles are called as *anchor piles* (Fig. 15.3).

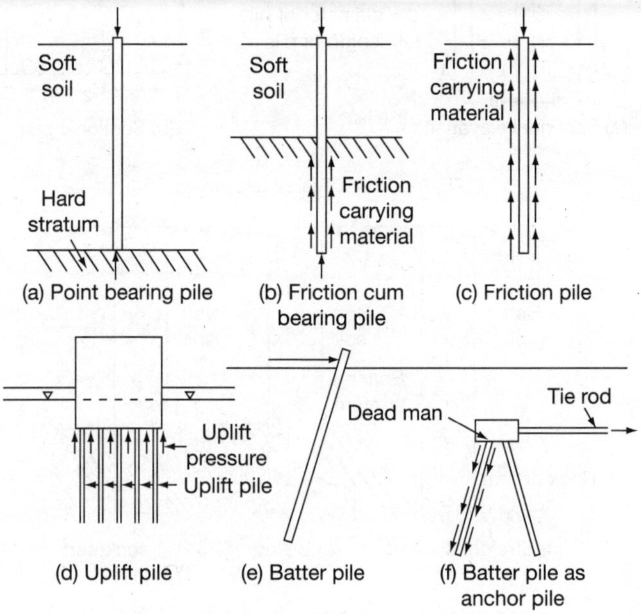

Figure 15.3 Classification based on function

2. *Pile Groups*

Where piles are used for foundation support, they are always used in a group. This requirement is essential so as to assure that the imposed structural load lies within the support area provided by the foundation. As per the building codes at least three piles should be used to support a major column and two piles to support a foundation wall.

A pile cap is provided near the ground encompassing all the top ends of piles. Pile caps are almost invariably made of reinforced concrete. The axial and the lateral load carrying capacity of a pile group is significantly affected by a pile cap. In order to keep the stresses in the pile cap to a minimum the piles should be arranged in the most compact geometric form. Typical arrangement of one pile group is shown in Fig. 15.4.

3. *Drilled Piers*

Drilled piers are structural members of relatively large diameter massive struts constructed of concrete placed in a pre-excavated hole.

They are also called bored piles, large-diameter piles, foundation piers and drilled caissons. The shaft can be enlarged at the base resulting in a belled or under-reamed pier.

The common type of drilled pier is the straight shafted type (Fig. 15.5(a)). The shaft is taken through the upper soil layers and the end is placed on the firm ground or rock. Drilled piers which are provided with a broad base (called a bell) at the bottom of the straight shaft are referred to as *belled piers*. The bell may have a shape of a dome or it may be angled

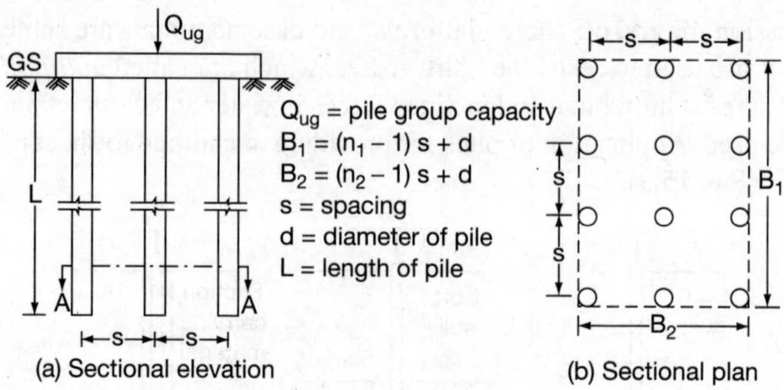

Figure 15.4 Typical pile group

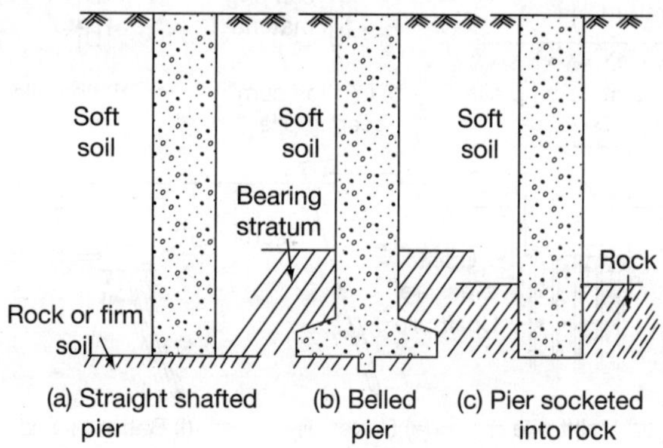

Figure 15.5 Types of drilled piers

(Fig. 15.5(b)). The third type is the extended straight shaft or socketed pier in which the straight shafts are extended into the underlying rock layer (Fig. 15.5(c)).

4. Caissons

Caissons are structural boxes or chambers that are sunk in place through ground or water. The sinking is systematically done by excavating below the bottom of the unit which thereby descends to the final depth. These have large cross-sectional area and hence provide high bearing capacity. Two types of caissons, viz., *open caisson* and *monolith caisson* are shown in Fig. 15.6.

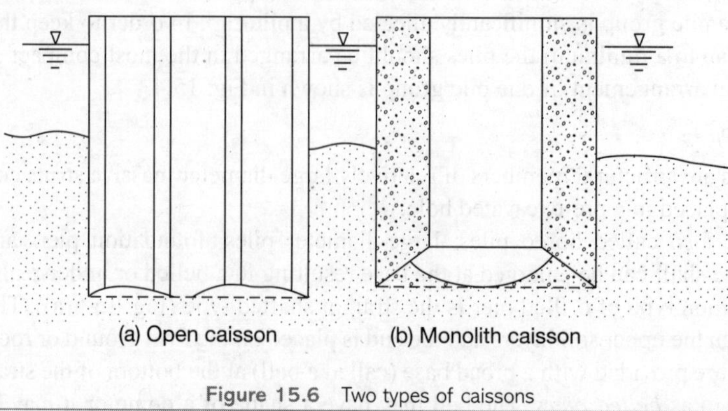

Figure 15.6 Two types of caissons

5. Well Foundation

Well foundation is a type of caisson. It is constructed either on dry ground or over an artificially formed island. The curbs are pitched in the current position and then sunk into the ground to the desired level by grabbing the soil through the dredge holes formed by the masonry or concrete the steining. In India, this procedure of initial sinking is referred to as the caisson method. Well foundations have all the advantages of open caisson. Figure 15.7 shows a typical cross section of a well foundation.

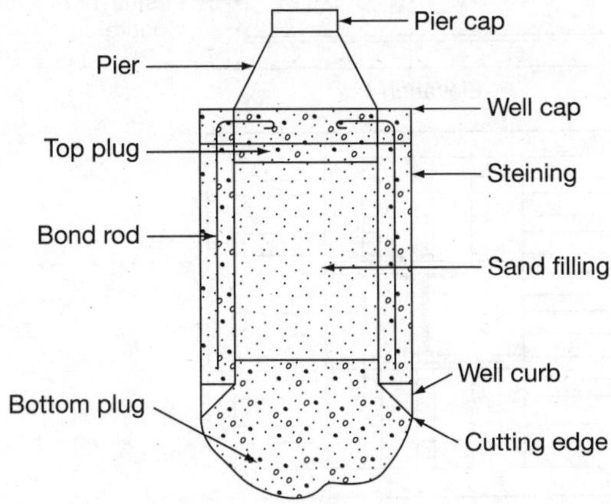

Figure 15.7 Typical section of a well foundation

15.6.3 Foundation under Special Conditions

For some structures or for soil conditions the routine method of providing foundations may not be suitable. Such foundations under special conditions are dealt below.

1. Grillage Foundation

Steel columns may be founded on concrete footings or footings of steel. Many a times the latter approach is preferred. This primarily consists of steel beams arranged in layers at right angles to one another and the beams are connected with each other by bolts in order to form a rigid unit. The entire assembly is embedded in concrete. This is known as Grillage foundation (Fig. 15.8).

The steel grillage foundations are adopted for structure having concentrated loads. Hence they are employed for the foundations of buildings such as theatres, factories, town halls, clock towers, etc.

(i) In this method the depth of foundation is fixed to 1–1.5 m and the width is increased to satisfy the safe bearing capacity and permissible settlements.
(ii) The beams are R.S.Js which are fully embedded in concrete so as to protect them from atmospheric actions.
(iii) The bed of concrete should have a minimum thickness of 15 cm and nowhere the depth of concrete is less than 80 mm. The concrete filling does not carry any load but it maintains the R.S.Js in proper position.

Sometimes timber beams are used for temporary grillage foundation.

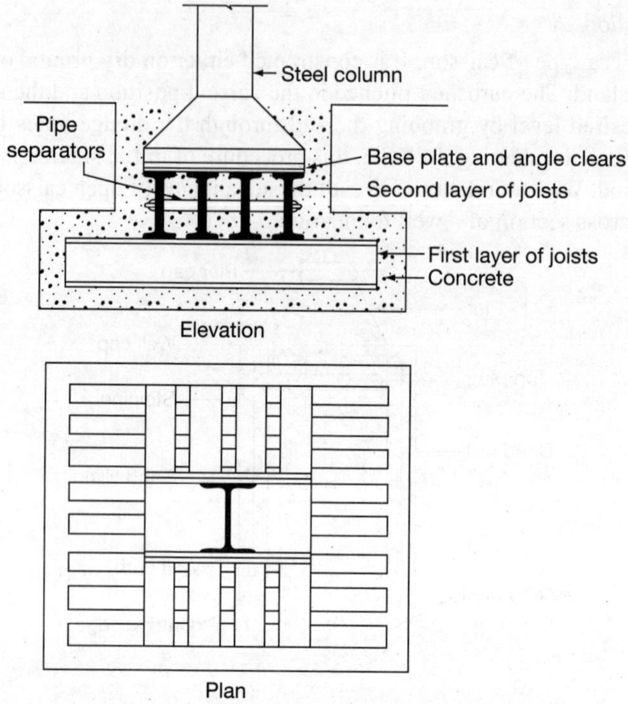

Figure 15.8 Typical grillage footing for steel columns

2. Stepped Foundation

In a sloping ground, it is uneconomical to provide the conventional type of foundation. In such cases, stepped foundation may be provided as given in Fig. 15.9.

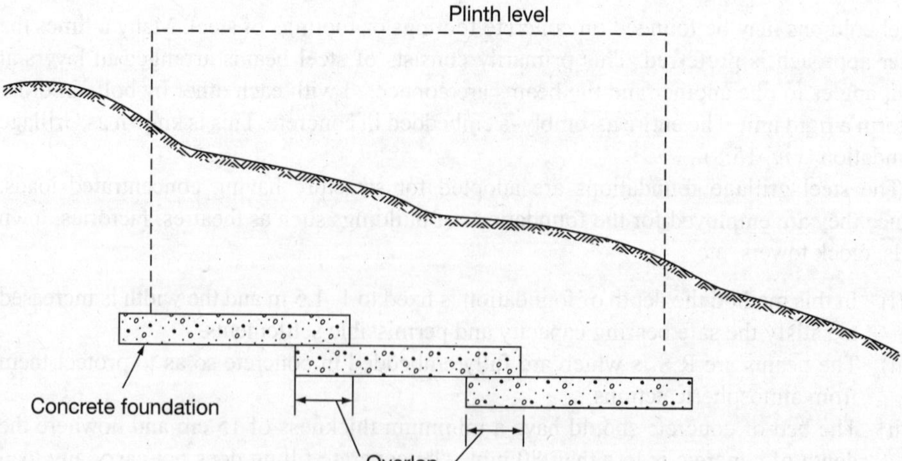

Figure 15.9 Stepped foundation

The following points are to be taken care of:

(i) The overlap between two layers of foundation concrete should be greater of the depth of foundation concrete or twice the height of the step.
(ii) In order to protect from weathering action a minimum depth of 80 cm should be provided at all points.
(iii) The depth of foundation concrete should be in even number.
(iv) The distance of the sloping surface from the lower edge point should be at least 100 cm for soils and 60 cm for rocks.
(v) Stability of slope has to be checked if heavy load is expected on the foundation.

3. Foundations near Adjacent Structures

The horizontal location of a footing is often affected by adjacent structures and property lines. The existing adjacent structure may be damaged due to construction of new foundation because of vibration and shock due to blasting, caving in due to nearby excavation, lowering of water table or increasing stress.

The Indian Standards (IS: 1904, 1986) recommends the following for footings placed adjacent to a sloping ground or when the bases of footings are at different levels.

When the ground surface slopes downwards adjacent to a footing, the sloping surface should not encroach upon a frustum of bearing material under the footing, as shown in Fig. 15.10(a) and (b) for granular soils and clayey soils respectively.

The following norms have to be adopted to avoid any damage to the existing structure:

(i) The footing should be founded at least at a distance *S* from the edge of the existing footing where *S* is the width of the larger footing.
(ii) The line from the edge of the new footing to the edge of the existing footing should make an angle of 45° or less.

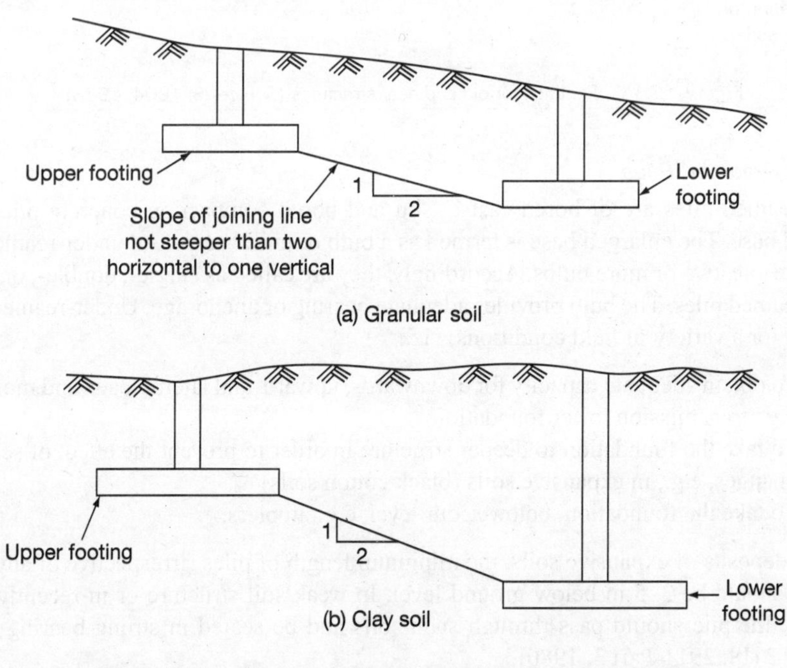

Figure 15.10 Footings at different levels (*Source:* IS: 1904, 1986)

(iii) When a new footing is constructed lower than an old footing, the excavation for the foundation must be carefully done with a suitable bracing system so as to prevent damage to the existing structure (Fig. 15.11).

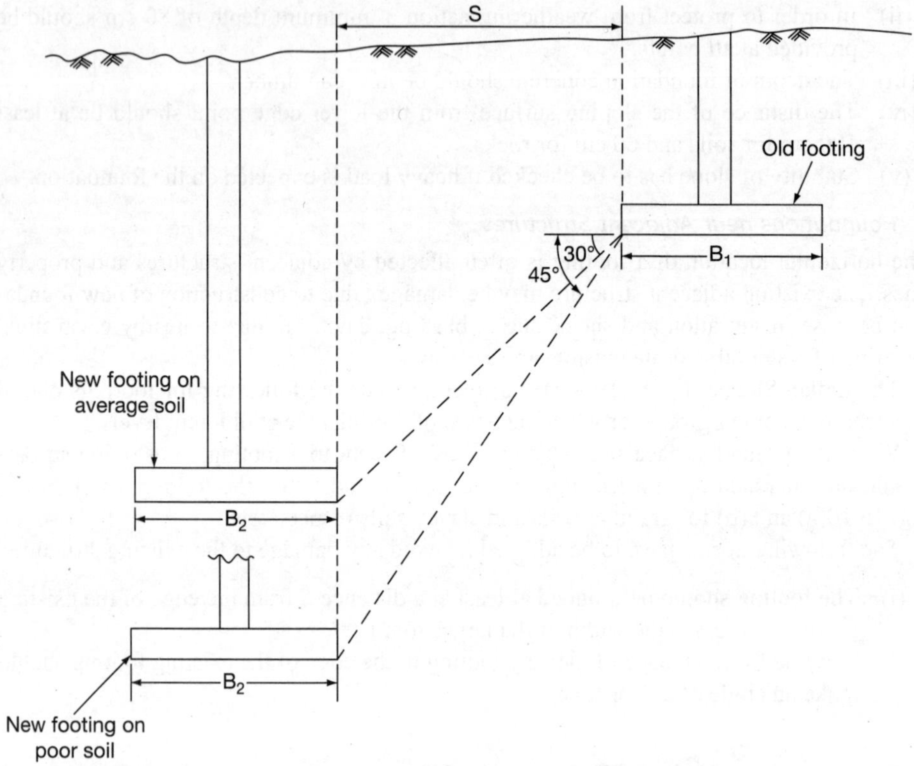

Figure 15.11 Footings for old and near structures (*Source:* IS: 1904, 1986)

4. Under-reamed Piles

Under-reamed piles are of bored cast-in-situ and bored compaction concrete piles with enlarged base. The enlarged base is termed as a bulb or under-ream. An under-reamed pile may have one, two or more bulbs. Accordingly, they are called as single-, double-, or multi-under-reamed piles. The bulb provides adequate bearing or anchorage. Under-reamed piles are used for a variety of field conditions, viz.,

(i) To obtain adequate capacity for downwards, upward, and lateral loads and moments, e.g., transmission tower foundation.
(ii) To take the foundation to deeper structure in order to prevent the effect of seasonal changes, e.g., in expansive soils (black cotton soils)
(iii) To take the foundation, below scour level, e.g., in piers.

In deep deposits of expansive soils, the minimum length of piles (irrespective of any other factors) should be 3–5 m below ground level. In weak soil structure or in recently filled grounds, the pile should pass through such soils and be seated in strong bearing strata, Fig. 15.12 (IS: 2911-Part 3, 1980).

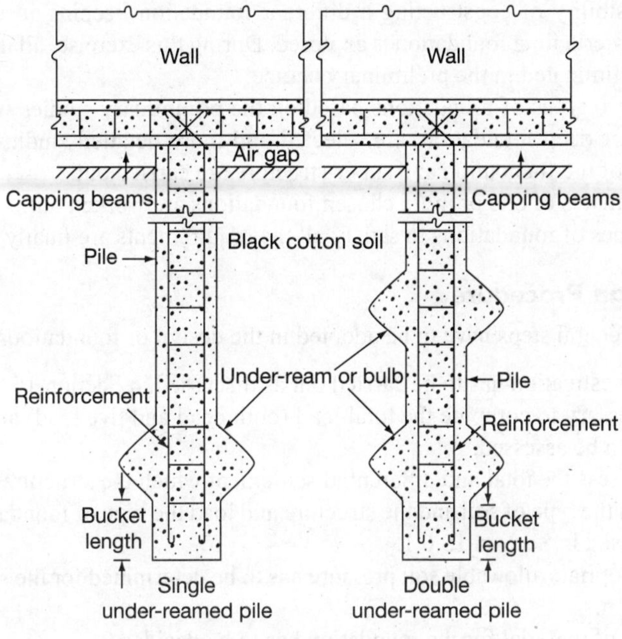

Figure 15.12 Single and double under-reamed Piles (IS: 2911–Part 3, 1980)

15.7 CAUSES OF FOUNDATION SETTLEMENT

Settlement of foundations may be caused due to the following reasons:

(i) Elastic compression of the foundation and the underlying soil.
(ii) Plastic or in-elastic compression of the underlying soil.
(iii) Ground water lowering is another major cause for settlement. This is more adverse in granular soils due to repeated raising or lowering of the ground water. In clayey soil prolonged lowering of ground water may cause settlement.
(iv) Vibrations caused by pile driving, machinery, blasting, etc. This is more adverse in granular soils.
(v) Other causes of settlement include volume change of soil, ground movement and excavation for adjacent structures, mining subsidence, etc.

15.8 SELECTION AND DESIGN OF SIMPLE FOUNDATIONS

15.8.1 Selection Procedure

The selection of a foundation suitable for the type of structure to be constructed or for a given size depends on several factors. Following are the general steps to be followed in choosing the type of foundation.

(i) Necessary data about the type of structure and the loads anticipated to be carried by the structure are collected.
(ii) Adequate information about the sub-soil condition through a suitable soil investigation is got.

(iii) The possibility of constructing a different foundation keeping in mind the basic design criteria for a foundation is explored. During this exercise, all unsuitable types may be eliminated in the preliminary choice.
(iv) One or two types of foundations based on the preliminary studies which may be a shallow or deep foundation, are selected and more detailed studies regarding the stability of the foundation and super-structure are carried out.
(v) Cost estimates of one or more chosen foundations are worked out.
(vi) Three types of foundations to satisfy all the requirements are finally decided.

15.8.2 Design Procedure

The following general steps have to be adopted in the design of foundations:

1. A soil investigation has to be carried out as discussed in Section 15.2.
2. It is necessary to compute the total load (both dead and live load) and the distribution has to be assessed.
3. It is to assess the total and differential settlement which the structure may under go.
4. Based on the type of soil and the structure and load the type of foundation is decided as discussed in Section 15.6.1.
5. The appropriate allowable soil pressure has to be determined for the selected type of foundation.
6. The type of material for the foundation has to be decided.
7. Alternate designs are to be made before finalization.
8. Cost estimate has to be made and any further modification may be made keeping in view economy and life of the structure.

15.8.3 Design of Shallow Foundations

Following guidelines may be taken while designing shallow foundation other than rafts and mats.

(i) In case of wall footing the width of foundation should be computed based on the allowable soil pressure.
(ii) In case no footings are to be provided to the walls the width of foundation should be equal to three times the width of the wall.
(iii) In case of piers the width of foundation is equal to square roots of total load of the pier divided by the allowable soil pressure.
(iv) For unreinforced strip footings the thickness should not be less than the projection from the base of the wall. It should not be less than 150 mm where the foundations are laid at more than one level.
(v) For unreinforced column footing the spread of footing may be 1 vertical to one horizontal.
(vi) As a general rule, the shallow foundation should be taken down to a depth where the allowable bearing capacity is adequate.
  (a) As for as possible the foundation should be kept above the ground water table.
  (b) In order to safeguard a against minor soil erosion, a minimum depth of 500 mm is provided for strip or column foundation
  (c) The depth of foundation can be also determined by plotting the pressure distribution lines (Fig. 15.13).

h_1, h_2 = Depth of footing, Depth of base concrete

h = Depth of foundation

Then $h = h_1 + h_2$

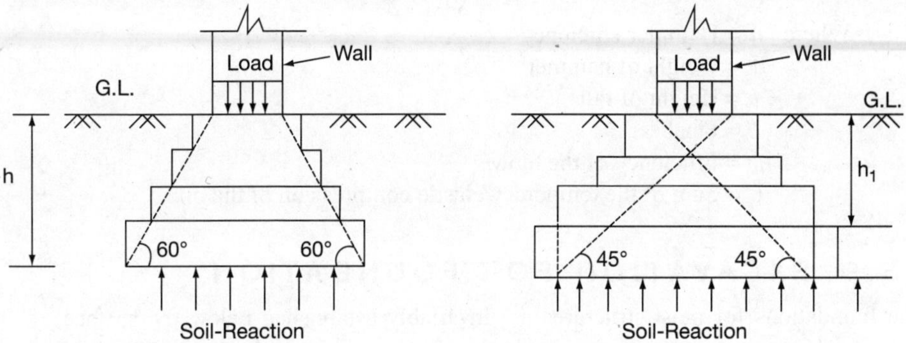

Figure 15.13

(d) Minimum depth of foundation for loose soils may be obtained from Rankine's formula, i.e.,

$$h = \frac{p}{w}\left(\frac{1-\sin\phi}{1+\sin\phi}\right)^2$$

Where h = Minimum depth of foundation in m

w = Weight of soil in kg/m^3

ϕ = Angle of repose

p = Load in soil kg/m^2

The depth of concrete block is given as $h_2 = \frac{1}{58}\sqrt{\frac{pa^2}{f}}$ in cm.

where a = offset of concrete in cm

f = safe modulus of rupture in kg/m^2

15.8.4 Design of Piles

Following guidelines may be considered in design of piles

(i) Direct vertical load coming on the pile should be considered.

(ii) In case of driven piles, the impact stresses induced due to pile driven is taken into account.

(iii) Bending stresses induced due to bending in piles and due to eccentricity to be accounted.

(iv) Lateral forces due to wind, waves, water, current, ice sheets, impact of ships are to be accounted

(v) Forces due to uplift may also to be considered.

(vi) If the area is earthquake – prone area necessary modifications have to be made.

(vii) Load carrying capacity of pile is computed based on the type of pile. Pile load tests can be done for all type of piles. For driven piles, pile driving formulas can be used.

One such formula is ENR formula which is derived on the basis of work-energy theory. The ENR formula has been modified by Hiley as the ultimate pile load, Q_u, is given as

$$Q_u = \frac{\eta W h \eta_b}{S + C/2}$$

Where η = Hammer efficiency
W = Weight of hammer
h = Height of fall
S = Final set
η_b = Efficiency of the blow.
C = Sum of the temporary elastic compression of the pile.

15.9 EXCAVATION FOR FOUNDATION

The foundations for most structures are invariably established below the surface of the ground. Thus they can not be constructed until the soil or rock above the base level of the foundations has been excavated. Open excavations are supported in some soils by lateral support called bracings. It is generally the engineer's duty to decide the construction procedure proposed by the builder and to check the design of bracing. In previous soils, excavation below the water table usually requires drainage of the site either before or during construction. The general aspects of excavating and providing support for the sides of the pits or cuts are discussed in the following sections.

15.9.1 Shallow Excavations with Unsupported Slopes

Shallow excavations can be made if there is enough space is available to establish slopes at which the material can stand. As a general rule construction slopes can be made as steep as possible although a few small slides is generally not serious. But the steepness of slope depends on the type of soil or rock, climate and weather conditions, the depth of excavation and the time to which the excavation should stand.

The steepest slope that can be used in a particular location are decided based on the experience. However, in sandy soils, slopes of about 1 vertical to 1½ horizontal are usually considered. The maximum slope in a clayey soil depends on the depth of cut and the shearing resistance of the clay.

15.9.2 Shallow Excavations with Sheeting and Bracing

Many a times, building sites extend to the edges of the property lines or are adjacent to other sites, over which some structures may already be existing. Under these conditions, it is mandatory that the sides of the excavation must be made vertical and should be usually supported. Several methods are available in such occasions. Two common and simple methods are explained below.

If the depth of excavation is less than 4 m, it is common practice to drive vertical planks knowing as sheeting around the boundary of the proposed excavation. The depth of sheeting is kept near to the bottom of the excavation in progress. The sheeting is held in position by means of horizontal beams called wales. These wales are in turn are commonly supported by horizontal struts extending from side to side of the excavation. The struts are usually are of timber for the excavation not more than 1.5 m wide. For wider excavation metal pipes called trench braces are commonly used (Fig. 15.14).

If the excavation is too wide, the wales may be supported by inclined struts known as rakes. Rakes can be used to provide the supporting soil is firm enough to withstand the forces (Fig. 15.15).

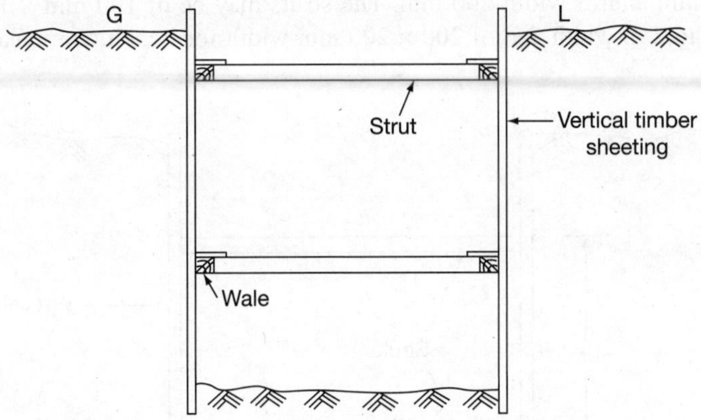

Figure 15.14 Sheeting of shallow excavation

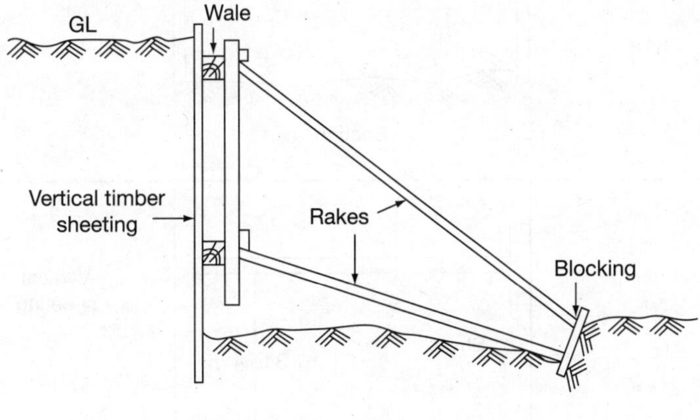

Figure 15.15 Shallow bracing

15.9.3 Deep Excavations

Excavation beyond depth of 1.5 m is generally categorised as deep excavation. The problems generally encountered in deep excavation are:

(i) The collapsing of the sides of the trench.
(ii) The prevention of water entering the trench from the sides or from the bottom of trench.

Only the first aspect is treated in the sections treated below. The secured aspect is dealt separately elsewhere. Following methods of bracing are commonly employed.

1. Stay Bracing

This arrangement is similar to that followed for shallow excavations. This type of bracing is used in moderately firm ground and when the depth of exaction does not exceed 2 m.

Here vertical sheets or poling boards are placed on opposite sides of the trench and they are held in position by one or two rows of struts. The sheets are placed at the spacing of 3–4 m and generally extend to the depth of trench. The thicknesses of poling boards are about 40–50 mm and of width 200 mm. The struts may be of 100 mm × 100 mm size for trench up to 2 m width and of 200 × 200 mm width for trench width exceeding 2 m (Fig. 15.16).

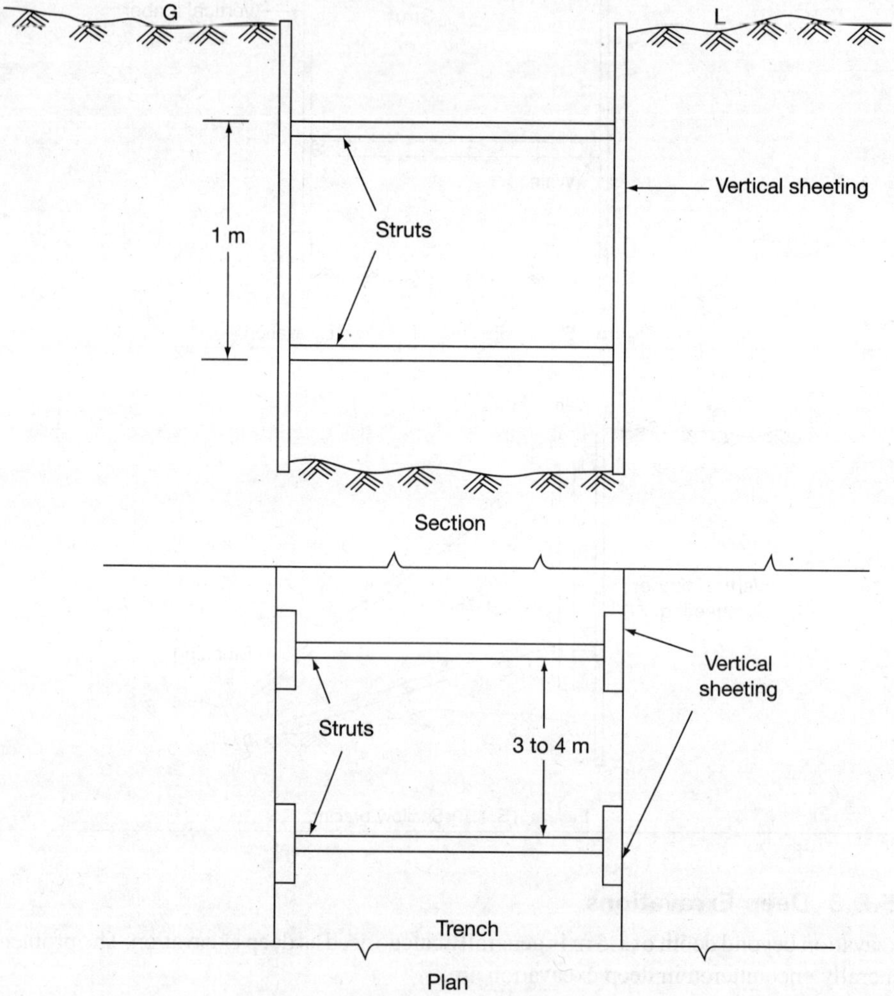

Figure 15.16 Stay bracing

2. Box Sheeting

This arrangement is made for loose soil and when the depth of excavation does not exceed 4 m. Sheeting planks, wales and struts are used to form box like structure as shown in Fig. 15.17. In this arrangement the planks are placed closer or sometimes touching each other. Tow longitudinal rows of wales keep the sheets in position. Struts hold the wales in position.

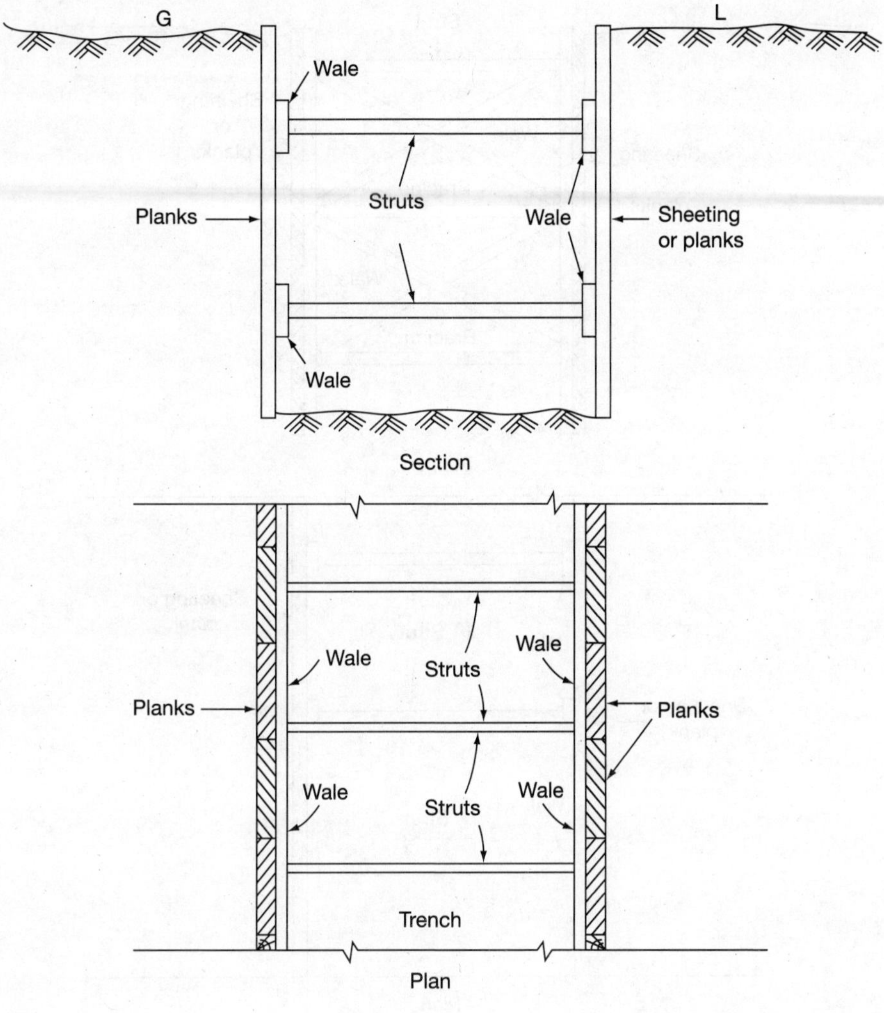

Figure 15.17 Box sheeting

In very loose soils additional bracings are provided. In this arrangement the planks are placed horizontally (in plan) and are supported by wales and struts as shown in Fig. 15.18.

3. *Vertical Sheeting*

In soft ground up to 10 m depth of trenches, the work is carried out in stages. This method is similar to box sheeting. Here at each stage of excavation one offset is provided for each stage separate vertical sheets, horizontal wales, struts and braces are provided. The offset is provided at 3–4 m depth and of 30–60 cm wide at each stage. Suitable working platform is provided (Fig. 15.19).

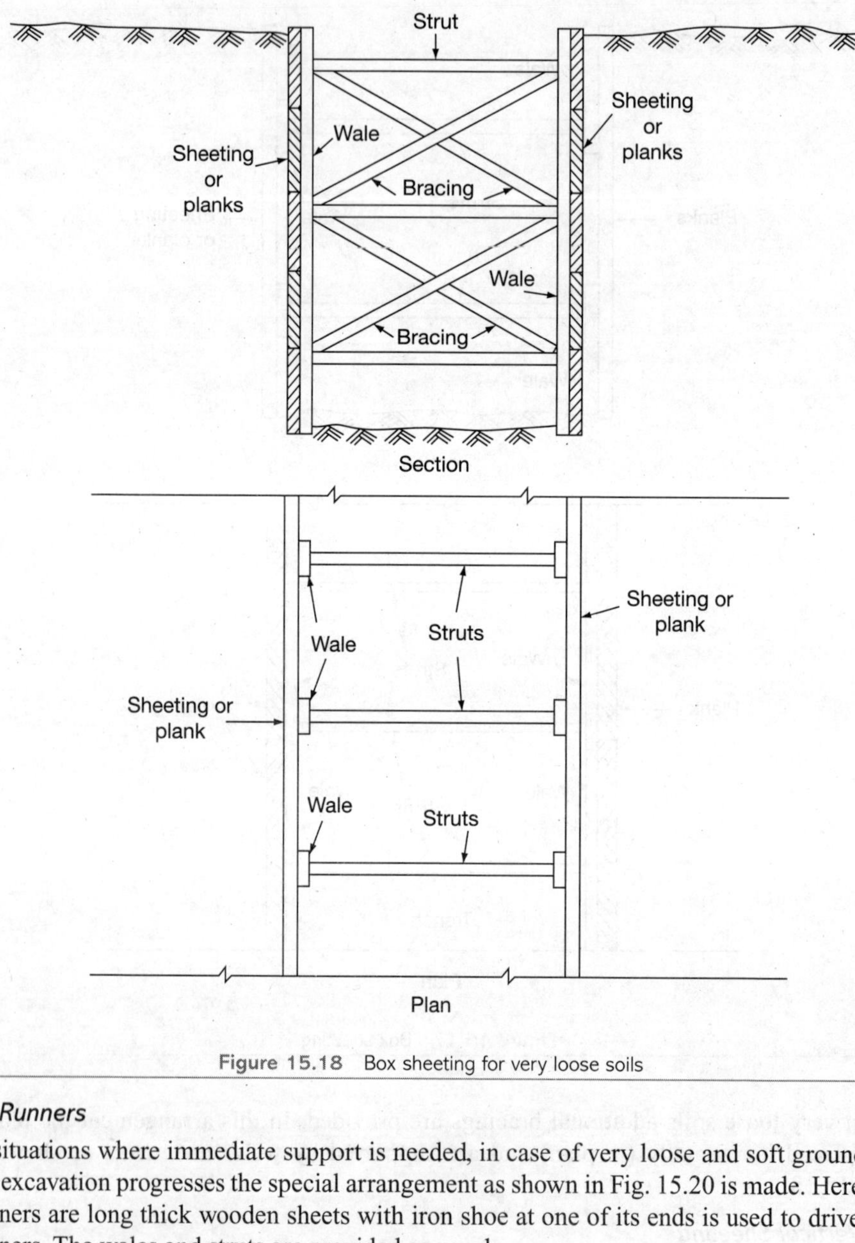

Figure 15.18 Box sheeting for very loose soils

4. *Runners*

In situations where immediate support is needed, in case of very loose and soft ground, as the excavation progresses the special arrangement as shown in Fig. 15.20 is made. Here the runners are long thick wooden sheets with iron shoe at one of its ends is used to drive the runners. The wales and struts are provided as usual.

5. *Sheet Piling*

When the depth of excavation exceeds 10 m the use of vertical timber sheeting becomes generally uneconomical. In such situations other methods of sheeting and bracing are commonly employed. One such procedure is driving of steel sheet piling around the boundary of the excavation. As the soil is removed from the enclosure wales and struts are inserted.

The types of sheet piles commonly used are shown in Fig. 15.21.

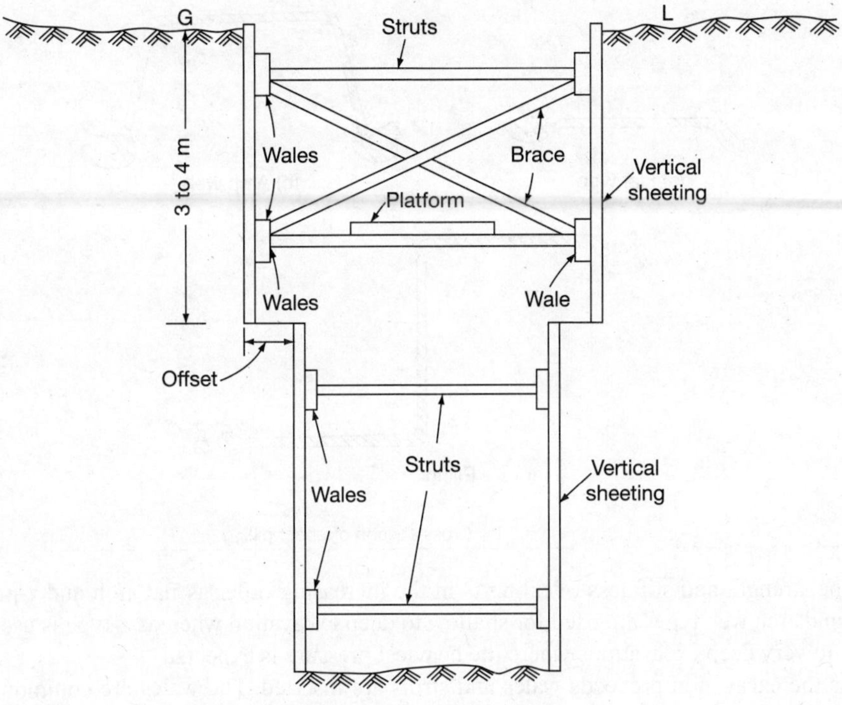

Figure 15.19 Vertical sheeting

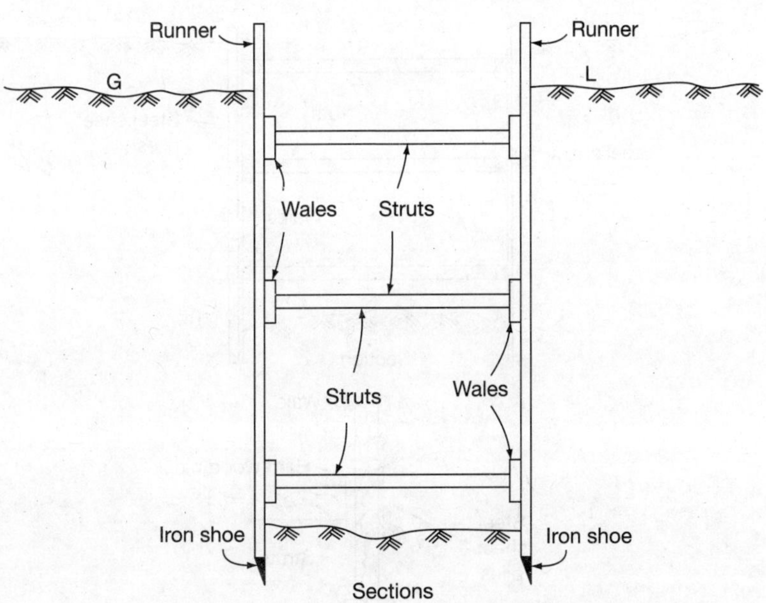

Figure 15.20 Arrangement of runners

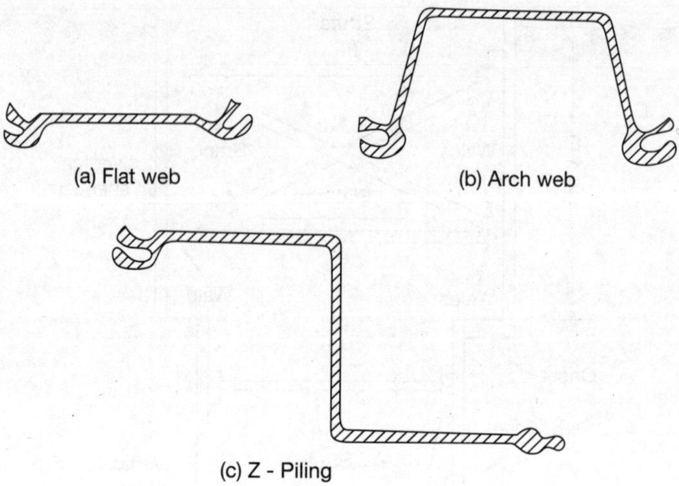

Figure 15.21 Cross-section of sheet pilling

The strength and stiffness of piling is in the increasing order as flat arch and z-piling. Flat and arch web types are used for shallow to deep excavation whereas z-type is used for deep to very deep excavations where the heaviest pressure is expected.

As the excavation proceeds wales and struts are inserted. The wales are commonly of steel, and the struts may be of steel or wood. Excavation is then proceeded to a lower level, and another set of wales and struts is installed. This process is continued until the excavation is completed. In order to prevent local heaves in most of the soils it is necessary to drive the sheet piles several cms below the bottom of excavation (Fig. 15.22).

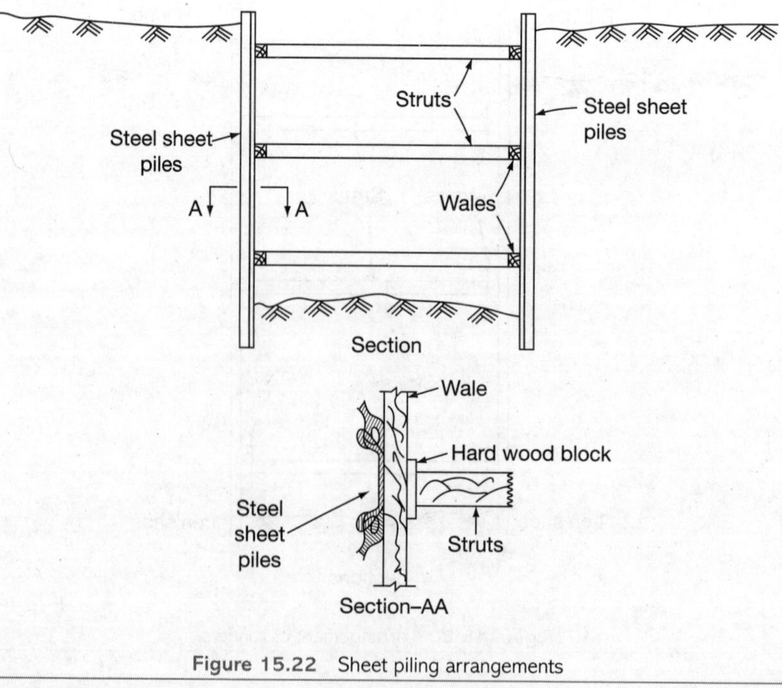

Figure 15.22 Sheet piling arrangements

15.10 CONSTRUCTION OF FOUNDATIONS FOR BUILDINGS

The construction procedure adopted in each of the building foundations are briefly explained below.

15.10.1 Construction of Spread Footing

Spread footing is called as isolated or column footing. They are used to support individual columns. They can be of stepped type or provided with projections in the concrete base. Main reinforcement is placed at the bottom. In case of heavily loaded columns reinforcement is provided in both the directions in the concrete bed. The concrete mix is based on the strength requirement. In general 1:2:4 mix is used. Generally 15 cm offset is provided on all sides of the concrete bed. In case of brick masonry columns an offset of 5 cm is provided.

15.10.2 Construction of Combined Footing

Combined footings are designed keeping the following aspects in view:

(i) The shape of the footing is so selected such that the centre of gravity of the column loads and of soil reaction remains in the same vertical line. Unusually a rectangular or trapezoidal shape of foundation will generally satisfy this requirement.
(ii) The area of the combined footing should be equal or greater than the ratio of the total load and the bearing capacity of the soil.
(iii) The combined footing is treated as an inverted floor, loaded by earth reaction and supported by columns.

Other procedure adopted for column footing may be adopted here too.

15.10.3 Construction of Continuous Footing

A wall footing is a typical case of continuous footing. This may have a base course of concrete or may be made of the same material as that used for the wall. This type of footing may be simple or stepped.

For light loads a simple base with a projection of 15 cm on either side is provided. As a general rule the base width of concrete bedding should be twice the width of the wall and the depth of the base concrete is at least twice the projections.

In another type of wall footing no base concrete is provided. In such a case in order to transmit the load gradually the width of the wall is gradually increased. This is adopted by projecting bricks regularly to a distance not greater than ¼ of a brick beyond the edge of the wall. Cement mortar is used in both the cases for walls. For foundation part a richer mix has to be used.

In another type of continuous footing series of columns in a line are provided with a footing. Here reinforced concrete slabs extends over the series of columns. In order to increase the stability a deep beam is constructed in between the columns. Such type of footings resist differential settlements.

15.10.4 Construction of Mat Foundation

As discussed earlier this type of foundation is used when the bearing capacity is low and total area of spread footing exceeds 50% of the total plinth area.

Mat foundation consists of rows of columns built monolithic with a continuous slab covering the entire foundation area, with or without depressions or openings.

A true mat is a flat concrete slab with uniform thickness throughout the entire area. This type is most suitable where the column spacing is fairly small and uniform and the column loads relatively small. For large column loads a portion of the slab under the column may be thickened. If bending stresses become large, thickened bands may be used along the column lines in both directions. Under extremely heavy column loads, two-way grid structure made of cellular construction may be used. Basement walls are also sometimes used as ribs or as deep beams.

The choice of mat type depends on one or more of the following factors:

(i) For fairly small loading and uniform column spacing and the supporting soil is not very much compressible a flat concrete slab with uniform thickness of mat may be provided.

(ii) In order to provide adequate strength against shear and negative bending moment for heavy loaded columns the slab is thickened.

(iii) For unequal column loading and wide spaced columns beam and slab type of raft is more economical.

(iv) For heavy structures, cellular rafts or rigid frames may be adopted.

Example 15.1

A residential building is to be constructed on a sandy soil with a safe bearing capacity of 1.65 kg/cm^2 and the angle of shearing resistance (angle of repose) is of 30°, and the unit weight of the soil 1580 kg/m^3. The thickness of wall is 30 cm. The total load transmitted is 11500 kg per metre length of the wall.

Solution:

Depth of foundation can be found using Rankine's formula

$$h = \frac{p}{w}\left(\frac{1-\sin\phi}{1+\sin\phi}\right)^2$$

$$= \frac{1.65\times100\times100}{1580}\left(\frac{1-0.5}{1+0.5}\right)^2$$

$$= 10.44/9 = 1.16 \text{ m say } 1.2 \text{ m}$$

Width of footing,

$$B = \frac{\text{Load}}{\text{SBC}} = \frac{11500}{1.65\times100\times100}$$

$$= 0.697 \text{ m} \approx 0.70 \text{ m}$$

$$= 70 \text{ cm}$$

From practical consideration:

$$B = 2W + 2\,(\text{offset})$$

$$= 2\times30 + 2\times15 = 90 \text{ cm}$$

Hence $B = 90$ cm is adopted.

Load on wall/metre length $= 11500$ kg

Assuming 10% of load–Self Weight $= 1150$ kg

Total load on the soil $= 11500 + 1150$

$= 12650$ kg/m length

Pressure on soil $= \dfrac{12650}{90 \times 100} = 1.41 \text{ kg/cm}^2$

This is less than SBC of the soil, hence satisfied.
Using 1:4:8 concrete the modulus of rupture f = 2.45 kg/cm^2 and taking offset of concrete = 15 cm,

$$h_2 = \frac{1}{58}\sqrt{\frac{1.41 \times 15^2}{2.45}} = 11.38 \text{ cm.}$$

As the structure is lightly loaded a bed thickness of 15 cm is considered. Design features are shown in Fig. 15.23.

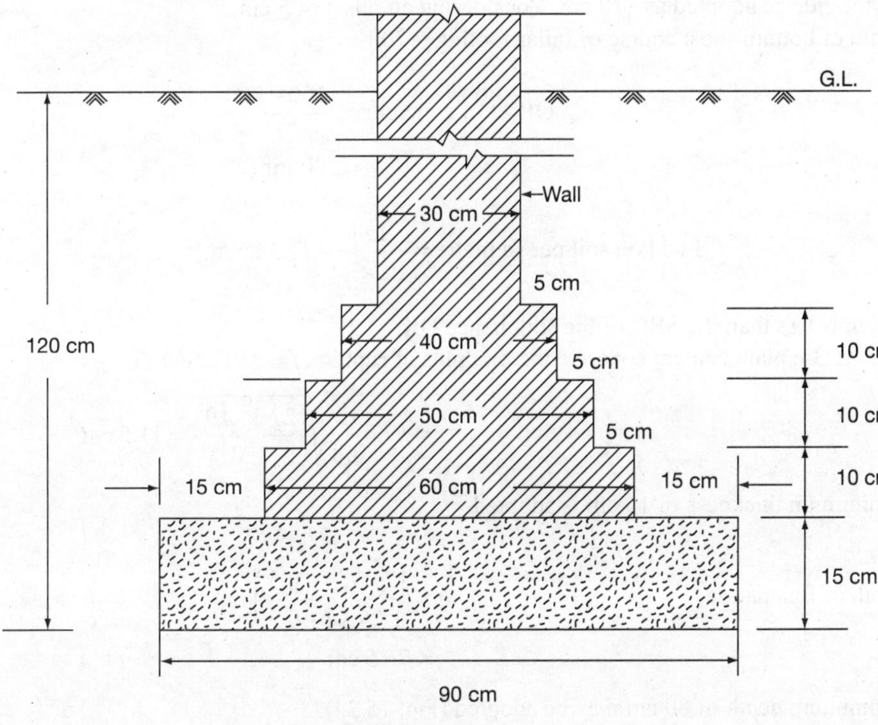

Figure 15.23

Example 15.2

Design an isolated footing to carry a brick-pillar of 300 mm square. The load transmitted at the top of footing is 140 kN. The bearing capacity of the soil at the location is 150 kN/m^2. The unit weight of soil is 18.5 kN/m^3. The angle of repose, ϕ is 30°. The base concrete is of 1: 3: 6 mix plain concrete.

Solution:

Load on the footing = 140 kN

Self weight (Taking 10% of the load) = $140 \times \frac{10}{100} = 14$ kN

$\therefore$ Total load on the soil = 140 + 14 = 154 kN

$$\text{Area of base concrete} = \frac{\text{Load}}{\text{SBC}} = \frac{154}{150} = 1.03 \text{ sq.m}$$

Considering a square base,

$$\text{One side of concrete base} = \sqrt{1.03} = 1.015 \text{ m}$$

Let the side be adopted as 110 cm. Considering an offset of 5 cm,
Width of bottom-most course of pillar footing = 700

$$\therefore \quad \text{Offset of concrete} = \frac{1100-700}{2} = 200 \text{ mm}$$

$$\text{Load on soil per sq.metre} = \frac{154}{1.1 \times 1.1} 127.3 \text{ kN/m}^2$$

Which is less than the SBC of the soil, hence OK.
Using 1:3:6 plain cement concrete the modulus of rupture, f = 350 kN/cm^2

$$= h_2 = \frac{1}{58}\sqrt{\frac{127.3 \times 19^2}{350}} = 11.5 \text{ cm}$$

A minimum thickness of 15 cm is adopted

Depth of foundation,

$$h = \frac{127.3}{18.5}\left(\frac{1-0.5}{1+0.5}\right)^2 = 76.5 \text{ cm}$$

A minimum depth of 80 cm may be adopted (Fig. 15.24).

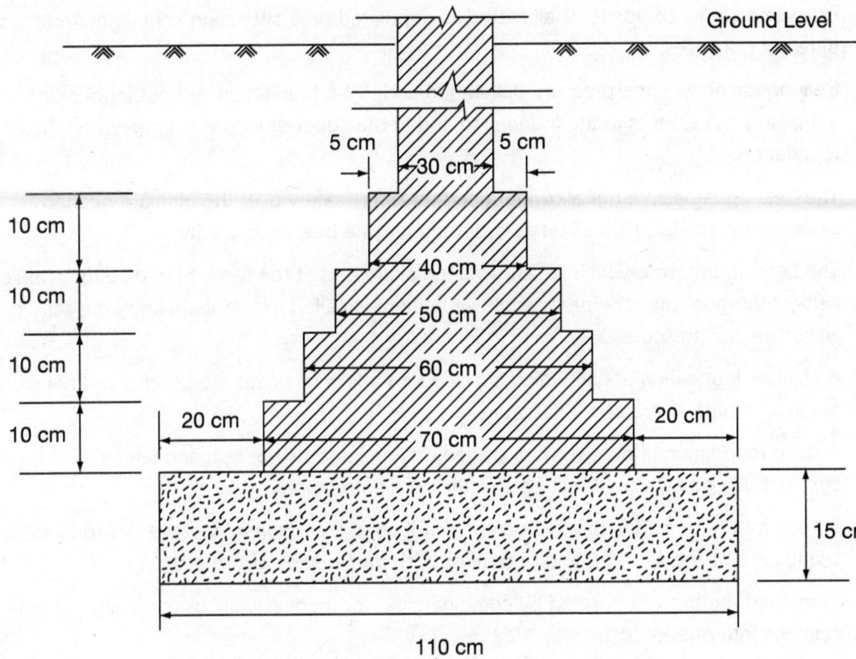

Figure 15.24

SALIENT POINTS

1. A foundation is that part of the structure which is in direct contact with the ground and transmits the load of the structure to the ground.
2. Information on surface and sub-surface conditions is a more critical requirement in planning and designing the foundations of structures, dewatering systems, shoring or bracing of excavation, the materials to be used in construction, and site improvement methods.
3. Planning the ground investigation comprises of (i) available information, (ii) reconnaissance and (iii) preliminary investigation and (iv) detailed investigation.
4. Methods of exploration are indirect methods, semi-direct methods and direct methods.
5. Indirect methods consist of geophysical and sounding methods. Semi-direct methods are common boring and drilling methods combined with intermittent sampling. Direct methods are boring and sampling methods.
6. Residual soils have formed from the weathering of rocks and practically remain at the location of origin with a little or no movement of individual soil particles.
7. Transported soils are those that have formed at one location (like residual soils) but transported and deposited at another location.
8. Regional soil deposits are marine deposits, black cotton soils, lateritic soils, alluvial deposits, desert soils and boulder deposits.

9. Re-claimed soils comprise of all materials deposited on a site using various methods for different purposes.
10. Bearing capacity is the pressure developed under the foundation without introducing damaging movements in the foundation and in the super-structure supported on the foundation.
11. The bearing capacity after allowing a certain factor safety over the ultimate bearing capacity against failure (collapse) is termed as safe bearing capacity.
12. The bearing pressure which is safe against failure and at the same time does not cause settlement more than the permissible settlement is called allowable bearing capacity or allowable soil pressure.
13. A shallow foundation is one which is placed on a firm soil near the ground and beneath the lowest part of the super-structure.
14. A deep foundation is one which is placed on a soil that is not firm and which is considerably below the lowest part of the super-structure.
15. Spread footing provides the function of distributing the column load over a wide area taking care of the strength and deformation characteristics of the soil.
16. Combined footings are formed by combining two or more equally or unequally loaded columns into one footing.
17. Continuous footings carry closely spaced columns or a continuous wall such that the load distribution is uniform and load intensity is low on the supporting soil. The footings are also named as strip footings or wall footings.
18. Mat or raft foundations are characterised by the feature that columns frame into the footing in two directions.
19. Piles are slender structural members normally installed by driving by hammer or by any other suitable means.
20. Bored cast in-situ pile is formed within the ground by excavation or boring with or without the use of temporary casing and subsequently filling it with plain or reinforced concrete.
21. Piles used to compact soils are called as compaction or displacement piles.
22. Driven piles installed in pre-drilled holes are called as non-displacement piles.
23. Where the top soil is soft or too weak to support the super-structure, piles are used to transmit the load to the underlying bed rock, such piles are called end-bearing or point-bearing piles.
24. If the bed rock does not exist at a reasonable depth below the ground surface, the load is transferred through friction along the pile shaft such piles are called friction piles.
25. Some structures are subjected to uplift pressure and piles are used to resist the uplift forces which are called uplift piles or tension piles.
26. In order to resist horizontal and inclined forces in water and earth retaining structures better piles are used.
27. Application of piles for providing anchorage to sheet piles is called as anchor piles.
28. Drilled piers are structural members of relatively large diameter massive struts constructed of concrete placed in a pre-excavated hole.

29. Caissons are structural boxes or chambers that are sunk in place through ground or water.
30. Well foundation is a type of caisson which is constructed either on dry ground or over an artificially formed island.

REVIEW QUESTIONS

1. List the purpose of ground investigation.
2. How do you plan a ground investigation for a multi-storeyed building?
3. What are the methods of exploration?
4. Explain the regional soil deposits.
5. Distinguish between safe bearing capacity and allowable soil pressure.
6. Discuss the factors affecting the bearing capacity of soil.
7. Mention the techniques adopted for improving the bearing capacity of the soils.
8. What factors determine whether a foundation type is shallow or deep?
9. Indicate the circumstances under which combined footings are adopted.
10. What precautions are to be taken while locating a footing (i) on a slope and (ii) adjacent to an existing structure?
11. How piles are classified based on materials?
12. Enumerate the different types of piles and describe each type briefly. Give the advantages and disadvantages of each type. (UPSC)
13. Briefly explain the probable causes of failure of pile foundations.
14. What is a grillage foundation?
15. Mention the precautions to be taken to ensure safety of foundations on expansive soil.
16. Under what ground conditions under-reamed piles are recommended.
17. Under the following field conditions, what type of foundation you would suggest:
    (i) Soil is of soft nature and the load is uniform.
    (ii) A cavity is met during excavation.
    (iii) A made – up ground.
    (iv) A new structure with five-stories to be constructed next to an old structure.
18. What are the factors you consider in the selection of a type of foundation?
19. Describe with sketches the method timbering in a trench of size 180 cm deep × 120 cm wide for laying a foundation in a moderately firm ground. (AMIE)
20. How temporary supports are made for deep foundation excavation?
21. How construction of a mat foundation is done?
22. What are the factors considered in the selection of mat foundation?
23. A six-storeyed RCC framed building has to be constructed on an old tank bed with loose soil over 12 m deep and water-logged up to 1.2 m from ground level. What type of foundation would you adopt? Explain in detail the process of constructing out such a foundation. (AMIE)

16 MASONRY CONSTRUCTION

16.1 INTRODUCTION

As per the old convention, masonry is a term used to indicate the art of building structures in stones using mortar and stone. But in a simplified form, it may be defined as construction of building units bonded together with mortar. The building units may be brick, stone, concrete, hollow blocks, etc.

The selection of the type of unit for the masonry is made keeping in view the requirements of strength, water proofing, thermal insulation, fire resistance, durability and economy. The binding strength of the mortar is usually disregarded as far as the strength of masonry is concerned.

The mortar is required to keep the building units in position. It is prepared by mixing lime or cement with sand after adding water which forms the mortar. Sometimes a portion of cement is replaced by lime and accordingly called as cement-lime mortar. If the mortar is to be strong, more plastic and workable lime-cement mortar is used.

The masonry is classified based on the basic unit used for the structure.

16.2 BRICK MASONRY

As bricks are comparatively small in size, they can be handled easily. Further, brick being uniform in size can be arranged systematically and bonded together with mortar to form a homogeneous mass, which is called brick masonry. The materials used for brick masonry are bricks, mortar and specially shaped bricks. Mortar helps to bind the individual bricks and to form a cushion to take up the inequalities in the brick and to distribute the pressure evenly and also to fill up the interstices in the bricks.

16.2.1 Terminology

Some of the terms which are used in brick masonry are defined below.

1. Header

A brick laid with its breadth or width parallel to the face or direction of a wall.

2. Stretcher

A brick laid with its length parallel to the face or direction of the wall.

3. Bed

The lower surface of a brick when laid flat.

4. Bat

A portion of a brick cut across the width.

5. Closer

A part of brick that is used to close up the bond at the end of brick course. The different closers are: (i) king closer, (ii) queen closer and (iii) bevelled closer.

6. Quoin

A corner of the external angle on the face side of a wall.

7. Squint

A brick cut to a special shape and used at oblique quoins.

Figure 16.1 shows courses of brickwork and Fig. 16.2 shows types of special bricks.

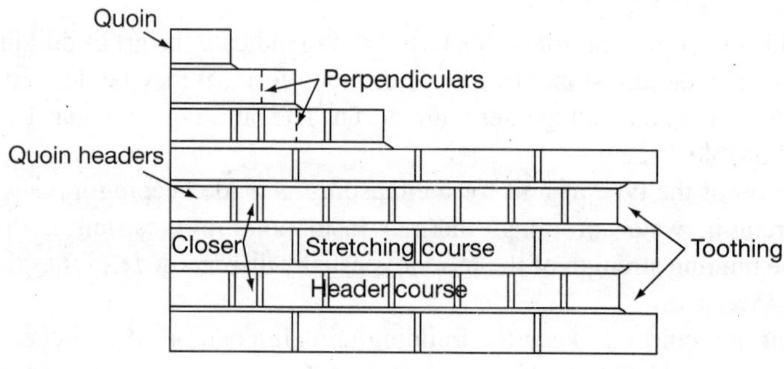

Figure 16.1 Courses of brickwork

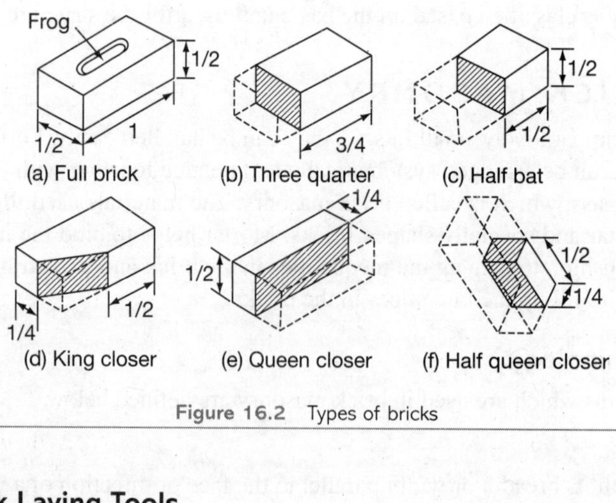

Figure 16.2 Types of bricks

16.2.2 Brick Laying Tools

Main tools used for brick laying are trowel, plumb bob, bolster, tri-square, lime and pins, sprit level, and brick hammer. Figure 16.3 shows the above tools and is explained below.

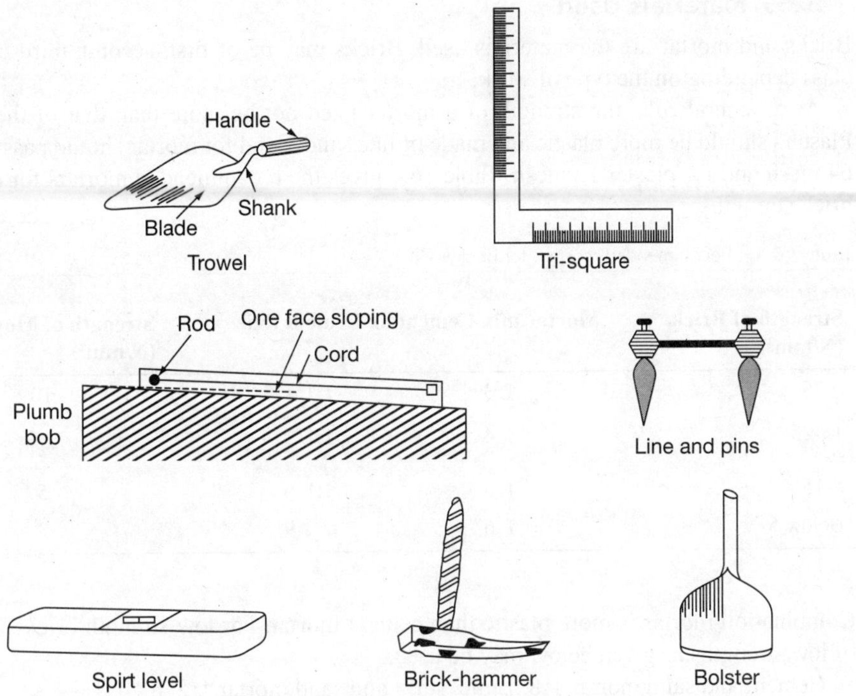

Figure 16.3 Common brick laying tools

1. Trowel

Brick trowel is usually about 30 cm long with a steel blade and a handle. The blade has one straight edge and one end curved edge.

2. Tri-square

It is used to set out a right-angled corner. It is like a large triangle having two legs. It should be used carefully to set the first course.

3. Plumb Bob

It consists of a piece of thick straight wood ruler with a piece of string which is fixed in the wood at one end and the metal plumb bob at the other end.

4. Line and Pins

The two pins used for brickwork have flattened and blunt point ends for inserting into the joints of brickwork as work proceeds.

5. Sprit Level

It is used to level the brickwork. It consists of a hardwood stock with anything from two to six bubbles. These bubbles are small, sealed inside a glass tube.

6. Brick Hammer and Bolster

Cutting of bricks to provide suitable shapes for the bonding is done with the club hammer and bolster.

16.2.3 Materials Used

Bricks and mortar are the materials used. Bricks may be of first, second, third or fourth class depending on the type of work.

As a general rule, the strength of a mortar need not be more than that of the bricks. Plasters should be more plastic and made of fine sand. Sand for mortar should pass through 64 mesh and for plaster 12 mesh. Table 16.1 gives the recommended mortars for different brick strengths.

Table 16.1 Recommended mortars for brickwork

Strength of Brick (N/mm^2)	Mortar mix Cement or Cement lime		Strength of Mortar (N/mm^2)
> 25	1:3	1:1/4:3	10
15-25	1:4	$1{:}1/2{:}4\frac{1}{2}$	7.5
5-15	1:5	1:1:6	5
Below 5	1:6	1:2:9	3

Combination mortar is more plastic than cement mortar. For low-strength bricks, mortars of low strengths as given below may be used.

Cement and sand mortar 1:8; Lime, surki and sand mortar 1:2:6
Lime and sand mortar 1:3; Cement, lime, surki and sand mortar 1:2:4:20
Cement, lime and sand mortar 1:4:14

16.2.4 Bonds in Brickwork

It is the process of arranging bricks in courses to ensure that vertical joints do not come one over the other. A wall built without any continuous vertical joint shall distribute the load properly and shall also be more strong and durable.

1. Stretcher Bond

In this type of bond, all courses are laid as stretchers. As only stretchers are visible in elevation, this bond is referred to as stretcher bond. This is used for partition walls (Fig. 16.4(a)).

2. Header Bond

In this type of bond, all courses are laid as headers. As only headers are visible in elevation, this bond is referred to as header bond. This is used for one brick and curved walls (Fig. 16.4(b)).

3. English Bond

This bond is produced by laying alternate courses of stretchers and headers. In order to break the joints vertically, it is essential to use a closer after the header quoin in the heading course. This is the most commonly used bond which is also the strongest. This type of bond is used for walls carrying heavy loads. Figure 16.4(c) shows formation of a wall adopting English bond.

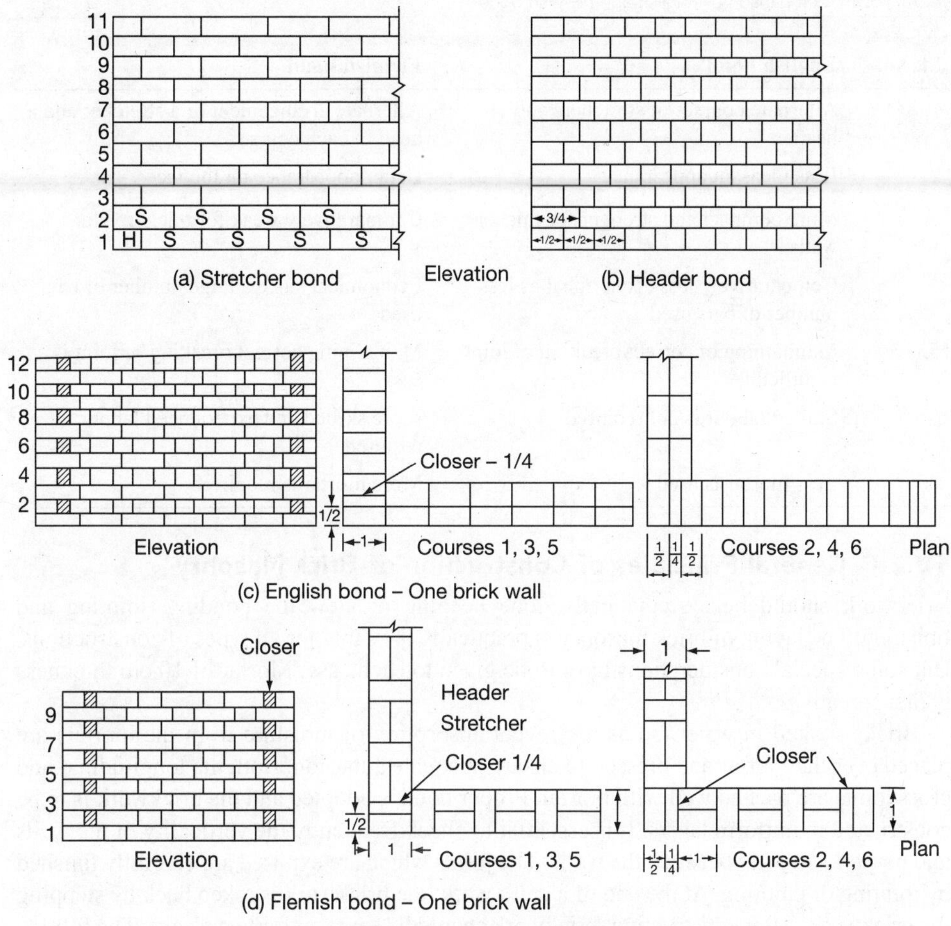

Figure 16.4 Bonds in brickwork

4. Flemish Bond

This bond is produced by laying alternate stretchers and headers in each course. The headers and stretchers appear in the same course alternately on the front and the faces. The queen closer is used next to the quoin header in alternate courses in order to break the continuity of the vertical joints. A header in any course is in the centre of a stretcher in the course above or below it. This bond is also used for walls to carry moderate loads. Figure 16.4(d) shows formation of wall adopting Flemish bond.

16.2.5 Comparison of English and Flemish Bonds

Table 16.2 gives a comparison of English and Flemish bonds.

Table 16.2 Comparison of English and Flemish bond

Sl. No.	English bond	Flemish bond
1.	Alternate courses of stretchers and headers	Alternate arrangement of a stretcher and a head in each course
2.	Less pleasant appearance	Better appearance on the face
3.	More compact and stronger for thicker walls	Comparatively weak for thicker value
4.	Comparatively less economical as less number of bats used	Economical since a large number of bats used
5.	Maintaining of correct breaking of joint is difficult	Maintaining correct breaking of joint is easy
6.	Skilled labour is not required	More skilled and experienced labour is required
7.	Less mortar is used	More mortar is required

16.2.6 General Principles of Construction of Brick Masonry

Brickwork should be systematically done keeping in view the bonding, jointing and finishing. The laying of brick masonry is practically the same for all types of constructions. But some special considerations have to be given to each case. Mortar of 10 cm thickness is first spread.

Bricks soaked in water (so as to prevent absorption of moisture from the mortar) are placed over the mortar and pressed to ensure adequate adhesion. Both the longitudinal and cross joints are packed well with mortar. Proper bond is adopted and the brickwork is to be constructed in uniform layers. It is essential to check frequently the verticality of the walls and perpendicularity between the walls. The joints which are exposed are carefully finished by jointing or painting. At the end of a day's work, the brickwork is raked back by stepping the brickwork. After construction, brickwork should be kept wet for two weeks. The following points should be observed while supervising the construction of brick masonry:

(i) Bricks should conform to the specification and requirements of the work.
(ii) Bricks should be saturated well with water before use.
(iii) Bricks should be laid with frog upwards, with mortar completely covering the bed.
(iv) A suitable bond has to be used to avoid continuous vertical joints anywhere in the work.
(v) A good quality mortar has to be used and the thickness of mortar should not be more than 10 mm.
(vi) The brickwork should be raised uniformly.
(vii) Brickbats should be minimised and to be used only in the required places.
(viii) In no case the height of masonry should be greater than 1 m in a day's work.
(ix) All vertical faces should be checked with a plumb bob and wall junctions should be checked for perpendicularity.
(x) After the construction is over, the brickwork should be watered well for two weeks.

16.3 STONE MASONRY

Stones are abundantly available in nature which can be properly shaped and used for construction of various parts of a building. Similar to brick masonry, stones also can be systematically arranged to form a homogeneous mass called stone masonry. The materials used for stone masonry are stones and mortar. The common types of stones used for stone masonry in India are: granites, sandstones, limestones, marbles and slates. Stone masonry is strong and durable.

Apart from the use in building construction, stones are used for structures like dams, piers, waterfront works, monuments and memorials. Stone masonry is not affected by dampness and does not require plastering. However, it will be cheap only in areas where it is abundantly available.

16.3.1 Terminology

1. Bed Surface

Surface of a stone perpendicular to the line of pressure.

2. String Course

A horizontal course of masonry continuously projecting from the face of the wall.

3. Corbel

A stone embedded in a wall to support a structural member.

4. Cornice

A moulded projection of masonry on the top of a wall.

5. Through Stone

Stone extending throughout the thickness of the wall.

6. Hearting

A filled-up core of a rubble wall.

7. Jam

A masonry portion forming the side of an opening.

16.3.2 Materials Used

Materials used for stone masonry are stones and mortar. Common types of stones which are used for stone masonry along with their crushing strength are presented in Table 16.3.

Table 16.3 Common types of stones used for stone necessary

Types	Crushing Strength (Minimum)
Limestone	20 N/mm^2
Sandstone	30 N/mm^2
Basalt	40 N/mm^2
Marble	50 N/mm^2
Granite	100 N/mm^2

Generally lime and cement mortars are used for stone masonry. Their function is to initially provide a workable matrix and ultimately a hard-binding material, which renders masonry into a monolithic unit. Hydraulic lime and Portland cement with river sand are used.

16.3.3 Cutting and Dressing of Stones

Dressing is the art of cutting the stones to shape required for use in structures. A brief discussion on dressing and purpose of dressing is made in Section 2.8.

Tools used for cutting and dressing stones is given in Fig. 16.5.

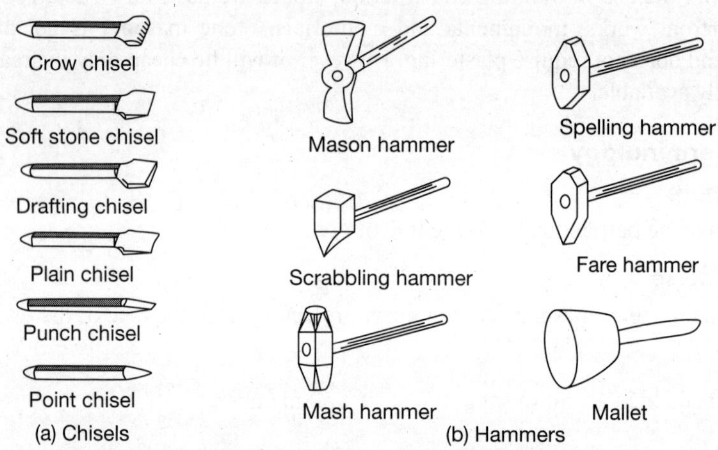

Figure 16.5 Tools for cutting and dressing stones

Stones for two major types of stone masonry, viz., rubble and ashlar masonry (discussed in Section 16.3.4) are dressed as follows.

1. Dressing for Rubble Masonry

(i) For this type of masonry, stones are shaped by means of hammer. Lower and upper surfaces of stones are almost dressed flat.

(ii) For this type of masonry, stones which have front face almost rectangular and remaining irregular are also used.

2. Dressing for Ashlar Masonry

(i) For this type of masonry, the faces of stones are finely dressed by means of chisels.

(ii) Further, chiselled faces are sometimes rubbed to give a smooth surface in case of special works like monumental works.

Different types of surface finishes adopted are presented in Fig. 16.6.

(i) Rough tooled surface
(ii) Tooled surface
(iii) Cut stone surface
(iv) Hammer dressed surface
(v) Polished surface
(vi) Rubbed surface

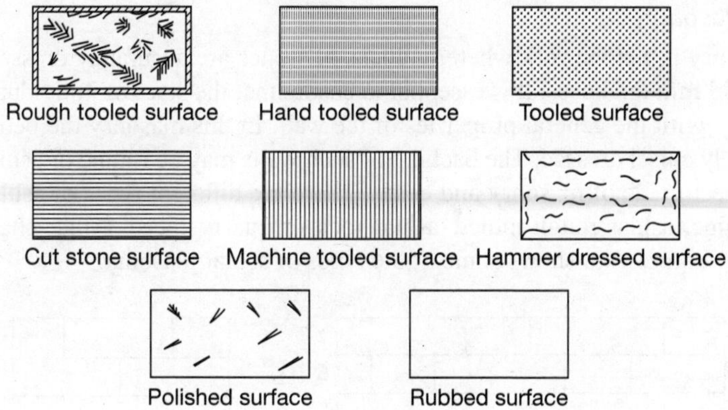

Figure 16.6 Types of surface finishes

16.3.4 Types of Stone Masonry

Stone masonry is classified based on the thickness of joints, continuity of courses and finish of face. Rubble Masonry and Ashlar Masonry are discussed below (IS: 1597, Parts 1 and 2, 1992).

1. Rubble Masonry

Rubble masonry is a stonework wherein blocks of stones are either undressed or roughly dressed and have wider joints. The stones used are not of uniform size and shape but generally of pyramidal in shape to some extent. Vertical and transverse bonds have to be attained. Through stones are provided from back to the face of wall to ensure better bonding. The strength of this masonry depends on the quality of mortar, use of through stones and filling of spaces. There are different types of rubble masonry, viz., random rubble, squared rubble and polygonal rubble. Figure 16.7 shows the various types of rubble masonry.

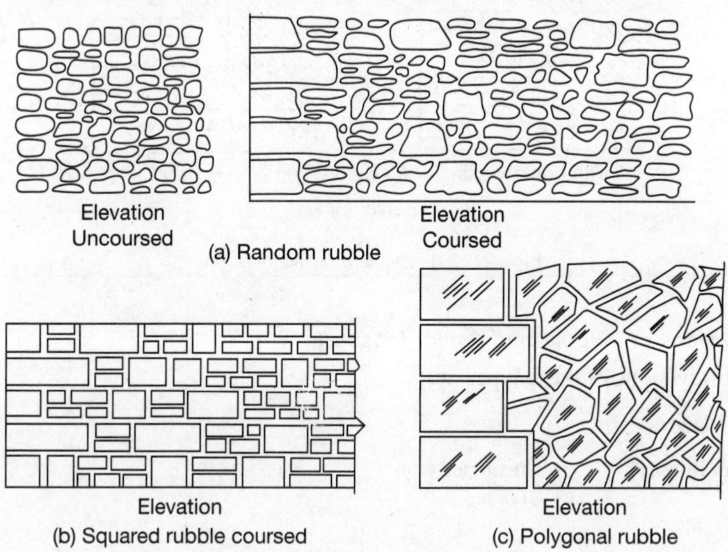

Figure 16.7 Types of rubble masonry

2. Ashlar Masonry

Ashlar masonry is a stonework wherein blocks of stones are accurately dressed with very fine joints of 3 mm thickness. It is essential to ensure that the sizes of individual stones are in conformity with the general properties of the wall. In this masonry the beds, sides and faces are finely chisel dressed. The backing of such walls may be rubble or ashlar masonry. It is the highest grade of masonry and costly. There are different types of ashlar masonry, viz., ashlar fine, ashlar rough tooled, ashlar rock or quarry faced, ashlar chamfered and ashlar facing. Figure 16.8 shows some of the types of ashlar masonry.

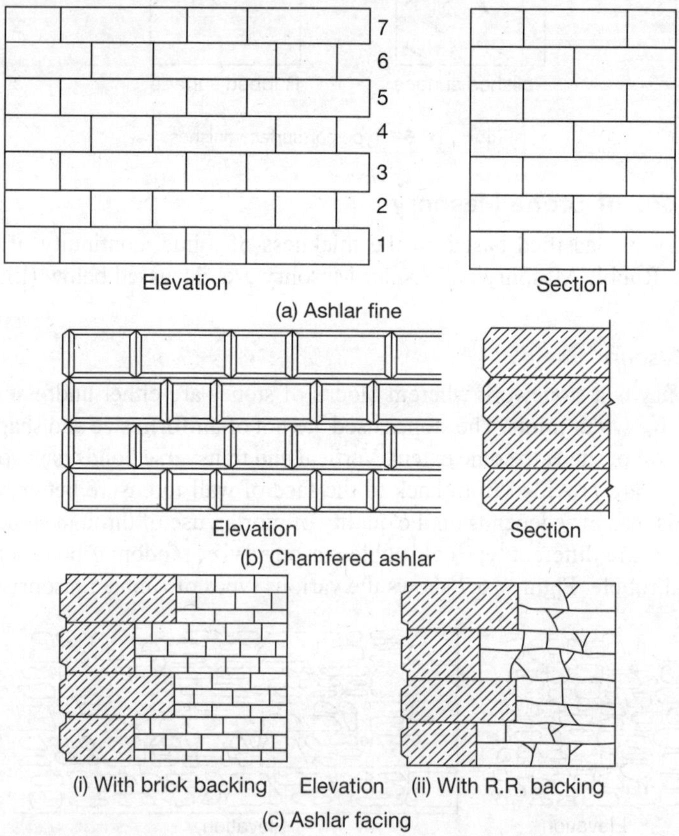

Figure 16.8 Types of ashlar masonry

3. Sub-divisions of Rubble and Ashlar Masonry

Various sub-divisions of these types are:

(i) Rubble Masonry

(a) Random Rubble (i) Uncoursed
(ii) Coursed

(b) Square Rubble
 (i) Uncoursed
 (ii) Coursed
 (iii) Built to regular courses

(c) Miscellaneous
 (i) Polygonal
 (ii) Flint

(ii) Ashlar Masonry

(a) Ashlar fine
(b) Ashlar rough tooled
(c) Ashlar rock or quarry faced
(d) Ashlar chamfered
(e) Ashlar facing
(f) Ashlar block-in-course

16.3.5 General Principles of Construction of Stone Masonry

The construction procedure for stone masonry is slightly different from that of brick masonry. Unlike in brick masonry, in stone masonry stones of different sizes and shapes are used and accordingly there is slight variation in construction. However, basically the requirements of vertical joints not being continuous, thickness of mortar joints and quality of mortar, verticality and perpendicularity of bonds, and the overall stability of wall are same in both. Even in stone masonry, procedures differ slightly between rubble and ashlar masonry constructions. However, the following general principles of construction should be observed while supervising the stony masonry construction:

(i) Stones should be hard, tough and durable.
(ii) Pressure acting on the stone should be perpendicular to the bedding planes.
(iii) Stone masonry work should not be designed to take any tension.
(iv) Stones should be well watered before use.
(v) Through stones should be used sufficiently such that they cover about 15–25% of the area in elevation.
(vi) Stones should be laid on its natural bed.
(vii) Extreme care has to be taken by providing proper bond to prevent formation of vertical joints.
(viii) Hearting of masonry should be properly packed with chips and mortar.
(ix) Verticality of the faces of the masonry walls should be checked with a plumb rule.
(x) Portions of the masonry should be raised uniformly throughout its length.
(xi) Cement mortar and sometimes rich lime mortar may be used.
(xii) Proper dressing of stones has to be done wherever needed.
(xiii) Surfaces of the stone masonry should be kept wet while the work is in progress and also till the mortar has set.
(xiv) All laid fine dressed stonework should be protected against damage during further construction.

16.3.6 Comparison of Brick and Stone Masonry

Brick masonry and stone masonry are compared in Table 16.4.

Table 16.4 Comparison of brick and stone masonry

Sl. No.	Brick masonry	Stone masonry
1.	Construction cost is less	Construction cost is more
2.	Easy to construct	Difficult to construct
3.	Semi- skilled labour is sufficient	Skilled labour is required
4.	Construction of openings and connections are easy	Difficult to make openings and connections
5.	Reasonably good quality mortar and thin mortar joints are needed	Only cement mortar or rich lime mortar of good quality and thick mortar joints are needed
6.	Plastering is required	Plastering is not required
7.	Compressive strength is weak	Compressive strength is high
8.	No special lifting devices are needed	Special lifting devices are needed
9.	Better resistance to atmospheric effects	Easily affected by atmospheric effects
10.	More moisture absorption from atmosphere	Less moisture absorption from atmosphere
11.	High fire resistance	Reasonable resistance to fire
12.	Elegant appearance	Massive appearance
13.	Reasonably durable and moderate long life	Highly durable and very long life

16.4 CONCRETE HOLLOW BLOCK MASONRY

Cement concrete hollow blocks have been in use for several masonry constructions. Several advantages have given room for rapid development and use of the same in place of traditional construction materials like stones and bricks.

Typical concrete masonry units are shown in Fig. 16.9 (Sharma, 1988), in addition to the concrete hollow blocks referred to in Chapter 9.

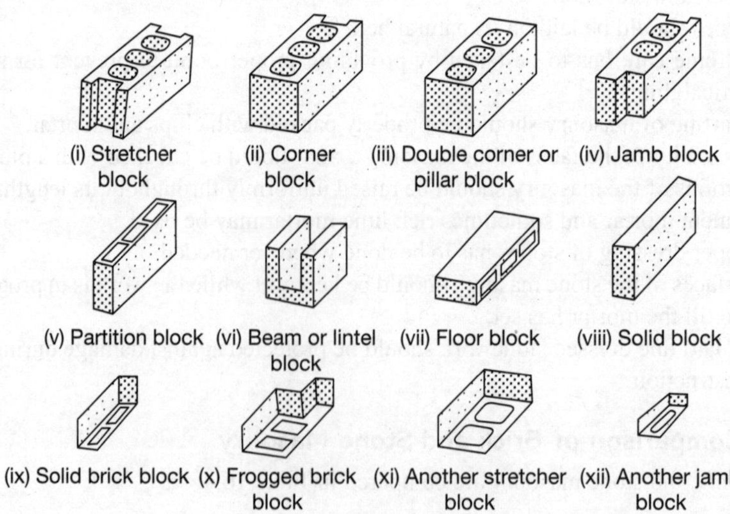

Figure 16.9 Typical concrete masonry units

Based on the job requirement, the concrete blocks may be made. There is no standard size of concrete blocks. However, Concrete Association of India (CAI) recommends that the face thickness should not be less than 5 cm and the net area should be atleast 55–60% of the gross area.

The common sizes for building blocks are given in Table 16.5.

Table 16.5 Common sizes of blocks

Sizes	Actual dimensions		
	Length (cm)	Breadth (cm)	Height (cm)
A	39	30	19
B	39	20	19
C	39	10	19
Tolerance	± 0.3 mm	± 1.5 mm	± 1.5 mm

16.4.1 Use of Hollow Concrete Blocks for Masonry

Following are the uses of concrete blocks for masonry construction:

(i) Large in size but easy to handle
(ii) Uniformity in design
(iii) Easy handling and placing
(iv) Adequate strength
(v) Attractive appearance
(vi) No need of plastering for outside work
(vii) Less number of masonry joints
(viii) Less cost
(ix) Easy method of manufacturing

16.4.2 Laying of Concrete Hollow Block Masonry

Different types of basic hollow blocks used in construction are discussed in Section 9.6.1. Additional hollow block units are given in Fig. 16.9 Now the construction features are explained.

1. Construction of Walls

A mortar bed is spread on the foundation concrete and levelled to have a uniform thickness everywhere. The corner block is first placed and positioned accurately. Mortar is applied to the other end and one block is positioned to the end and aligned. The level of the course is checked after placing a few blocks. If necessary, the blocks are tapped with additional mortar such that the mortar thickness is 2 cm below and on the ends. The first course is checked to be in plumb before planning second and additional courses.

As done in brick laying, the successive courses are laid in such a manner so as to break the joints vertically. For vertical joints the mortar is applied to the projection at the sides of the block. As followed for the first course, the courses are built starting from the corners only. Every time, the verticality and horizontality are checked. All the four vertical edges of the final block and the edges of the opening are covered with mortar and pushed in position. The face of the masonry may be pointed by running a tool. The type of joint recommended are weathered, V-shaped or concave, such that the joints shed off water easily.

2. Construction of Columns

Columns are used wherever a large pressure to be transferred through large bearing surface. The columns may form an integral part of the wall or it can be a separate unit. The columns are made of standard stretcher and corner blocks or other special shapes are used. For better stability, the hollows within the blocks may be filled with plain or reinforced concrete (Fig. 16.10(a)).

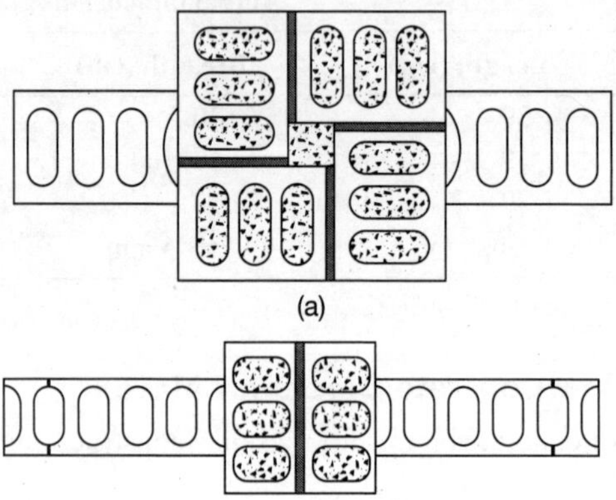

(a)

(b)

Figure 16.10 Pilasters and piers

3. Construction of Window and Door Openings

Blocks with hole (jamb) near the opening should be filled with concrete and wooden plugs (Fig. 16.10(b)). The door or window frames are screwed to the wooden plugs. It is also fixed in the lintel with small dowels of mild steel. Under the base of the window or door, a course of solid concrete block masonry is laid which is extended into the adjacent walls up to a distance of at least 30 cm on either side. Lintels are also of hollow channel-shaped sections which can be filled with concrete and provided with steel reinforcement at their bottom.

4. Construction of Reinforced Walls

This is made by providing vertical reinforcement in the hollow with concrete (Fig. 16.11). In order to increase the strength of the wall, reinforcement is provided at the horizontal joints. Because of this provision, expansion cracks which may occur due to moisture and change of temperature may be reduced. Further, two horizontal bars of 6 mm diameter are placed on each on the face of the wall. Instead of steel rods, welded steel mesh may be used wherever needed (Fig. 16.11).

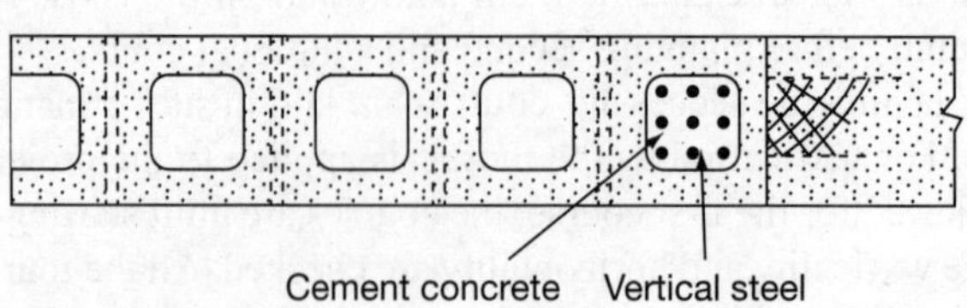

Figure 16.11 Jamb details for 200 mm thick hollow block wall

16.4.3 Concrete Blocks with Concrete Filling

Concrete hollow blocks constructed with two through holes to form continuous vertical hole are ideal for concrete filling. These holes can be completely or partly filled with concrete during the process of construction itself. For this purpose, mixes with large size aggregates may be used. This provision will add greater strength to walls. It has been found from the tests that filling of hollow blocks with plain concrete will give basic strength approximately halfway between these for unfilled blocks and the solid blocks with the same quality of concrete.

16.4.4 Reinforced Concrete Hollow Block Masonry

In the conventional hollow block masonry, the vertical holes may be made to accommodate steel reinforcement and can thus be made of reinforced concrete construction. Similarly special concrete blocks can be made to accommodate horizontal reinforcements. Such a horizontally reinforced hollow block can be used as lintels and horizontally strengthened reinforced concrete bonds can be built up with these blocks. This type of construction is known as reinforced concrete hollow block masonry construction.

16.4.5 Special Features of Concrete Block Masonry

The important features to be incorporated, compared to regular brick masonry, are control joints, band beams and joint reinforcements. These factors are to be provided failing which the block masonry is bound to result in random cracking due to shrinkage and moisture movements. These factors are discussed below.

1. Control Joints

Concrete has higher coefficient of expansion than brickwork. In order to control thermal movements, these joints are built as continuous and vertical. Further, to minimise cutting of units, they are to be located at change of sections. These joints may be shrinkage joints and expansion joints. The shrinkage joints are continuous vertical joints approximately 18 mm thick. These joints are later raked and caulked. Expansion joints are located at spacing of 45–60 m in suitable locations. These joints are filled with joint filers, which are compressible materials that close the gap but allow movement.

2. Bond Beams

Bond beams are similar to lintels which are to be provided on top of openings or top of walls such as compound walls. This is constructed using special channel-shaped units which are reinforced and filled with concrete. It serves as a structural element as that of a lintel to control cracks on top of openings.

3. Joint Reinforcement

For crack control, horizontal joint reinforcements in the form of minimum two No. 9 gauge wires at intervals can be used. As they are expensive, they are not generally provided. When used as load-bearing walls with RCC slabs resting on the walls, the top three courses at least should be provided with joint reinforcement (Fig. 16.12).

16.4.6 Compound Walls in Hollow Block

Compound walls are nowadays preferred to be of unplastered concrete blockwork. The construction is cheap and no maintenance is needed. There are many variations in the

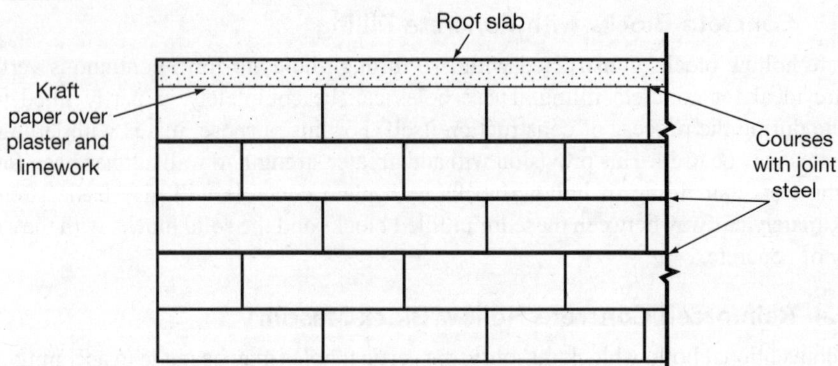

Figure 16.12 Load-bearing wall

construction of these walls. The convenient method is to install under-reamed or cast-in-situ piles at stipulated intervals (of 3–4 m) with grade beam on top of the piles and the block-work built on top of the grade beams.

Vertical reinforcements are provided in the hollow blocks at regular intervals from the piles. This provision strengthens the wall along its length giving lateral resistance to wing loads or earth pressure. The top of these walls are finished with a coping cast in a channel-shaped block on top (Varghese, 2012).

16.5 REINFORCED BRICK MASONRY

Ordinary brick masonry can not take tensile stresses as the mortar joints separate out and the bricks pulled out. In order to increase the load-carrying capacity, in particular tensile stresses, of ordinary brick masonry steel reinforcement is introduced in between mortar joints. This type of construction can withstand tensile and shear stresses for moderate loading.

Reinforced brick masonry is recommended in the following cases:

(i) When excessive compressive loads are to be supported
(ii) When the brickwork has to withstand tensile and shear stresses
(iii) When the supporting soil may be subjected to differential settlement
(iv) When the brick masonry is to be provided over openings of doors, windows and passages
(v) When the brick masonry has to sustain lateral stress
(vi) When it is intended to strengthen the longitudinal bond
(vii) When constructions in earthquake-prone areas are to be made

The following points should be kept in mind:

(i) All the bricks should be structurally sound and of uniform size.
(ii) In order to bond the brickwork, a cement mortar of 1: 3 should be used.
(iii) The joints have to be constructed carefully such that the reinforcement gets sufficient cover and is also prevented from corrosion.

Typical reinforced brick masonry are discussed below.

16.5.1 Reinforced Brick Masonry Walls

Here the reinforcement may be iron base or expanded metal mash provided at every third or fourth course. Flat bars of section about 25 mm × 2 mm may be used as hoop iron reinforcement for walls. They are hooked at corners and junctions. In order to increase the resistance against rusting, the bars are dipped in tar and sanded immediately. Reinforcement in vertical direction is provided by using special bricks or blocks. Mild steel bars of 6 mm diameter may also be used as longitudinal reinforcement in walls. A detailed discussion is made in Chapter 7.

16.5.2 Reinforced Brick Masonry Columns

These columns are made with special-purpose bricks and vertical bars running through them. In order to keep the bars in position and to strengthen the brickwork, steel plates of 6 mm thick are introduced. Bent bars of small diameters of about 12 mm can be used in the horizontal joints (Fig. 16.13).

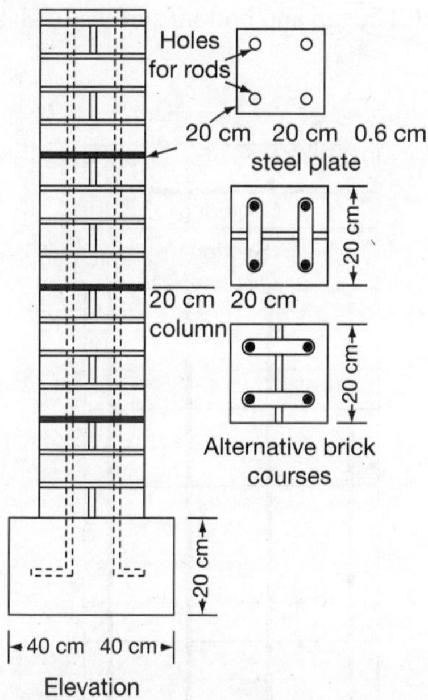

Figure 16.13 Reinforcement of brick masonry column

16.5.3 Reinforced Brick Masonry Lintel

In lintels, bars are provided longitudinally in between the vertical joints and extending from joint to joint. Further, to resist the vertical shear, additional 6 mm diameter steel stirrups may be used (Fig. 16.14).

16.5.4 Reinforced Brick Masonry Slab

Floor slabs of brickwork with reinforcement is also done. The conventional centring is made. The centring is covered with earth for a depth of about 20–25 mm and well compacted and sand sprinkled on the top.

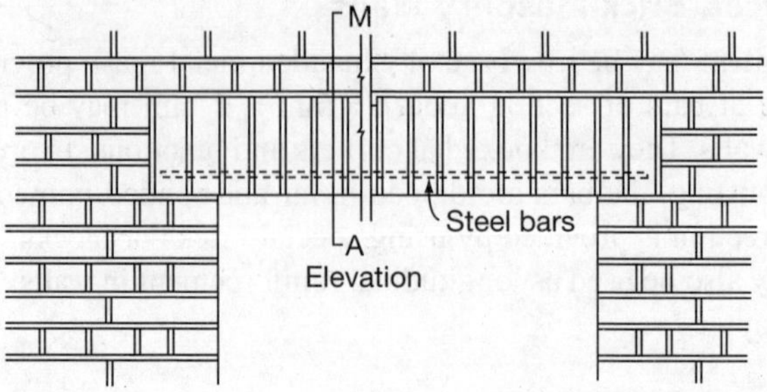

Figure 16.14 Reinforcement in brick masonry lintels

Then the reinforcement is placed in the correct direction. The bricks are laid in one or two directions. The joints are then filled with mortar such that the reinforcement is fully closed. The work is kept moist for a period of two days and then kept fully wet for 14 days. The centring is then removed. The top and bottom surface of slab are then suitably finished (Fig. 16.15).

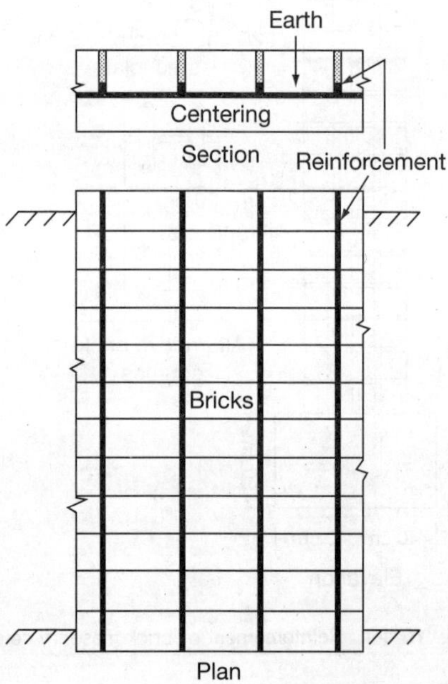

Figure 16.15 Slab with reinforced brickwork

SALIENT POINTS

1. Masonry may be defined as construction of building units bonded together with mortar.
2. The selection of type of unit for the masonry is made keeping in view the requirements of strength, water proofing, thermal insulation, fire resistance, durability and economy.

3. The mortar is required to keep the building units in position. The binding strength of the mortar is usually disregarded as far as the strength of the masonry is concerned.
4. Main types of masonry are brick masonry, stone masonry, concrete hollow block masonry and reinforced brick masonry.
5. Bond is the process of arranging bricks in courses to ensure that vertical joints do not come one over the other.
6. In stretcher bond all courses are laid as stretchers. As only stretchers are visible in elevation, this bond is referred to as stretcher bond.
7. In header bond, all courses are laid as headers. As only headers are visible in elevation, this bond is referred to as header bond.
8. English bond is produced by laying alternate courses of stretchers and headers, so as to break the joints vertically.
9. Flemish bond is produced by laying alternate stretchers and headers in each course. The headers and stretchers appear in the same course alternately on the front and the back faces.
10. Brickwork should be systematically done keeping in view the bonding, jointing and finishing.
11. Stones also can be systematically arranged to form a homogeneous mass called stone masonry.
12. Dressing is the art of cutting the stones to the shape required for use in structures.
13. Stone masonry is classified based on the thickness of joints, continuity of courses and finish of face. The two major types are Rubble masonry and Ashlar masonry.
14. Rubble masonry is a stonework wherein blocks of stones are either undressed or roughly dressed and have wider joints.
15. Ashlar masonry is a stonework wherein blocks of stones are accurately dressed with very fine joints of 3 mm thickness.
16. Cement concrete hollow blocks have been in use for several masonry constructions. Several advantages have given room for rapid development and use of the same in place of traditional construction materials like stones and bricks.
17. In order to increase the load-carrying capacity, in particular tensile stresses, of ordinary brick masonry steel reinforcement is introduced in between mortar joints. Such a structure is called reinforced brick masonry.
18. In reinforced brick masonry walls, iron bars or expanded metal mesh are provided at every third or fourth course.
19. Reinforced brick masonry columns are made with special-purpose bricks and vertical bars running through them.
20. In reinforced brick masonry lintel, bars are provided longitudinally in between the vertical joints and extending from joint to joint.

REVIEW QUESTIONS

1. What is meant by efflorescence in brick work? List the sources which cause efflorescence.
2. How does English bond differ from Flemish bond? Explain with a neat diagram.
3. Draw neat sketches, in plans of successive courses and elevation, to illustrate the construction of a brick wall, 2½ brick thick in English bond at a right-angled corner. (AMIE)
4. Discuss the general principles to be observed during the construction of brick masonry work.
5. What is the difference between English bond and Flemish bond in brick masonry construction? (AMIE)
6. Explain with neat sketches the various types of joints used in stone masonry for placing stones together.
7. Enumerate the various classes of rubble masonry.
8. What are the common defects which are to be checked in a stone before it is used in a building?
9. What do you understand by dressing of stones? Describe various methods of dressing.
10. Explain the usual types of finishes used in stone masonry.
11. What are the various types of Ashlar masonry? Explain any one.
12. What are the factors on which a particular type of finish is selected in stone masonry work?
13. Explain the differences between brick masonry and stone masonry with reference to their stability, cost and method of construction.
14. State the advantages of hollow block construction over the stone masonry.
15. Explain the advantages and disadvantages of concrete hollow block construction.
16. What type of bonding is used in hollow concrete block construction?
17. What are the precautions to be kept in view while constructing concrete block masonry.
18. Explain three types of construction possible with hollow concrete blocks.
19. What is reinforced brick masonry?
20. Explain the special advantages of Reinforced Brick Masonry.
21. Explain the reinforced brick pier with a neat sketch.
22. What is reinforced brick slab? Explain with neat sketches.

17 WALLS

17.1 INTRODUCTION

A wall is a vertical member with width greater than four times its thickness. If this condition is not satisfied it is a column. Based on functional aspects, walls may be classified under the following two broad groups:

(i) Load-bearing walls
(ii) Non-load-bearing walls

The masonry types explained in the previous chapter may be used for load-bearing structures and as separators or partitioners.

17.2 LOAD-BEARING WALLS

In general, masonry used for exterior walls, central main walls and some cross walls in a building are of load-bearing walls. These walls should have adequate thickness such that it will sustain a slight eccentricity in loading. The mortar used for such walls should give adequate bonding to hold the structure even when there is a slight buckling. This load-bearing wall, apart from supporting the loads, subdivides the space, provides thermal and acoustic insulation, and affords fire and weather protection.
Load-bearing walls are of the following five types:

(i) Solid wall
(ii) Solid wall with piers (Pilaster)
(iii) Cavity wall
(iv) Faced wall
(v) Veneered wall

17.2.1 Solid Wall

It is a wall built of solid bricks or perforated bricks and designed to carry an imposed dead loads and live loads, including its self-weight (Fig. 17.1).

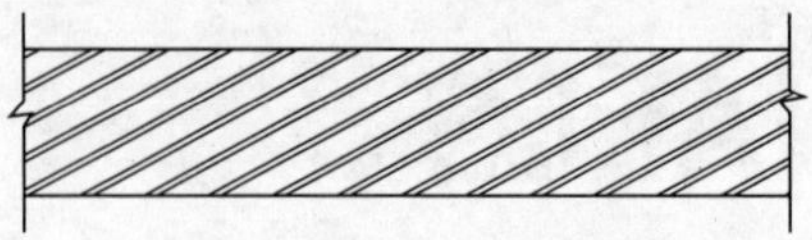

Figure 17.1 Solid wall

17.2.2 Solid Wall with Piers or Pilasters

These walls are similar to solid walls but the thickness of wall at intervals is increased (Fig. 17.2). The thickened portions are called as piers or pilasters. These walls are used for the following purposes:

(i) To carry concentrated loads from roof or floor beams
(ii) To provide lateral support
(iii) To reduce the slenderness ratio by stiffening the walls.

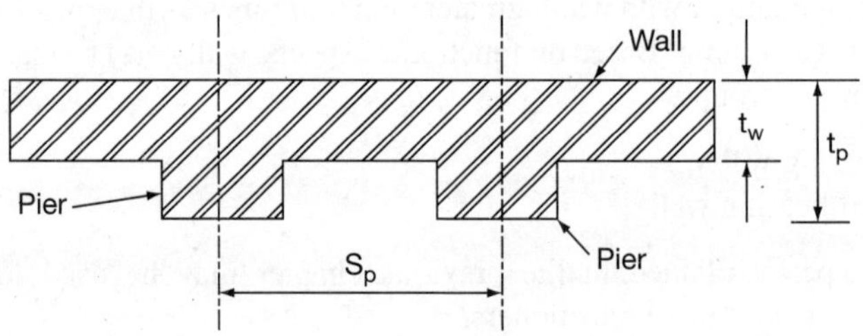

Figure 17.2 Solid wall with piers

17.2.3 Cavity Wall

Cavity wall consists of two structural leaves separated by an uniform continuous space called cavity. The two leaves are interconnected by metal ties (Fig. 17.3). The provision of cavity forms a barrier against penetration of dampness through the internal wall and also keeps the room cool because of air column in cavity. In such walls, the inner leaf will be of load-bearing and the outer wall carries its weight only. But bending moment is carried by the stiffness of both the leaves.

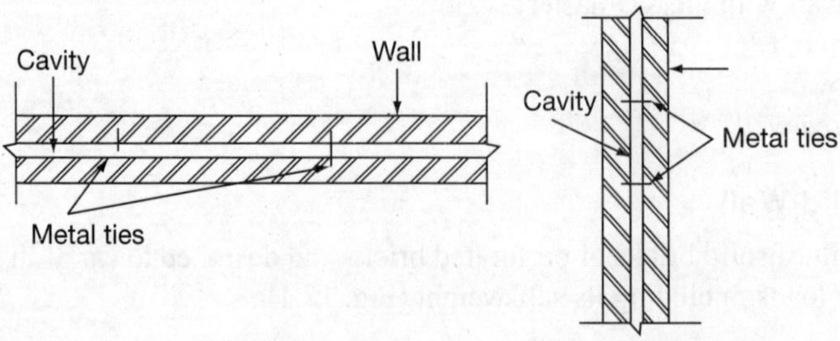

(a) Plan (b) Vertical section

Figure 17.3 Cavity wall

17.2.4 Faced Wall

It is similar to a solid wall but with a different facing material. These two materials are bonded together such that both take the load. The facing material may be of a different quality such that it may give a better aesthetic view. Generally dressed stone facing is done on brick walls (Fig. 17.4).

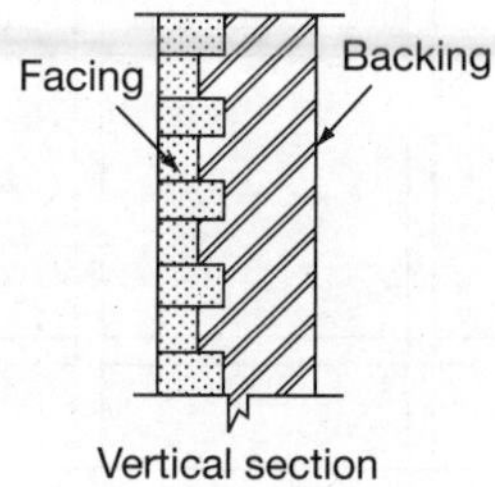

Figure 17.4 Faced wall

17.2.5 Veneered Wall

In veneered wall the facing is attached to the backing but need not be bonded. The entire load is taken by the backing. The facing is mostly for decoration purposes or to meet aesthetic needs (Fig. 17.5).

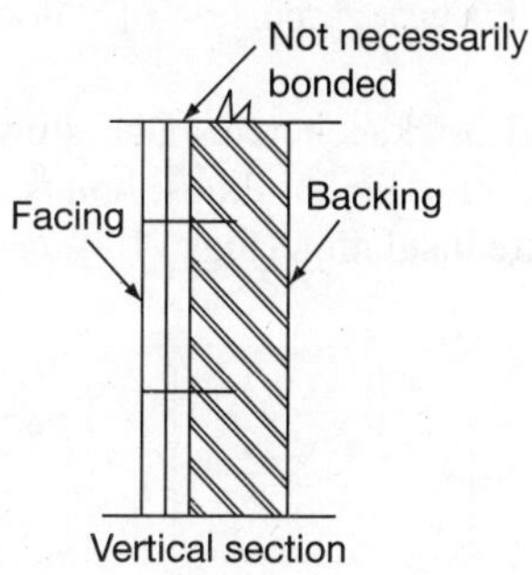

Figure 17.5 Veneered wall

17.3 NON-LOAD BEARING WALLS

In general, non-load bearing walls have adequate strength, stability, sound insulation and fire resistance. Non-load bearing walls may be interior walls or exterior walls. Different types of materials are used for non-load bearing walls, viz., brick, wood, hollow block, metal lath, corrugated sheet, etc. Following are some of the important non-load bearing walls:

1. Panel wall
2. Partition wall
3. Curtain wall
4. Free-standing wall

17.3.1 Panel Wall

It is an exterior wall in a structural frame construction. It forms into a unit in each storey (Fig. 17.6).

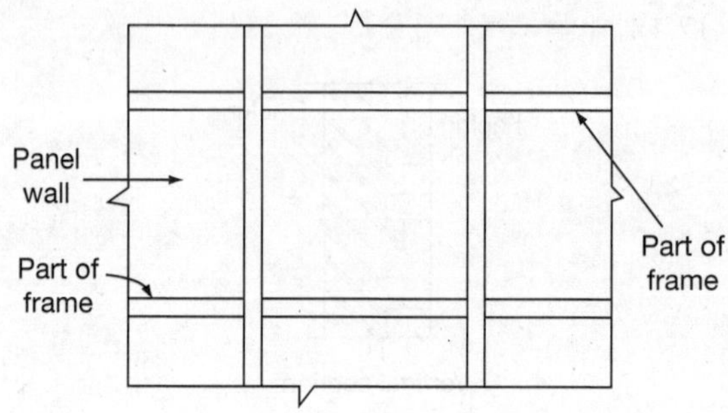

Figure 17.6 Panel wall

17.3.2 Partition Wall

Partition wall is an interior wall whose main function is to divide the space within a building to rooms and other areas of varied use.

Sometimes partition walls are required to support girders. In such a case, it is called a load-bearing partition wall. In this case, a portion of a floor is transferred to the partition wall through the girders.

In residential buildings, wood, brick, concrete and hollow block partitions are commonly used. Glass partitions are used in public buildings, hotels, recreation centres, etc. Hollow block partition provides adequate insulation (Fig. 17.7).

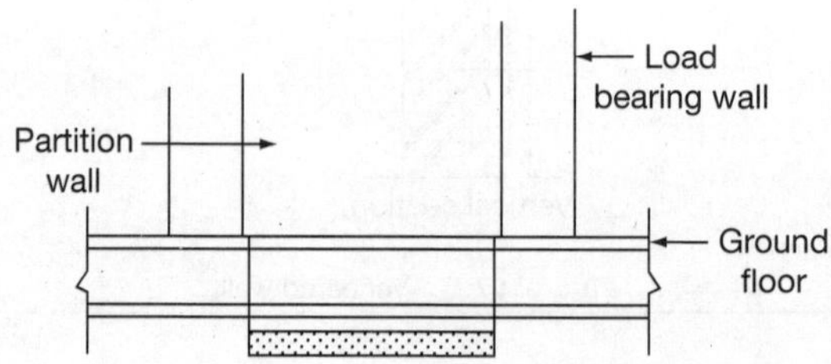

Figure 17.7 Partition wall

17.3.3 Curtain Wall

It is a wall carrying its self-weight but subjected to lateral loads. But it may be laterally supported by horizontal structural members wherever necessary (Fig. 17.8).

17.3.4 Free-Standing Wall

Parapet wall, compound wall, shear wall, buttress wall, counter fort wall are the examples of free-standing walls. These walls are expected to carry their self-weight and horizontal force due to wind and while supporting other structures.

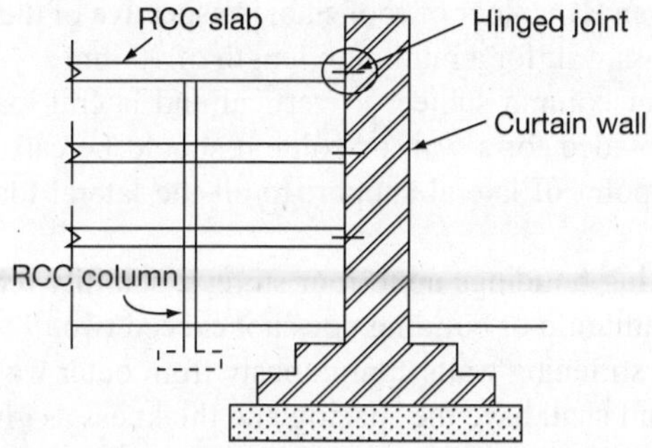

Figure 17.8 Curtain wall

17.4 LOADS ON WALLS

Loads on walls may be classified based on the load–wall reaction. The load–wall interaction may be divided into the following two major groups:

(i) Vertical
(ii) Lateral or Transverse

17.4.1 Vertical Loads

Vertical loads may be uniformly distributed load or concentrated load. Loads acting parallel and along the axis of wall cause axial stress. Loads may act eccentrically. In such cases, these loads will cause bending stress in addition to the axial stress. Thus, a uniformly distributed load or concentrated load may act axially or eccentrically. The design of such structural elements is different from that normally used in the case of walls with lateral or transverse loads.

17.4.2 Lateral Loads

Lateral or transverse loads may act parallel to the face of the wall or on the surface-perpendicularly. This type of loads may be concentrated, uniformly distributed or triangularly distributed. The lateral or transverse loads cause bending stresses in addition to in plane or transverse shear force.

17.5 LATERAL SUPPORTS AND STABILITY

Masonry structures gain stability from support offered by cross walls, floors and roof. Load-bearing walls are structurally sound as long as the load is applied axially without any eccentricity.

Lateral support for load-bearing walls or columns limit the slenderness of the structure. Further the lateral supports reduce the possibility of buckling of member due to vertical loads and to resist horizontal forces. Thus in total the lateral support ensures stability against sliding and overturning.

It is mandatory that an RCC floor or roof slab, irrespective of the direction of space, has to bear on a wall or cross wall for a minimum length of 90 mm.

Stability of a wall or column subject to vertical and lateral loads should be ensured. The lateral support provided for a wall or column should be capable of resisting simple static reactions at the point of lateral support to all the lateral loads, plus 2.5% of total vertical load.

In case of load-bearing buildings up to four storeys, stability requirements are ensured when the height-to-width ratio of building does not exceed two.

Cross walls used as stiffening walls continuously from outer wall to outer wall or outer wall to load-bearing wall shall have the spacing and thickness as given in Table 17.1.

Table 17.1 Thickness and spacing of stiffness walls

Sl. No.	**Thickness of load-bearing wall to be stiffened (mm)**	**Height of storey not to exceed (m) (centre to centre dimension)**	**Stiffening wall (centre to centre dimension)**		
			Thickness		**spacing (m)**
			1–3 storeys (mm)	**4–6 storeys (mm)**	
1.	Above 300	5.0	$\nless$100	$\nless$200	8.0 (max)
2.	300	3.4	$\nless$100	$\nless$200	8.0 (max)
3.	200	3.2	$\nless$100	$\nless$200	6.0 (max)
4.	100	3.2	$\nless$100	–	4.5 (max)

Halls exceeding 8 m span have to be adequately laterally supported.

For basement walls the following stability requirements are needed:

1. Bricks should have a minimum crushing strength of 5 N/mm^2.
2. Mortar used in masonry should be of grade M1 or better.
3. Clear height of ceiling in basement should not exceed 2.6 m.
4. Adequate cross walls.
5. Thickness of basement should be 300–400 m for spans up to 1.75 m and 2.5 m, respectively.

17.6 EFFECTIVE HEIGHT OF WALLS

If both lateral and rotational restraints are offered by a support, then the wall is said to be fully restrained at the support. It is said to be partial, if only lateral restraint is provided. Combination of these two restraint cases yields different boundary conditions depending on location. The effective height of a wall is based on the boundary conditions. Table 17.2 presents the condition of supports and the corresponding effective heights and H is the distance between the supports.

Figure 17.9 shows the effective height of standing in different situations.

Table 17.2 Effective height of walls

Sl. No.	Type of restraint		Effective height
	Top	Bottom	
1.	Full	Full	0.75 H
2.	Full	Partial	0.85 H
3.	Partial	Full	0.85 H
4.	Partial	Partial	1.00 H
5.	No	Full	1.50 H
6.	No	Partial	2.00 H

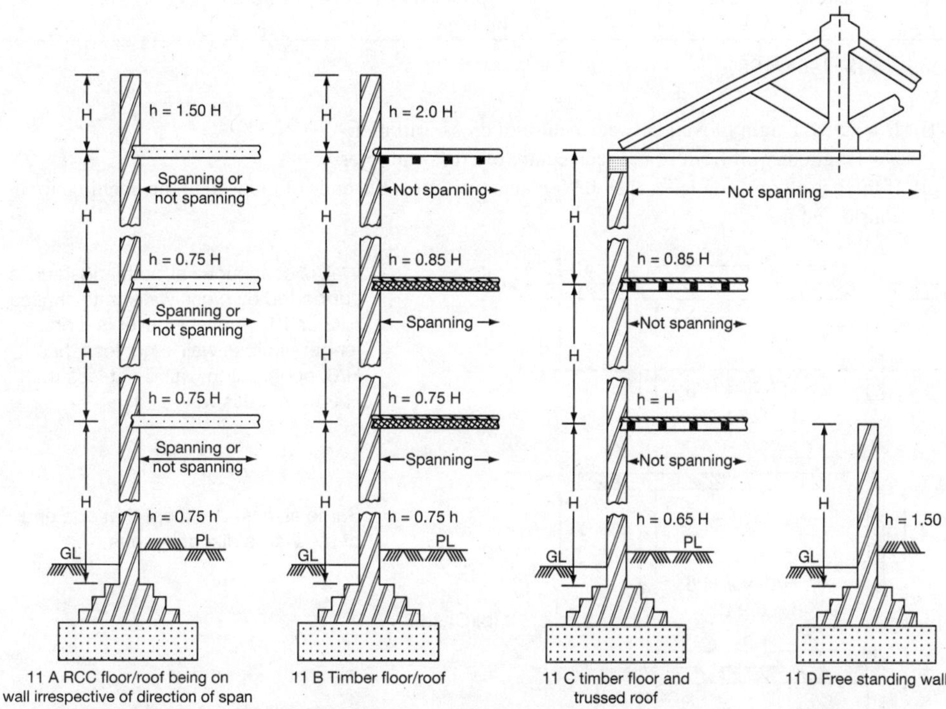

Figure 17.9 Effective height of walls

17.7 EFFECTIVE LENGTH OF WALLS

While deciding the length of walls, the following end support conditions are considered:

1. Free end of the wall
2. Continuity of the wall
3. Support from cross walls or piers or buttresses
4. Openings

Various combinations of the above conditions and the effective length of a wall are presented in Table 17.3 (Fig. 17.10).

Table 17.3 Effective length of walls

Sl. No.	Condition of support		Effective length
1.	End 1	End 2	
2.	Continuous wall supported by cross wall/piers buttress		0.8 L
3.	Supported by a cross wall/pier/ buttress	continuous with cross wall/pier/ buttress supporting	0.9 L
4.	supported by cross wall/piers buttress	supported by cross wall/piers buttress	1.0 L
5.	Free	continuous with across wall/pier/ buttress	1.5 L
6	Free	supported by a cross wall/pier/ buttress	2.0 L

Source: IS: 1905, 1987.

Note:

(i) H = actual height of wall between centre of cross wall/pier.
L = length of wall from or between centre of cross wall/pier.

(ii) If there is an opening taller than 0.75 H in a wall, then the ends of the wall at the opening are considered free.

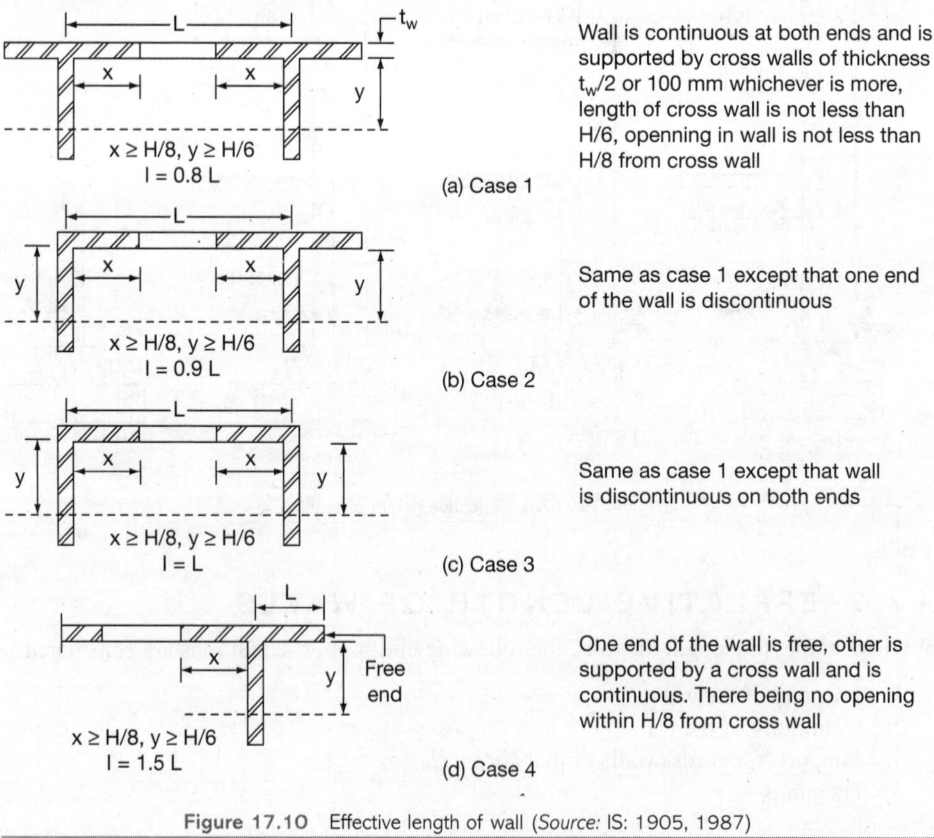

Figure 17.10 Effective length of wall (*Source:* IS: 1905, 1987)

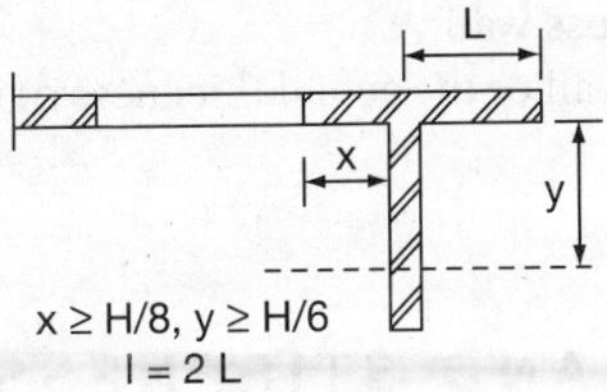

Same as case 4 but opening is within H/8 from cross wall and thus that end is taken as discontinuous

(e) Case 5

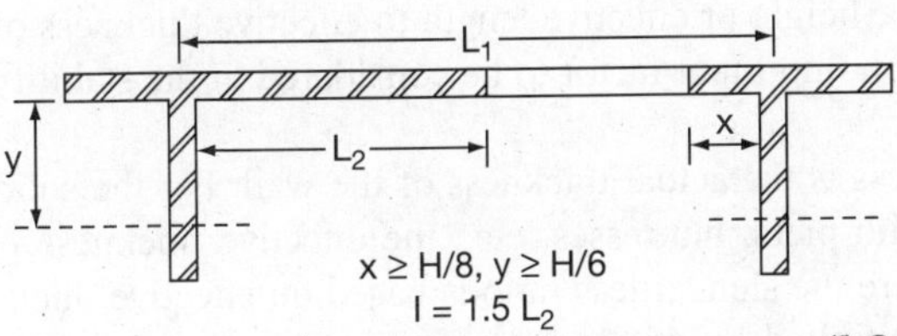

This illustration is with an opening which is within H/8 from cross wall

(f) Case 6

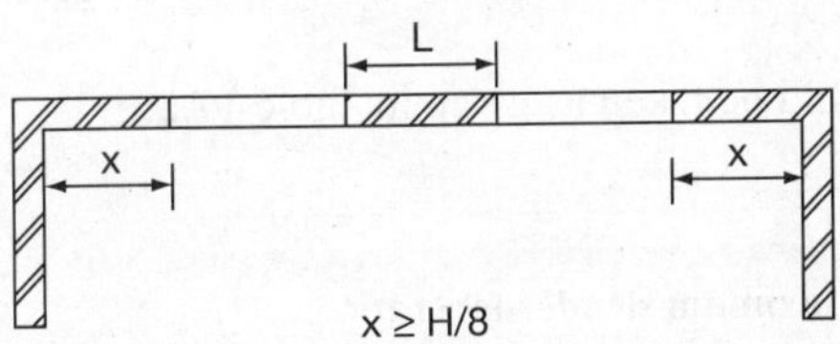

Wall length is between two openings which are closer than H/8 from cross walls

(g) Case 7

Figure 17.10 *(Continued)*

17.8 EFFECTIVE THICKNESS OF WALLS

Effective thickness of a wall is an idealised thickness which reflects the behaviour of the wall. Effective thickness is determined as detailed below.

1. **Solid Walls and Faced Walls:** Effective thickness is the same as actual thickness.
2. **Cavity Walls with Uniform Leaves:** Effective thickness is two-thirds of the sum of actual thickness of both the walls.
3. **Solid or Faced Walls Stiffened by Piers or Cross Walls:** Effective thickness is obtained by multiplying the actual thickness by a stiffening coefficient as given in Table 17.4.

Table 17.4 Stiffness coefficient

Sl. No.	S_p/w_p	Stiffness coefficient		
		$t_p/t_w = 1$	$t_p/t_w = 2$	$t_p/t_w = 3$ or more
1.	20 or more	1.0	1.0	1.0
2.	15	1.0	1.1	1.2
3.	10	1.0	1.2	1.4
4.	8	1.0	1.3	1.7
5.	6	1.0	1.4	2.0

where S_p = Centre to centre spacing of pier or cross wall
w_p = Width of pier in the direction of the wall or the actual thickness of cross wall
t_p = Thickness of pier
t_w = Thickness of wall proper

17.9 SLENDERNESS RATIO AND STIFFNESS

Slenderness ratio is the ratio of effective height or effective length to effective thickness of the masonry unit. Slenderness ratio is an important factor to be considered in the stability of a wall.

For solid walls, the effective thickness is the actual thickness of the wall. For the solid walls which are adequately bonded with piers, buttresses, etc., the effective thickness is determined using slenderness ratio. Here the slenderness ratio is based on effective thickness which is the actual thickness multiplied by stiffness coefficient values as given in Table 17.5.

Maximum slenderness ratio for walls should be taken as given in Table 17.5.

Table 17.5 Maximum slenderness ratio

Number of storeys	Maximum slenderness ratio	
	Cement mortar use	Lime mortar use
Not exceeding 2	27	20
Exceeding 2	27	13

17.10 REINFORCED BRICK WALLS

Ordinary masonry walls are reinforced with iron bars or expanded metal mesh and such walls are called reinforced brick walls. Here, the reinforcement, iron bars or expanded metal mesh are provided at every third or fourth course (Fig. 17.11).

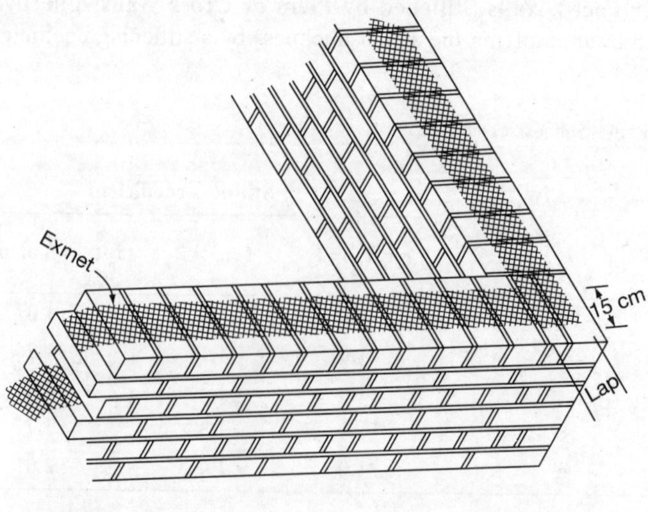

Figure 17.11 Reinforcement of brick wall with metal mesh

Alternately flat bars of sections about 25 mm × 15 mm may be used as loop iron reinforcement for walls (Fig. 17.12). They are hooked at corners and junctions. In order to increase the resistance against rusting, the bars are dipped in tar and sanded immediately.

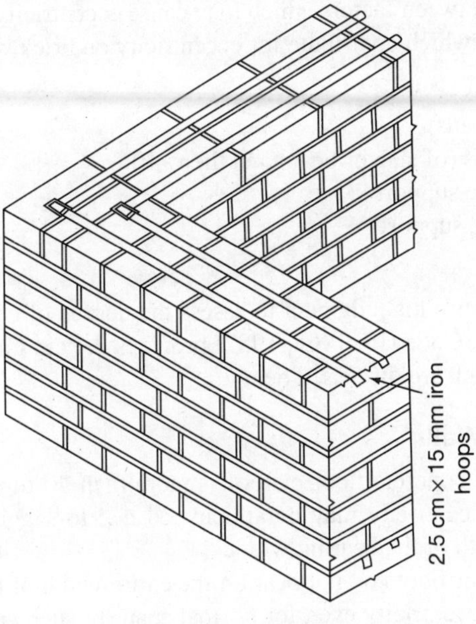

Figure 17.12 Reinforcement of brick wall with loop iron

Reinforcement in vertical direction is provided by using special bricks or blocks as shown in Fig. 17.13. Mild steel bars of 6 mm diameter may also be used as longitudinal reinforcement in walls.

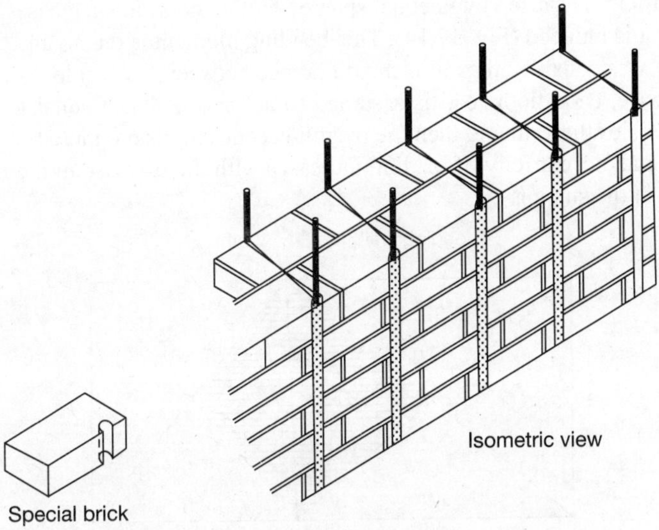

Figure 17.13 Reinforcement of wall by vertical rods

17.11 ECCENTRICALLY LOADED BRICK WALL

In general, walls in buildings are commonly loaded with some eccentricity. Eccentricity may be caused due to one reason or another. Thus, there is a little possibility of establishing an exact relationship between factors which may cause eccentricity.

Some of the factors which contribute for eccentricity on brick walls are:

(i) Long floor edges
(ii) Magnitude of loads
(iii) Relative stiffness (of slab or beam and the wall)
(iv) Flexibility of the support
(v) Geometry of the support
(vi) Unequal spans

Thus a designer has to use his judgment to assess the degree of eccentricity based on the situation. However, I.S. Code (IS: 1905, 1987) provides certain guidelines for determination of eccentricity which are discussed below.

17.11.1 Exterior Walls

(i) When a span of concrete floor or roof is more than 30 times the thickness of the wall, then all eccentricity may be anticipated due to sagging. The eccentricity is given as one-sixth of the bearing width.
(ii) When the roofs or floors do not bear on the entire width of the wall, then there is a possibility for eccentricity even for normal span. In such cases, the eccentricity is taken equal to $1/12^{th}$ the thickness of the wall.
(iii) For timber and other light weight floors, eccentricity is assumed one-sixth the thickness even for full-width bearings.

17.11.2 Interior Walls

(i) Eccentricity is caused by unequal span of roof or floor. In such cases a net bending moment is induced (Fig. 17.14). This bending moment is due to an eccentric load. The load is considered axial if the difference between the two loads is within 15%. Otherwise, each floor load is assumed to act at a distance equal to one-sixth the thickness of the wall and then the overall eccentricity is computed.
(ii) In general, eccentricity of loading increases with the increase in the fixity of slabs/ beams at the supports.

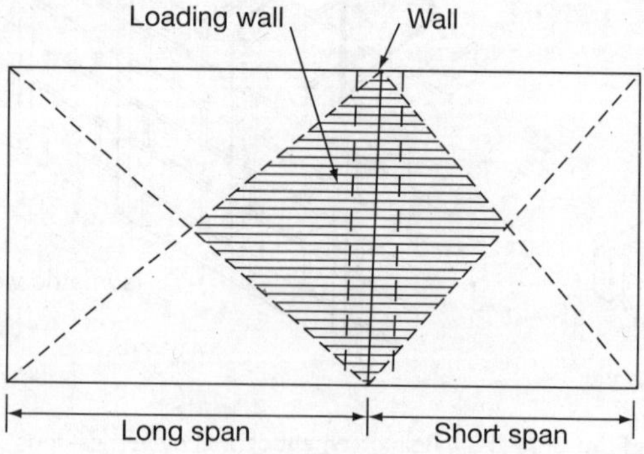

Figure 17.14 Eccentricity due to unequal span in interior wall

17.11.3 Stress Distribution Under Eccentric Loads

In an eccentrically loaded wall, there is an axial load and a bending moment. These two may be combined into a single resultant load acting at a distance. This is known as equivalent eccentricity (Fig. 17.15).

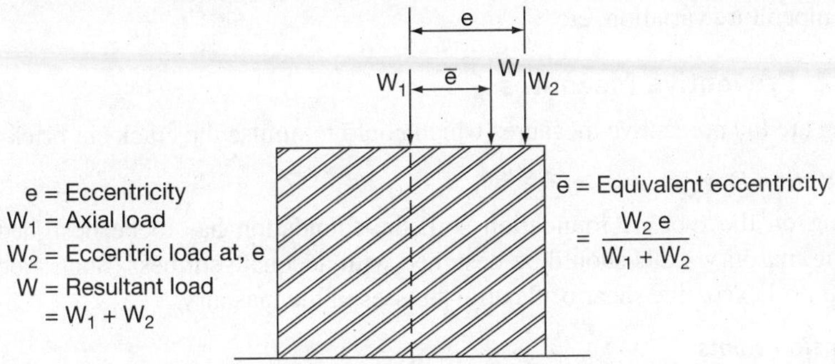

Figure 17.15 Equivalent eccentricity

The stress distribution due to axial load and the bending moment are combined to get the stress distribution due to the resultant load. The stress distributions for various eccentricities are shown in Fig. 17.16.

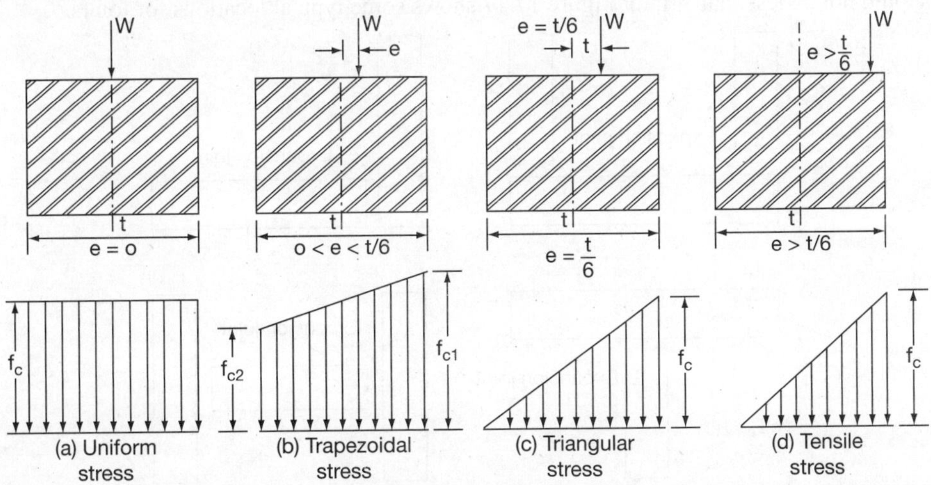

Figure 17.16 Variation of stress distribution (*Source:* IS: 1905, 1987)

It can be observed that with an increase in eccentricity, the net compressive stress in the tension face decreases. That is, the tensile stress due to bending moment decreases.

17.12 CRACKING IN WALLS

17.12.1 Causes

Cracks are frequently found in brick masonry walls due to some of the reasons given below.

(i) Brick masonry behave differently when constructed in conjunction with concrete foundations and concrete framing.

(ii) Combination of brick masonry with other members having greater deflections and strains.
(iii) Effect of deflection and shrinkage of concrete slabs resting on walls.
(iv) Due to introduction of new types of construction.
(v) Restraint of stresses developed inside the brick masonry due to moisture absorption, temperature variation, etc.

17.12.2 Preventive Measures

Following are the preventive measures which could minimise the cracks in brick masonry.

1. Foundation Design

Depending on the type of foundation soil, the foundation has to be designed and the supporting masonry walls should be designed with adequate stiffness. Such a design will help to control excessive shear of flexural stresses in the masonry.

2. Expansion Joints

Providing horizontal and vertical expansion joints in walls helps to reduce the cracks to a considerable extent. Horizontal and vertical expansion joints absorb vertical and horizontal movement respectively. In general, expansion joints have to be provided for every 15 m. The sealant used for joints are natural or cellular rubber, bitumen, expanded plastics, coconut pith, etc. The depth of sealant should not be more than half the width of joint and should not be less than 4 mm. Figure 17.17 shows some typical locations for joints.

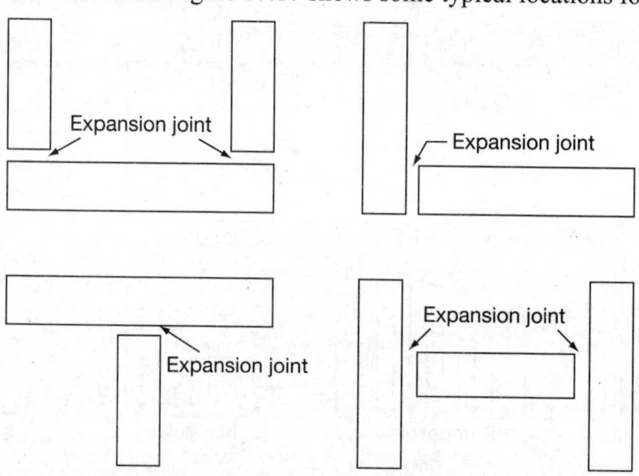

Figure 17.17 Plan of locations of expansion joint

Typical expansion joints in brick masonry provided at different locations to avoid cracks are shown in Fig. 17.18 to 17.20.

3. Isolation Joint

Isolation joint is similar to expansion joint but provided under the following conditions:

(i) When it is desired to separate the foundation of machines from the rest of the structure.
(ii) When one portion of a building is higher than the other.
(iii) When one portion of a building rests on rock and the adjacent portion on compressible clayey soil.

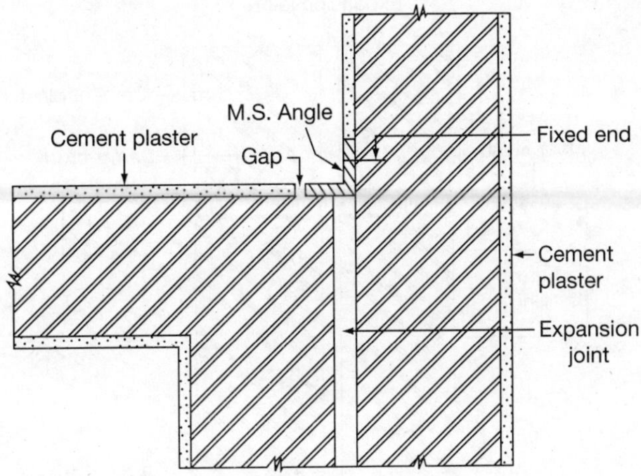

Figure 17.18 Expansion at corner of walls

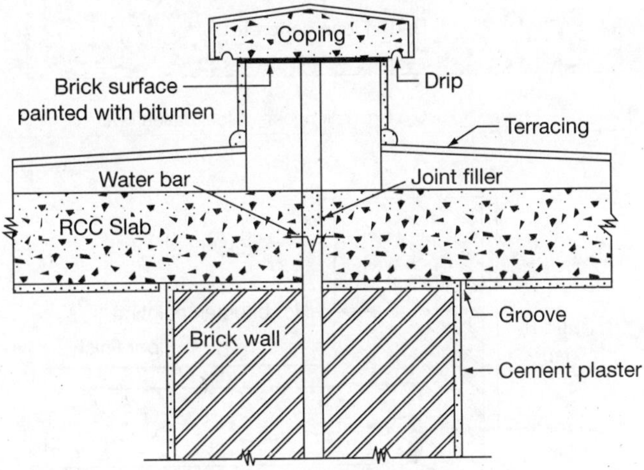

Figure 17.19 Expansion joint at roof level

4. Sliding Joint

Sliding joint is provided when one part of a structure has a tendency to slide over the other due to variations in temperature and moisture content. Figure 17.21 shows the details of a sliding joint at floor level.

5. Slip Planes

Smooth slip planes are provided between the roof slabs and brick walls. Because of this arrangement, cracks will be developed only at the re-entrant corners. These cracks can be easily covered after the complete construction is over.

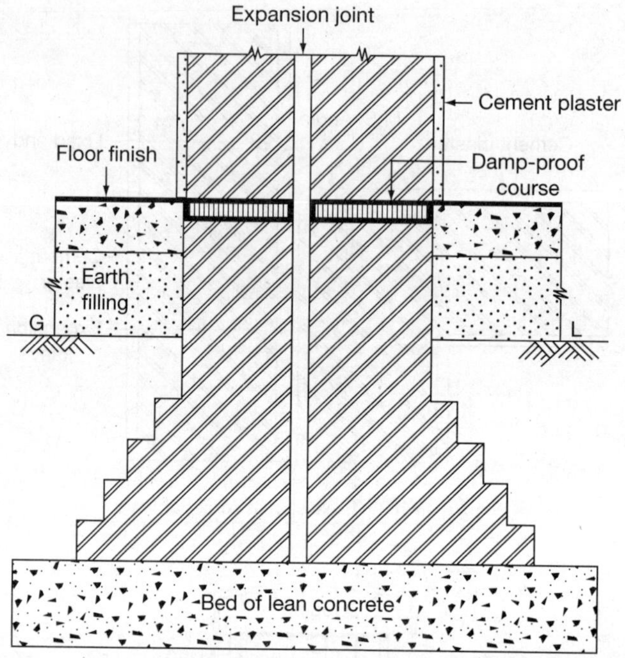

Figure 17.20 Expansion joint at foundation level

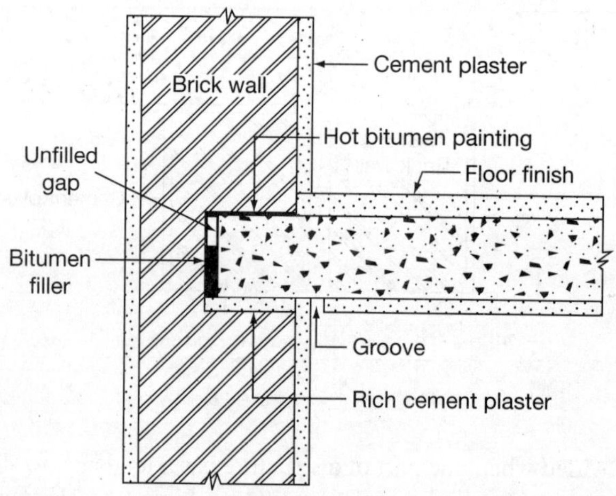

Figure 17.21 Sliding joint

6. *Spans*

It is recommended to provide short spans for the floor slabs.

7. *Quality of Concrete*

For floor and roof slabs, it is desirable to use concrete of low shrinkage characteristics.

SALIENT POINTS

1. A wall is a vertical member with width greater than four times its thickness.
2. Walls are classified as load-bearing walls or non-load bearing walls.
3. Solid wall is a load-bearing wall which is built of solid bricks or perforated bricks and designed to carry an imposed dead load and live loads including its self-weight.
4. Solid wall with piers or pilasters is a load-bearing wall with thickness of wall being increased at intervals. The thickened portion is called a pier or pilaster.
5. Cavity wall is a load-bearing wall consists of two structural leaves separated by a uniform continuous space called cavity. The two-leaves are connected by metal ties.
6. Faced wall is a load-bearing wall similar to solid wall with a different facing material. These two materials are bonded together such that both take the load.
7. Veneered wall is a load-bearing wall in which the facing is attached to the backing but need not be bonded. The entire load is taken by the backing.
8. Panel wall is a non-load bearing wall and an exterior wall in a structural frame construction. It forms into a unit in each storey.
9. Partition is a non-load bearing and an interior wall whose main function is to divide the space within a building to rooms and other areas of varied use.
10. Curtain wall is a non-load bearing wall which carries its self-weight but subjected to lateral loads. It may be laterally supported by horizontal structural members wherever necessary.
11. Free-standing wall is one which carries its own weight and also the horizontal force due to wind. This is a non-load bearing wall.
12. Walls are subjected to vertical and horizontal forces. Vertical loads may be uniformly distributed load or concentrated load. Lateral or transverse loads may act on the face of the wall in any direction. This type of loads may be concentrated, uniformly distributed or triangularly distributed.
13. Masonry structures gain stability from supports offered by cross walls, floors and roof.
14. Load-bearing walls are structurally sound as long as the load is applied axially without any eccentricity.
15. Lateral support for load-bearing walls limit the slenderness of the structure.
16. If both lateral and rotational restraints are offered by a support, then the wall is said to be fully restrained at the support. It is said to be partial, if only lateral restraint is provided.
17. Length of a wall is decided based on the following conditions: (i) free end of the wall, (ii) continuity of the wall, (iii) support from cross walls or piers or buttresses and (iv) opening.
18. Effective thickness of a wall is an idealised thickness which reflects the behaviour of the wall.
19. Slenderness ratio is the ratio of effective height or effective length to the effective thickness of the masonry unit.

20. Ordinary masonry walls are reinforced with iron bars or expanded metal mesh and such walls are called reinforced brick wall.
21. Factors contributing for the eccentricity in walls are: (i) long floor edges, (ii) magnitude of load, (iii) relative stiffness, (iv) flexibility of the support, (v) geometry of the support and (vi) unequal span.
22. In an eccentrically loaded wall, there is an axial load and a bending moment. These two are combined into a single resultant load acting at a distance. This distance is known as equivalent eccentricity.
23. Cracks in walls may be minimised by adopting the following preventive measures: (i) foundation design (ii) providing expansion joints, isolation joints, sliding joint, slip planes, (iii) providing short spans and (iv) quality control of concrete.

REVIEW QUESTIONS

1. Distinguish between load-bearing and non-loading bearing wall.
2. What are the advantages and disadvantages of cavity wall construction?
3. Distinguish the difference between a bearing and non-bearing portion?
4. Briefly discuss the types of load-bearing walls.
5. What are partition walls? List the materials used for partition walls.
6. What are the requirements of partition walls?
7. What are the advantages of concrete partitions?
8. Explain different types of wooden partitions commonly used.
9. How later support helps in the stability of a wall?
10. How the effective height of a wall is decided?
11. Discuss the end support conditions in deciding the effective length of walls.
12. How the thickness of a wall is designed?
13. Why damp-proofing courses are provided?
14. What are the ill-effects of dampness in building?
15. How damp-proof surface treatment is done?
16. What are the causes for cracking in walls?
17. Discuss the preventive measures to be taken to prevent cracks in walls.

18 FRAMED STRUCTURES

18.1 INTRODUCTION

New technologies for vertical transportation and high socio-economic levels of urban growth have given room to tall buildings. Further, large-scale industrialisation and exhaustive land cost resulted in the expansion of building programme.

Framed structures are comprised of series of frames with horizontal and vertical components. These framed structures concept are effectively used in the construction of multi-storey buildings.

Although timber, steel and RCC can be used for framed structures, RCC plays the top-most role in space frame technology. The idea of space frame was originally developed for airplane hangers in 1940s.

Apart from construction of framed structures using the conventional construction techniques, the technique of pre-fabrication has also been used to economise the cost and time.

These tall buildings and pre-fabricated buildings which are discussed in the following sections with introduction of framed structures.

18.2 CONCEPT OF FRAMED STRUCTURES

18.2.1 Principles of Framed Structures

Framed structures are the structures which are formed by connecting a series of horizontal and vertical structural members in suitable positions. The framed structures primarily comprise of columns and beams. Columns are connected by beams at floor and roof levels. The floors are divided into rooms and passages of required sizes by walls.

The walls may be load-bearing walls or partition walls. The materials used for framed structures, in general, may be wood, steel or RCC. Light framed structures are constructed of wood, steel or RCC, whereas multi-storeyed framed structures are constructed of mild steel.

The entire load of floors, roofs and partition walls are supported by the beams which transmit these loads to the columns. These columns finally carry the whole weight of the structure to the foundation.

18.2.2 Materials Used for Framed Structures

As discussed earlier, the materials used for framed structures, viz., wood, metal and RCC, are discussed below.

1. Timber

Use of timber for framed structures offers several advantages like beauty, versatility, durability, low cost, high strength-to-weight ratio, excellent electrical insulation and high shock absorption capacity. Variety of range of finishes can be provided to satisfy decorative and protective purposes.

In wooden frames, the walls are conventionally built with slender studs spaced about 40 cm centre to centre. In a similar way, joints and rafters which are supported on the walls are also spaced 40 cm apart. Timber studs are usually placed in walls and partitions keeping the wide faces perpendicular to the face of the wall or partition. For attachment to studs, joints and rafters, facings and decking for the required sizes are available. Wood bearing wall constructions are generally either braced to balloon frame or platform frame.

2. Steel

In general, steel framework supports all loads. In this type, all columns, beams and girders in the framed structure are of steel section. Adequate bracing is needed for these structures in order to resist the wind and earthquake forces.

For these structures, fire-proof and other light materials are used for partitions and exterior walls if it is a building. Steel framed structures are capable of sustaining greater load in a given space. In steel framed structures, circular or curved work should be avoided.

3. RCC

Reinforced cement concrete (RCC) forms the best material for framed structures, provided they are properly designed and constructed. As discussed earlier, columns and beams are interconnected with each other so as to form a grid of the beams and girders. In order to carry various floor levels, the slabs are built monolithically. The foundations may be either isolated or combined footing or it may be a raft or mat depending on the soil and the load on the foundation.

18.2.3 Advantages of Framed Structures

Following are the advantages of framed structures:

1. If the framed structure consists of several floors, for example, in a building, it is possible to carry out different activities simultaneously. For example, finishing of lower floors can be done while the construction of framework of upper floors can be taken up.
2. Based on the requirement, the panel walls may be changed to a different position. This brings in greater freedom in planning.
3. In a building, this helps to divide the components into two distinct categories, viz., load-bearing and non-load bearing. This enables to construct the non-load bearing components using low cost materials.
4. The framed structure construction requires thin panels which in turn leads to the larger floor area. However, outer walls are off a type which shall offer heat and sound insulation and shall withstand weather conditions.

5. Framed structures can highly resist vibrations and ideal for seismic zones.
6. Very suitable in filled-up or soft ground compared to ordinary masonry constructions.

Tall buildings are generally of framed structures which are discussed in the next section.

18.3 TALL BUILDINGS

Large-scale industrialisation have resulted in great expansion of building programme. Prohibitive land cost in urban areas and demand to meet large population in urban areas have made way for construction of tall buildings called as multi-storeyed buildings. Thus, we have reached a stage now that multi-storeyed construction is essential and inevitable in urban areas. Buildings with more than five stories are called as multi-storey buildings. Most of the tall buildings in cities have five to twelve stories. But in metros like Kolkata, Delhi, Mumbai and Chennai, 20- to 25-storey buildings have started coming up.

18.3.1 Advantages of Tall Buildings

Following are the advantages of tall buildings:

1. Economy in use of less land for construction.
2. Gives room for large proportion of open space for creating natural environment.
3. Enables better day-lighting and greater flow of air.
4. Freedom from street noises.
5. Provides a pleasant panoramic view of the city.
6. Provides amenities like common car parking, recreation centres, park, swimming pool, etc.

18.3.2 Disadvantages of Tall Buildings

Following are the disadvantages of tall buildings:

1. Density of population is high in a small area.
2. Prevention of congestion is difficult.
3. Excessive and imbalanced load on municipal services like water supply, sewage, electricity, etc.
4. Difficult to prevent accidents due to fire, earthquake disasters, etc.
5. Poses a number of social and human problems.

18.3.3 Construction of Tall Buildings

Important aspects which are to be considered in the construction of RCC tall buildings are given below:

1. RCC frames are monolithic construction of columns, beams and slabs. Because of monolithic construction, deflection and bending moments are reduced which results in economical construction. Further adequate safety is ensured.
2. An RCC frame essentially consists of beams, columns and slabs (as floor or roof). In case of large spans for better distribution of load, secondary beams spanning across main beams can be introduced.
3. A typical frame of an RCC multi-storeyed building is shown in Fig. 18.1. It shows the monolithic construction of columns, beams, slabs and girders.

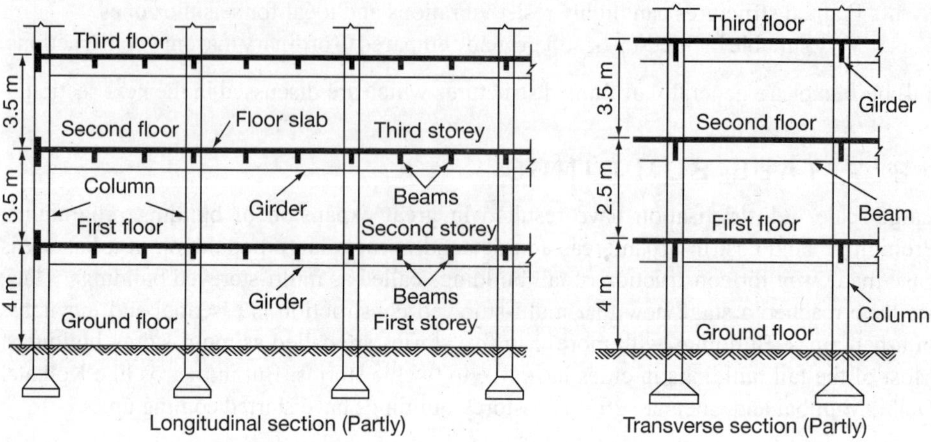

Figure 18.1 Details of RCC multi-storeyed building

4. The concreting procedure is the same as for other structures. However, a sequential procedure has to be followed. Here the formworks for different members to be cast are first installed or erected in position. Necessary reinforcement is then placed and concreted. The formwork is removed after the concrete has attained adequate strength.
5. The next member to be formed is taken up and formwork is fixed and concreting done. The general sequence is construction of columns followed by beams, cross beams and slabs together.
6. As the entire construction can not be done in one stretch, construction joints are required to be provided at intervals. Further, in framed construction the joints should be at the point of minimum shear.
7. The height of the columns is concreted so as to provide proper lap with the sides of the beams and columns in upper storeys.

18.4 PRE-FABRICATED BUILDINGS

A general discussion on different building units which could be pre-fabricated is made in Chapter 9. Pre-fabricated buildings are constructed based on the articulated structure concept.

Articulated structure means, the separation of a structure into two or more elements and join the entire structural elements such that it functions as a monolithic structure. The elements are pre-fabricated and are assembled and erected. This technique is applied to framed structures also.

18.4.1 Design and Manufacturing

While designing pre-fabricated buildings, manufacturing and effective installation techniques should be taken into account. That is, the design of structural parts, utilisation of structural parts and their joints should be installed with minimum use of materials and manpower for manufacture and erection.

In fully pre-fabricated construction, it is the practice to use larger elements while simultaneously reducing the relative mass. This is achieved by using more efficient design, light-weight concrete, synthetic heat insulation and other efficient materials.

At present, pre-fabricated concrete factories not only manufacture structural components but also assemble buildings from fabricated blocks and perform the whole complex of construction work. As pre-fabricated elements grow larger and taller, pre-fabrication makes possible speedier construction.

The erection should involve minimum consumption of labour, time and other means. Effectiveness in erection depends on efficient pre-assembly of structures relatively, equal weights of section unit, high degree of pre-fabrication and accuracy of manufacture, and simplicity of the butt-joints and provision of fastening devices.

18.4.2 Delivery and Storage of Pre-fabricated Structures

Structural units are delivered to erection zones by most effective mechanised procedures and allowing for haulage distance, availability of approach roads and conditions of in-site roads. Depending on the character of application of structural units, they are transported from the place of manufacturing to construction sites and unloaded at a pre-assembly area. Pre-assembly area is a site storage area or a zone identified in an erecting area.

Pre-fabricated structural units are generally transported by trucks with two-axle trailers, tractor trucks with semi-trailers and panel transporters. Delivered units to the construction site should correspond to assembly lists which specify the name, the type and the number of pre-fabricated elements intended for placement in a specified area of the building.

18.4.3 Pre-assembly of Pre-fabricated Concrete Structural Units

It is difficult to haul large-size buildings because of their bulkiness and large mass. Thus, these structural units have to be made as transportable components which could be assembled by erection units at construction site.

Reinforced cement concrete structural units are manufactured at areas within the radius of truck haulage. They are generally transported as complete units by means of special transport vehicles. Sometimes bulky reinforced concrete structures are delivered to construction sites as sub-components. These are then assembled into heavy columns, beams, arches, shells and other large constructions.

Reinforced concrete trusses are pre-assembled in a horizontal position on special racks with the use of jigs securing elements in precise positions or in the design position on special stands. The assembly of trusses from two half-trusses in the design position is carried out on special purpose universal stands.

Pre-assembly can also be done in the zone of erection at ground level with the use of movable stands or along the line of erection work. Structural units may be cast for assembling frame structure to form a multi-storey building. Erection of structural units to form multi-storeyed pre-fabricated building is discussed in the next section.

18.5 ERECTION OF PRE-FABRICATED MULTI-STOREY FRAME BUILDINGS

Multi-storey buildings may be built of standardised construction with column spaces based on some modular grid (say 6 m × 9 m or 6 m × 6 m). Based on dimensions, multi-storey buildings are erected by means of tower or derrick cranes placed on one or both sides of the frame (Fig. 18.2).

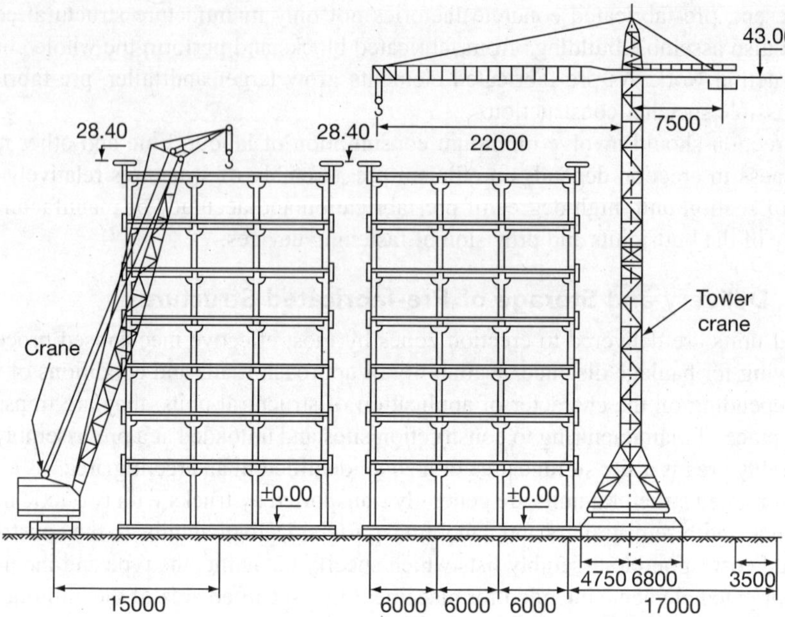

Figure 18.2 Diagram of locating crane for erecting high-rise building

When two cranes are put into use, they are to be arranged such that there is no dead area (i.e., areas not covered by the cranes). Further, another requirement is that the booms or loads they hoist should not interfere with one another. This is achieved by positioning the cranes such that they are spaced more than two radii of reach apart so as to provide a space sufficient for their safe operation. Sometimes the cranes operate in sequence with one lagging behind the other.

When cranes are placed outside the building, erection should be carried out one storey after another. Based on this procedure, the erection of a storey is started only after the completion of all the structures of the preceding storey have been completed. Further, the erection braces are placed so as to ensure longitudinal stability of the building. When cranes are located within the building cross-section, the vertical break between adjacent cells (units) of the framework should not exceed one-tier.

Columns at the ground floor are placed on heads of foundation columns or in foundation pockets. Columns of subsequent storey are mounted using group jigs which are intended for the erection of four or six columns (Fig. 18.3). A group jig consists of a box-type metal structure with collars for securing columns and a wooden working platform for the erectors. Further, a jig carries three collars for each column. The bottom collar is attached to the jig projecting caps of the underlying storey columns. The jig is aligned in the centre lines with the aid of a special frame. Further, it is secured to erection parts of intermediate floors of bracing, using screw jacks and the jig is levelled.

After the jig has been placed and secured to the caps of columns of all underlying storeys, all four new columns are installed, secured and trued by adjacent screws.

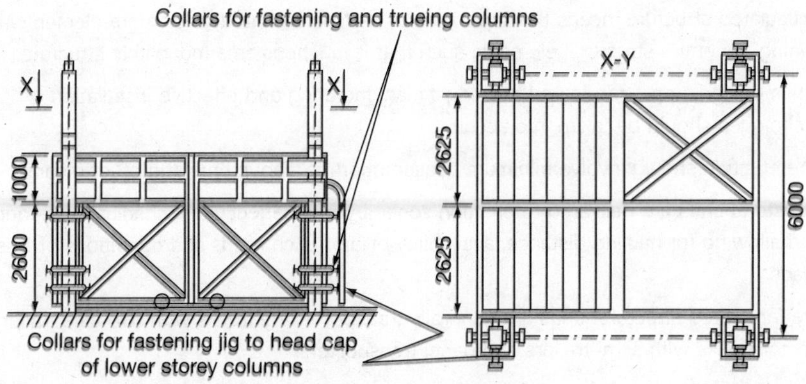

Figure 18.3 Single-tier jig for trueing four columns

Using a theodolite, the verticality of the columns are checked. The admissible deviation of longitudinal axes of columns is ±10 mm for columns less than 4.5 m and ±15 mm for columns 4.5–15 m high. Once columns are trued and secured to jigs, their joints are then grouted with concrete (Fig. 18.3).

Crossbars of the ground floor are placed then. Once the crossbar has been positioned correctly, its cast-in fittings are then welded to column brackets. After securing the crossbar throughout the width of the building, reinforcement projections are welded. The cast-in fittings of crossbars and column brackets are permanently welded. The joints are then grouted with concrete.

SALIENT POINTS

1. Framed structures are the structures which are formed by connecting a series of horizontal and vertical structural members in suitable positions.
2. The framed structures primarily comprise of columns (vertical structural member) and beams (horizontal structural member).
3. Columns are connected by beams at floor and roof level. The floors are divided into rooms and passages of required size by walls (load-bearing or non-load bearing).
4. The entire load of columns, beams and floors are transmitted to the foundation.
5. Timber, used for framed structures, offers several advantages like beauty, versatility, durability, low cost, high strength-to-weight ratio, good electrical insulation and high shock absorbing capacity.
6. If steel is used as the material, all columns, beams and girders are of steel. Adequate bracing is needed for these structures in order to resist the wind and earthquake forces.
7. RCC forms the best material for framed structures, provided they are properly designed and constructed.
8. Buildings with more than five stories are called as multi-storey buildings.

9. Articulated structure means the separation of a structure into two or more elements and joining the entire structural elements such that it functions as a monolithic structure.
10. While designing pre-fabricated buildings, manufacturing and effective installation techniques should be taken into account.
11. The erection should involve minimum consumption of labour, time and other means.
12. Structural units are delivered to erection zones by most effective mechanical procedures and allowing for hauling distance, availability of approach roads and conditions of in-site roads.
13. Pre-fabricated structural units are generally transported by trucks with two-axle trailers, tractor trucks with semi-trailers and panel transporters.
14. Structural units have to be made as transportable components which could be assembled by erection units at the construction site.
15. Reinforced concrete trusses are pre-assembled in a horizontal position on special racks with the use of jigs securing elements in precise position or in the design position on special stands. The assembly of trusses from two half-trusses in the design position is carried out on special purpose universal stands.
16. Pre-assembly can also be done in the zone of erection at ground level with the use of movable stands or along the line of erection work.
17. Based on dimensions, multi-storey buildings are erected by means of tower or derrick cranes placed in one or both sides of the frame.
18. The verticality of columns is checked using a theodolite. The admissible deviation of longitudinal axes of columns is ±10 mm for columns less than 4.5 m and ±15 mm for columns 4.5 to 15 m high. Once columns are trussed and secured to jigs, their joints are then grouted with concrete.
19. Crossbars of the ground floor are placed then. After securing the crossbar throughout the width of the building, reinforcement projections are welded.

REVIEW QUESTIONS

1. What are framed structures? Describe their advantages briefly.
2. List the differences between load bearing and non-load bearing partitions in framed structures.
3. Briefly explain the materials used for framed structures.
4. Enumerate the advantages and disadvantage of tall buildings.
5. Enumerate the problems encountered in tall buildings.
6. Write a brief note on design considerations of tall buildings.
7. Describe the important aspects to be considered in the construction of RCC tall buildings.
8. What is an articulated structure?

9. Discuss the concept of modular co-ordination in precast construction.
10. How delivery and storage of pre-fabricated structures are dealt?
11. Describe the various structural members of a building which can be prefabricated.
12. What is pre-assembly? Why is it needed?
13. List the import equipment which are used in the erection of multi-storeyed framed buildings.
14. Explain the erection procedure of multi-storey buildings.

19 ARCHES AND LINTELS

19.1 INTRODUCTION

Arch and lintel are provided above doors, windows and passages. The function of an arch and a lintel is to carry the weight of the structure above the opening. In an arch because of the shape, the blocks support each other by the mutual pressure of their own weight and the structure remains in position by the resistance from the supports. Special types of centring devices are incorporated.

A lintel is a horizontal member used to span openings in walls. It functions similar to that of an arch. Lintels provided over windows are connected to a sunshade to provide shade and shelter. Centring for lintels are much simpler than that of arches.

Materials such as wood, stone, steel, brick and concrete are used both for arches and lintels. Now-a-days, reinforced cement concrete (RCC) lintels are preferred for its simplicity and easy construction.

In walls, at certain levels, beams are constructed to have uniform pressure and even out settlements if any. Such beams are called plinth beams and grade beams.

Arches, lintels, and plinth and grade beams are discussed in this chapter.

19.2 ARCHES

An arch is a mechanical arrangement of wedge-shaped blocks of stones or bricks mutually supporting each other and in turn supported at the ends by piers or abutments. However, arches made of steel and RCC are built as a single unit without the use of wedge-shaped units.

19.2.1 Stability of Arches

Basically the stability of an arch depends on the friction developed between the wedge-shaped blocks and the cohesion of the mortar.

The stability is disturbed by one of the following reasons:

1. Crushing of arch material
2. Sliding of wedge-shaped blocks
3. Rotation about an edge
4. Differential settlement of supports

1. Crushing of Arch Material

Crushing of arch material can be safeguarded by taking the following precautions:

(i) The size of the wedge-shaped blocks should be adequate enough to withstand the pressure exerted by the adjacent blocks.
(ii) The thickness of the arch ring is accordingly adopted depending on the span. As a general rule, the thickness of the ring may be taken 1/12th of the span.
(iii) High quality stones or bricks should be used for the arch. In case of large spans, steel reinforcement may be used to increase the strength.

2. Sliding of Blocks

In order to prevent sliding of blocks, the following points are considered:

(i) All the bed joints are made perpendicular to the line of least resistance. To satisfy this, they are made normal to the curve of the arch.
(ii) The depth of wedge blocks should be adequate to resist the tendency of the joints to open and slide upon one another.

3. Rotation About an Edge

This is prevented by adopting the following aspects:

(i) The thrust on any block should be within the middle third of the arch.
(ii) The thickness of arch is such that the line of resistance atleast falls within the section and crosses each joint away from the edge.

4. Differential Settlements of Supports

This is taken care of by observing the following points:

(i) The supports are to be adequately strong to withstand the thrust of the arch due to self-weight and the super-imposed loads.
(ii) Irrespective of the type of arch, the arch should be symmetrical.

19.2.2 Terminologies in Arch Work

Following are the terminologies used in arch work (Fig. 19.1).

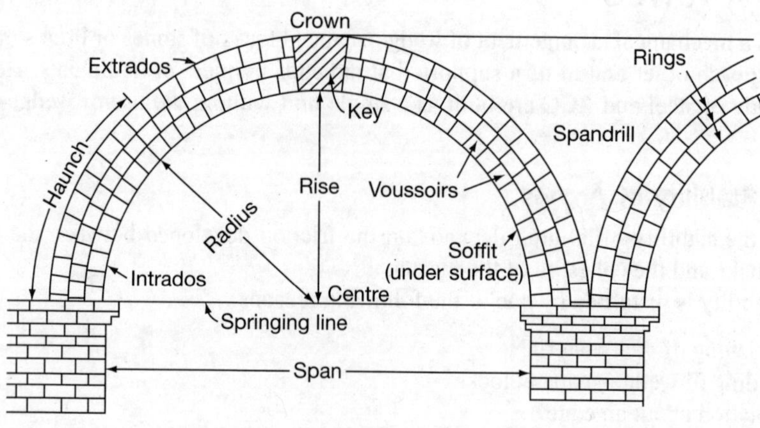

Figure 19.1 Various parts of an arch

1. Abutments

These are the outermost supports of an arch from which the arch starts. That is, this is the wall which supports the arch.

2. Springing Line

It is the horizontal line joining the two springing points.

3. Span

This is the width of the opening covered by the arch.

4. Intrados

The inner curve of the arch.

5. Extrados

The outer curve of the arch.

6. Arch ring

This is the curved portion of masonry forming the arch.

7. Voussoirs

Wedge-shaped blocks of stone or brick masonry of which the arch ring is made.

8. Key Stone

The central voussoir at the central point (crown) of the arch.

9. Crown

The highest part of the extrados.

10. Soffit

The inner or under surface of the arch.

11. Skewbacks

This is the springing point of an arch at the support.

12. Springing Points

The points at the intersection between the skewbacks and the intrados.

13. Springer

The lowest voussoir next to the skewback.

14. Haunch

The lower half of the arch between skewback and crown.

15. Rise

The vertical distance between the highest point of intrados and springing line.

16. Spandrill

The triangular space between the back of the arch ring and a horizontal plane tangent to it at the crown.

19.2.3 Types of Arches

Intrados is a combination of arcs of various radii. Based on the number of centres, arches may be classified as follows:

1. One-Centred Arch

Under this category, semi-circular arch, segmental arch (which is less than a semi-circle), horse-shoe arch (which includes more than a semi-circle) and stilted arch (which consists of semi-circular arch with two vertical portions at the springings) are included. Figure 19.2(a–d) schematically represent these arches.

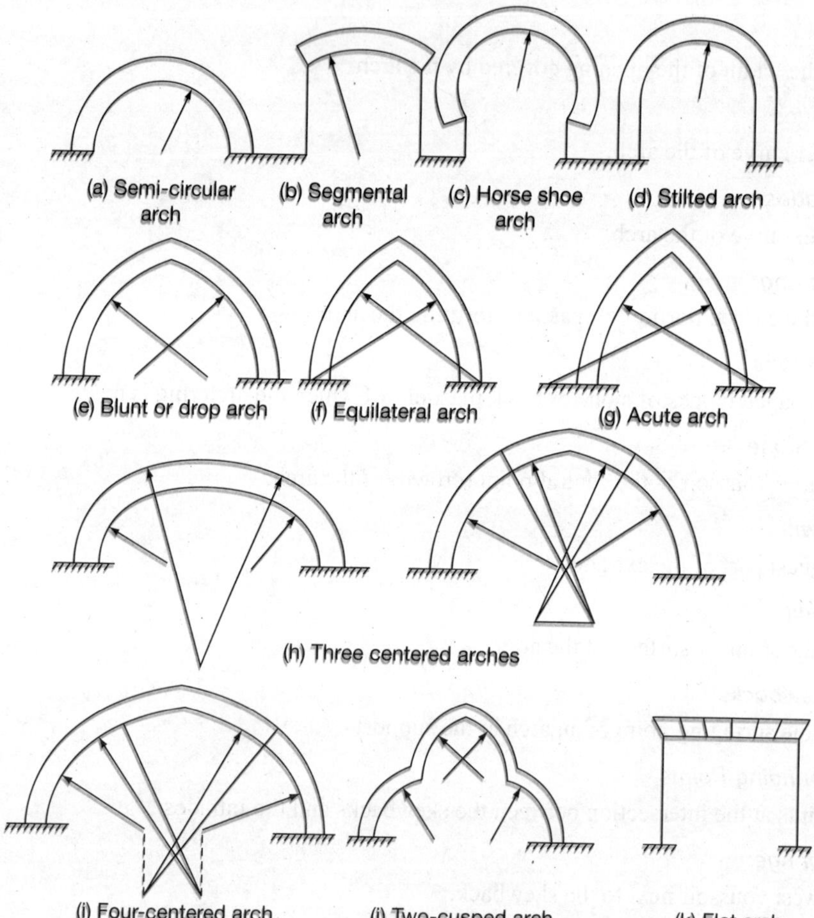

Figure 19.2 Types of arches

2. Two-Centred Arches

There are different types of two-centred arches. Three of them are shown in Figs. 19.2(e), (f) and (g). They are as follows:

(i) Blunt Arch

The two centres of the arch are within the arch itself.

(ii) Gothic Arch

This is also called as equilateral arch. Here the radius of intrados are equal to the span length and the centres are on the springing line.

(iii) Acute Arch

In this type, the centres of the arch are outside the arch.

3. Three-centred Arches

Here there are two types. In the first type, the central portion of the arch is drawn with one centre and the other two centres are used for making up the ends of the arch (Fig. 19.2(h)). In the other type, the procedure is reversed (Fig. 19.2(k)).

4. Four-centred Arches

Here the centres for the lower section do not coincide as in the case of three-centred arches (Fig. 19.2(i)).

(i) Two Cusped Arch

It is used mainly for decorative purposes (Fig. 19.2(j))

(ii) Elliptical Arch

It has an elliptical shape for its intrados.

(iii) Flat Arch

In this type of arch there is no size (Fig. 19.2(k))

19.2.4 Materials Used for Arches

Materials used for arches are stone, brick and concrete. A classification based on number of centres is discussed in Section 19.2.3. A classification of arches based on materials is also feasible which are discussed below.

1. Stone Arches

Ashlar and rubble masonry are best for arch construction.

Ashlar arches are constructed of stones which are cut and dressed to their true wedge shapes. Ashlar arches are used for superior work and have a good appearance. Ashlar arches can also be made to give a flat arch appearance (Fig. 19.3). These flat arches are strengthened by related or jogged joints. For excessive loads, relieving arches can be built.

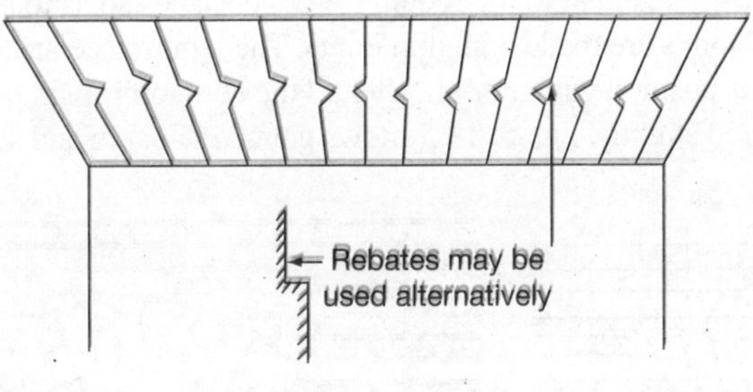

Figure 19.3 Flat arch of ashlar masonry

For rubble masonry arches, roughly dressed stones are made. They are arranged and fitted into a definite arch shape using cement mortar. Because of unevenness of individual stones, the points are thicker. These arches comparatively have less strength than that one made of ashlar masonry (Fig. 19.4).

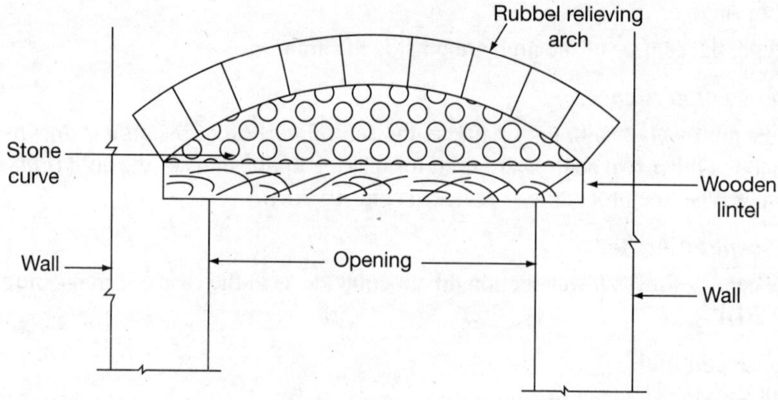

Figure 19.4 Relieving rubble masonry arch

2. Brick Arches

The following four types of brick arches are in use:

(i) Gaughed arches
(ii) Axed or rough cut arches
(iii) Rough brick arches
(iv) Brick flat arches

The above classification is based on the type of bricks used in the arch construction. The different types of bricks which can be used for arches are as follows:

(a) Ordinary standard bricks.
(b) Ordinary bricks cut to a wedge shape.
(c) Special bricks of different sizes and shapes called as purpose-made bricks.
(d) Soft bricks which are made of diatomaceous earth. They can be sawn and rubbed to the required shapes.

In gaughed arches, bricks are cut to the required wedge shapes so as to suit the various voussoir sizes. True radii joints are made with thin joints. The required centring is erected and each voussoir is fitted in its position with mortar. After all the voussoirs are in position, the key stone is fitted and hammered slightly. Figure 19.5 shows gauged flat arch and semi-circular arch.

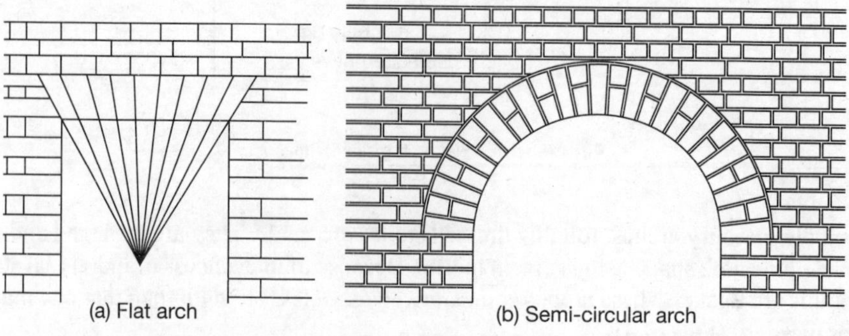

Figure 19.5 Gauged brick arches

For axed arches, bricks are cut with a bricklaying axe to make wedge-shaped bricks. They show rough joints. Construction procedure is same as that of gauged arches.

Rough brick arches are constructed using ordinary bricks. Here, the length of extrados is greater than that of the intrados. Thus the joints near the extrados are wider. These arches do not give pleasing appearance.

Flat brick arches are constructed using cut bricks made to wedge shape and arranged such that to have a flat under surface (Fig. 19.5(a)). All the joints are made to radiate from one point. One brick or half brick may be used. In case of large loads, segmental or semi-circular arches can be built above the flat arches in order to reduce the load on the latter (Fig. 19.6).

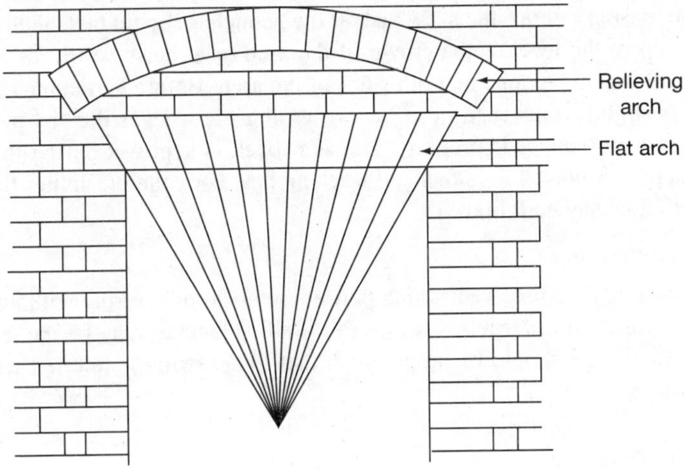

Figure 19.6 Relieving arch over flat arch

3. Concrete Arches

Now-a-days, only RCC arches are used. Here, the blocks are similar to stone voussoirs and of precast type. These arches are used only for ornamental purposes.

19.2.5 Methods of Construction of Arches

Construction of arches is done under the following three operations:

1. Installation of centring
2. Laying the arch
3. Removal of centring

1. Installation of Centring

Arches are generally built over a centring made to the shape of the arch needed. The centring is a temporary structure which has to support the arch till it develops strengths. For small spans, mud masonry is used. The masonry is made to the shape of the soffit of the arch to be built and plastered, over which the arch is constructed. After attainment of adequate strength, the masonry can be dismantled.

Timber centring is preferred generally as it is easier to erect and dismantle. Further, the material can be re-used. Thick wooden planks are used to get the shape of the soffit and are supported by vertical timber posts. Wedges are used to tighten or loosen the centring.

For heavy arches, the shape is got by suitable braces and struts. In such cases it is necessary to support the posts on boxes containing sand. A hole is kept plugged into the box. The post sinks slowly when the plug is removed and the sand flows out, thereby relieving the arch centring without any shock.

2. Laying of Arch

Skewbacks at the springing of the arch are prepared first and placed very carefully in the correct angles. Voussoirs are then laid on either side of the crown. In case of brick arch, bricks may be placed as showing stretcher-face or header-face. More than one ring may be used. Joints are to be filled with mortar not less than 5 mm and not more than 15 mm thick. Finally the key stone is fitted in to lock the whole arch in position. In order to counteract the thrust, the masonry above the skewback at the haunches should be loaded by filling up the spandrels up to the level of the crown of the arch or at least two-thirds of the height of the arch. This has to be done on both sides of the arch. Before hardening of the mortar, the centring is slightly slackened by means of easing the wedges thereby permitting the voussoirs to rest on their beds properly. The skewback is kept wet during the work. The above procedure is followed for stone and brick arches. For concrete arches the procedure followed for lintels may be followed.

3. Removal of Centring

Sandbox method may be followed, which permits removal of centring without any shock. Releasing of wedges in order may also be followed. Whatever may be the material used for centring, the design should be proper such that the centring is lowered with ease and without any shock.

19.3 LINTELS

A lintel is a horizontal structural member placed to span openings for doors, windows and corridors. Lintels are usually rectangular in shape. Lintels may be made out of several materials such as wood, stone, brick, reinforced brickwork, RCC or rolled steel sections embedded in cement concrete.

Reinforced cement concrete lintels are mostly used these days for all spans as other materials have some drawbacks. For example, wooden lintels are liable to decay, stone lintels are costly and not available for long lengths without defect, and steel sections become loose when exposed to sun due to the high rate of expansion.

19.3.1 Purpose of Lintels

The purpose of providing a lintel is just the same as that of an arch. Lintels take the load of the wall above the opening and transfer to the walls. The effective load causing on a lintel is shown in Fig. 19.7.

The lintels are easy and simple in construction and do not need special form work. The ends of lintels are built into the masonry and thus the load carried by lintel is transferred to the masonry in jambs. The bearing of lintel should be either 10 cm or 40 cm for every 30 cm of span, whichever is greater. As a general rule, the depth of the lintel can be adopted as 1/12th of the span or 15 cm, whichever is greater.

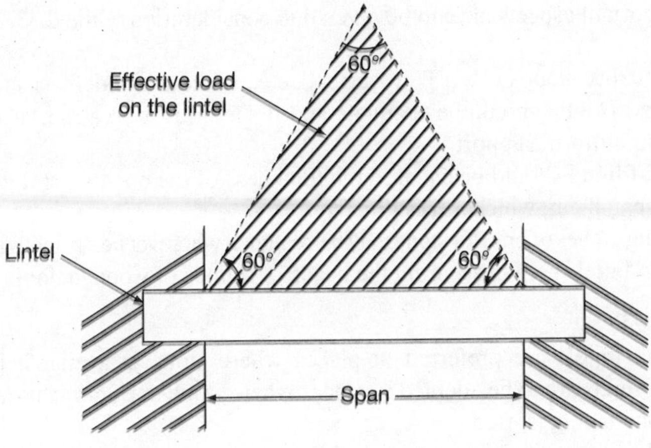

Figure 19.7 Effective load acting on lintel

19.3.2 Types of Lintels

Based on the type of material used, lintels are classified as follows:

1. Wooden lintels
2. Stone lintels
3. Brick lintels
4. Steel lintels
5. RCC lintels and
6. Reinforced brick lintels

1. Wooden Lintels

Wooden lintels are the oldest type of lintels. It is even now used in areas where timber is abundantly available. However, the risk of using wooden lintels is that they are susceptible to catch fire and liable to decay by rot or termite.

Single piece of timber is used for smaller spans. For larger spans, built-up sections are to be used. Built-up sections may be formed by using two or three timber pieces and bolted to serve as a lintel. Figure 19.8 shows a wooden lintel composed of two timber pieces and two distance or packing pieces. Bolts are provided through the packing pieces. Sometimes, in order to strengthen the wooden lintels, mild steel plates are placed at the top and bottom and such lintels are called as flitched lintels.

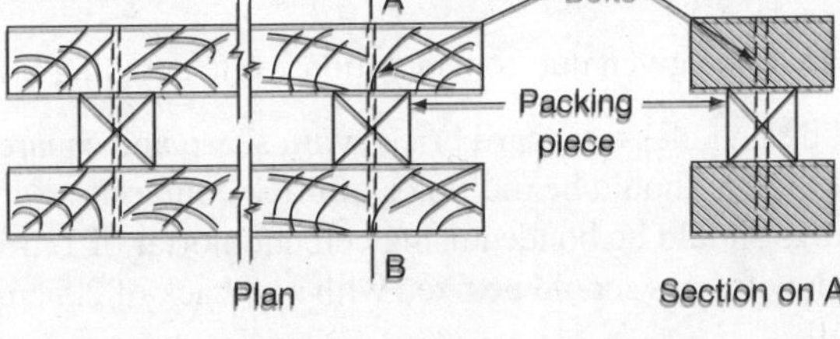

Figure 19.8 Wooden lintel

Following important aspects have to be given due consideration while designing a wooden lintel:

(i) Good quality woods like teak, rosewood, etc., without any defects should be used.
(ii) Bearing of lintel should be adequate and it should rest on a mortar base to have a firm and uniform support.
(iii) Width of lintel should be equal to the thickness of the opening and the depth 1/12th of the span 8 cm, whichever is greater.
(iv) Relieving arches of brick or stone may be designed wherever heavy loads are anticipated.
(v) Wooden lintels help in securing the heads of frames of wooden doors and windows.

2. Stone Lintels

Basically stone lintels are preferred in places where stone is available in abundance. Further, stone lintels are to be adopted in places where stone masonry is used and in buildings where stone facing is used.

Stone lintels consist of slabs of stones which may be one or more pieces. Stone lintels are not recommended for the following reasons:

(i) As stone is weak in tension, it can not withstand the transverse stresses.
(ii) Only spans up to 1 m may be tried with stone slabs. For large spans, relieving stone arches may be used.
(iii) It is extremely difficult to get long stone slabs without defects.

3. Brick Lintels

Brick lintels are used for short spans up to 1 m with light loadings. Brick lintels generally consist of bricks laid on end points or edges (Fig. 19.9).

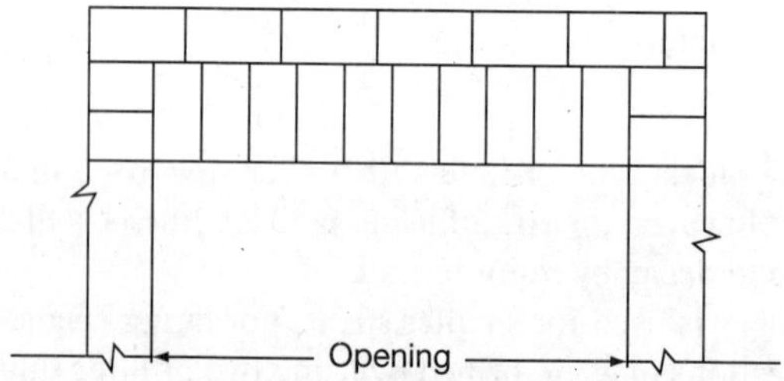

Figure 19.9 Brick lintel

Brick lintels are constructed over temporary wooden supports. Bricks with frogs are generally preferred as they form juggles when bounded. Such construction possesses more shear resistance at the end joints. Lintel formed by using frog bricks is termed as *joggled brick lintel.*

Following points are to be given due consideration while deciding brick lintels:

(i) Bricks should be first class standard bricks with sharp and square edges
(ii) Depth of brick lintels should be multiples of brick courses.
(iii) Bricks with frogs should be bonded using cement mortar of 1 : 3 ratio.
(iv) Door and window frames should be fixed with a setback of 2.5 cm from the external face of the wall.
(v) Steel angles or flats may be provided at the bottom of the lintel.
(vi) Brick lintels should not be recommended for supporting heavy loads.

4. Steel Lintels

Rolled Steel Joints (RSJs) are used individually or embedded in concrete. Particularly for large openings and to support heavy loads, embedded RJS lintels are preferred. These lintels are also called as *bressummers* (Fig. 19.10).

For light loadings and short spans steel angles or channel sections in concrete are used. For large spans and heavy loads, built-up sections are used. Built-up sections comprise of three RSJs or channel sections which are held together at proper distances by tube-separators or cross-bolts. In order to protect these built-up sections from corrosion and fire they are embedded in concrete. In order to increase the bond or grip between the sections, a strip of metal lathing or wire netting is placed at the bottom of the steel section before concreting. The whole system is cured like an RCC lintel.

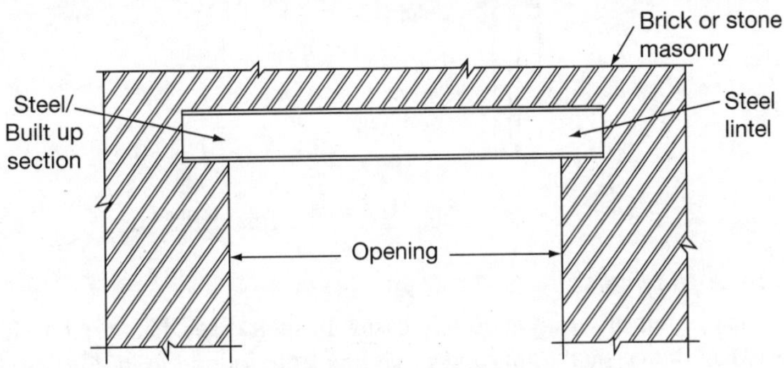

Figure 19.10 Steel lintel

5. RCC Lintels

Reinforced cement concrete lintels have practically replaced all other materials for the lintels. The RCC lintels have plenty of advantages over other materials as stated below.

(i) They are fire-proof, durable, strong, economical and simple in construction.
(ii) Practically there is no restriction on length of span and loading.
(iii) There is no need to go in for relieving arches.
(iv) Use of precast units shall increase the speed of construction.

A concrete mix of 1:2:4 with mild steel reinforcement is usually adopted. Precast RCC lintels are preferred for a span of 2 m. While laying precast RCC units, special care has to be taken to see that the marked face of the lintel is on the top or upper side.

For large spans, lintels should be cast in place. Construction of RCC lintels is similar to any other RCC construction. Figure 19.11 shows the position of RCC lintel over a door or a passage. Adequate bearing is given for the lintel to sit well on the supports and the width is equal to the width of the wall.

Lintels with sunshades have to be planned sometimes. As stated earlier, the bearing for individual lintels should be at least one brick on each side. Window sunshades are also to be extended of this length of 23 cm on either side of the external opening. Sunshades should be projected for a length not less than that required to protect the window or door leaf which may have to be opened out from rains. As sunshade is a shading device, it is taken as a minimum of 60 cm for the windows of residence. For entrance doors, the length may be up to 1 m (Fig. 19.12).

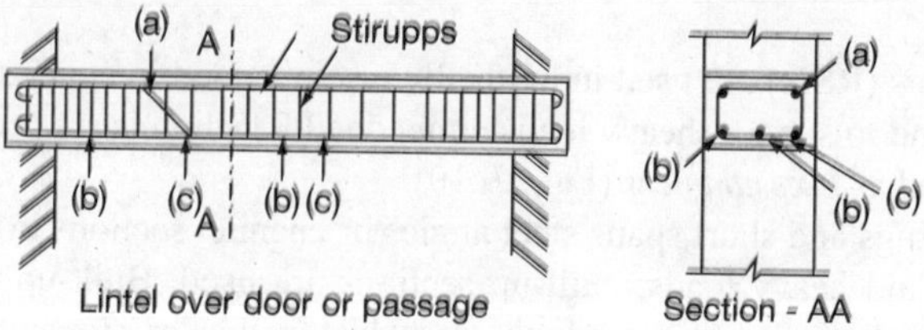

Figure 19.11 Details of RCC lintel over door or passage

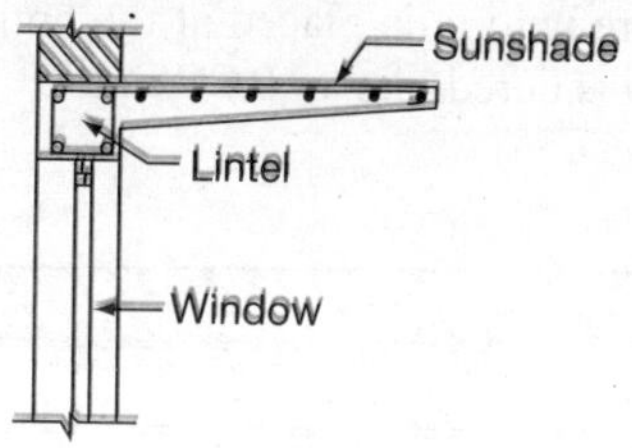

Figure 19.12 Lintel over a window with sunshade

6. Reinforced Brick Lintels

Reinforced brick lintels are used in certain cases. In such cases, bars are provided longitudinally in between the vertical joints and extending from joint to joint. Further, to resist the vertical shear additional 6 mm diameter steel stirrups may be used.

19.4 PLINTH BEAMS AND GRADE BEAMS

Two types of beams, viz., plinth beam and grade beam are constructed in walls for load distribution which are discussed below.

19.4.1 Plinth Beams

Plinth beams are constructed at plinth level in load-bearing wall for uniform load distribution and to even out settlement of buildings. They also allow to provide Damp Proof course at this level. Such beams are very much helpful to buildings in seismic areas. Similar types of beams are provided at lintel level and roof level. Through beams provided at lintel level and roof level are very much helpful for any alternations to be made in future. Further, they are also helpful in any underpinning work to be taken at a later date (Fig. 19.13).

The minimum depth of plinth beams is 10–15 cm and should be provided to the full width of the wall above the plinth beam. The minimum reinforcement recommended is three numbers of 6 mm MS bars at top and bottom and longitudinal bars with 6 mm stirrups at 23 cm spacing.

19.4.2 Grade Beams

Grade beams are provided in between isolated footings. Thus under reamed piles and column on isolated footing are connected by grade beams and then the wall is constructed over it. These beams are stronger than the plinth beams. The recommended minimum depth of grade beam is 150 mm and minimum three numbers of 8 mm bars of Fe415 should be provided at top and bottom (Fig. 19.14).

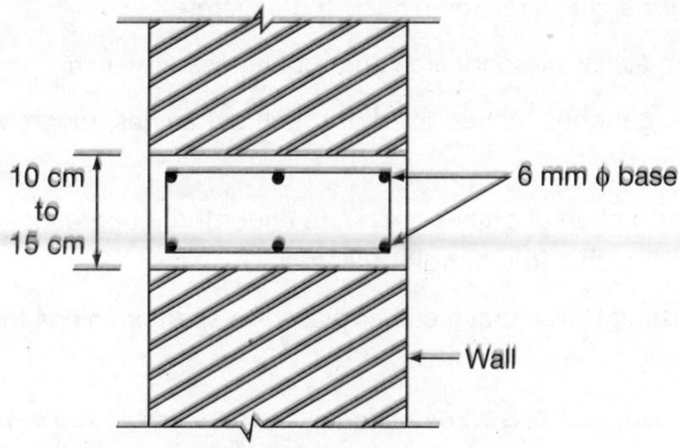

Figure 19.13 Details of plinth beam

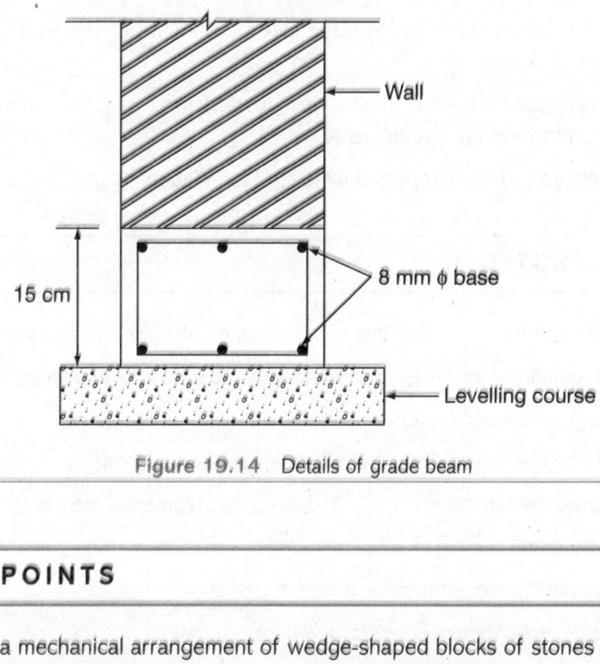

Figure 19.14 Details of grade beam

SALIENT POINTS

1. An arch is a mechanical arrangement of wedge-shaped blocks of stones or bricks mutually supporting each other and in turn supported at the ends by piers or abutments.
2. Basically the stability of an arch depends on the friction developed between the wedge-shaped blocks and the cohesion of the mortar.
3. Stability of an arch is disturbed by one of the following reasons:
   (i) Crushing of arch materials
   (ii) Sliding of wedge-shaped blocks
   (iii) Rotation about an edge
   (iv) Differential settlement of supports.
4. Arches are classified based on number of centres.
5. One-centred arches are: Semi-circular, segmental, horse shoe and stilted.
6. Two-centred arches are: blunt or drop, equilateral and acute.

7. Materials used for arches are stone, brick and concrete.
8. Stone arches are ashlar masonry arch and rubble masonry arch.
9. Brick arches are: gaughed arches, axed or rough cut arches, rough brick arches and brick flat arches.
10. Methods of construction of arches are made under three operations, namely installation of centring, laying of arch and removal of centring.
11. A lintel is a horizontal structural member placed to span openings for doors, windows and corridors.
12. Lintels may be made out of several materials such as wood, stone, brick, reinforced brickwork, RCC or rolled steel sections embedded in cement concrete.
13. The purpose of providing a lintel is just the same as that of an arch. Lintels take the load of the wall above the openings and transfer to the walls.
14. Types of lintels are: wooden lintels, stone lintels, brick lintels, steel lintels, RCC lintels and Reinforced brick lintels.
15. Plinth beam are constructed at plinth level in load-bearing walls for uniform load distribution and to even out settlements.
16. Grade beams are provided in between isolated foundation.

REVIEW QUESTIONS

1. What is an arch? Sketch roughly five common types of brick arches used in building.
2. Under what situations you would recommend brick arches in a building?
3. Differentiate the load distribution of a stone arch with an RCC arch.
4. Describe briefly the construction procedure of masonry arches.
5. Briefly describe the construction of a 3 metre span segmental arch for 2 brick thick wall.
6. Where do you prefer a steel arch? Justify.
7. Discuss the stability requirement of different arches.
8. Why are stone lintels generally not favoured?
9. Differentiate between stone lintels and brick lintels.
10. Draw neat sketches of steel lintel and wooden lintel.
11. Explain the detail of an R.C.C. lintel having weather shed.
12. Bring out the advantages of RCC lintels.
13. Sketch a reinforced concrete lintel to span an opening 2 metre clear in a masonry wall.
14. Where do you use a plinth beam and a grade beam?
15. Discuss the difference between a Plinth Beam and a Grade Beam.

20 DOORS AND WINDOWS

20.1 INTRODUCTION

Doors, windows and ventilators are important and main fixtures in a building which play vital role in the comfortable usage of a building.

Door is a framework with shutters made of wood, steel, aluminium or a combination of these materials (with glass wherever needed), which is secured in an opening left in the wall for the purposes of providing access to the users of the structure. For small openings, door is provided with one leaf or shutter and for larger openings two or more shutters may be provided.

Window is a framework with shutters fixed in an opening left in the wall for the purpose of providing daylight, vision and ventilation. The window shutters may be of wood or glass. All materials used for construction of door may be used for window also.

Ventilator is a window having small heights with shutters pivoted and fixed at the top of the door or window.

Details of doors and windows are discussed in this Chapter.

20.2 TECHNICAL TERMS

Technical terms used in connection with doors and windows are discussed below (Fig. 20.1).

1. Frame

This consists of a group of members which form a support for a door or window.

2. Style

This is the outside vertical member of the shutter of a door or window.

3. Head

This is the horizontal uppermost part of a frame.

4. Sill

This is the bottom horizontal part of a window.

5. Top Rail

This is the topmost horizontal member of a shutter.

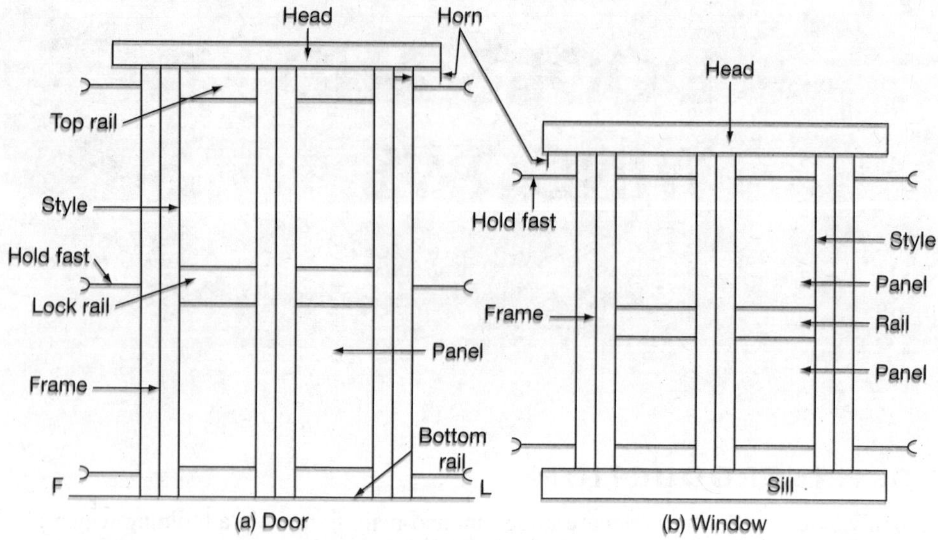

Figure 20.1 Different parts of door and window

6. Lock Rail

This is the middle horizontal member of a shutter where the locking arrangements are made.

7. Bottom Rail

This is the bottom most horizontal member of a shutter.

8. Cross Rail

This is also called as intermediate rail which is an additional horizontal rail fixed between the top and bottom rails.

9. Panel

This is the portion of the shutter enclosed between the adjacent rails.

10. Hold Fast

These are used to keep the frame of door or window in position. Three numbers of such hold fasts are fixed on each side of a door frame and two on window frames. They are of mild steel of size 30 mm × 6 mm and of 20 cm length.

11. Horn

This is a horizontal projection of head or sill of the frame beyond the vertical face of the frame. It facilitates the fixing of the frame on the wall opening and is of 10–15 cm length.

12. Shutter

The entire assembly of styles, panels and rails is known as the shutter.

13. Mullion

This is a vertical member which sub-divides a window or a door opening vertically.

14. Transom

This is a horizontal member which is used to sub-divide a window opening horizontally.

15. Louver

This is a piece of timber which is fixed in an inclined position within a frame.

16. Jamb

This is the vertical wall face of an opening which supports the frame of door and window.

17. Rebate

This is the depression or recess made inside the door frame to receive the door shutter.

18. Architrave

When doors and windows are placed flushed with the masonry, a strip of wood usually moulded or splayed is fixed around the sides closing the joint. This gives a decent appearance to the joint of the frame and masonry.

20.3 DOORS

Doors are to be located such that there shall be freedom of movement in and out of the room. Further it should ensure the maximum use of accommodation in the room and at the same time privacy of the occupants.

In general, it is preferred to place the door at one end of the room. If there happens to be two doors, they are to be located in the opposite walls of the room. This provides good ventilation and also creates least interference with the inside arrangement of the room.

20.3.1 Standard Sizes of Doors

Dimensions of doors and windows are measured inside to inside of openings. Door is designated as 8 DS 20 in which D denotes door and S denotes single shutter. The first number indicates the width of opening in module of 100 mm. The last number denotes the height of the opening in module of 100 mm.

As discussed above, 8 DS 20 refers to Door with single shutter of width 800 mm (= 8 × 100) and height 2000 mm (= 20 × 100). Standard sizes of doors as per Indian Standards are shown in Table 20.1.

Table 20.1 Indian standard dimensions of timber door frames

Sl.No.	Designation	Size of opening (mm)	Size of wooden frame (mm)		Size of door shutters (mm)
			Width	Height	
1.	8 DS 20	800 × 2000	790	1900	700 × 1905
2.	8 DS 21	800 × 2100	790	2090	700 × 2005
3.	9 DS 20	900 × 2000	890	1990	800 × 1905
4.	9 DS 21	900 × 2100	890	2090	800 × 2005
5.	10 DT 20	1000 × 2000	990	1990	900 × 1905
6.	10 DT 21	1000 × 2100	990	2090	900 × 2005
7.	12 DT 20	1200 × 2000	1190	1990	1100 × 1905*
8.	12 DT 21	1200 × 2100	1190	2090	1100 × 2005*

* Doors with two shutters 560 mm each shutter and 20 mm overlap when closed.

20.3.2 Materials Used

Wood, metals and concrete are used for door frames where plywood and glass are used for door shutters/panels.

1. Wood

Wood is used in several forms for the construction of doors. As wood is largely available in India, it is mostly used in India. Doors may be made of solid planks or built up of small individual pieces. Further, it can take a good polish and can be moulded to produce an attractive appearance. One set back of wood is that it is easily attacked by vermin.

2. Metals

Metal doors, particularly aluminium, have been in use in different forms. Cast metal doors are generally preferred. Hollow metal doors made up of a metal frame and covered with sheet metal have high resistance, life and fire resistance.

3. Concrete

In situations where there is a danger of vermin, reinforced cement concrete frames are preferred.

4. Plywood

This is used for flush doors which have a wide variety of finishes.

5. Glass

In order to permit light, glasses are used in panels. Further, to get a rich appearance, big openings in doors are fitted with thick glass sheets. Coloured glass, frosted glass and beaded glass can be used for ornamental appearance.

20.3.3 Types of Doors

Based on the type of materials, positioning of different components, method of construction, intended purposes, etc., the common types of floors used in the building are discussed below.

1. Ledged Doors

A ledged door is formed by joining the vertical boards (battens) and horizontal supports (ledges) securely by suitable joints (Fig. 20.2). The battens are of 10–15 cm wide and 20–30 mm thick, whereas the ledges are of 10–20 cm wide and 30 mm thick. Top, middle and bottom ledges are used. Shutter is hung on T-hinges which are fixed to the ledges. This is the simplest form of door. It is used where strength and appearance are not very much important.

2. Ledged and Braced Doors

These doors are similar to the ledged doors discussed above but for the difference that diagonal members, known as the braces, are provided as shown in Fig. 20.3. The braces are generally of 10–15 cm wide and 30 mm thick. The braces provide rigidity and for this reason such doors are preferred for wide openings. The braces are usually housed in the ledges. It is to be noted that the braces should work in compression and not in tension.

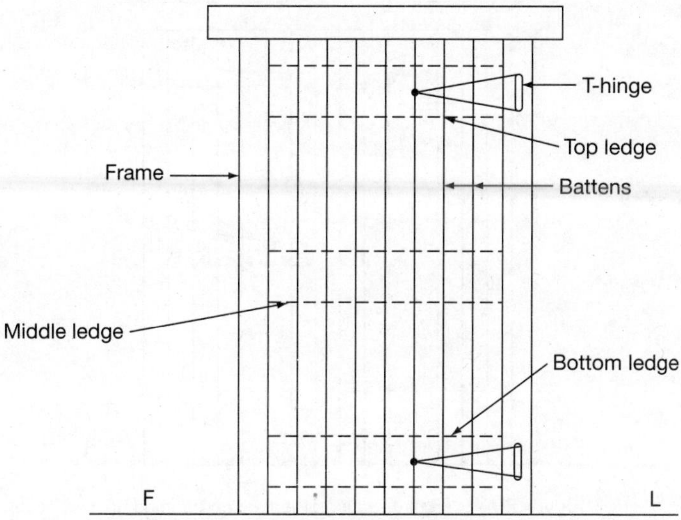

Figure 20.2 Front elevation of ledged door

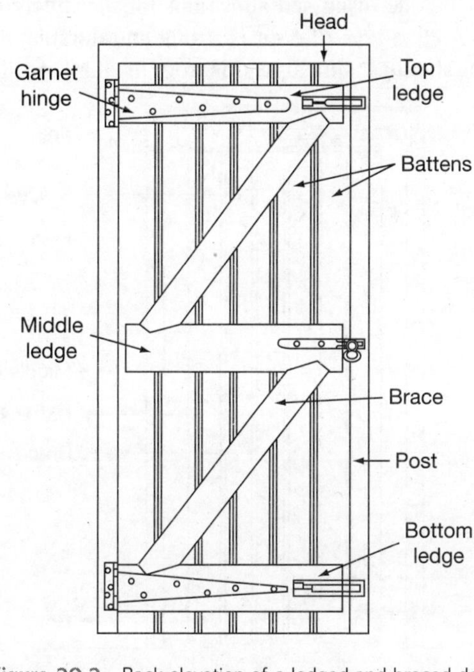

Figure 20.3 Back elevation of a ledged and braced door

3. Ledged and Framed Doors

In this type of doors, a framework for shutter is provided. This makes the door stronger and better in appearance as shown in Fig. 20.4. Styles are of 10 cm wide and 40 mm thick. As usual the battens and ledges are provided.

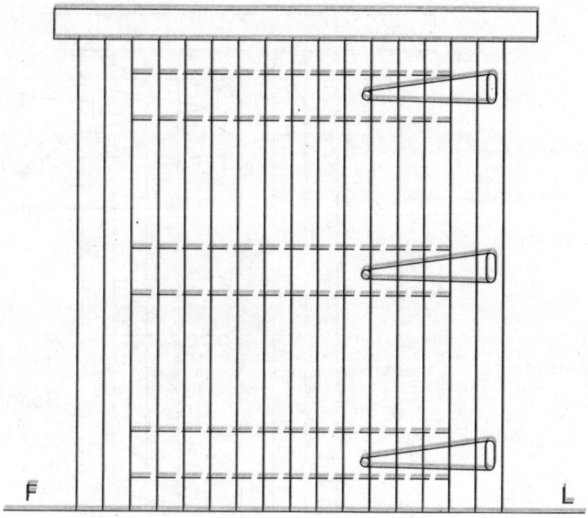

Figure 20.4 Front elevation of ledged and framed door

4. Ledged – Framed and Braced Doors

This type is similar to the one discussed above but for the difference that the braces are introduced (Fig. 20.5). As this type of door is strong and durable, this can be adopted for external use. The braces should incline upwards from the hanging side.

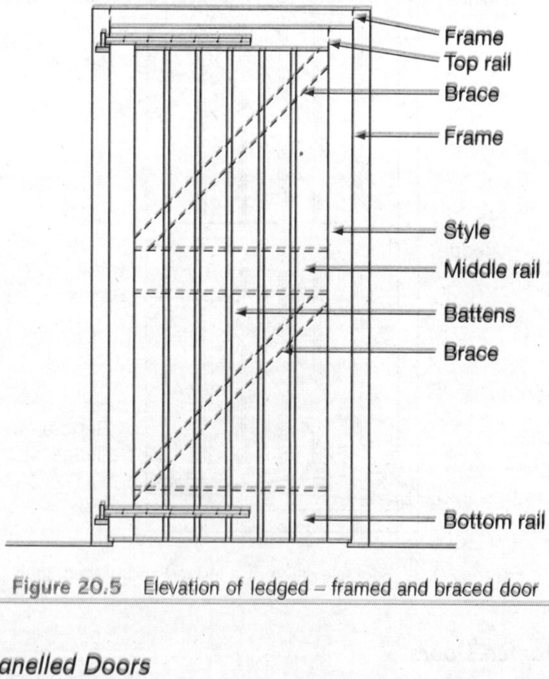

Figure 20.5 Elevation of ledged – framed and braced door

5. Framed and Panelled Doors

Panelled doors are mostly used. It consists of a framework in which panels are fitted. The styles are continuous from top to bottom and rails are joined to the styles. Mullions, if any, are joined to rails. The thickness of the shutter depends on various factors such as type of work, situation of the door, number of panels, etc. The thickness of the shutter is about 30–40 mm and the thickness of panels is about 20 mm. Panels are secured in position by the grooves inside the edges of the framework. Figure 20.6 shows panelled doors with panels. Moulding of panels adds to the beauty. Number and size of panels depend on the desire of the owner.

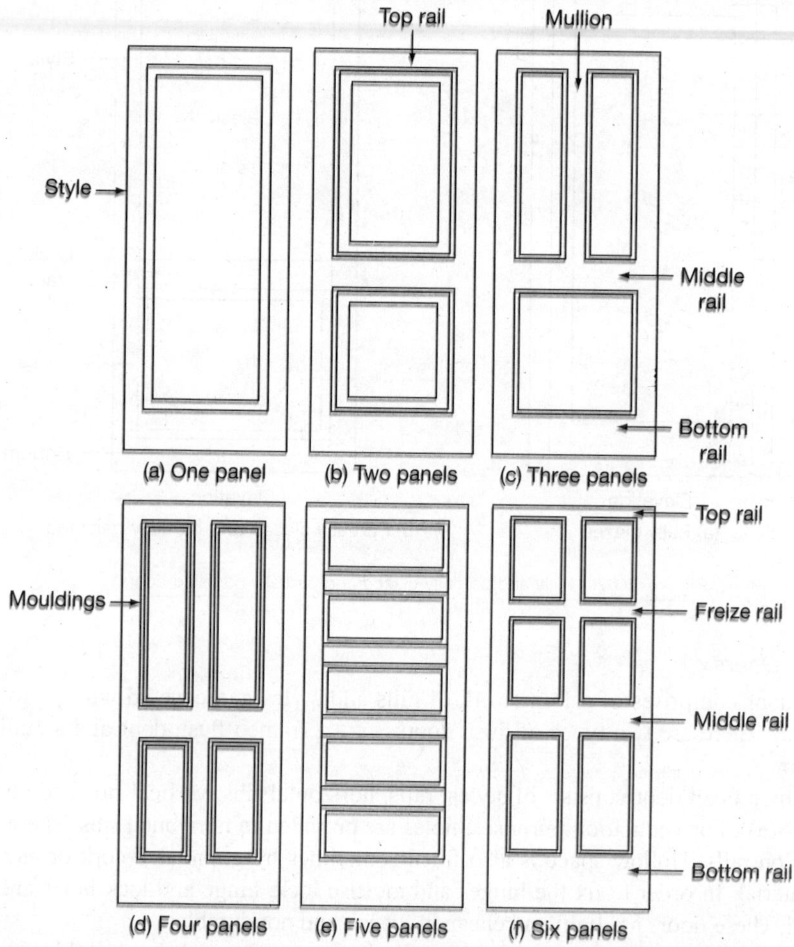

Figure 20.6 Elevation of framed and panelled doors

For small openings, single shutter may be adequate while double-leafed shutters are used for large openings.

6. Glazed or Sash Doors

When there is a need for more light in a room, glazed doors may be preferred. Partly panelled and partly glazed doors may be considered while some light is already available through glazed windows.

Usually the ratio of glazed portion to the panelled portion is 2 : 1. The glass is fixed into the rebates made in the wooden sash bars and secured by nails and putty or by wooden beads (Fig. 20.7).

In order to increase the area of the glazed portion, the width of styles above lock rail is decreased as shown in Fig. 20.7. These are known as the diminished styles or gunstock styles. Glazed or sash doors are preferred in hospitals, offices, libraries, show rooms, etc.

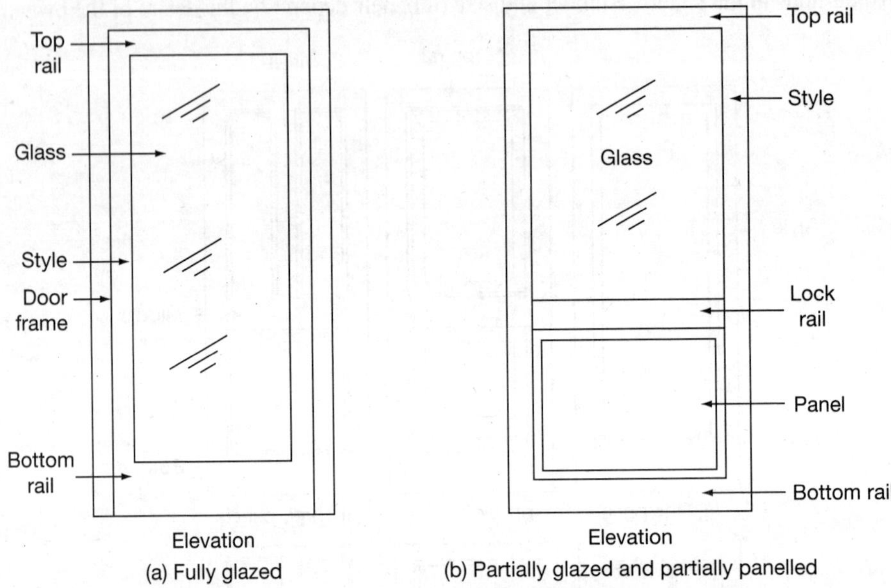

Figure 20.7 Fully and partially glazed and partially panelled doors

7. Flush Doors

A flush door comprises of a framework of rails and styles and covered with plywood or hardboard. There are two types of flush doors, viz., a framed flush door and a laminated flush door.

A framed flush door consists of styles, rails, horizontal ribs, vertical ribs and plywood (Fig. 20.8(a)). For ventilation purposes, holes are provided in horizontal ribs. The vertical ribs rest on rails. Hollow space is also filled sometimes by granulated cork or any other light material. In order to fix the hinges and mortise lock, hinge and lock holes are to be provided. These doors are light and cheap but weak and not durable.

A laminated flush door consists of styles, rails, laminated core and plywood (Fig. 20.8(b)). Under great pressure the laminae of wood are glued. Along with the plywood, sheets on either side are also glued to the laminated core under great pressure. Evidently these doors are heavy and durable.

8. Louvered Doors

Louvers are provided in the shutters which may be fully or partially louvered. Further the louvers may be movable or fixed. Figure 20.9 shows a movable type partially louvered door. Louvers are arranged at such an inclination that the horizontal vision is obstructed. That is the upper back edge of any louver will be higher than the lower front edge of the louver just above it. In the movable louvers, a vertical piece of timber is provided to which the louvers are attached through hinges. With this vertical piece, the movement of louvers can be performed. Louvers may be of glass or wood (Fig. 20.9).

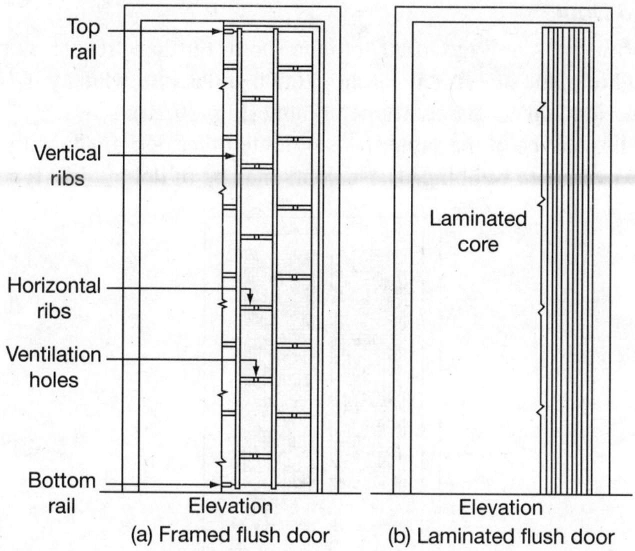

Figure 20.8 Types of flush doors

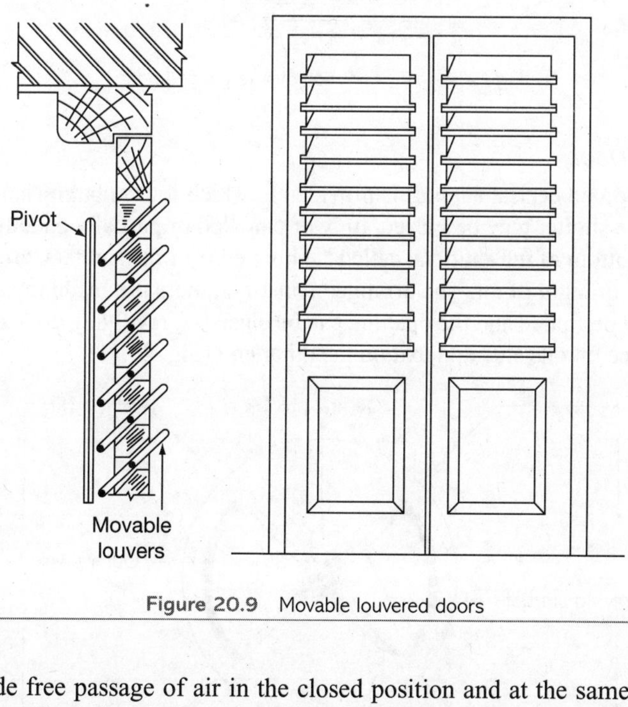

Figure 20.9 Movable louvered doors

Louvers provide free passage of air in the closed position and at the same time maintain sufficient privacy. Such doors are preferred in sanitary blocks of public buildings, residential buildings, etc. However, the louvers should be cleaned often failing which dust may accumulate.

9. Wire-gauged Doors

These doors permit easy passage of air into the room. Further prevent movement of flies or insects. They are made of vertical and horizontal styles into which wire gauge is fixed. They may be provided on a separate frame or hung (Fig. 20.10).

Styles and rails shall be of the same size similar to other doors. The wire gauge shall be of best quality woven wire webbing. These doors may be of double-leaf type.

Figure 20.10 Elevation of wire-gauged door

10. Revolving Door

In a revolving door, a central support is provided to which four radiating shutters are fixed (Fig. 20.11). The shutter may be glazed, fully or panelled or partially panelled, or partially glazed. At the bottom of the shutters, called rubbing ends of the shutters, are rubber pieces fixed to prevent draught of air. The arrangements of shutters are made in such a way that the shutters may be folded and the opening can be closed. A revolving door simultaneously provides entrance on one side and exit on the other end (Fig. 20.11).

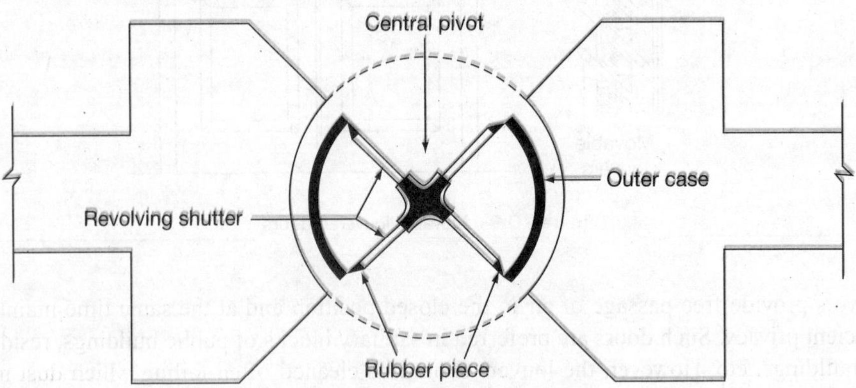

Figure 20.11 Revolving doors

20.4 WINDOWS

Windows made of materials as that of doors and fixed to the wall openings for admission of light, free circulation of air, etc. They comprise of a frame which is fitted in the opening in the wall and small shutter frames containing glass, wood, etc., are attached to the frame by hinges. Window shutters have different types of movements.

20.4.1 Window Shutter Movements

Windows have different types of movements of shutters which are discussed below (Fig. 20.12).

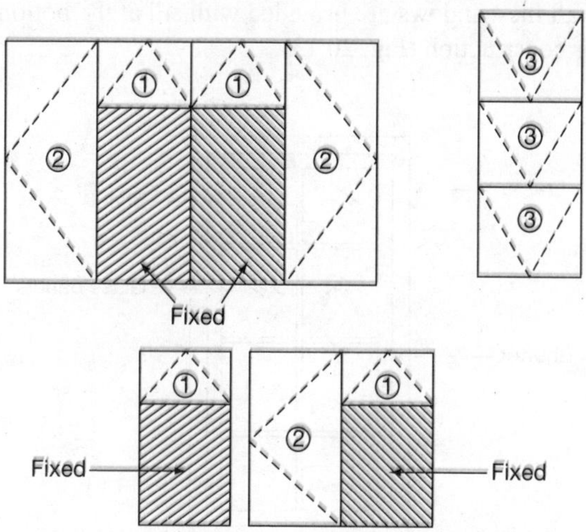

Note: 1 – Top Hing; 2 – Side Hing; 3 – Bottam Hing

Figure 20.12 Type of window shutter movements

1. Fixed Type

In this type, the framework is fitted with a glass and the framework can not move.

2. Opening Outside

In this type, the shutters open outside of a building.

3. Opening Inside

In this type, the shutters open inside a building.

4. Top-hinged

The shutter is hinged at the top and can be moved outside.

5. Bottom-hinged

Here the shutter is hinged to the bottom of the frame such that it can be moved inside.

6. Pivoted

In this type, the shutters are made to rotate about a pivot which may be fixed to the window frame. The windows can rotate vertically or horizontally based on the position of the pivot.

7. Sliding

In this type, the shutters can move inside the walls vertically or horizontally.

8. Double Hung

In this type, one shutter goes up and the other moves down.

20.4.2 Types of Windows

Different types of windows are discussed below.

1. Casement Windows

Casement window is one in which the shutters open like doors. This window comprises of a frame consisting of styles, top rail and bottom rail. Window frames are made similar to frames of doors but all the windows are provided with sill at the bottom. The construction is similar to the door construction (Fig. 20.13).

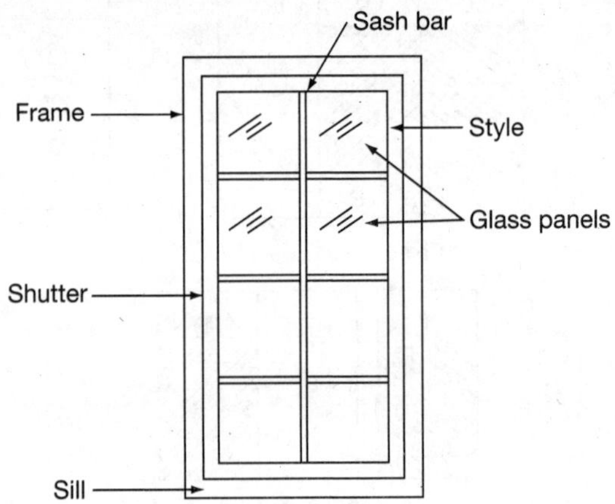

Figure 20.13 Casement window

2. Double-hung Windows

This window comprises of a pair of shutters which can slide within the grooves provided in the frame. A pair of metal weights is provided for each sash with connected cord or chain over a pulley. When the weights are pulled, the shutters open to the required level and thus provide ventilation.

3. Pivoted Windows

This is similar to casement window except that no rebates are provided and the shutter movement is different. Here, the shutters are allowed to swing round pivots. The windows may be pivoted horizontally or vertically (Fig. 20.14). These windows admit more light than side hung windows and easy to clean.

4. Louvered Windows

In this type of windows, louvers are provided as in case of louvered doors. They provide adequate privacy apart from allowing free passage of air. Here, the shutter consists of top rail, bottom rail and two styles which are grooved to receive the louvers. Angle of inclination of louvers is 45° and are fixed in position. If it is expected to move the louvers some mechanism has to be provided to operate. In order to prevent the entry of rain water, the louvers should slope downward to the outside.

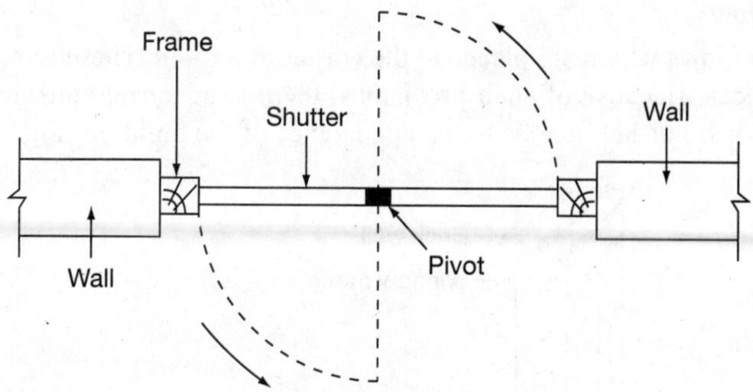

Figure 20.14 Vertically pivoted window

5. Sliding Windows

Here the shutters move on roller bearings, either horizontally or vertically. Suitable opening are provided in the walls to receive the shutters when windows are opened out. Such windows are preferred in trains, buses, bank counters, shops, etc.

6. Glazed Windows

These windows are fully glazed casement windows. In order to receive glass panels, sashes are rebated. The glass is kept in position either by putty or by small fillets, known as glazing beads.

7. Metal Windows

Metal windows, made of mild steel, bronze, aluminium or alloys, are widely used these days. Steel windows are manufactured in standard sizes. The metal frame may be fixed directly to the wall (Fig. 20.15) or it may be fixed on a wooden frame. In the former case, the hinges and plugs are provided to keep the frame in position and in the latter case the screens are used to fix the metal frame with the wooden frame.

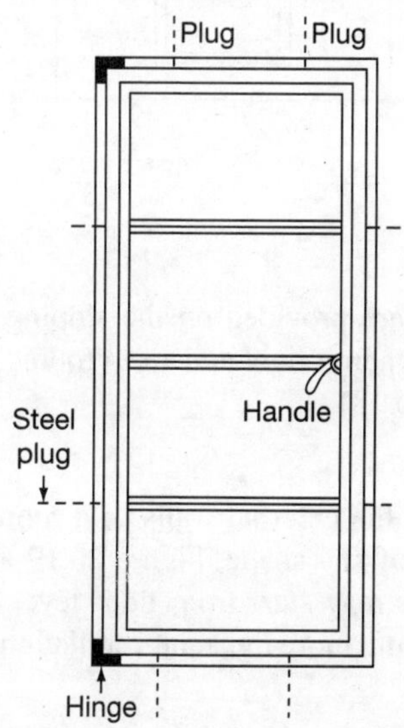

Figure 20.15 Metal window fixed to wall

8. Corner Windows

These are the windows which are placed at the corner of a room. These windows have two perpendicular faces. Because of such provisions, there is adequate entry of light and air from two directions. Further, it adds to the appearance of the building (Fig. 20.16).

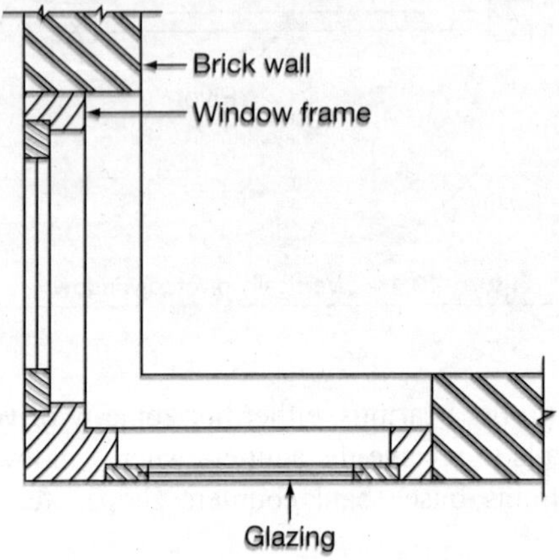

Figure 20.16 Corner window

9. Gable Windows

These windows are provided at the gable ends of a roof as shown in Fig. 20.17.

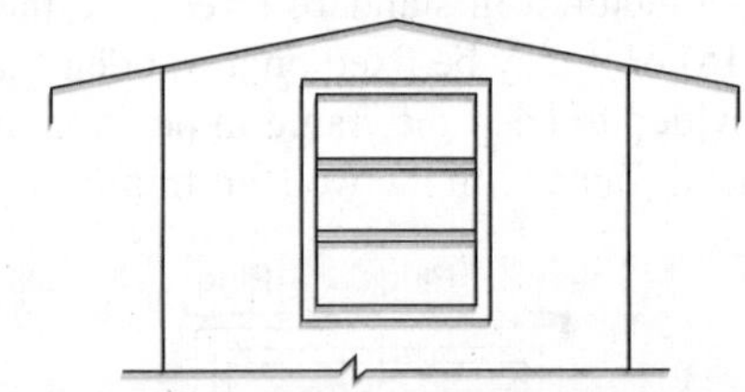

Figure 20.17 Gable window

10. Dormer Windows

These are the windows which are provided on the sloping roofs. The main purpose of providing dormer windows is to admit light and air to rooms which are constructed within or below the roof slopes (Fig. 20.18).

11. Bay Windows

These windows project outside the external walls of a room. The shape may be square, splayed, circular, polygonal or of any shape. Figure 20.19 shows a splayed bay window. The projection of bay windows may start from floor level or sill level. These windows increase the opening area and admit more light and ventilation. Also adds to the appearance.

12. Clerestory Window

These windows are fixed near the top of the main roof (Fig. 20.20). Pivoted windows are used for this purpose. These windows provide ventilation to the inside of the room which

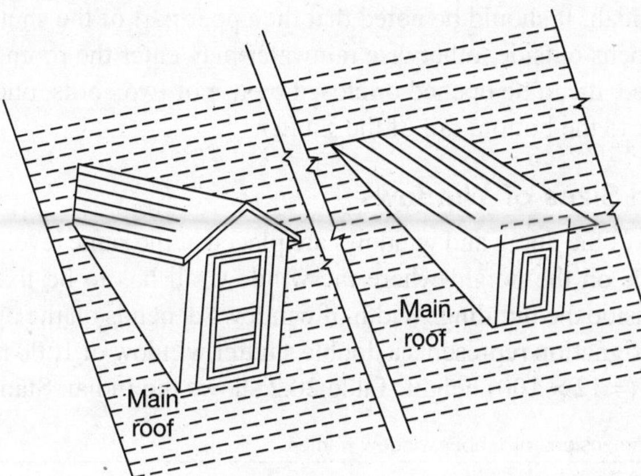

(a) Gable type (b) Shed type

Figure 20.18 Dormer windows

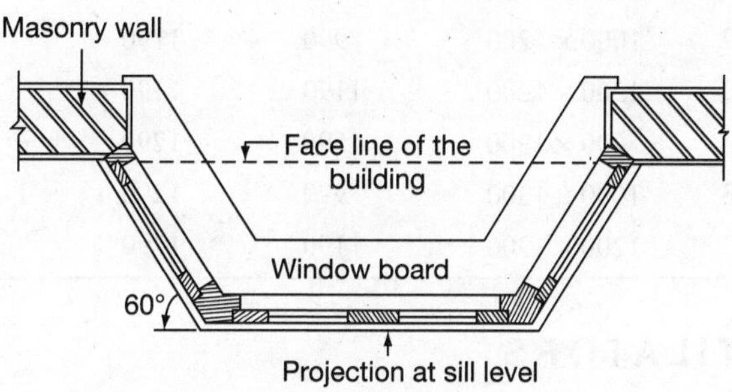

Figure 20.19 Bay window

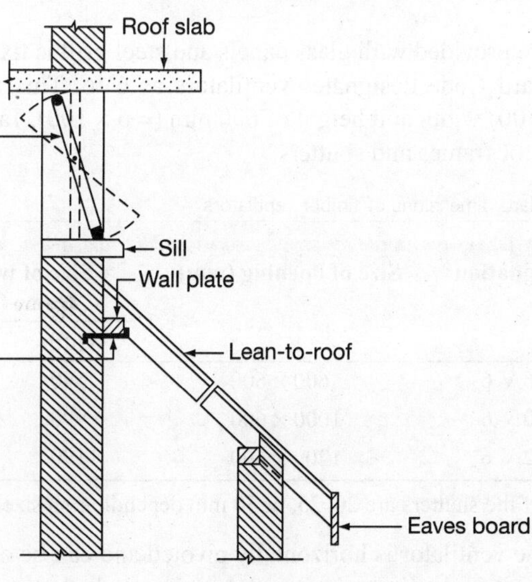

Figure 20.20 Clerestory window

is blocked by a verandah. It should be noted that the upper part of the shutter opens inside and the lower part opens outside, otherwise rainwater may enter the room. As the window is horizontally pivoted, it can be opened or closed by use of two cords, one attached to the top rail and the other to the bottom rail of the shutter.

20.4.3 Standard Sizes of Windows

In a building, the top of the doors and windows are placed at the same level. Thus the height of a window depends on the height where the window sill has to be fixed based on the circulation needed inside the building. Width of windows depends on the opening required.

The 10 WT 12 designation represents a double shutter window of 1000 mm (= 10 × 100) width and 1200 mm (= 12 × 100) height. Table 20.2 shows the Indian Standard Windows.

Table 20.2 Standard dimensions of timber window frames

Sl. No.	Designation	Size of opening (mm)	Size of wood frame (mm)		Size of wind shutters (mm of wood)
			Width	Height	
1.	6 WS 12	600 × 1200	590	1190	500 × 1100
2.	10 WT 12	1000 × 1200	990	1190	460 × 1100
3.	12 WT 12	1200 × 1200	1190	1190	560 × 1100
4.	6 WS 13	600 × 1300	590	1290	500 × 1200
5.	10 WT 13	1000 × 1300	990	1290	460 × 1200
6.	12 WT 13	1200 × 1300	1190	1290	560 × 1200

20.5 VENTILATORS

A ventilator is a narrow window of small height which is generally fixed at the top of door or window or near the roof of a room for providing ventilation in the room. Ventilator may be fixed which provides only light, whereas the movable ones provide light and air circulation.

Ventilators are provided with glass panels and steel grill is fixed for safety conditions.

Indian Standard Code designates ventilators as 6V6. This represents a ventilator of 600 mm (= 6 × 100) width and height of 600 mm (= 6 × 100). Table 20.3 shows the Indian Standard ventilator frames and shutters.

Table 20.3 Standard dimensions of timber ventilators

Sl. No.	Designation	Size of opening (mm)	Size of wooden frame (mm)		Size of shutter (mm)
			Width	Height	
1.	6 V 6	600 × 600	590	590	500 × 500
2.	10 V 6	1000 × 600	990	590	900 × 500
3.	12 V 6	1200 × 600	1190	590	1100 × 500

Note: Thickness of the shutters are 20, 25, or 30 mm depending on size of the opening.

The shutter of the ventilator is horizontally pivoted and can be opened or closed by use of the chords, one attached to its top rail and the other to the bottom rail.

SALIENT POINTS

1. Doors, windows and ventilators are important and main fixtures in a building which play vital role in the comfortable usage of buildings.
2. Door is a framework with shutters made of wood, steel, aluminium or a combination of these materials (with glass wherever needed), and is secured in an opening left in the wall for the purposes of providing access to the uses of the structure.
3. For small openings, a door is provided with one leaf or shutter and for large openings two or more shutters may be provided.
4. Doors should ensure the maximum use of accommodation in the room and at the same time privacy of the occupants.
5. Dimensions of doors and windows are measured inside to inside of openings.
6. A ledged door is formed by joining the vertical boards (battens) and horizontal supports (ledges) securely by suitable joints. This is the simplest form of door.
7. Ledged and braced door is similar to ledged door but for the difference that diagonal members, known as braces, are provided. It is to be noted that the braces should work in compression and not in tension.
8. Ledged and framed door is one in which a framework for shutters is provided. This makes the door stronger and better in appearance.
9. Ledged-framed and braced door is similar to ledged and framed one but for the difference that additional braces are introduced diagonally.
10. Framed and panelled door is one in which panels are fitted to the framework. Moulding of panels adds to the beauty of the doors.
11. When there is a need for more light in a room, glazed doors may be preferred. Partly panelled and partly glazed doors may be considered while some light is already available through glazed windows.
12. A flush door comprises of a framework of rails and styles and covered with plywood or hardboard. There are two types of flush doors, viz. a framed flush door and a laminated flush door.
13. Louvered door is one in which louvers are provided in the shutters. Louvers may be movable or fixed type and the door may be partially or fully louvered.
14. Wire-gauged door permits easy passage of air into the room. Further it prevents movement of flies or insects.
15. In a revolving door a central support is provided to which four radiating shutters are fixed. The shutters may be fully glazed, panelled or partially panelled and partially glazed.
16. Window is a framework with shutters, fixed in an opening left in the wall for the purpose of providing daylight, vision and ventilation. The window shutters may be wood or glass.
17. Windows have different movement of shutters, viz., fixed type, opening outside, opening inside, top-hinged, bottom-hinged, pivoted, sliding and double hung.

18. Casement window is one in which the shutters open like doors. The window comprises of a frame consisting of styles, top rail and bottom rail.
19. Double-hung window comprises of a pair of shutters which can slide within the grooves provided in the frame.
20. Pivoted window is similar to casement window except that no rebates are provided and the shutter movement is different. Here, the shutters are allowed to swing around pivots.
21. In louvered windows, louvers are provided as in the case of doors. They provide adequate privacy apart from allowing free passage of air.
22. Sliding windows shutters move on roller bearings, either horizontally or vertically. Suitable openings are made in the walls to receive the shutters when windows are opened out.
23. Glazed windows are fully glazed casement windows. In order to receive glass panels sashes, are rebated.
24. Metal windows made of mild steel, bronze, aluminium or alloys are widely used at present. Steel windows are manufactured in standard sizes. The metal frame may be directly fixed to wall or it may be fixed on a wooden frame.
25. Corner windows have two perpendicular faces placed at the corner of a room.
26. Gable window is provided at the gable end of a roof.
27. Dormer window is provided on the sloping roof. Main purpose is to admit light and air into the room which is constructed below the roof.
28. Bay window is projected outside the external wall of a room. The shape may be square, splayed, circular, polygonal, etc.
29. Clerestory window is fixed near to the top of the main roof where the ventilation is blocked by the verandah.
30. Ventilator is a narrow window of small height which is generally fixed at the top of door or window or near roof of a room for providing ventilation in the room.

REVIEW QUESTIONS

1. How are doors and windows designated by Indian Standards?
2. Write a note on the materials used for making doors and windows.
3. What are the different types of ledged doors? Explain with neat sketches.
4. Name the important components of a panelled door. Why would you recommend the bottom sill for external doors?
5. What are the advantages of flush doors?
6. Under what situations louvered doors are preferred? Explain.
7. It is intended to provide a partially panelled and partially glazed door for a 1.20 m wide and 2.40 m high opening. Suggest a suitable door and sketch the door with all details.
8. Name the various types of windows. Draw an elevation and vertical section of a casement window.

9. Draw an elevation and a vertical section of a casement window.
10. Why the metal windows are preferred to wooden windows?
11. Bring out the necessity of providing dormer window and bay window.
12. What are the standard sizes of windows? Explain.
13. List the precautions to be taken in case of aluminium windows.
14. Explain the various factors to be considered while locating the location of doors and windows.
15. What are ventilators? Explain their uses. Mention the standard dimensions of timber ventilators.
16. What are the points to be observed in case of natural ventilation?
17. Explain the requirements of a good ventilating system.
18. Explain the meanings of the following designations as recommend by Indian Standards.
    (i) 8 DS 20
    (ii) 6 WS 12
    (iii) 10 WT 12
    (iv) 12 DT 21
    (v) 12 V 6

21 STAIRS AND ELEVATORS

21.1 INTRODUCTION

A stair is a sequence of arrangement of steps which is provided as a means of easy ascent and decent between the floors or landings.

Stairs are of different types which are designed and used based on the location and usage. Stairs are provided in almost all types of buildings starting from residential building to multi-storeyed buildings.

Different materials are used for the construction of stairs, viz., stone, wood, brick, metal and concrete (both plain and reinforced).

In order to have a comfortable climb and descend, proper design has to be stipulated as regards to size of stairs, slope, landing space, etc.

Other modes of transportation between floors are lifts, ramps and moving stairs.

In this chapter, salient features of stairs and other transportation facilities are discussed.

21.2 COMPONENTS OF A STAIRCASE

The space occupied by the stairs is called a staircase or stairwell. Various components of a stair are shown in Fig. 21.1 and also defined below.

1. Baluster

It is a vertical member supporting the handrail.

2. Flight

It is a series of steps without any break such as a platform, landing, etc.

3. Tread

It is an upper horizontal portion of a step used to keep the foot while ascending or descending.

4. Riser

It is a vertical portion of a step which provides support to the tread.

5. Step

It is a portion of the stair comprising of tread and riser.

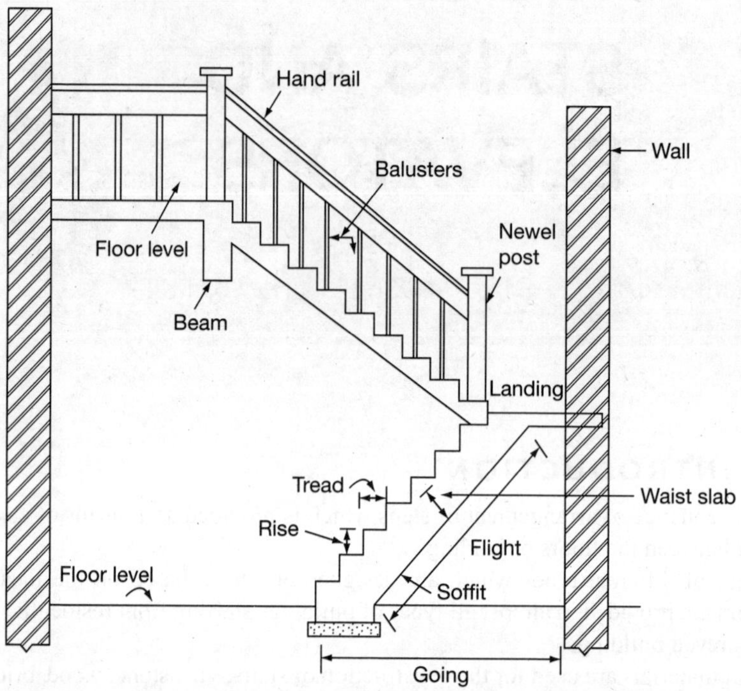

Figure 21.1 Components of a staircase

6. Rise

It is a vertical length between the upper faces of any two consecutive steps.

7. Landing

It is a flat platform provided between the flights.

8. Winders

These are the tapering used for a change of direction in a stair.

9. Soffit

It is the under surface of a stair.

10. Going

It is the horizontal projection of the flight.

21.3 REQUIREMENTS OF A GOOD STAIR

A well-planned and designed stair should be comfortable, quick, safe and easy to use. Different aspects which are to be considered in the planning and design of stairs are discussed below.

21.3.1 Location

The location of stairway should be so chosen such that sufficient ventilation and light should be available. If it is feasible, it may be located centrally so as to be of easily accessible from different parts of a building.

21.3.2 Width of Stair

Width of a stair, in general, should be adequate for the users without much crowd. It depends very much on the type of building and the number of users. In general, 1.0 m width is adopted for residential building and about 1.5 m width for public buildings.

21.3.3 Pitch of a Stair

The pitch or slope of a stair should prevent undue exortion to the user and at the same time there should not be any waste of space. Normally the slope of a stair should not exceed 40° and should not be flatter than 20°.

21.3.4 Length of a Flight

In order to make the ascend comfortably, the stairway should not have steps more than 12 and a minimum of 3.

21.3.5 Headroom

The headroom or the clear distance between the tread and the soffit of the flight immediately should be adequate. This should be at least 2.14 m.

21.3.6 Materials of a Stair

Construction of a stair should be carried out with sound and fireproof materials. Further, it should be adequately durable so as to have long life as that of the other materials used for the construction of the building.

21.3.7 Landing

The width of landing should not be less than the width of a stair.

21.3.8 Winder

As far as possible, the introduction of a winder should be avoided. They are liable to be dangerous. If it is inevitable, it may be provided near the lower end of a flight. Thus instead of quarter-space landing, three winders may be used. Similarly for half-space landing five winders or four radiating risers may be adopted.

21.3.9 Step Proportions

All the risers and treads should be of uniform dimensions. The steps should neither be too wide or too short. Further high rise may cause inconvenience and fatigue to the user.

The following empirical rules may be followed:

(i) Rise + Tread : not $<$ 40 and not $>$ 45
(ii) 2 Rise + Tread : not $<$ 58 and not $>$ 63
(iii) (Rise) × (Tread) : not $<$ 400 and not $>$ 500

For residential buildings, the common size of step is 15 × 28 cm. Rise greater than 20 cm and tread less than 22 cm should not be used. In general, (2 Rise + Tread) approximately equal to 60 will give a comfortable flight.

21.4 TYPES OF STAIRS

Different types of stairs which are in general use are discussed below.

21.4.1 Straight Stairs

In this type of stairs, there is no change in direction of any flight between successive floors. These stairs may have (i) straight run with a single flight or (ii) straight run with a series of flights with one or two landings in between. If the staircase hall is long and narrow, only straight stairs can be thought of (Fig. 21.2).

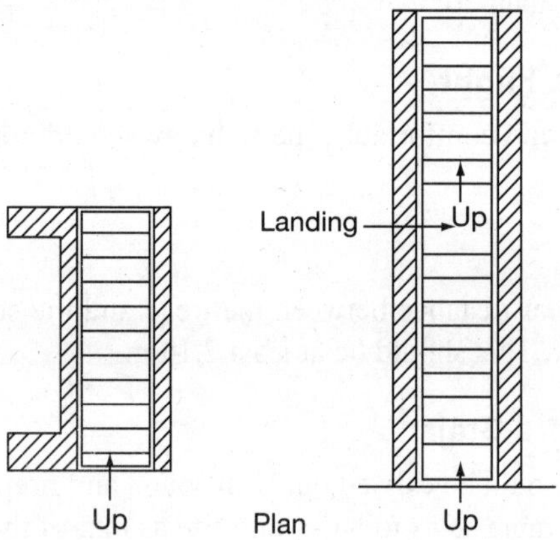

(a) Single flight (b) Two flights

Figure 21.2 Straight stairs

21.4.2 Quarter-turn Stairs

This type of stairs is preferred when there is a need for change of direction by 90°. The change in direction may be accompanied by providing either by introducing quarter-space landing or by providing winders (Fig. 21.3).

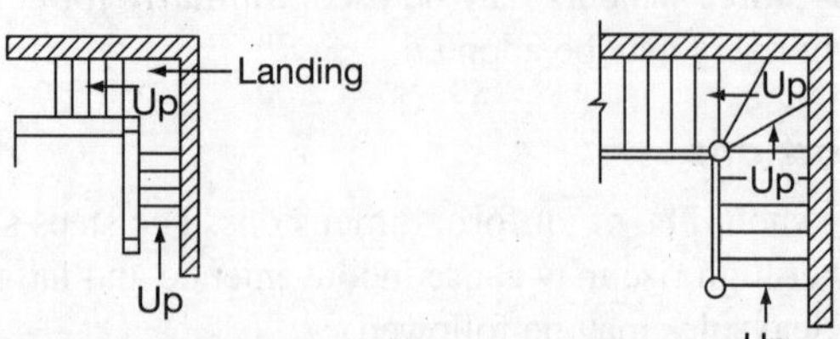

(a) Quarter-turn with landing (b) Quarter-turn with winders

Figure 21.3 Quarter-turn stairs

21.4.3 Dog-legged Stairs

In this type of stairs, the direction of flight is changed through 180° by introduction of landings and or winders. A half-space landing may be used. It is also called as dog-legged stairs. This type of stair is useful where the width of the stair-case is just sufficient to accommodate two widths of stair (Fig. 21.4).

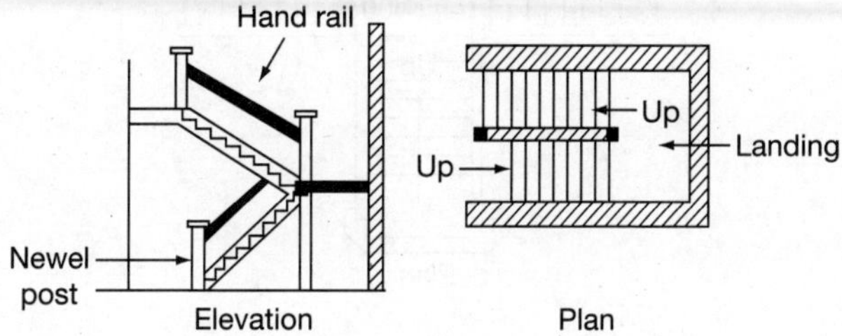

Figure 21.4 Dog-legged stairs

21.4.4 Three-quarter Turn Stairs

This type of stairs is provided with change in directions by 270°. That is, the direction is changed three times. In this case, an open-well is provided (Fig. 21.5). It is also called open newel stair.

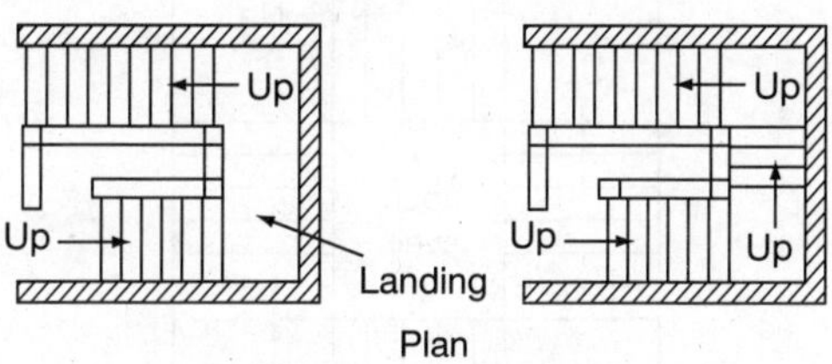

Figure 21.5 Three quarter-turn stairs

21.4.5 Geometrical Stairs

These stairs have no newel post and may be of any geometrical shape. Here the strings and hand rails are continuous and are set out in accordance with geometrical principles. Thus these stairs are also called as continuous stairs. They may be winding stairs, half-turn stairs (with landing or newels) and spiral stairs (Fig. 21.6).

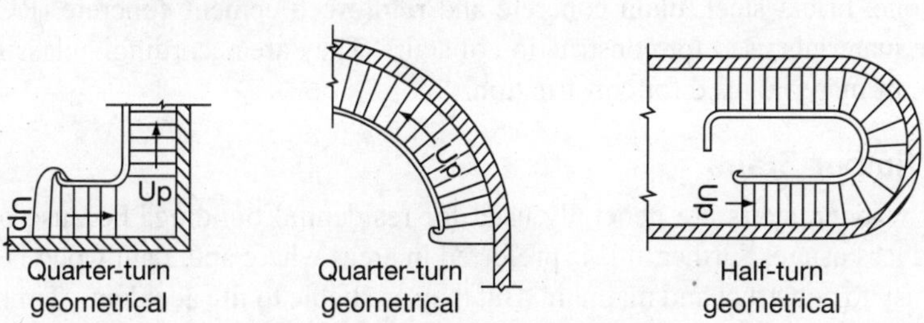

Figure 21.6 Geometrical stairs

21.4.6 Bifurcated Stairs

In these stairs, the bottom flight is wide which is divided into narrow flights from either side of the landing (Fig. 21.7). This type of stair is provided in modern public buildings.

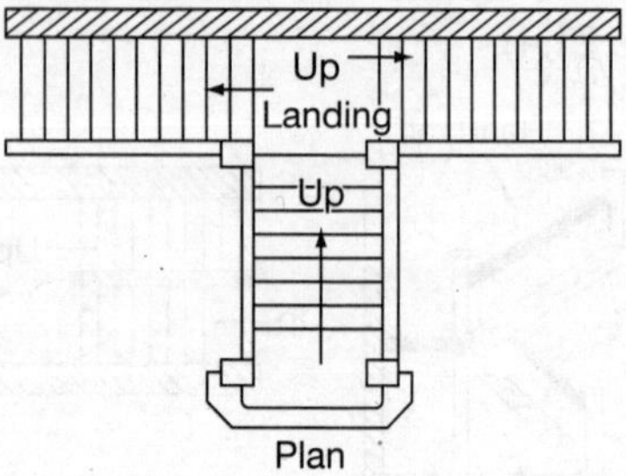

Figure 21.7 Bifurcated staircase

21.4.7 Multiple Flight Stairs

It is an open-well type of stair which has four flights each with flight turn of 90°. It has an excellent view in a building (Fig. 21.8).

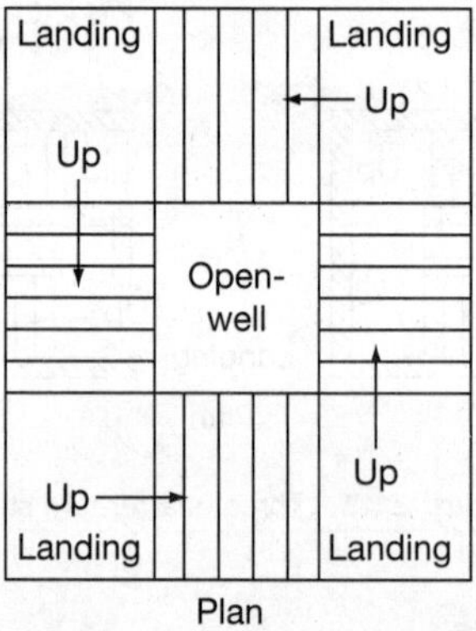

Figure 21.8 Multiple flight stairs

21.5 MATERIALS USED FOR STAIRS

Timber, stone, brick, steel, plain concrete and reinforced cement concrete (RCC) are in general the materials used for construction of stairs. They are accordingly classified based on the type of material used for construction.

21.5.1 Timber Stairs

Timber or wooden stairs are generally used for residential buildings because of its light weight and less usage. Further, this is preferred in areas where abundant wood is available. They are easy to construct and maintain. But they are liable to fire accidents. Timber should be free of fungal decay, insect attacks and other defects.

A typical detail of a wooden stair is shown in Fig. 21.9.

Straight flight stairs, half-turn stairs and open-well stairs may be constructed of wood.

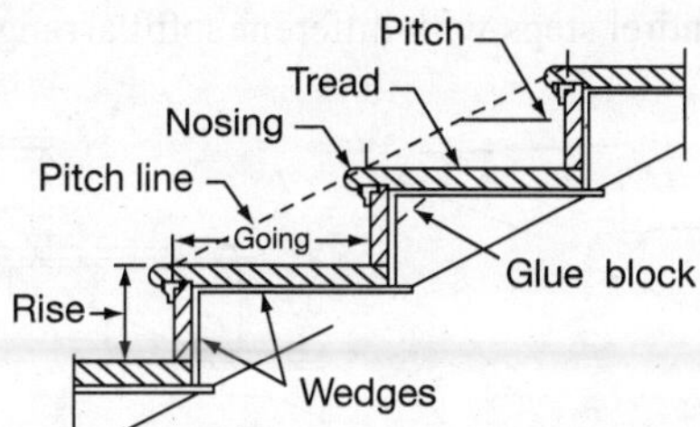

Figure 21.9 Typical details of a wooden stair

21.5.2 Stone Stairs

Stone stairs are heavy and need substantial supports. They are preferred for outside stairs and for threshold, approach to basements, heating chambers, etc. Due to regular use, the steps become slippery and dangerous.

Stones are in general hard, durable and weather resistant. They are widely used at places where ashlar stones are readily available. Different types of stone steps adopted are as follows:

(i) Rectangular steps
(ii) Spandril steps
(iii) Built-up steps

1. Rectangular Steps

Typical rectangular steps are shown in Fig. 21.10.

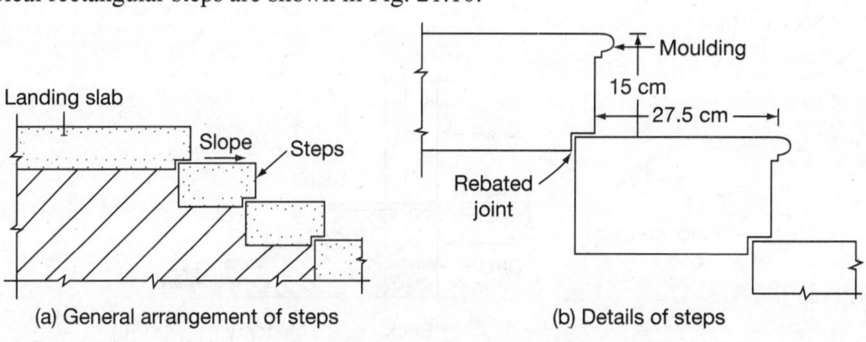

(a) General arrangement of steps (b) Details of steps

Figure 21.10 Typical rebated stone steps

In rectangular steps, the lower edge of one step is supported on the top back edge of the other. In order to strengthen the joint, rebated or checked joints are used. To provide a better appearance, the front edges have moulded noses.

2. Spandril Stone Steps

Here the steps are cut so as to provide a plain soffit. In general, these steps are nearly triangular in shape except at the ends which are built into the wall. This arrangement is used where the headroom is desired. The soffit affords a nice appearance and the weight of steps is also reduced.

Figure 21.11 shows spandrel steps with different soffit arrangements.

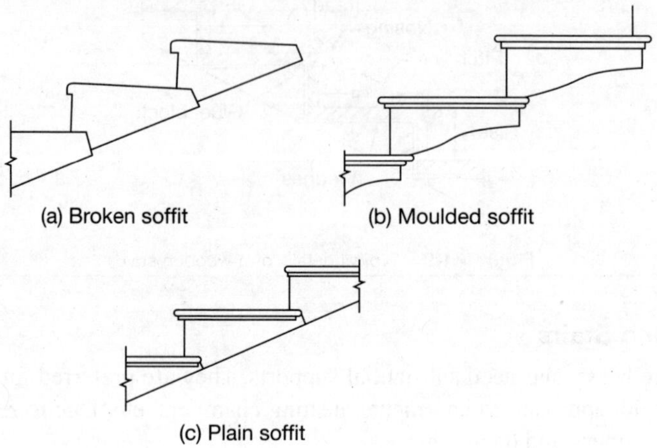

Figure 21.11 Spandril steps with different soffit arrangements

3. Built-up Steps

In built-up steps, the tread and riser of each step are made of thin swan slabs similar to timber steps. The stone slab treads and risers are connected by dowels. In case the treads are supported at the ends only, then the thickness of the tread should be more than 5 cm. The stone slabs are placed over concrete or brick steps. This type of steps will give an ornamental look where granite or marble stones are used. Figure 21.12 shows a built up steps at an entrance.

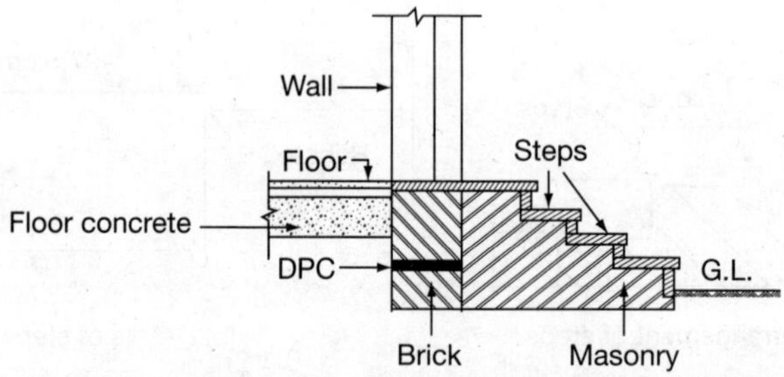

Figure 21.12 Built-up step at entrance

21.5.3 Brick Stairs

Brick stairs may be of solid masonry construction or provided with arches in the lower portion. When arches are provided, the total masonry work is reduced. This space may be used as a cupboard or as a passage. Here the treads and risers are generally made equal to length of $1\frac{1}{2}$ bricks and height of two layers of bricks respectively. Figure 21.13 shows a brick stair. The treads and risers of brick stair are finished with suitable flooring material.

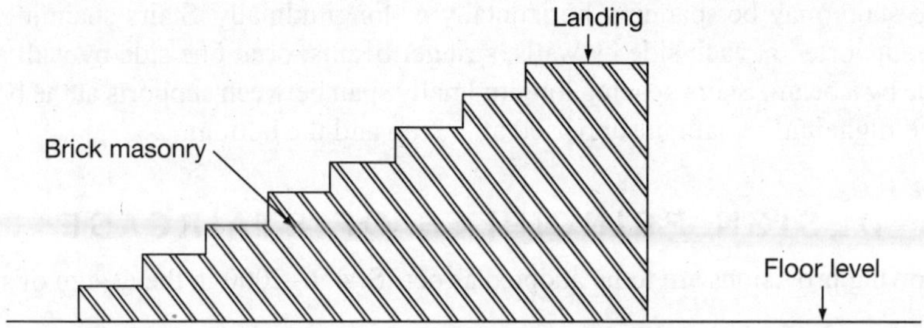

Figure 21.13 Brick stair with arch opening

21.5.4 Steel Stairs

Steel stairs are of special type which are fire resistant and are preferred in factories, workshops, godowns, etc. These are made of pre-stressed sheet steel. The steel sheets are formed into risers or pans to which the treads are inserted. The treads and risers are supported on angles which are in turn connected to the stringer. Treads may be of stone-concrete composition or metal.

21.5.5 Concrete Stairs

Plain concrete stairs are used in place of stone stairs these days. They are mostly used for the entrance to the buildings. These steps are mostly of spandril type and are supported as in the case of stone stairs.

Now-a-days, RCC stairs are mostly used in all types of buildings. Reinforced concrete stairs derive a lot of advantages compared to other types. They are as follows:

1. It can be easily moulded into any required geometrical shape.
2. It has high resistance to wear and fire better than any other material.
3. It needs less bulky sections, thereby more headroom may be attained.
4. It is less noisy.
5. It gives highly attractive appearance with appropriate finishes.
6. It can be kept clean.
7. It can be easily rendered non-slippery.
8. It can be designed to accommodate greater widths and longer spans.

A typical RCC slab stair is shown in Fig. 21.14.

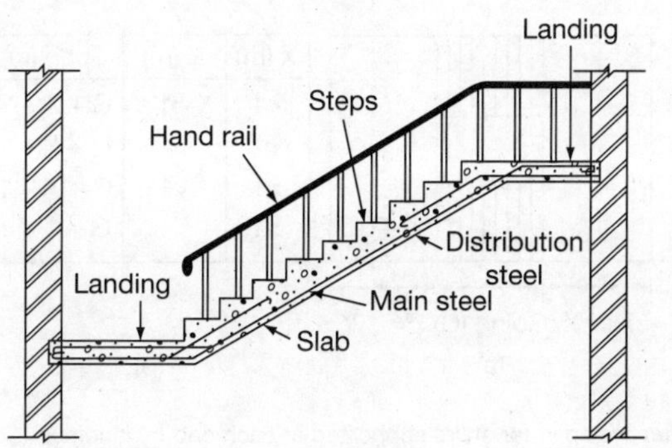

Figure 21.14 Typical RCC slab stair

Concrete stairs may be spanning horizontally or longitudinally. Stairs spanning horizontally are supported at each side by walls, stringer beams, or at one side by wall and at the other side by a beam. Stairs spacing longitudinally span between supports at the bottom and top of the flight and remain unsupported at the top and the bottom.

21.6 DESIGN PRINCIPLES OF STAIRCASE

The following provisions are to be adopted as per IS: 456–2000 in the design of staircases.

21.6.1 Dead Loads

Dead loads to be considered for staircase design comprise of self-weight of the waist slab, treads and risers, and self-weight of finishes. Finishes include hand rail, balusters and newel post.

21.6.2 Live Loads

For residential buildings, a uniformly distributed load of 2–3 kN/m^2 may be adopted depending on the users. For public buildings, a uniformly distributed load of 5 kN/m^2 is specified.

21.6.3 Effective Span of Stairs

1. When the flight of the stair is supported at the ends of landing beams, the effective span is taken as the projected horizontal distance between the centre lines of landing beam.
2. When the stairs are not provided with stringer beam, then the effective span shall be taken as the following horizontal distances:
   (i) Where supported at top and bottom risers by beams standing parallel with the risers, the distance centre-to-centre of beams.
   (ii) Where spanning on to the edge of a landing slab, which spans parallel with the risers (Fig. 21.15) a distance equal to the going of the stairs plus at each one, either half the width of landing or 1 m, whichever is smaller.
   (iii) Where the landing slab spans in the same direction as the stairs, they shall be considered as acting together to form a single slab and the span is determined as the distance centre-to-centre of the supporting beam or walls, the going being measured horizontally.

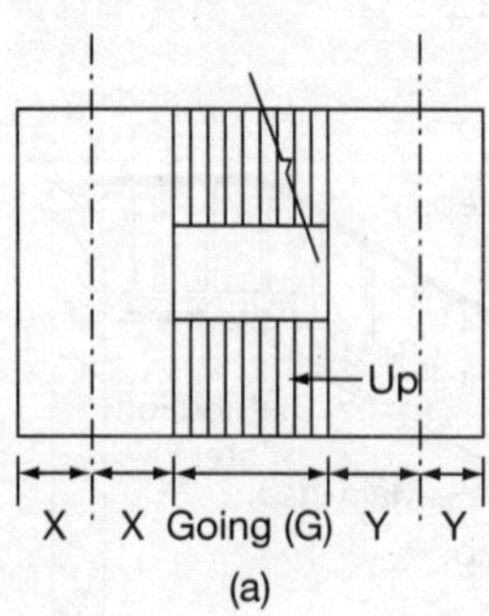

(a)

X (m)	Y (m)	Span (m)
<1	<1	G + X + Y
<1	>1	G + X + 1
>1	<1	G + Y + 1
>1	>1	G + 1 + 1

(b)

Figure 21.15 Effective span for stairs supported at each end by landings (*Source:* IS: 456–2000)

21.6.4 Rise and Tread

Following are the norms for the size of rise and tread:

	Rise (mm)	Tread (mm)
(i) Public buildings	150	300
(ii) Residential buildings	150–200	200–300

The following guidelines may be followed while deciding the size of rise and tread of a stair:

(i) 400 mm < (Rise + Tread) < 450 mm
(ii) 580 mm < (Rise + Tread) < 630 mm

21.6.5 Distribution of Load

Distribution of load for two conditions are given below:

1. In the case of stairs with open-well where spans partly crossing at right angles occur, the load on areas common to any two such spans may be taken as one-half in each direction (Fig. 21.16).

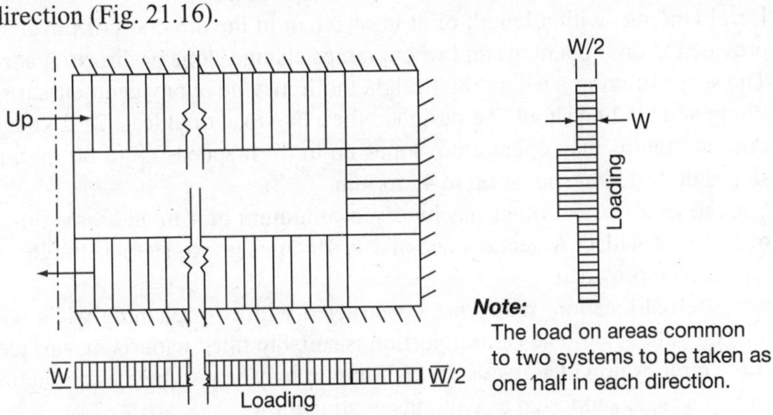

Figure 21.16 Distribution of loads on stairs (*Source:* IS: 456–2000)

2. Where flights or landings are embedded into walls for a length not less than 110 mm and are designed to span in the direction of the flight a 150 mm strip may be deducted from the loaded area and the effective breadth of the section increased to 75 mm for the purpose of design (Fig. 21.17).

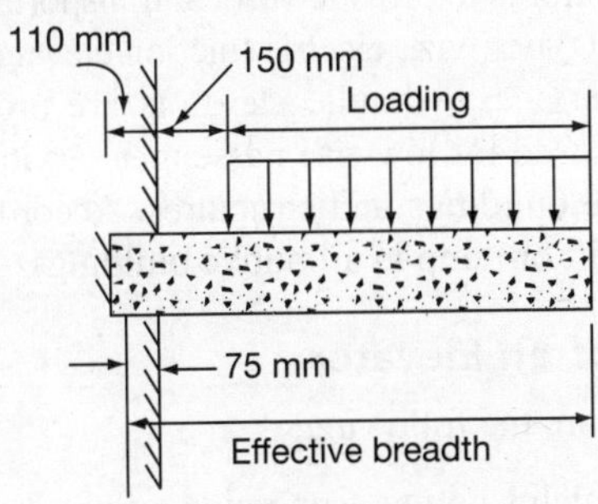

Figure 21.17 Loading on stairs built into walls (*Source:* IS: 456–2000)

21.7 RAMPS

Ramps are sloping surfaces that join two floors. These are provided where large numbers of people or vehicles or equipment have to be moved from floor to floor. Now-a-days, ramps are specially provided in all public places like hospitals, railway stations, etc., for movement of aged and physically challenged persons. Multi-storey car parks that are generally provided at the heart of the cities have ramps.

Ramps should be provided with a non-slippery surface. They occupy much larger space than stairs. It should be provided with a uniform sloping surface.

Ramps may be located either externally from the general building line or internally (i.e., inside the building). Based on the intended use, it can be located internally or externally. For movement of heavy equipment's, it may be preferred outside.

The ramp size evidently depends on its use. However, the following factors are taken into account in designing a ramp:

(i) The general slope of ramp is 15%. But a slope of 10% is preferred.
(ii) The width of ramp depends on the use. Minimum width of pedestrian ramps is 75 cm and a minimum slope of 1 in 10 cm is adopted.
(iii) Level landings with a length of at least 1.1 m in the direction of travel should be provided at door openings and where ramps change slope or direction abruptly.
(iv) The shape of ramp need not be straight but it may be of any geometrical shape.
(v) Ramp and landing should be designed for a live load of at least 21.2 kg/cm^2.
(vi) Powered ramps may operate on slopes up to 8° at speeds up to 60 m/min. and/or slopes up to 15° at speeds up to 47 m/min.
(vii) For carrying car and other machinery, a minimum of 4 m and a maximum width of 8 m is provided. A gentle slope of 1 in 10 cm is given. Here a straight or curved type can be provided.
(viii) For external location, ramps are constructed with two retaining walls with rising tops on either side. The central portion is suitably filled with rising surface.
(ix) If the ramp is provided inside the building, it may be designed as an inclined RCC slab or as slab supported as columns or stingers.
(x) Ramp surfaces may be rough cement or granolithic surface. Surface patterns or transverse grooves may be formed to improve slip resistance.

21.8 ELEVATORS

Buildings which are having more than four storeys are provided with elevators or also called as lifts. The function of an elevator is to provide vertical transportation of passengers or freight. The elevators may be of two types, viz., electric traction elevators or hydraulic elevators.

For a low-rise freight service, hydraulic elevators are provided which can rise up to about six storeys. It is also used for low-rise passenger service. Electric traction elevators can be provided for all rise buildings and structures. According to building regulations, it is mandatory to provide lift or ramp in all public buildings.

21.8.1 Components of an Elevator

Components of an elevator are the following:

1. Lift car is the one which is to move on guide rails.
2. A machine room with electric motor, winding machine and other necessary machines.

3. Suspension rope to hold and move the car.
4. Counterweight on pulleys to balance the car with about 50% of maximum live load.
5. Buffers for the car in the lift pit floor. For low velocity lifts, spring buffers are used and oil buffers for velocity lifts.
6. Special landing facility for entry and exit.
7. Capacity of a passenger lift is based on the weight of person taken as 68 kg.

21.8.2 Structural Component of an Elevator

Civil engineering construction components needed for an elevator operation are given below:

1. A lift well of suitable size. It is usually extending up to 1600–2000 mm below the bottom landing.
2. At every floor level for entry of people an opening of height of 2 m is to be provided.
3. An upper machine room on the top of the lift to suit the lift size is chosen.

21.8.3 Types of Elevators

Elevators are used for different purposes and accordingly they are classified as follows:

1. Passenger lifts
2. Goods lifts
3. Hospital lifts
4. Service lifts
5. Fireman's lifts

Separate specifications are available for each type of elevator. The relevant IS code has to be referred for specifications.

21.9 ESCALATORS

These stairs are also known as moving stairs or even moving flights. By a revolving drum, the escalator is kept in motion. A few steps at top and bottom are kept level through moving individually. A person has to occupy a step (preferably the first step) of the escalator for the upward or downward motion.

Essential features of the escalators are discussed below.

21.9.1 Essential Parts of Escalator

An escalator comprises of three parts, viz., steel trussed framework, hand rails and an endless belt with steps. Accurately arranged tracks are attached to the steel trusses and the steps move on these tracks.

21.9.2 Speed and Slope of Escalator

Speed of the escalator is normally 45 cm per second. The pitch or slope of the inclined bridge is kept at 30°. The moving stair is in the form of an inclined bridge between two successive floors.

21.9.3 Location of Escalator

It is necessary to make a careful study of flow of traffic before deciding the position of the moving stair. In a new structure the escalator has to be fixed at a position where the traffic is likely to be the heaviest.

21.9.4 Design Aspects of Escalator

The anticipated load on the escalator has to be properly assessed before design. The main factor which affects the design is the floor-to-floor height. The stair way should be kept independent by providing a structural framework around the stair well. The structural frame is designed to take care of the load of floor, hand rail, etc.

21.9.5 Installation

The various parts of the moving stair are fabricated in a workshop and installed at the location. Adequate care has to be exercised while installing each part. This arrangement ensures smooth working of the morning stair. Generally, escalators are fixed in pairs, one for upward movement and the other for downward movement. The escalators may be positioned diagonally opposite to each other or parallel to each other.

Example 21.1

The staircase room available in a residential building is 1.4 m × 4.8 m. The head room available is 3.2 m. Draw a lay-out plan of the stairs.

Solution:

As the space available is only 1.4 m, the width of staircase be taken as 1.4 m and no turn can be provided. Further the length of the stair is 4.8 m only and a landing can not be provided. Thus a straight-flight staircase be provided.
A rise of 20 cm is assumed

$$\text{No. of risers } (R) = \frac{3.2\times 100}{20} = 16$$

$$\therefore \quad \text{No. of treads } (T) = R-1 = 16-1 = 15$$

Width of tread is assumed as 25 cm

$$\text{Spaced occupied by 15 treads} = 25\times 15 = 375 \text{ cm}$$
$$= 3.75 \text{ m}$$

Space for passage = 4.8 – 3.75 = 1.05 m
The layout is shown in Fig. 21.18

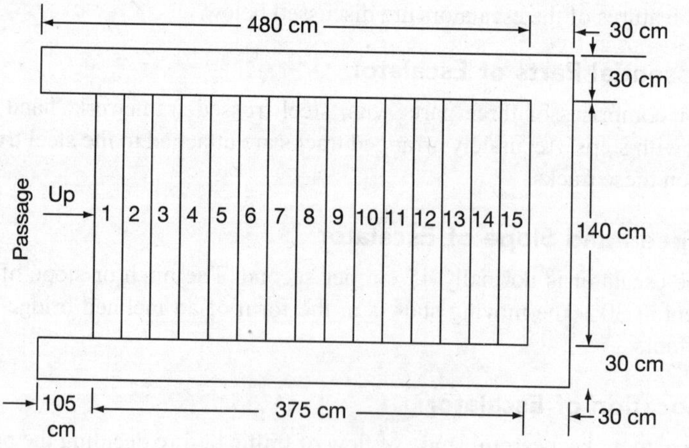

Figure 21.18

Example 21.2

A dog-legged staircase has to be provided in an office building. The staircase room is of 3.0 m × 6.0 m. The distance between the floors is 8.5 m. Design the stair-case and draw the lay-out plan.

Solution:

Width of stair is assumed as 1.4 m. As the width of the staircase is 3 m, a half space landing is provided with two flights. The width of landing is made equal to width of the stair.

Space available between the stairs $= 3.0 - 2\times 1.4$

$= 0.2$ m

Height of each flight $= \dfrac{3.5}{2} = 1.75$ m

A riser of 15 cm is assumed

No. of risers $(R) = \dfrac{175}{15} = 11.67 \approx 12$

Exact rise of each step $= \dfrac{175}{12} = 14.58$ cm

No. of treads to be provided in each flight

$$T = R - 1 = 12 - 1 = 11$$

Tread is taken as 30 cm

Here 15 cm × 30 cm step satisfies all the rules.

$$\text{Space occupied by 12 treads} = 30\times 12 = 360 \text{ cm} = 3.6 \text{ m}$$

Space left for passage = 6 – 3.6 = 2.4 m.

The lay-out plan is shown in Fig. 21.19.

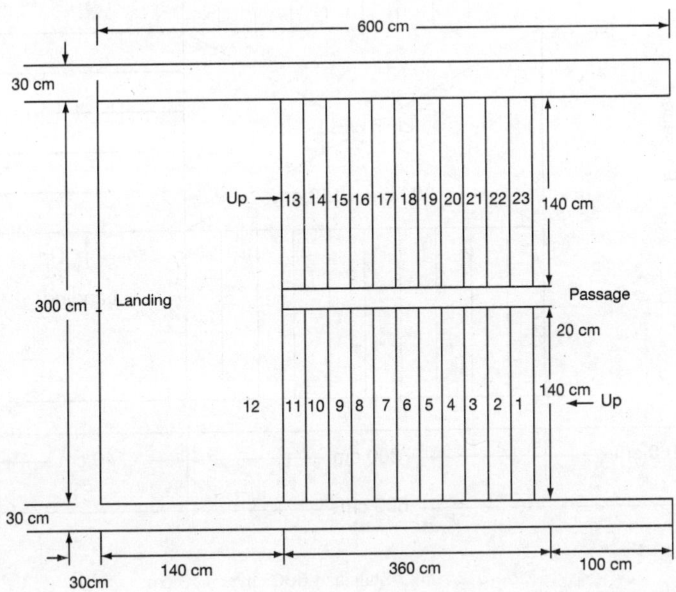

Figure 21.19

Example 21.3

It is intended to provide a suitable staircase for a hospital which consists of ground floor and first floor. The details are

Staircase room size = 6.0 m × 4.3 m

Height of each floor = 3.5 m

Thickness of floor slab = 15 cm

Width of stair = 140 cm

Provide a provision for lift.

Solution

An open-newel stair may be designed. The space of the well may be utilized for provision of a lift.

As in Example 21.2, adopt a rise of 15 cm and a tread of 30 cm.

Then,

$$\text{No. of risers} = \frac{\text{Total height of floor}}{\text{Hight of riser}}$$

$$= \frac{350+15}{15} = 25$$

Provide 10 steps in the first flight, 5 steps in the second flight and 8 steps in third flight. Figure 21.20 shows the lay-out plan of the staircase.

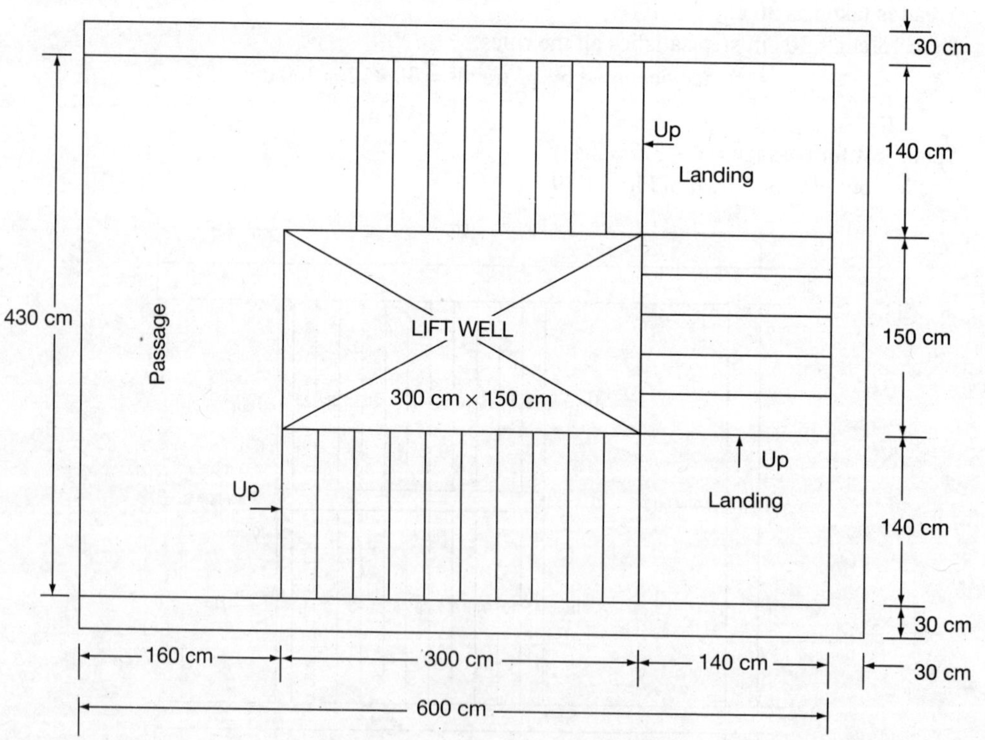

Figure 21.20 Staircase 600 cm × 430 cm

SALIENT POINTS

1. A stair is defined as a sequence of steps which are provided to ascend or descend between the floors or landings.
2. The space occupied by the stairs is called a staircase or stairwell.
3. A well-planned and designed stairway should be comfortable, quick, safe and easy to use.
4. The location of a stairway should be so chosen such that sufficient ventilation and light should be available.
5. Width of a stair in general, should be adequate for the users without much crowd. It depends on the type of building and the number of users.
6. Pitch or slope of a stair should prevent undue exhaustion to the user and at the same time there should not be any waste of space.
7. In order to make the ascend comfortable, the stairway should not have steps more than 12 and a minimum of 3.
8. The headroom or the clear distance between the tread and the soffit of the flight immediately should be adequate. This should be at least 2.14 m.
9. As far as possible, the introduction of a winder should be avoided. If it is inevitable, it may be provided near the lower end of a flight.
10. All the risers and treads should be of uniform dimensions. The steps should neither be too wide or too short. The following empirical rules may be followed:
    (i) Rise + Tread: not $<$ 40 and not $>$ 45
    (ii) 2 Rise + Tread: not $<$ 58 and not $>$ 63
    (iii) (Rise) $\times$ (Tread): not $<$ 400 and not $>$ 500
11. In straight stairs there is no change in direction of any flight between successive floors.
12. Quarter-turn stirs are preferred when there is a need for change of direction by 90°.
13. In dog-legged stairs the direction of flight is changed through 180° by introduction of landings and or winders.
14. Three-quarter turn of open newel stairs are provided with change in direction through 270°.
15. Geometrical stairs also called as continuous stairs have no newel post and may be of any shape.
16. In Bifurcated stairs the bottom flight is wide which is divided into narrow flights from either side of the landing.
17. Multiple flight stairs is an open-well type of stairs which has four flights each flight turn 90°.
18. Timber, stone, brick, steel, plain concrete and RCC are in general the materials used for construction of stairs.
19. Timber or wooden stairs are generally used for residential buildings because of its light weight and less usage.
20. Stone stairs are preferred for outside stairs and for threshold, approach to basements, heating chambers, etc.

21. Brick stairs may be of solid masonry construction or provided with arches in the lower portion.
22. Steel stairs are of special type which are fire resistant and are preferred in factories, workshops, godowns, etc.
23. Plain concrete stairs are used mostly for the entrance of the buildings. These steps are of spandrel type.
24. RCC stairs are used mostly in all type of buildings now-a-days.
25. Design principles to be considered are: Dead load, Live load, Effective span of stairs, Rise and Tread proportion and Distribution of load.
26. Ramps are sloping surfaces which join two floors. These are provided where large numbers of people or vehicles or equipment have to be moved from floor to floor. Ramps are specially provided in all public places like hospitals, railway stations, bus terminals, airports, etc., and for movement of aged and physically challenged persons.
27. Buildings which are having more than four storeys are provided with elevators or lifts. Lifts are operated hydraulically or by electric traction.
28. Elevators are provided for different purposes and accordingly classified as Passenger lifts, Goods lifts, Hospital lifts, Service lifts and Fireman's lifts.
29. Escalators are the moving stairs which moves by using revolving drum. A few steps at top and bottom are kept level through moving individually.
30. Different factors to be considered in the design of an escalator are: Speed and slope of escalator location.

REVIEW QUESTIONS

1. Discuss the requirements of a good stair.
2. List the components of a staircase.
3. What are the various types of stairs used in buildings? Illustrate your answer with sketches.
4. Draw a sectional elevation and plan of a dog-legged stair connecting two floors of a building.
5. Enumerate and justify the essential requirements of a domestic staircase.
6. Distinguish between quarter-turn and half-turn staircases used in buildings.
7. What are the types of stone stairs? Illustrate with sketches.
8. How are the stone stairs constructed and supported?
9. Bing out the advantages of concrete stairs compared to other stairs.
10. Explain the design principles of a staircase.
11. (a) State briefly the requirements of a good staircase
    (b) How are treads and risers proportioned?

(c) Where do you see:
   (i) Dog-legged stair
   (ii) Open-newel-stair, and
   (iii) Geometrical stair. (AMIE)

12. A suitable stair has to be located in a staircase 7.67 m long, 4.54 m wide and 4.54 m high, with a door 1.22 m wide in each of the longitudinal walls. The doors face each other and are located with their centres at a distance of 1.07 m from the respective corners of the staircase. Assuming that it is in a public building, design the stair and draw a dimensional sketch plain and a dimensional sketch section to illustrate your proposals (AMIE)
13. Indicate the situations where you would recommend (i) a ramp and (ii) an escalator.
14. When it is necessary to provide elevators in a building? Give the components of a lift.
15. What is an escalator? Discuss the essential requirements for its provision.
16. Where do ramps are preferred?

22 TEMPORARY SUPPORTING STRUCTURES

22.1 INTRODUCTION

Temporary supporting structures are those structures which are required in the routine building construction works or repair works. During the process of concreting work, temporary structures called formworks are needed till the concrete attains its full strength.

Similarly for laying of masonry work, plastering, painting and for temporary repair, some temporary structures are used which are called scaffolding.

During the process of excavation, or temporarily supporting a weak structure some temporary structures are constructed which are called shoring.

Sometimes there may be some necessity to strengthen the existing foundation or to undertake some repairs or to increase the number of floors or introducing a basement floor, some temporary structures are constructed which are called underpinning.

In this chapter such temporary structures are discussed.

22.2 FORMWORK

Concrete initially is in a plastic state and has to be kept within an closure of a desired shape by proper temporary supports till it gains adequate strength. This temporary enclosure is known as formwork or *shuttering* or simply as *forms*.

For circular works such as arches, doors, etc., the term centring is used generally instead of formwork or shuttering. Similarly the term *moulds* is used to indicate formwork of relatively small units such as lintels, cornices, cubes for testing, etc.

22.2.1 Requirements of Formwork

Irrespective of the type of material used the formwork should satisfy the following requirements:

1. Adequate strength
2. Smooth inner surface
3. Enough rigidity
4. Quality
5. Less leakage
6. Economy
7. Easy removal
8. Supports

1. Adequate Strength

As the formwork has to bear the weight of concrete which exerts a lot of pressure on all sides when it is wet, strength is an important and essential requirement. Resistance to outward push is achieved with the help of bracings and the wire fastened on the outer side of the formwork. Further the temporary load during placing of concrete by men and machinery should also be withstood by the formwork.

2. Smooth Inner Surface

It is an essential requirement that the inner surface of the formwork to have smooth surface on all concrete structures. This is specially desired, in case, the concrete is to be exposed without plastering. All the projections, if any, are to be smoothened before concreting. Steel sheet formworks give a smooth surface. Further it is the practice to apply crude oil or soft soap solution to the inside surface of the formwork. This also further makes the removal of formwork easy.

3. Enough Rigidity

The formwork should be rigid enough to retain the shape without undue deformation. For visible finished surfaces, the allowable deformation is 1/300 of span and 1/150 of span for hidden surface. In order to use the formworks repeatedly, they should be rigid, robust and stiff.

4. Quality

Formwork should be designed correctly such that the required size, shape and finish are obtained. Particularly the verticality of the side formwork should be made. In order to avoid leakages through the formwork, the timber used for formwork should not have any cracks or holes.

5. Less Leakage

Adequate care is taken to see that the formwork should be so placed that there is minimum joint. Sometimes joints open once some concrete is placed. This should be avoided and the leakage should be corrected by adjusting the vertical supporting poles of the formwork.

6. Economy

The cost of the formwork should be minimum as the formwork is only a temporary structure and not going to contribute to the stability of the structure. This demands a careful selection of formwork material and an optimum design.

7. Easy Removal

Formwork should be placed in such a manner that it can be removed easily without causing any damage to the structure. During removal of the formwork, the nails and screws should be removed with little hammering. Such a careful removal prevents the possibility of injury to the concrete. Further such smooth removal enables the formwork to be re-used with less expense.

8. Supports

In general formwork should rest on strong, hard and non-yielding supports. For this the ground should be hard or the supports should rest on cured base concrete. The vertical poles supporting the formwork should be strong and should not due to the weight of concrete.

22.2.2 Materials Used

Materials generally used are timber and steel. Sometimes plywood and aluminium are also used. The choice of material to be used for the preparation of formwork depends on the nature of the job and economy.

1. Steel Formwork

If it is intended to re-use the framework material for several times it is preferred to use steel or aluminium. The initial cost of steel is very high.

Based on the requirements, the sizes of steel section are decided. For example, if formworks are prefabricated into panels or sections it is to be done as big as the handling will permit or the size of concrete. It should be noted that individuals can carry a maximum load of 35 kg.

Steel formworks are preferred based on the following advantages:

(i) As it can be easily installed and dismantled, there is less labour cost.
(ii) It is feasible to use several times.
(iii) It is of high strength and hence durable.
(iv) Precise design can be done.
(v) As it does not absorb water from concrete, the formation of honeycomb surface is avoided.
(vi) Concrete surface exposed in uniform without any undulations.
(vii) High degree of accuracy is possible, as it does not shrink or expand.

2. Timber Formwork

Timber formwork can not be used repeatedly as that of steel formwork. Timber intended to use as formwork should be well seasoned and should be neither too dry nor too wet. The components of timber formwork depend on the design load and the type of timber available. Number of nails used in the formwork should be less and the heads projecting out for easy removal.

Timber formwork claims the following advantages:

(i) Initial cost is less and hence cheap.
(ii) With less modification it can be easily altered.
(iii) It is used for small works requiring less repetitions.
(iv) In buildings it is economical as the formwork can be used for some other building work with minimum variations.

3. Plywood Formwork

For light loads, plywood's are used nowadays. Following are the advantages compared to timber formwork:

(i) Like steel it can be re-used for a few times.
(ii) No finishing treatment is needed, as it gives surfaces which are plain and smooth.
(iii) As large sizes of plywood are available, it is possible to cover large area and less labour cost is involved.

Typical formwork for columns and for beams and slab are shown in Figs. 22.1 to 22.3.

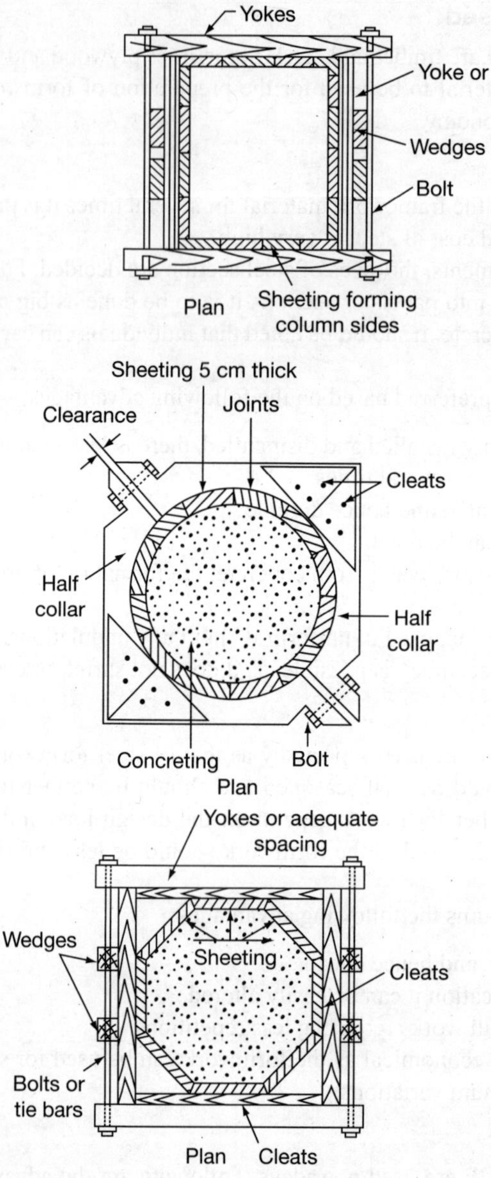

Figure 22.1 Formwork for different shapes of columns

22.3 DE-SHUTTERING OF FORMWORKS

Under no circumstance the forms are removed until the concrete has hardened sufficiently. Under normal weather conditions, the concrete is left in place as long as possible up to a maximum of 28 days.

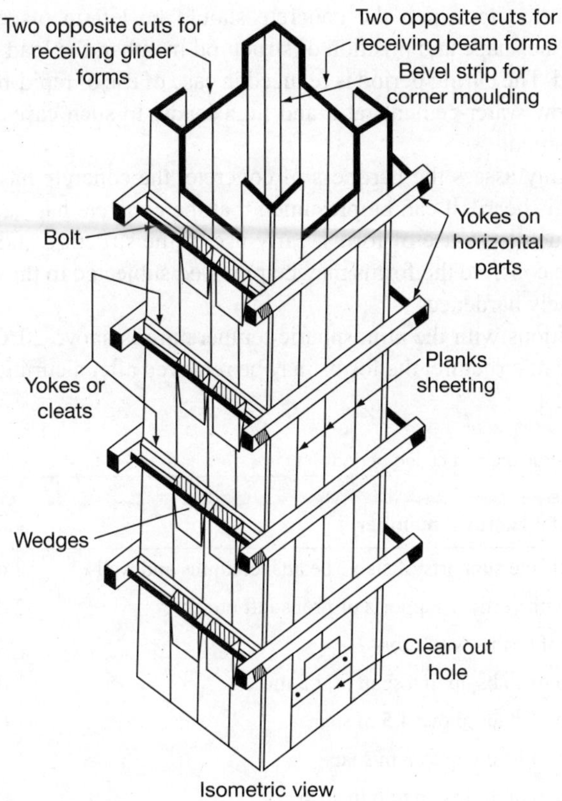

Figure 22.2 Formwork for square column

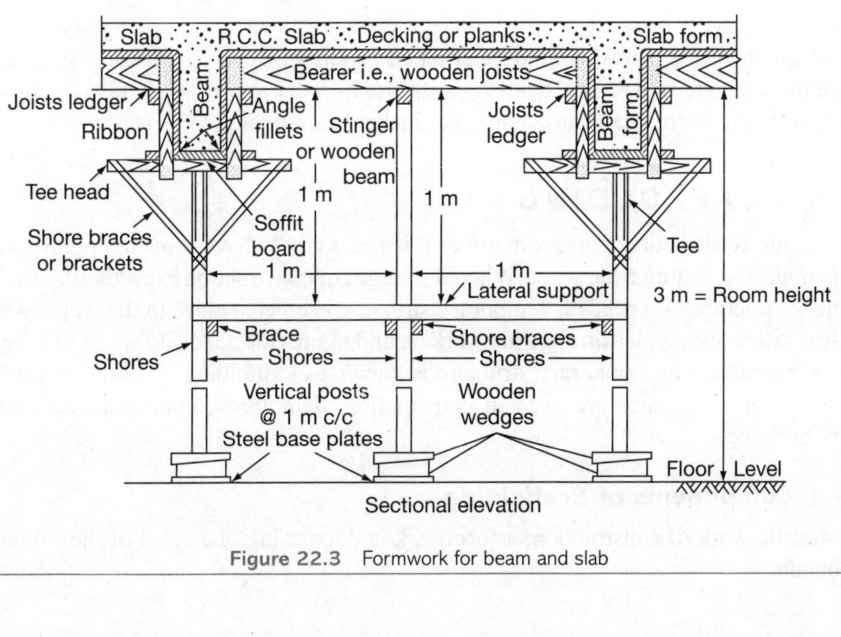

Figure 22.3 Formwork for beam and slab

However, the period to which the concrete should be left in place depends on the temperature of air, the shape and position of structural member, the load condition and the type of cement used. The curing period is reduced in case of use of rapid-hardening cement, low temperature, low water-cement ratio and high loads. In such cases early removal of forms may be permitted.

In order to roughly assess the hardness of concrete, the concrete has to be struck and if a metallic sound is heard, it can be presumed that the concrete has hardened. However, for other reasons such as re-use of forms, early use of the structure and need to cool the concrete in massive concrete the forms may be removed subjected to the condition that the concrete is adequately hardened.

In normal conditions with the atmospheric temperature is above 20°C, normal cement is used and the type of structure, the forms may be removed after a curing period as recommended in Table 22.1.

Table 22.1 Type of structure and curing period

Sl. No.	Type of structural member	Curing period
1.	Vertical side supports of slabs, beams, columns and walls	2 days
2.	Slabs with vertical supports or props left under	7 days
3.	Soffits of beams with props	7 days
4.	Bottom of slabs up to a span of 4.5 m	7 days
5.	Bottom of slabs above 4.5 m span	14 days
6.	Bottom of beam up to 6 m span	14 days
7.	Bottom of arch ribs up to 6 m span	14 days
8.	Bottom of beams over 6 m span	21 days
9.	Bottom of arch ribs over 6 m span	21 days

After adequate curing, forms should be stripped or removed with great care so as not to damage the concrete wedges, vertical supports etc. In order to prevent sudden application of load on the structure, the forms should be slackened gradually.

22.4 SCAFFOLDING

In the routine construction work, sometimes it is necessary to have some temporary structure or support to continue the work. When the height of construction exceeds about 1.5 m, temporary structures are needed. Temporary structure is erected close to the work so as to provide a safe working platform for the workers and to provide adequate space to keep the working materials. This temporary structure is known as scaffolding or simply a scaffold. Such temporary structures are used in construction, demolition, maintenance or repair work of buildings.

22.4.1 Components of Scaffolding

For a specific work of construction, a formwork is done using one or all of the following components.

1. Standards

These are the vertical members of a scaffold which are either supported on the ground or embedded in to the ground or rested on sand-filled bags.

2. Ledgers

These are the horizontal members of the scaffold.

3. Putlogs

These are transverse pieces which are placed on ledgers in perpendicular direction and supported on the wall.

4. Transoms

There are putlogs whose both ends are supported on the ledgers.

5. Bridges

These are used to bridge an opening in a wall and support one end of the putlog at the opening.

6. Braces

These are the cross or diagonal pieces fixed on the standards.

7. Guard Rail

This is a horizontal member provided like a ledger at the working level.

8. Toe Board

This is a board placed parallel to the ledgers and supported between the putlogs. This facility is made as a protective measure to work on the working platform.

9. Raker

This is an inclined support.

22.4.2 Requirements of Scaffolding

Scaffolding has to satisfy the following requirements irrespective of the material used:

(i) Method of erection should be easy with less wastage of material.
(ii) As persons have to use the scaffolding at all heights, it should possess adequate strength.
(iii) Materials needed for scaffolding should be available in all the required sizes and lengths.
(iv) It should be feasible to interchange the material for other works with minimum wastage.
(v) Initial cost should be comparatively less and should have high scrap value.
(vi) Fire-resistant scaffolding should be preferred.
(vii) For erection purposes no skilled labour should be depended upon.

22.4.3 Types of Scaffolding

Following are the different types of scaffolding:

1. Single scaffolding or bricklayer's scaffolding
2. Double scaffolding or mason's scaffolding
3. Cantilever or needle scaffolding
4. Suspended scaffolding
5. Trestle scaffolding
6. Steel scaffolding
7. Patented scaffolding

1. Single Scaffolding or Bricklayer's Scaffolding

This type of scaffolding is commonly used and particularly in the construction of brickwork. It consists of a single row of standards placed at a distance of about 1.20 m from the wall. The spacing between the standards is about 2–2.5 m. The ledgers are fixed at a vertical distance of 1.20–1.80 m on the standards.

The putlogs are placed at a horizontal spacing of 1.20–1.80 m. The details of the scaffolding is shown in Fig. 22.4. This type of scaffolding is also called as *putlog scaffolding*.

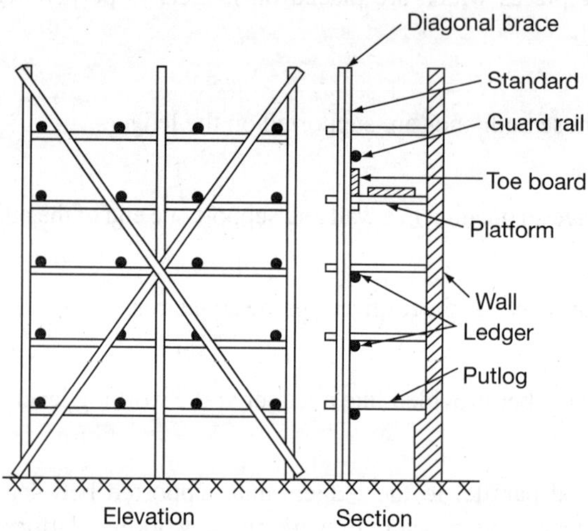

Figure 22.4 Single Scaffolding

In order to extend the work vertically, the platform is raised by extending, the standards by adding extra pieces. The scaffolding is removed, generally, after the completion of plastering and pointing works. After the removal of the scaffolding, the holes in the walls used by putlogs are filled immediately.

2. Double Scaffolding or Mason's Scaffolding

This type of scaffolding is stronger than the single scaffolding. This is similar to that of single scaffolding except two rows of standards are used. Out of these two standards, one is close to the wall and the other is 1.2–1.5 m away from the face of the wall, Fig. 22.5.

Here, no holes are made in the wall for putlogs and the putlogs are supported at both ends on ledgers. Sometimes diagonal bracings and inclined supports called rakers are provided. This type of scaffolding is particularly used for stone masonry construction.

3. Cantilever or Needle Scaffolding

In this type of scaffolding the general framework may be of single or double type of scaffolding. But the standards are supported by needles or ties. These ties are projected out at floor levels or through openings or through holes provided in the masonry. Two types cantilever scaffolding are shown in Fig. 22.6.

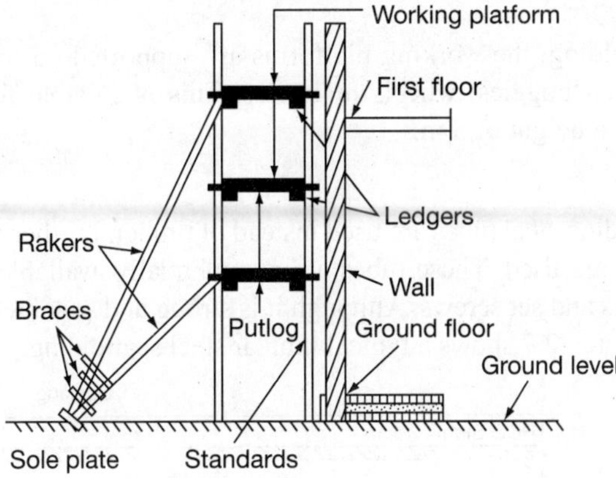

Figure 22.5 Double scaffolding

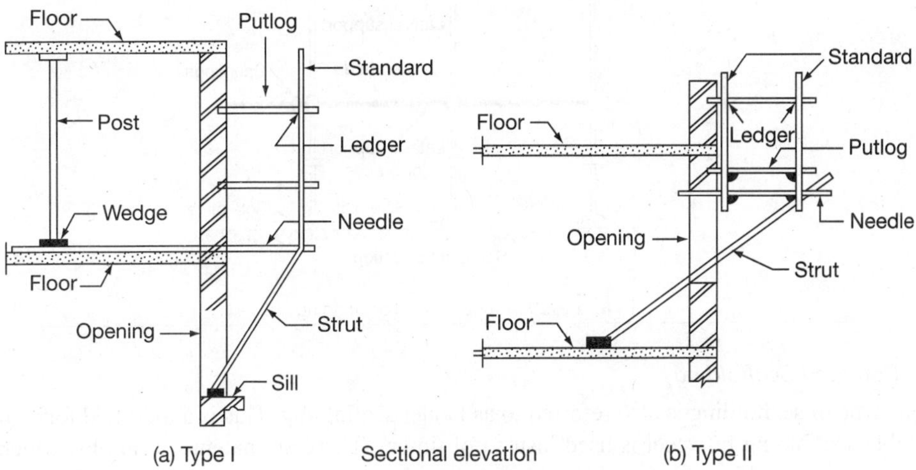

Figure 22.6 Cantilever scaffolding

This type of scaffolding is useful under the following conditions:

(i) In situations where it has to be provided on a busy street so as not to disturb the traffic.
(ii) In situations where it is difficult to fix the standards on the ground.
(iii) In construction of tall buildings.

4. Suspended Scaffolding

This is a light type of scaffolding used only for maintenance works like pointing, white washing, etc. The working platform is suspended from the roofs. Special arrangements are made with pulleys, ropes, etc., to suspend the platform from the roof and to raise or lower based on the need. As it does not create any obstruction on the ground and only a minimum space is required, this arrangement is preferred.

5. Trestle Scaffolding

In this type of scaffolding, the working platforms are supported on tripods, ladders, etc., which are mounted on boggies, wheels or lorries. This is suitable for minor repairs or painting work within a height of 5 m.

6. Steel Scaffolding

In this type of scaffolding steel tubes are used instead of timber. Further 40–50 mm diameter tubes of 5 mm thick are used. These tubes are commercially available in suitable lengths with special couplings and set screws. Although it is strong and suitable it can be used up to a specific height. Figure 22.7 shows a typical tubular steel scaffolding.

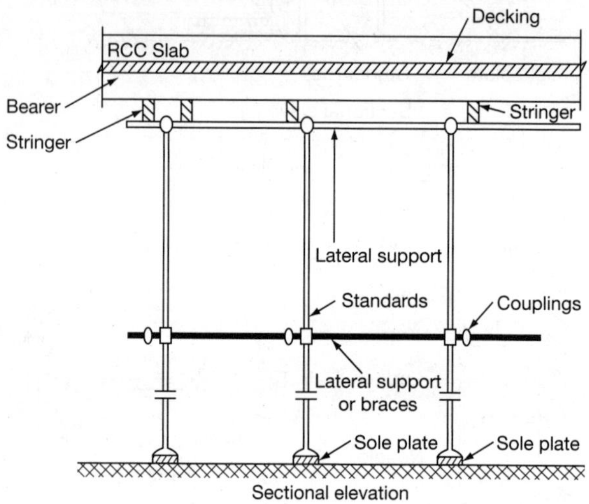

Figure 22.7 Tubular steel scaffolding

7. Patented Scaffolding

This type of scaffolding is also referred to as ladder scaffolding. This is a modified form of double scaffolding but steel is used instead of timber. There are patented scaffolds which are readily available in the market with special types of couplings and frames. Here, the working platform is supported on a bracket which can be adjusted to any suitable height. Such patented scaffoldings are used to light works like painting or decoration, Fig. 22.8.

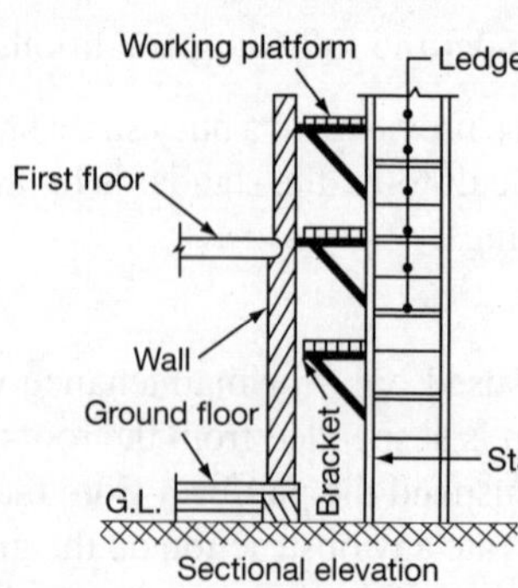

Figure 22.8 Patented steel scaffolding

22.5 SHORING

Shoring is the means of providing support to get stability of a structure temporarily under certain circumstances during construction, repair or alteration.

Such a circumstance may arise when

(i) The suitability of a structure is a endangered due to removal of a defective portion of the structure.
(ii) The stability of a structure is endangered due to the unequal settlement during construction.
(iii) Certain alterations are required to be done in the present structure itself (e.g., re-modelling of walls, changing position of windows, introducing a beam, etc).
(iv) Alterations are carried out in adjacent building for re-modelling, strengthen foundation, etc.

22.5.1 Installation of Shoring

Following are the points to be kept in view while installing shoring:

(i) For shoring, timber or steel tubes may be used. Sometimes both are used in combination. If timber is used its surface should be coated with a preservative so as to protect from wet rot.
(ii) The shoring should be designed based on the load it has to sustain and the duration of load.
(iii) Shoring may be installed internally or externally depending on the requirement and in certain cases they may be provided on either side of the wall to produce additional stability.
(iv) Over-turning forces, if any, must be resisted by the shoring.
(v) Shoring should be installed only after getting the necessary permission, if necessary, from the local authorities.
(vi) There is no time limit to which the shoring has to be kept. It may range from weeks to years depending on the case.

22.5.2 Types of Shoring

Shorings are classified under the following categories:

(i) Raking or Inclined shores
(ii) Flying or Horizontal shores
(iii) Dead or Vertical shores

1. Racking or Inclined Shores

In the type of shoring inclined members are adopted to provide temporary support to the external walls from the ground. These inclined members are called as rakers. An inclined shore primarily consists of rakers, braces, wall plate, needles and plank.

The wall plate is fastened to the wall by means of square needles. The needles penetrate into the wall for a depth of about 15 cm and prevent the wall plate from sliding against the wall. The wall plate distributes the pressure evenly (Fig. 22.9). Further the needles in turn are strengthened by providing wooden cleats. The inclined rakers are interconnected by timber braces. The feet of the rakers are tied together by braces and hoop iron. They are in turn connected to sole plate by means of iron dogs or dog spikes.

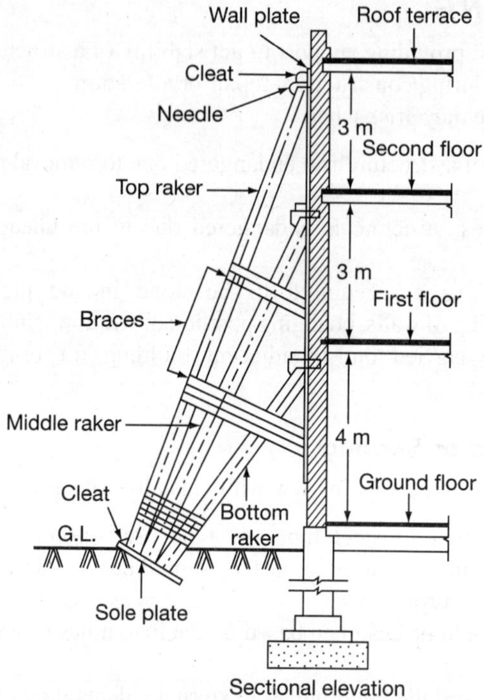

Figure 22.9 Raking or Inclined shoring of a wall

The details shown in Figure 22.9 is a simple raking shore for a building height of 10 m. For multi-storeyed building and building on road side, special rakers have to be used. While dealing with road side buildings, the bye-laws in force of the locality have to be followed.

2. Flying or Horizontal Shoring

In this type of shoring horizontal supports are provided parallel to the walls which have become unsafe due to some reason (Fig. 22.10). This arrangement is called a simple flying shore.

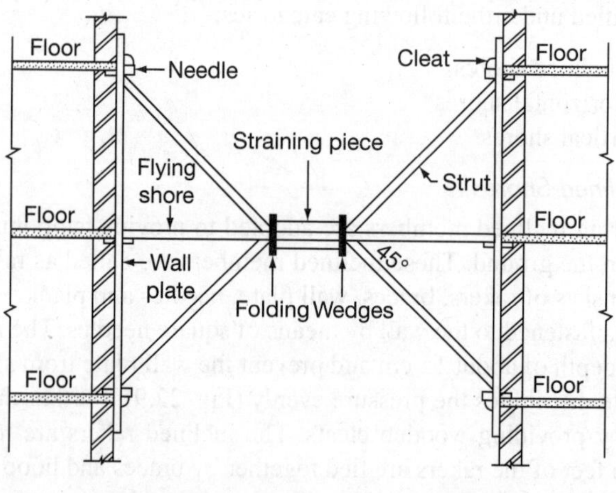

Figure 22.10 Details of single flying shore

This type of flying shore comprises of wall plate, needles, cleats, struts, straining pieces and folding wedges. As in inclined shores, here also the wall plates are secured against the wall by means of needles and cleats. The horizontal shore is kept in the required position by means of wedges, needles and cleats to the wall plate. The inclined struts are supported by the needles at there one end and straining sill at the other end. In turn the straining is fixed to the horizontal shore.

This type of simple flying shoring can be adopted for a maximum distance of about 9 m between the adjacent parallel walls. When the distance is from 9 to 12 m, a compound or double flying shore, shown in Fig. 22.11, may be provided. It is to be noted that both the horizontal shores are symmetrically placed with respect to floor levels.

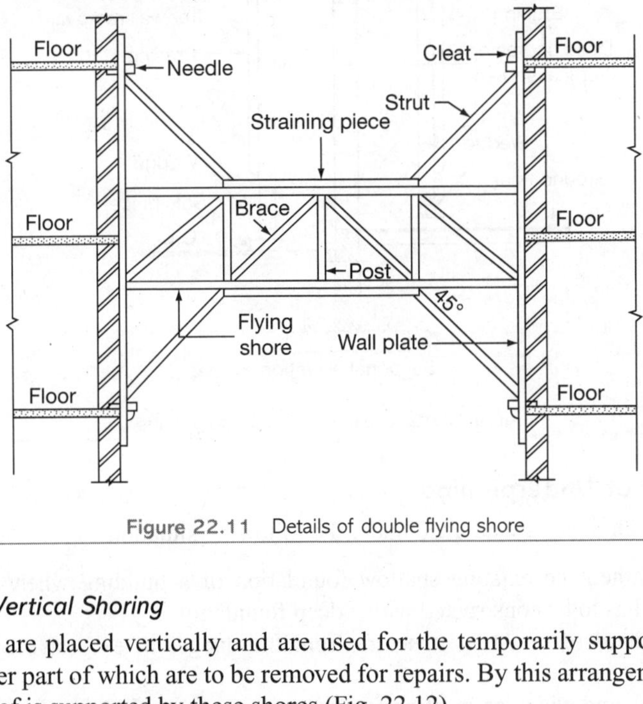

Figure 22.11 Details of double flying shore

3. *Dead or Vertical Shoring*

These shores are placed vertically and are used for the temporarily supporting the wall while the lower part of which are to be removed for repairs. By this arrangement the whole load of the roof is supported by these shores (Fig. 22.12).

Needles are used to transfer the load from the walls. These are first inserted into small wall opening which is made just sufficient to fit. Horizontal beams are laid along the floors. This beam supports the dead or vertical shore and distributes the load evenly on the base. The dead shores are installed in between the beams and the needles by means of wedges. In order to avoid any damage, the usual walls are supported with props before removing the desired portions. In the same way, the windows or other openings are duly strutted.

22.6 UNDERPINNING

Underpinning is the method of supporting structures while providing new foundations or strengthening the foundation or carrying out repairs and alterations without affecting the stability of the existing structures.

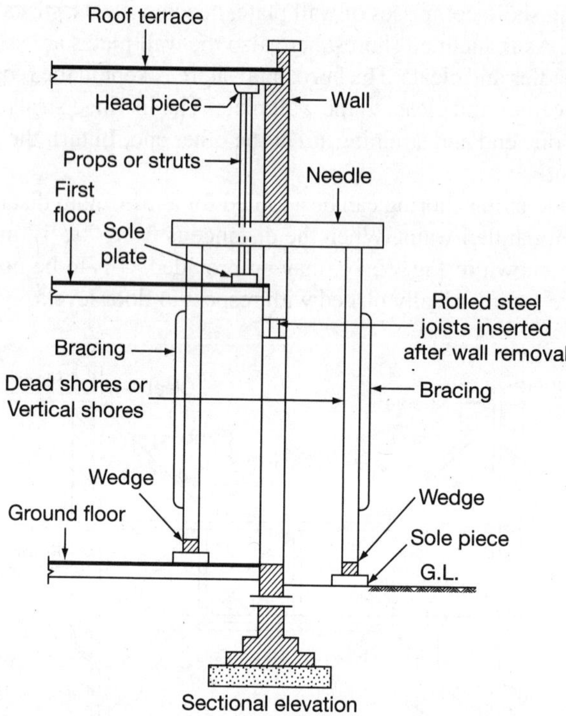

Figure 22.12 Dead-shores used for repairs

22.6.1 Uses of Underpinning

Underpinning techniques are adopted under the following situations:

(i) To strengthen the existing shallow foundation of a building when an adjoining building has to be constructed with a deep foundation.
(ii) To safeguard the existing structure from the danger of excessive or differential settlement.
(iii) To deepen and widen an existing foundation to increase the bearing capacity of the foundation soil.
(iv) To build a basement floor to an existing building.
(v) To lift a building fully or partly to alter the foundation so as to prevent from waterlogging.

22.6.2 Precautionary Measures Before Underpinning

The following general measures should be undertaken before starting the underpinning operations:

(i) The existing strength of the building should be ascertained before resorting to underpinning (e.g., inferior material used or use of construction method which is not standard or poor workmanship).
(ii) Temporary support should be provided by adequate shoring and strutting wherever needed.

(iii) During underpinning of buildings, a watch on possible movements should be checked and rectified then and there.
(iv) No damage should be caused to the adjoining structures.
(v) It should be ensured that no obstruction is created to the passage of people or vehicles in the adjoining areas.

22.6.3 Methods of Underpinning

Several methods are available for underpinning foundations but four routine methods which are used in general are:

1. Pit Method
2. Pile Method
3. Pier Method
4. Chemical Method

1. Pit Method

Generally underpinning by pit method is carried out by excavating slowly in stages. For this the existing wall is divided into suitable width of about 1.20–1.50 m. Holes are made in the existing wall in the required points. Needles with bearing plates are then inserted through these holes and supported on jacks, Fig. 22.13. This is followed by excavation, first up to the existing foundation level and then to the required depth.

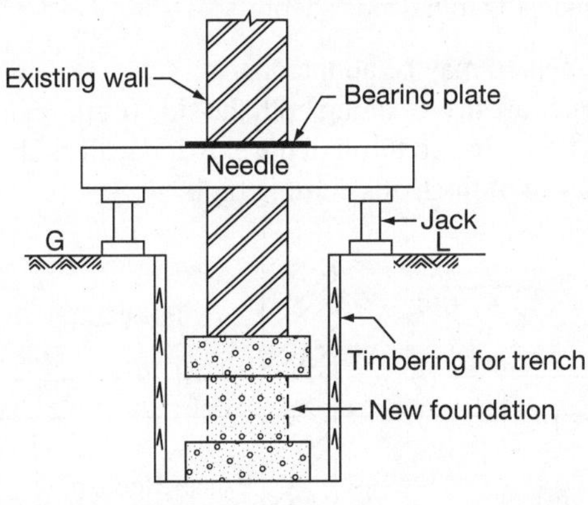

Figure 22.13 Pit method

During the process of excavation, the following precautions are to be taken to avoid formation of crakes or settlement:

(i) Excavation should be done one at a time preferably starting at the middle and progressing sideways.
(ii) Proper timbering is to be provided to the excavated trenches.
(iii) Instead of central needle beam, cantilever needle beam may be adopted as shown in Fig. 22.14.
(iv) After completion of excavation, foundation concrete is laid.
(v) A suitable connection between the old foundation and new foundation has to be made using vertical mild steel bars.
(vi) After the new foundation is completely set, then only the needle beams and raking shores are removed.

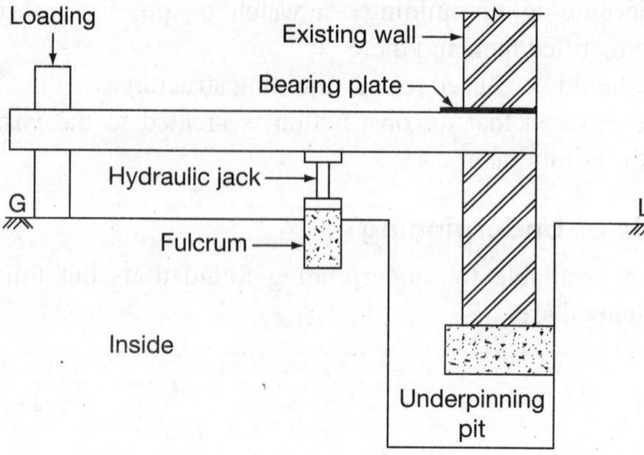

Figure 22.14 Support by cantilever needle

2. Pile Method

Pit method may be impracticable or uneconomical in the following cases:

(i) Waterlogged area
(ii) Heavy loads on existing structures
(iii) Loads to be transferred to a deeper depth

In such situations pile method may be adopted.

In this method, the piles are driven along both the sides of the existing wall. Then needles in the form of pile caps are provided through the existing wall as shown in Fig. 22.15. Thus, the existing wall is relieved of the loads coming on it.

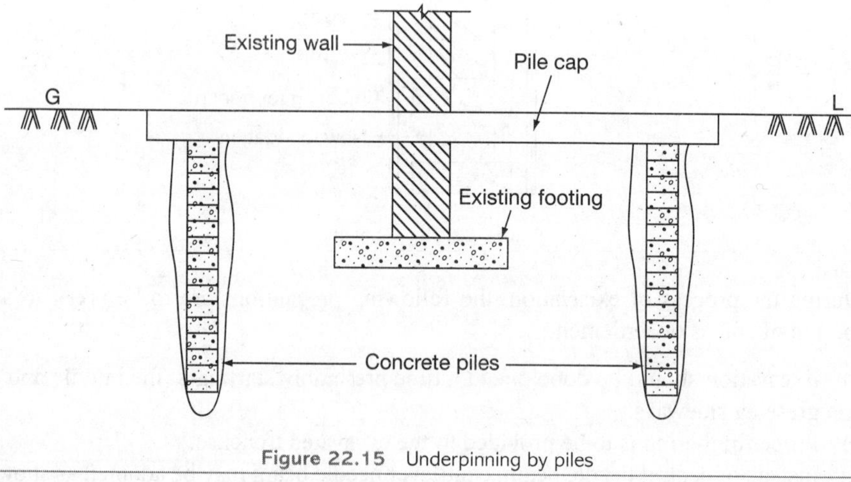

Figure 22.15 Underpinning by piles

This method is particularly useful in clayey soils and for waterlogged areas and for walls bearing heavy loads. In case of light structures piles are driven along the structure and then brackets or cantilever needles are provided.

In another approach, pre-cast piles are installed beneath the existing foundation using jacks and digging approach pits. The materials surrounding the pit is removed by water jets or air lifts and then filled with cement concrete.

3. *Pier Method*

This method is generally resorted to in dry ground. This is a simple method of underpinning. It involves the installation of piers under foundations of structures through the approach pits. Filling the pits with concrete and finally wedging up to transfer the loads to the new piers, Fig. 22.16.

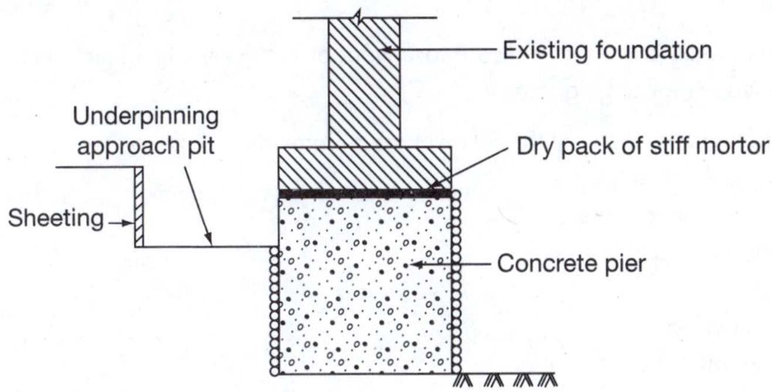

Figure 22.16 Underpinning by pier

4. *Chemical Method*

In this method the foundation soil is consolidated by employing chemicals.

Perforated pipes are driver in an inclined direction beneath the foundation, Fig. 22.17. The slopes are provided such that the entire area under the existing footing corners under the area used to be strengthened.

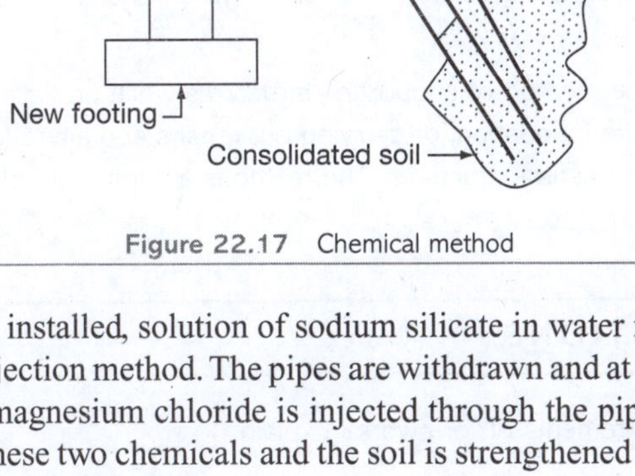

Figure 22.17 Chemical method

After the pipes are installed, solution of sodium silicate in water is injected through the pipes. This is a two-injection method. The pipes are withdrawn and at the time of withdrawal of pipes, calcium or magnesium chloride is injected through the pipes. Chemical reaction takes place between these two chemicals and the soil is strengthened by consolidation. This method is suitable for granular soils.

SALIENT POINTS

1. Temporary supporting structures are those structures which are required in the routine construction of building or repairs works.
2. Concrete is placed in a timber or steel casing during the period of curing. Such a casing is known as shuttering, formwork or forms.
3. Generally for circular works, such as arches, domes, etc., the term centring is used instead of formwork.
4. The term moulds is used to indicate formwork of relatively small units such as lintels, cornices, cubes for testing, etc.
5. Formwork should satisfy by the following requirements:
   (i) Adequate strength,
   (ii) Smooth inner surface,
   (iii) Enough rigidity,
   (iv) Quality,
   (v) Less leakage,
   (vi) Economy,
   (vii) Easy removal and
   (viii) Supports.
6. Materials used for the formwork are steel, timber, aluminium and plywood.
7. Temporary structure erected close to the work so as to provide a safe working platform for the workers and to provide adequate space to keep the working materials is known as scaffolding or scaffold.
8. The component parts of scaffolding are: standards, ledgers, putlogs, transoms, bridges, braces, guard rail, toe board and raker.
9. Types of scaffolding are: single scaffolding or bricklayer's scaffolding, double scaffolding or mason's scaffolding, cantilever or needle scaffolding, suspended scaffolding and trestle scaffolding.
10. Shoring is the means of providing support to get stability of a structure temporarily [illegible] certain circumstances during construction, repair or alteration.
11. Types of shoring are: raking or inclined shores, flying or horizontal shores and dead or vertical shores.
12. Underpinning is the method of supporting structures while providing new foundations or strengthening the foundation, or carrying out repairs and alterations without affecting the stability of the existing structures. The methods are: pit method, pile method and tier method.

REVIEW QUESTIONS

1. What are the requirements of formwork? Explain.
2. Discuss the materials used for formwork.

3. Enumerate the general rules to be observed to avoid the failure of formwork.
4. Explain the circumstances which are favourable for timber centering and steel centering.
5. Explain the requirements of shuttering with respect to quality and leakage.
6. How de-shuttering of forms are done?
7. What are the essential requirements of scaffolding?
8. Name different types of scaffolding. Explain any one in detail.
9. Detail the method of scaffolding provided to plaster the outer walls at first floor level.
10. What is meant by shoring?
11. It is proposed to make an opening of 6.5 m × 4.0 m height in front wall of an existing two storey building. The floor to floor height is 3.6 m. Suggest a suitable temporary structure to execute the work.
12. Distinguish between raking and flying shores.
13. What is underpinning? Briefly explain any one method of underpinning.
14. It is intended to add are more floor on an existing two-storey building. The foundation was designed only for two floors. Explain with sketches the operation of carrying out new construction and strengthening of the existing foundation.
15. Explain how will you handle the following field situation. Draw neat sketches wherever needed.
    (i) A basement floor has to be provided for an existing three-storied building.
    (ii) In a public building a large opening has to be made to use as a passage.

23 GROUND AND UPPER FLOORS

23.1 INTRODUCTION

Floors are the horizontal units of a building structure which divide the building into different levels. The purpose of separation or division is to create more accommodation within a restricted space one above the other for different purposes, viz., to provide support for the occupants, to allow space for furniture and equipment, etc. Strictly, the division of vertical occupation is called floors and the exposed top surfaces of floors are termed as floorings. But in usage, both the terms reflect the same meaning.

Thus the floor of a building immediately above the ground is known as ground floor. The construction below the ground level or the building has basement, the floor is termed as basement floor. The top portion of a ceiling forms the first floor or an upper floor. Any subsequent top portions of ceilings form upper floors.

The major problems of ground and basement floors are dampness removal and protection from heat. Generally there will be no problem regarding strength and stability for ground and basement floors as they get the full support from the ground. The major problems of upper floors are the strength and stability.

23.2 FUNCTIONS OF A FLOOR

A floor consists of a sub-floor and a floor covering.

Sub-floor may be a base-course or a sub-grade. This is required to impart strength and stability to support a floor covering and all other super-imposed loads including live loads.

Floor covering or paving or simply as flooring is the one constructed over the sub-floor and is intended to provide a hard, clean, smooth, impervious, durable and impressive surface to the floor.

The functional requirements of floors in general are as follows:

(i) Strength and stability
(ii) Damp prevention
(iii) Heat insulation
(iv) Sound insulation
(v) Fire resistance
(vi) Durability

23.3 SELECTION OF FLOORS

In order to select a suitable type of floor construction in a building, certain factors have to be considered. Although certain points may be common for ground, upper and basement floors, some distinct different points are to be observed in certain cases.

23.3.1 Ground Floors

For ground floors, the selection of the type of the wearing surface is important and the other factors which need consideration are given below:

(i) Initial cost
(ii) Appearance
(iii) Durability
(iv) Cleanliness
(v) Thermal insulation
(vi) Dampness
(vii) Indentation
(viii) Noiselessness
(ix) Maintenance
(x) Fire resistance

23.3.2 Upper Floors

Selection of a suitable type of construction for upper floors of a building depends on the following main factors:

(i) Initial cost
(ii) Floor loads
(iii) Type of construction
(iv) Plan of the building
(v) Function of the building
(vi) Fire resistance
(vii) Sound insulation
(viii) Type of ceiling
(ix) Wearing surface
(x) Weight and position of floors

23.3.3 Basement Floor

It is not a routine type of floor provided in every building. It is provided for a particular type of buildings like apartments, hotels and restaurants, cinema halls, etc. Selection of basement floors depends on the following factors:

(i) Initial cost
(ii) Availability of ventilation
(iii) Drainage of water from the floor
(iv) Adequate safety against fire
(v) Ground water level

23.4 TYPES OF GROUND FLOORS

As the ground floor directly rests on the ground, there is no need for a sub-floor. In order to drain the water outside completely, adequate drainage arrangements have to be made beneath the floor.

In normal practice, the space above the ground, up to a height of about 25–30 cm below the plinth level, called the basement, is filled with some inert materials like sand, gravel, crushed stone, cinder, etc. Over this course, a damp-proof course if needed is laid. Otherwise the floor covering is laid directly on the uniform bed.

The materials used for ground floor construction are: bricks, stones, concrete, hollow concrete blocks or wooden blocks.

The materials generally used for floor coverings are bricks, concrete, terrazo tiles, marbles, stones, mosaic, wood, etc.

23.4.1 Stone Floors

Usual sizes of stones of 30 cm × 30 cm, 45 cm × 45 cm or 60 cm × 60 cm with a thickness of 2–4 cm are used. Generally square stone slabs of the above sizes are used but the slabs can be of rectangular or oblong in shape with square edges. Stone slabs are laid on concrete bedding.

Before laying the slab a base is prepared after excavating to the required depth and the earthen base is levelled, rammed and watered. A layer of lime concrete of thickness 10–45 cm is spread over which the concrete bed or sub-grade is laid.

After setting the stone floor with a slope of 1 in 40, the mortar joints are raked out to a depth of 2 cm and flush pointed with cement mortar of 1 : 3.

Following aspects are kept in mind while laying stone slabs, Fig. 23.1:

(i) To begin with two stone slabs from diagonally opposite corners are laid. A string is stretched touching these corner slabs so as to fix top level of the floor.

(ii) Intermediate slabs are placed from both the sides taking care that the tops of stone slabs are touching the string which ensures the provision of proper grade.

(iii) All stone slabs should be of uniform thickness, strength and well drained on edges for residential and public buildings.

(iv) Width of joints between the slabs should not exceed 0.5 cm.

(v) Rough surface stone, may be used for godown, sheds, stores, etc.

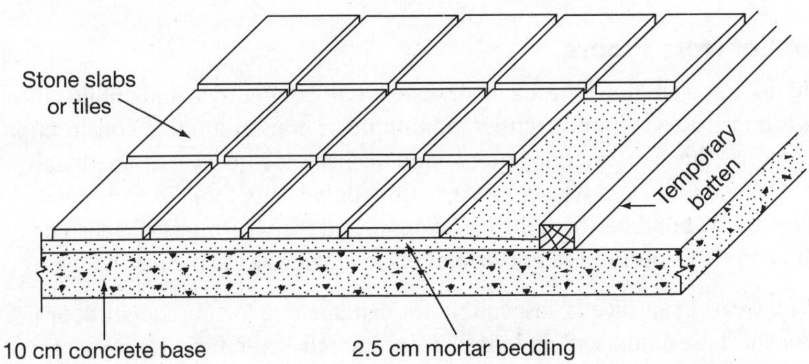

Figure 23.1 Stone flooring

23.4.2 Brick Floors

This type of flooring is used in case of warehouses, stores and godowns. This is a cheap construction and used in areas when stones are not available but good quality bricks are available.

This flooring may be laid flat or on edge. They may be arranged in herring bone pattern or at right angles to the walls. Brick on edge is preferred compared to bricks laid flat as the brick on edge is less liable for crack under pressure because of the higher depths. The bricks, in both the cases, are laid on ordinary mortar and pointed with cement or set in hydraulic mortar.

The construction of brick flooring is done as explained below, Fig. 23.2:

(i) An excavation of about 40 cm depth below the intended level of the floor is made.
(ii) The earth surface is levelled, watered and well rimmed until it is dry and hard.
(iii) Over the earth, above a sub-grade of 25 cm depth consisting of rubble or brickbats is laid.
(iv) Over this, a 10–15 cm thick layer of lime concrete or lean cement concrete (1:3:6) is laid.
(v) Upon this prepared sub-grade, bricks are laid in the desired shape.

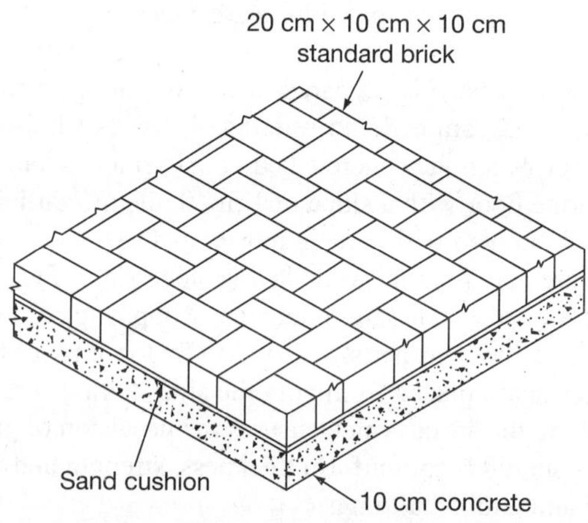

Figure 23.2 Brick flooring

23.4.3 Concrete Floors

Concrete floorings are mostly used in all residential, commercial and public buildings. The flooring is constructed adopting either monolithic or non-monolithic construction.

In the monolithic construction, after laying the base course layer, immediately a concrete topping is provided. In this type of construction, only a small thickness is needed for wearing surface as the bond between the base course and the wearing surface is good.

But this type of flooring has got several disadvantages as explained below:

(i) The wearing surface is susceptible for damage due to subsequent floor use.
(ii) As the base course if undergoing only small settlement immediately, there is a possibility for hair cracks.
(iii) Repairing the damaged top surface is difficult.
(iv) The progress is slow as the base course has to settle.

Non-monolithic construction is sound as the wearing surface is laid only after adequate drying of base course. The floor finish generally used is ordinary concrete finish of $1{:}1\frac{1}{2}{:}3$ (Fig. 23.3). Under controlled conditions, a mix of 1:2:4 ratio with carefully selected aggregates may be used.

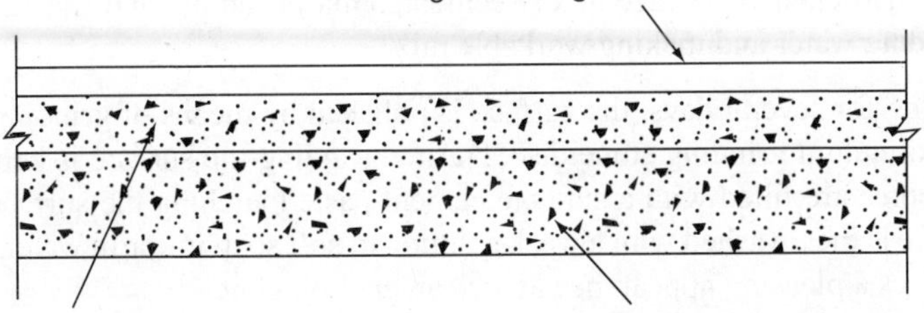

Figure 23.3 Concrete flooring

For non-monolithic construction, the surface of the base concrete is brushed with a stiff broom and cleaned thoroughly. The surface is wetted and excess water removed. The floor is laid in rectangular panels not greater than 2 × 2 m. Generally alternate bays are concreted so as to avoid initial shrinkages. When the concrete layer is even, the surface is rapidly compacted by remaining or beating and screeded to a uniform level. Then trowelling is done to give a level smooth surface. Adequate curing is done for 7 days by spreading a layer of wet sand or special membrane may be used.

23.4.4 Granolithic Floors

It is a concrete flooring with a different type of floor finish called granolithic. Granolithic finish is a concrete made of special selected aggregate. The thickness of layer varies from 1.25 to 4 cm. When it is greater than 4 cm this may be laid monolithically or after the base concrete has hardened.

This flooring is made with very hard and tough quality aggregate in rich concrete of 1:1:2. Generally hard fine grained granite, basalt, limestone and quartzite stones are suitable for coarse aggregate. In order to get a better granolithic finish, aggregate may be crushed and used. The fine aggregates are the natural or crushed sands with a suitable gathering. Non-slippery surfaces can be obtained by adding suitable abrasives. In case of non-monolithic construction, the base course may be prepared as done in concrete flooring. Granolithic flooring has all the advantages of concrete flooring.

23.4.5 Terrazo Floors

Terrazo is a special type of concrete flooring containing marble chips as aggregates. Any designed colour and designs can be obtained by using marble chips of different shades and colour cement.

Terrazo mix of 1:2 or 1:3 (1 cement to 2–3 marble chips) is used depending on the size of marble chips. Terrazo finish is of 10 mm thick. The terrazo finish is laid over the concrete base course (as explained in concrete flooring) by two methods.

(i) The cement concrete base is covered uniformly by a 6 mm thick sand cushion and a tar paper is placed on this. Over this paper a layer of rich mortar (1 : 3) about 30 mm is placed uniformly.
(ii) A thin coat of cement is spread over the wet concrete base. This layer is cleaned and a layer of cement mortar 12 mm thick is spread evenly over it. When the mortar bed has hardened the terrazo mix (1 cement: 3 marble chips) of 6 and 12 mm is laid after adding water and making workable mix.

After curing for several days, the surface is polished by means of grinding machine fitted with carborundum grinding stone disc. During grinding the surface is kept wet and small holes or pores are filled with a suitable cement paste matching the surface configuration. The surface is then washed with a weak solution of soft soap in warm water. Such a finished surface gives a pleasing appearance and clean environment.

This type of floor is used in public buildings like banks, hotels, offices, etc., because of its decorative appearance and excellent wear-resisting properties.

23.4.6 Mosaic Floors

For construction of this type of flooring, a hard concrete base is laid first. When the base is wet, a 2 cm thick layer of cement mortar (1:2) is evenly laid. Over this layer small pieces of broken tiles are arranged in different pattern. After this the inner space between tiles is filled with coloured pieces of marble in the desired fashion.

Following this cement or coloured cement is sprinkled at the top to get a complete floor without pores. Then the surface is rolled by light stone roller till an even surface is obtained. After 24 hours of drying the surface is rubbed with a pumice stone of 20 cm × 20 cm × 7 cm fitted to a long wooden handle. This polished surface is allowed to dry for 2 weeks before put into use.

23.4.7 Tiled Floors

In this type of flooring tiles either of clay or cement concrete, manufactured in different shapes, are used. A 15 cm thick layer or lime or cement concrete is laid over the levelled ground. In order to receive the tiles at 25 mm thick layer of lime mortar (1:3) or cement mortar (1:1) is laid. A cement slurry is spread over the hardened mortar. Tiles are laid flat on this bed and a cement paste is applied on the sides. Joints are rubbed with carborundum stone after allowing 2–3 days for setting. The entire surface is polished with a pumice stone (Fig. 23.4).

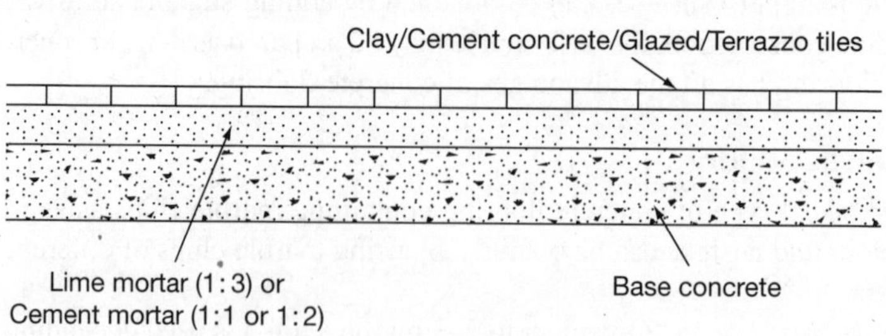

Figure 23.4 Tiled flooring

Tiled floorings are used in residential buildings, hotels, offices and other public buildings. These floors can be constructed in very short time with pleasing appearance and good durability.

23.4.8 Asphalt Floors

Asphalt floorings are of two types, viz., (i) using asphalt tiles and (ii) using mastic asphalt. Asphalt tiles are made from asphalt, asbestos fibres and other materials and pressed under pressure. Asphalt mastic is a mixture of fine aggregates and natural or artificial asphalt.

Asphalt tiles are used to cover wooden or concrete floors. These tiles are resilient, non-absorbent, moisture proof and cheap. They are used in school, offices and hospitals, etc.

Asphaltic mastic can be mixed hot and laid in continuous sheets or pressed into blocks which can be used for flooring. As an alternative it may be mixed with oil and asbestos and applied cold. An ordinary concrete or wood base may be used for laying this mixture.

23.4.9 Marble Floors

High quality marble slabs in the required sizes and colours are available in the market. The construction procedure is same as that of mosaic flooring except that marble slabs or pieces are used instead of mosaic tiles or pieces. This type of flooring is preferred when sanitation and cleanliness are required as in the case of hospitals, temples, theatres and other superior type of works. Now-a-days some quality marbles are used for residential buildings also.

23.4.10 Timber Floors

Timber flooring is not generally preferred for ground floors. If it is used as a ground floor, the prevention of dampness is most important. Timber floorings can be provided in any one of the following methods. Strip flooring consists of narrow and thin strips of wood jointed to each others by tongue and groove joint. Planked flooring is one type in which wider planks are used and they are also of tongued and grooved type. Wooden block flooring consists of small square or rectangular blocks with tongue and groove joints, Fig. 23.5.

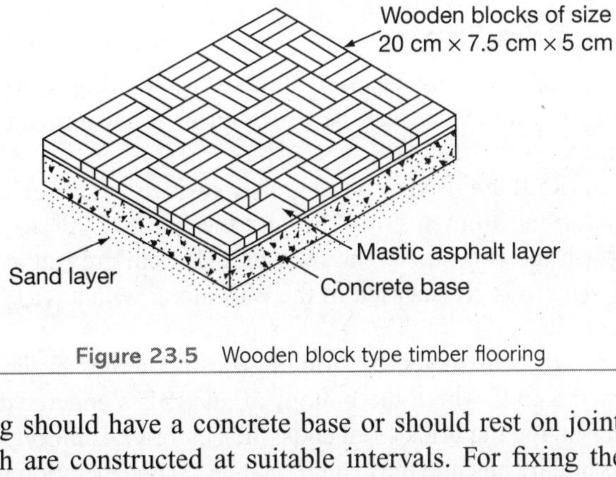

Figure 23.5 Wooden block type timber flooring

Timber flooring should have a concrete base or should rest on joints spanning across dwarf walls which are constructed at suitable intervals. For fixing the timber floors on concrete slabs, longitudinal nailing strips are provided. Planked flooring should be laid with spaces of metals spaced 1 mm apart temporarily for providing expansion joints. Strip flooring is used in thickness of 2–2.5 cm and width of 6–10 cm.

23.5 TYPES OF UPPER FLOORS

Upper floors should be strong to take heavy loads, should have sound insulation and fire resistance and also have a good wearing resistance. The upper floors are generally classified based on the materials of construction arrangement of beams and girders or materials used.

Commonly used floors are explained below.

23.5.1 Timber Floors

Timber floors are used only in areas where more timber is available at reasonable cost. Further they are used for residential buildings where the span is less and load on the floor is less.

The floors may be of single joist, double joist or triple joist. Joists of floors should be strong enough and not deflect too much causing any cracking of plaster of ceiling. Further long joists need strutting to avoid buckling. The planking consists of wooden boards of 4 cm thickness and 10–15 cm wide. Figure 23.6 shows the details of single joist timber floor, and the details of the joist's joint is shown separately.

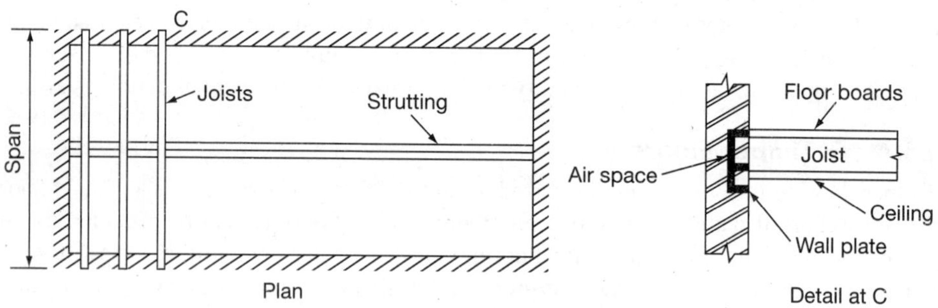

Figure 23.6 Details of single-joint timber floor

23.5.2 Jack Arch Brick Floors

Bricks and plain concrete can not be used directly as flooring system without proper strengthening. Either reinforcement can be provided or provision can be made so as to develop arch action.

Rolled steel joists (RSJ) are used to build arches so as to form jack arch floorings. The arches have spares varying from 1.25 to 2 m and the rise is 1/12th to 1/16th of the span. In order to accomplish the end thrusts the rods of 2–2.5 cm are run at 2–2.75m apart as shown in Fig. 23.7. The rods are encased in the wall along with a RSJ. The spandulus are filled with concrete.

Brick jack arches are constructed by bringing up the arches by laying the bricks on edges starting from the ends where the bottom of an RSJ is concreted. Lime or cement mortar is used. After laying the brick from ends, the key brick is introduced to set the arch action. Next layer is laid by pushing the centering ahead. This operation is continued till the work is completed.

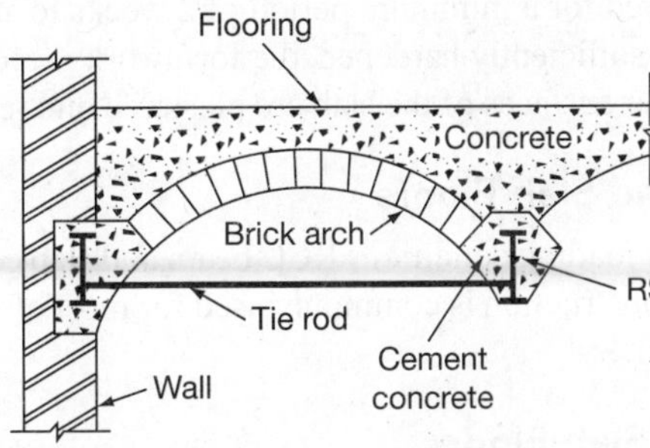

Figure 23.7 Brick jack arch floor

23.5.3 RCC Slab Floors

All modern buildings are invariably constructed with reinforced cement concrete. For small spans a simple RCC slab floor is generally suitable. For rooms, with the ratio of length of the room to its width is greater than 1.5, slabs are designed to span along the shorter width. That is, the main reinforcements are placed to the shorter width. The thickness of the slab depends on the type of concrete used, the span, floor loads, etc. These slabs are considered along with a frame then it is made monolithically with the supporting frames (Fig. 23.8).

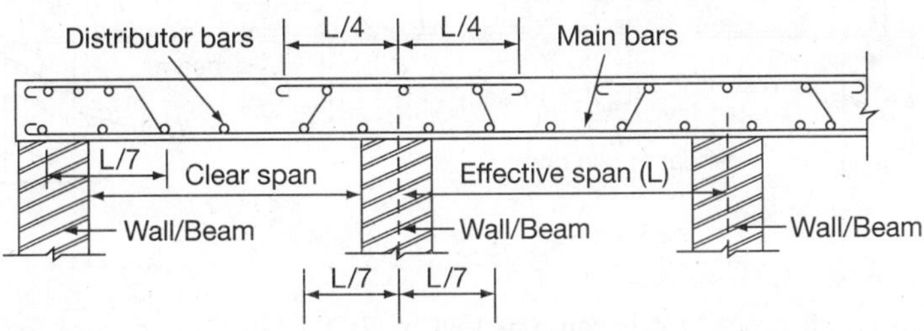

Figure 23.8 RCC slab floor

If the spans of rooms are approximately equal then the slab is designed as a two-way slab (i.e., main reinforcements are placed parallel to both the spans). At corners suitable reinforcement is provided at top and bottom to take the stresses due to partial fixity.

Reinforced concrete slabs are laid adopting the routine mixing, laying, finishing and curing. The slab provides a very smooth surface at the bottom and a pleasing appearance. It also accommodates all lighting arrangements. The RCC slabs are restricted up to 4 m span beyond which beam and slab construction has to be adopted.

The following is the construction procedure:

(i) A centring of steel or timber is erected to support its own weight and the superimposed load.
(ii) The reinforcement is placed on the centring before coating a thin layer (2–5 cm) of cement concrete.
(iii) The required mix of cement concrete is then prepared and poured around the reinforcement up to the required thickness of the slab.
(iv) The concrete is well consolidated by ramming or using a vibrator.

(v) The concrete is cured for a minimum period of 2 weeks to attain its full strength.
(vi) After concrete has sufficiently hardened, the formwork is recovered.
(vii) The upper and lower surfaces of the slab are plastered and required finish are given.

23.5.4 RCC Beam and Slab Floors

For larger spans and heavy loading conditions RCC beam and slab construction is generally resorted to. This type of construction is commonly used for most of the important buildings (Fig. 23.8).

23.5.5 RCC T-Beam Slab Floors

In this type of floor, beams and slabs are designed as rectangular sections and the slabs are supported on beams. It is a monolithic construction both the beams and slabs are cast together. The beam used in monolithic construction is called a T-beam (i.e., a part of slab acts as a flange of the T-beam).

The main reinforcement of the slab runs parallel to the short span. However, some reinforcement is provided on the other, as in RCC slab floor, as distributors. In case of equal spans two-way slabs may be constructed with reinforcement provided on both directions. Sometimes the projecting beams are covered by providing a false ceiling underneath it (Fig. 23.9).

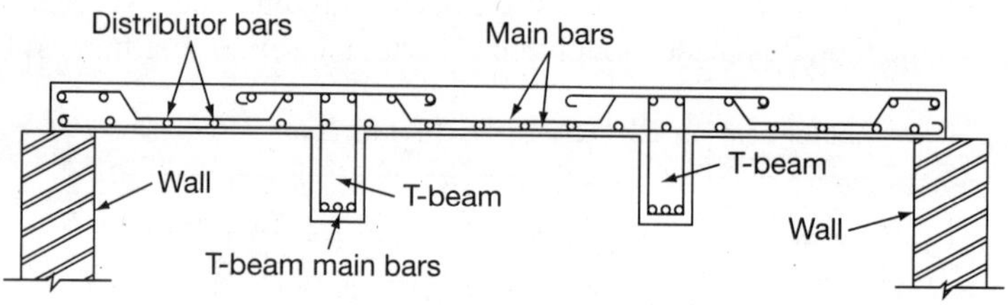

Figure 23.9 RCC T-beam slab floor

The construction procedure is same as that of RCC slab floor, except for the type of centring or formwork required for the floor.

23.5.6 Flat Slab Floors

This type of floor is called as beamless slab floor. This flat slab floor is directly supported on columns without any intermediate beams. This type of floor is preferred where heavy loads are anticipated and where there is head room restriction. The columns supporting the floor are invariably circular in cross-section and tops of the column are flared or tapered, which is called as *capital*. Sometimes certain portion of the slab, symmetrical with the column, is thickened which is called *drop panel* (Fig. 23.10).

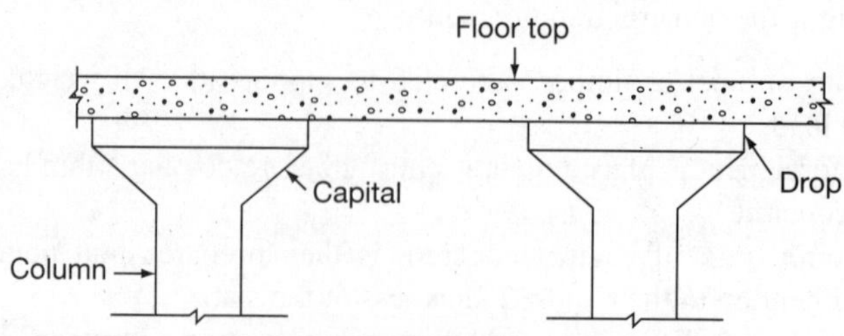

Figure 23.10 Flat slab floor

The flat slab floors have several advantages which are given below:

(i) As there is no projection of beam, there is no need for additional ceiling.
(ii) More clean head room is available.
(iii) Better ventilation and lighting is available.
(iv) Even for heavier loads only thin slab thickness is needed.
(v) The construction is comparatively easier.

The flat slab floors are preferred under the following conditions:

(i) When large equal number of panels are required.
(ii) Where clear large spans are required.
(iii) Where there is limitation in head room.
(iv) The ratio of length to breadth of panels is to be more than 4 : 3.
(v) The difference between the length and breadth of any two adjacent panels should not be more than 10% of the greater of the two.

The following two types of systems of reinforcement are in use: (i) two-way system and (ii) four-way system.

In the two-way system of reinforcement, the reinforcement is provided between the columns and perpendicular to the columns. The area left is considered as supported on four sides.

In the four-way system of reinforcement, the reinforcement is placed in four directions. That is two rows of main steel from column to column and other two rows placed diagonally across the panel from column to column.

23.5.7 Resilient Floors

Floors made from materials such as PVC, linoleum, rubber, etc., are called resilient floors. Resilient floors are sound proof and dust free and are used in libraries, offices, computer rooms, show room, etc. The three important resilient floors are: (i) PVC (Vinyl floors), Linoleum floor and Rubber floor, which are briefed below.

1. PVC Floor

The material in PVC floors is the thermoplastic binder which can be vinyl chloride polymer or vinyl chloride copolymer or both. The floor covering is backed with hessian or other woven fabric. Vinyl asbestos tiles have 40% of chrysotile asbestos fibre together with powdered mineral fillers and pigments. The thickness of normal PVC floor coverings are 1.5, 2.0, 2.5, 3.0 and 4.00 mm and the thickness of *back floor coverings* come in 2.0, 2.5, 3.0, 4.0 and 5.0 mm sizes. In roll form it is usually available in 1.0, 1.5 and 2.0 m and lengths of 10 m.

2. Linoleum Floor

Linoleum is a covering materials generally laid over wooden or concrete floors. Linoleum material is lubricated by mixing oxidised linseed oil with gum, resin, pigments, wood floor, cork dust and other filler materials. It is available as rolls of 2 or 4 m width with 2–2.5 m thick and both in plain and printed forms. Linoleum coverings are fixed to the sub-floor by means of suitable adhesive in order to have adequate bond and high durability. In some cases, it is nailed at the edges.

3. Rubber Floor

It is composed of natural rubber with various filling compounds. As the cost of rubber is high, their type of floor is expensive. They are produced in sheets and also in tile form. A sub-floor is essential.

23.5.8 Hollow Block and Rib Floors

In this type of floors, to reduce the total weight of the floor hollow blocks of clay or concrete are used. In one type, the blocks are placed at 10 cm apart and in this space, mild steel bars are placed (Fig. 23.11). Suitable flooring at top and ceiling finish are provided. In order to develop enough bond with the concrete, the blocks are provided with rough or grooved surface.

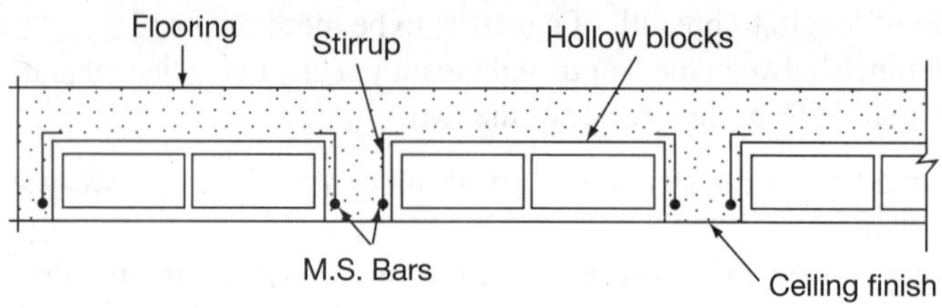

Figure 23.11 Hollow block and rib floor

The structural hollow clay floor tiles should be free from cracks and of uniform colour and texture. Projections of tiles are kept on one of the external face of such tiles so as to facilitate the application of plaster. The dimensions of length, 45–90 cm, width, 19–24 cm and height 70–90 mm. The weight per metre length of structural hollow clay tiles does not exceed 140 N.

SALIENT POINTS

1. Floors are the horizontal units of a building structure which divide the building into different levels.
2. The floor of a building immediately above the ground is known as ground floor. The top portion of a ceiling forms the first floor or an upper floor.
3. The construction below the ground level or the building has basement, the floor is termed as basement floor.
4. The major problems of ground and basement floors are dampers removal and protection from heat.
5. Sub-floor may be a base-course or a sub-grade. This is required to impart strength and stability to support a floor covering and all other super-imposed loads including live loads.
6. Floor covering or paving or simply as flooring is the one constructed over the sub-floor and is intended to provide a hard, clean, smooth, impervious, durable and impressive surface to the floor.

7. The functional requirement of floors in general are:
   (i) Strength and stability
   (ii) Damp prevention
   (iii) Heat insulation
   (iv) Sound insulation
   (v) Fire resistance
   (vi) Durability
8. Factors to be considered for the ground floor selection are as follows:
   (i) Initial cost
   (ii) Appearance
   (iii) Durability
   (iv) Cleanliness
   (v) Thermal insulation
   (vi) Dampers
   (vii) Indentation
   (viii) Noiselessness
   (ix) Maintenance
   (x) Fire resistance
9. Factors to be considered for the selection of upper floors are as follows:
   (i) Initial cost
   (ii) Floor loads
   (iii) Type of construction
   (iv) Plan of the building
   (v) Function of the building
   (vi) Fire resistance
   (vii) Sound insulation
   (viii) Type of ceiling
   (ix) Wearing surface
   (x) Weight and position of floors
10. Types of ground floors are: stone floors, brick floors, concrete floors, granolithic floors, terrazo floors, mosaic floors, tiled floors, asphalt floors, marble floors and timber floors.
11. Types of upper floors are: timber floors, Jack Arch brick floors, RCC slab floors, RCC beam and slab floors, RCC T-beam slab floors, flat slab floors, resilient floors and hollow block rib floors.

REVIEW QUESTIONS

1. Mention the factors to be considered in the selection of ground floors.
2. What factors you would consider while selecting an upper floor?
3. In what situations would you use stone floors?
4. A room 4 m × 5 m is to be covered by a wooden flooring. Describe the procedure to be followed.

5. Explain the construction procedure of a concrete floor giving its relative merits and demerits.
6. Under what circumstances cement concrete flooring is recommended.
7. Describe how a marble floor is laid with ordinary mortar joints.
8. How timber floors are used for ground floor and for upper floor?
9. Distinguish between granolithic and terrazo floors.
10. How a flat slab floor is constructed? Bring out the advantages of the same.
11. Describe four commonly adopted type of floorings that are recommended in modern buildings giving their merits and demerits.
12. What are resilient floors? Discuss.
13. Explain hollow block and rib floor.
14. Name the various types of floor construction suitable for the upper floors of a building.
15. What are the flooring materials you would recommend for the structures given below? Justify your recommendation.

    (i) Hospital
    (ii) Art Gallery
    (iii) Bathroom
    (iv) Hotel
    (v) School Building
    (vi) Bus Shelter

24 STRUCTURAL STEEL WORKS

24.1 INTRODUCTION

Steel is used in large extent in different forms in buildings. The materials generally used for structural purposes are cast iron, wrought iron and steel. As cast iron can withstand only compression, it is used limitedly. Wrought iron could resist tensile stresses. But mild steel can withstand both. Hence, it has replaced cast iron and wrought iron almost completely.

Various types of sections and shapes are used in building works. Most of the sections are made by rolling. Different types of sections are combined and as a unit is called built-up section, which is used in building work.

Mild steel in combination with cement forms the versatile material, the reinforced cement concrete (RCC). Apart from their structural steels are used for frame work and also to act as beam, column and a truss. Construction in steel work is more economical for buildings of more number of floors. Due to ease in erection and capable to take heavy loads, steel is commonly adopted for factory buildings.

This chapter discusses the various forms of use of steel in building industry.

24.2 STEEL SECTIONS

In order to use in construction works, steel should be available in certain forms which are known as steel sections in commercial forms or market forms. Some of the common sections or forms are given below.

24.2.1 Bars

Bars are of round or square cross-sections. Square sections of size 5–32 mm are commonly used. Round bars of diameter 6–32 mm are available. Both are available in lengths of 10–12 mm. Square bars are used for railings and of grill work and round bars are used in reinforced concrete and brick works.

24.2.2 Plates

Rolled plates with thickness of 5–28 mm and with a maximum area of 30 m^2 are available. Thin plates with 5 mm thickness are called sheets. These are used in flanges of beams and columns and as column bases.

24.2.3 Flats

Flats are plates with longer length with shorter width. The width varies from 18 to 500 mm and thickness from 3 to 80 mm. Flats are comparatively costlier and used in grill works and railings.

24.2.4 Angle Sections

The angle section may be of equal or unequal legs (Fig. 24.1(a)) and available in different sizes. Angle sections are used in steel roof trusses, steel columns, beams and as stiffeners in girders.

24.2.5 Channel Sections

It has two equal flanges and a web (Fig. 24.1(b)). A channel section is designated by the height of web and the width of flange and they are available in different sizes. These are used in steel framed structures, girders and steel bridges.

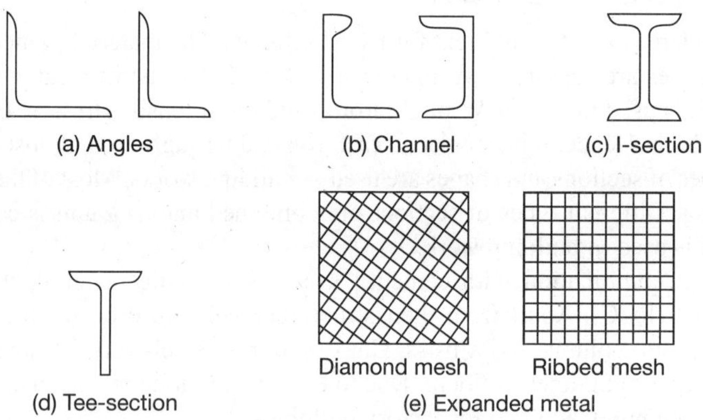

Figure 24.1 Commercial sections of steel

24.2.6 I and T Sections

These sections are named based on their shapes (Fig. 24.1(c) and (d)). I-section is usually referred to as rolled steel joist (RSJ) or beam. RSJs are used for floor beams, lintels, columns, etc. T-sections are used in steel roof trusses, in built-up sections, in steel water tanks and in chimneys and bridges. The economic use of material is derived by concentrating the material on the flanges where the bending stresses are maximum.

24.2.7 Expanded Metal

Plain or ribbed steel sheets are cut and expanded to form expanded metal. These are used in ferro-cement concrete works. Welded fabric with ribbed mesh has a rectangular or square mesh is another form of mesh used in construction (Fig. 24.1(e)).

24.2.8 Ribbed TOR Steel Bars

These are the new type bars produced from the ribbed TOR steel which is a deformed high strength steel. These bars have ribs or projection on their surface. The ribbed TOR steel bars are available in sizes varying from 6 to 50 mm diameter.

These bars are largely used as reinforcement in concrete structures. These bars have several advantages compared to conventional MS round bars.

24.3 METHODS OF STEEL CONNECTIONS

Steel members are fastened together by means of rivets, bolts or by welding. Among the three approaches welding has been used in most of the works as they have large advantages.

24.3.1 Riveted Connections

Members are connected by rivet which can not be removed unless the rivet is destroyed. Rivets are either fitted at site or in the fabrication shop. The former is called field rivets and the latter as shop rivets.

Rivets used in building construction are of soft steel with a tensile strength of the order of about 4000 kg/cm^2. Sometimes wrought iron rivets are favoured as it has the property of retaining malleability even longer than that of steel after being forged.

Generally rivets have a hemispherical button-shaped head and a cylindrical shank. Other shapes are flattered head and counter sunk head. Semi-circular button head is used commonly. Wherever more clearance is needed flattered head is used. Counter-sunk heads are used when least projections are needed.

The process of driving a rivet is first to heat the rivet red hot and to place in a hole through the members which are to be connected. The projecting end is then upset to form the head of required shape with the aid of a pneumatic hydraulic riveter. While driving, the other end of the rivet is held in place by a dolly. The rivet contracts as it cools, and thereby tightening the grip. At times cutting of rivets may become necessary. In such cases, this is accomplished by chipping off the head with a pneumatic chipper and then the rivet is driven out with a pin-maul.

Centre-to-centre distance between the adjacent rivets on the same rivet line is called pitch. Minimum pitch is generally three times the diameter of the finished rivet. However, the absolute minimum pitch is twice the diameter of the rivet plus one cm.

24.3.2 Bolt Connections

Bolt joints are used for temporary connections. For bold joints, the diameter of the hole is kept 1 mm larger than the external diameter of the bolt threads. In difficult situations, such as lesser number of rivets or difficulty in connecting, bolt connections are resorted to.

24.3.3 Welded Connections

There are two principal forms of welding of steel, viz., electric arc welding and oxyacetylene gas welding.

The electrical welding is further divided into resistance arc, shielded arc and atomic hydrogen welding. In both the types of welding, the members to be welded are placed in contact and edges are melted so that the metal flows from both the members together. When allowed to cool, the members get welded with the aid of the weld. In order to get a perfect joint, additional metal is supplied with the aid of a metallic rod which can be used as an electrode.

In general, the defects noticed in welding are: undercut, porosity, incomplete penetration, lack of fushion at edges, slag inclusions, hot crack, gas pores and surface imperfections.

Undercut occurs when a groove is melted into the member adjacent to the arc formed by the arc action which has not subsequently included in the weld metal.

Porosity is the presence of cavities in the weld metal caused by the gas entrapment. This happens because of the presence of moisture, scales, oil and other contaminations. This is overcome by striking the arc and travelling for 15 mm and then going back over the initial arc to melt out the start porosity.

Incomplete penetration invariably occurs when larger diameter electrodes are used in fillet welds. In butt welding, incorrect, 'back gauging' can result in lack of penetration.

Lack of fushion at edges happens in fillet welds and is due to the failure to fuse together adjacent surfaces of the weld and the material of the member.

Slag is derived from the flux used with welding process. If it is entrapped in the weld it is harmful. This can be prevented by correct plate preparation.

Typical hot crack in a fillet weld is a longitudinal crack characterised by a blue appearance along the crack due to surface oxidation at high temperatures.

Gases are formed inside the weld. Main surface imperfections can happen due to (i) edge of the plate melt off and (ii) overlap.

24.4 BUILT-UP SECTIONS

In order to make steel structures, for example, beams and columns, the standard sections are combined in a suitable way. Such a combined section is called a built-up section.

Figure 24.2 shows built-up sections composed of I-sections only. Likewise Fig. 24.3 shows built-up sections formed by combining I-section and plates.

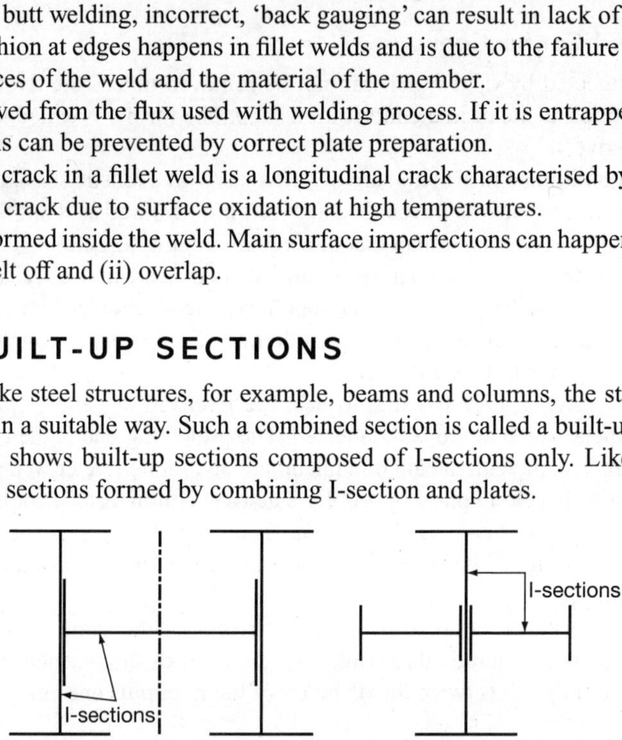

Figure 24.2 Built-up sections formed by combination of I-sections

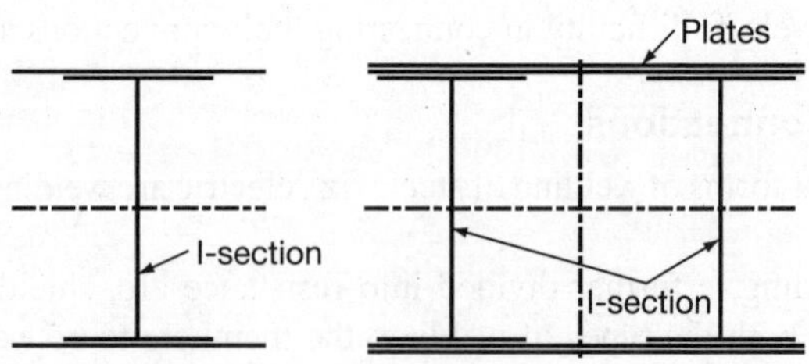

Figure 24.3 Built-up sections formed by combination of I-sections and plates

Figure 24.4 shows the combination of I and channel sections, whereas Fig. 24.5 shows combination of channel and plates.

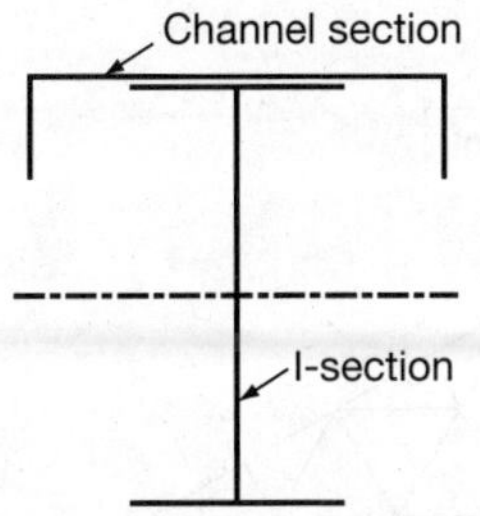

Figure 24.4 Built-up section formed by combination of I-section and channel

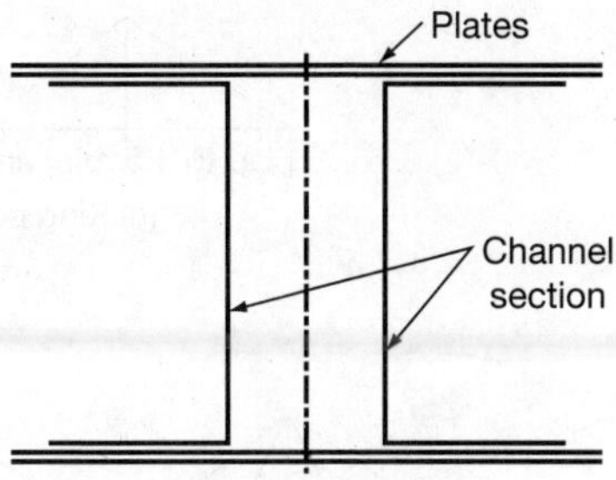

Figure 24.5 Built-up section formed by combination of channels and plates

Figure 24.6 shows combination of angle sections and plates.
Built-up sections are used for columns, beams, girders and other structural needs.

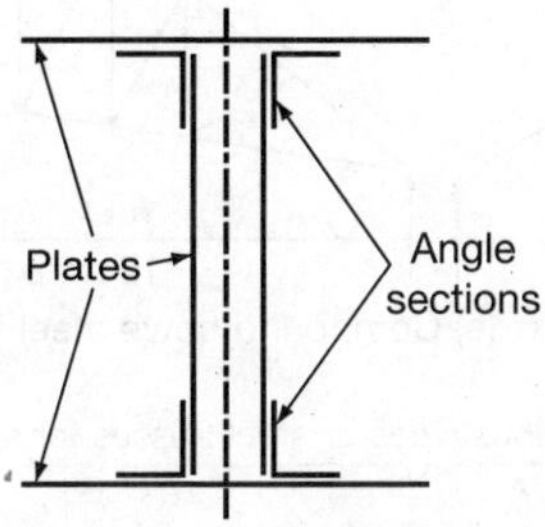

Figure 24.6 Built-up section formed by combination of angle sections and plates

24.5 STRUCTURAL STEEL MEMBERS

Different steel sections are used to form trusses, beams and columns. Individual steel sections or built-up sections are used.

24.5.1 Steel Trusses

Various types of steel roof trusses are shown in Fig. 24.7(a–e) and 24.8(a–d). Steel roof trusses are designed in such a way that the members are either in compression or in tension only. Members of a truss which form the roof base depends upon the span, roof slope, covering materials, centre to centre of trusses, etc.

Tee-section is the most suitable section for principal rafter. For struts either angle iron or channel section may be used. Round or flat sections are used for tension members. Built-up sections are also used in certain cases. Members of a truss are connected by bolts, rivets and thin plates called gussets.

1. Trusses for Small Spans

Small trusses (span up to 15 m) are rested on bed plates at the ends. The bed plates may be of stone or concrete. The ends are bolted down with rag bolts which hold the truss down.

Small trusses generally consist of angles connected with gusset plates. In order to seat the foot of the truss on the bed plate short angles are fitted. For spans up the 7 m, 15 mm diameter rivets are used (Fig. 24.7(a–e)).

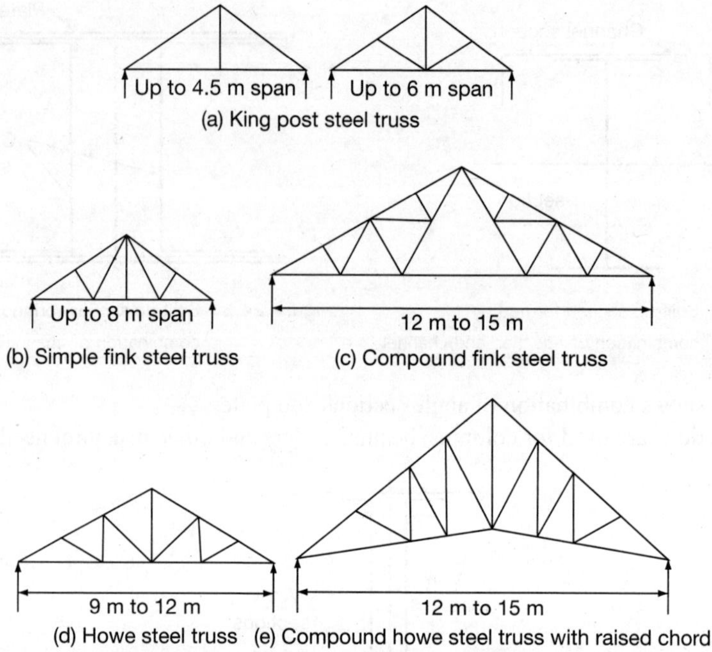

Figure 24.7 Various types of steel trusses for spans up to 15 m

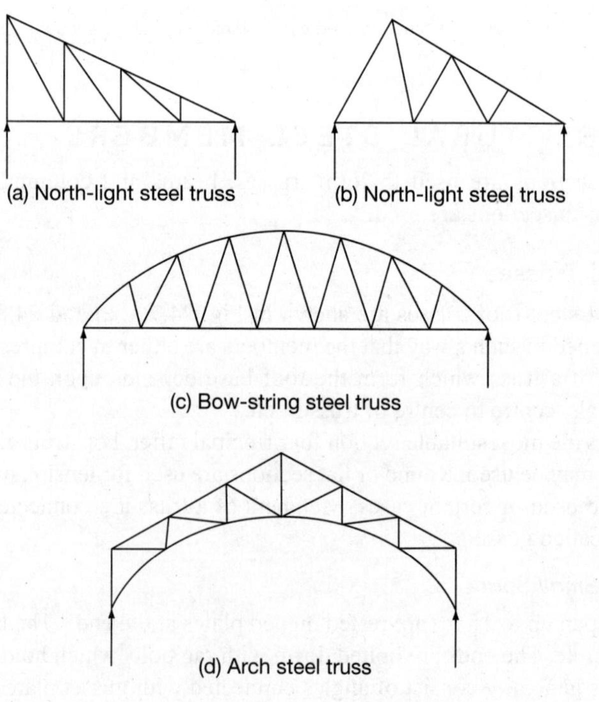

Figure 24.8 Various types of steel trusses for span more than 15 m

2. Trusses for Large Spans

In large-span trusses the members may consist of more than one section. Two angles or channels or flats may be connected in the gusset with other members at the joints. As the truss is hung, handling may be difficult. Thus the trusses are riveted into two portions at the fabrication centre and gusset plates at the connected ends are riveted to them. The two halves are erected at the site and riveted. Bigger bearing plates are used for these trusses. Figure 24.8(a–d) show different types of trusses grouped under large span trusses.

24.5.2 Steel Beams

Steel beams include girders, lintels, etc. The simplest form of steel beam would be a single RSJ or an angle section for carrying small loads. In case of large loads, compounds (built-up) sections are used.

Compound beams comprise of two or more single RSJs connected together through bolts and separated suitably with separators which hold them in required position. Separators are placed $1\frac{1}{2}$ to 2 m apart (Fig. 24.9).

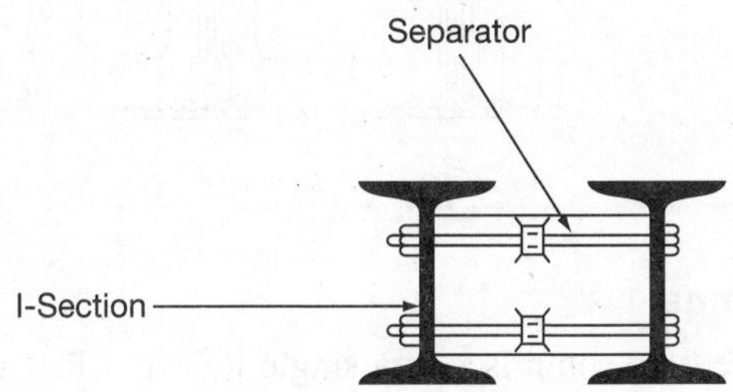

Figure 24.9 Compound beam

For still heavier loads, girders are formed by two channels placed back to back and their flange area increased by the addition of one or more plates at top and bottom (Fig. 24.10).

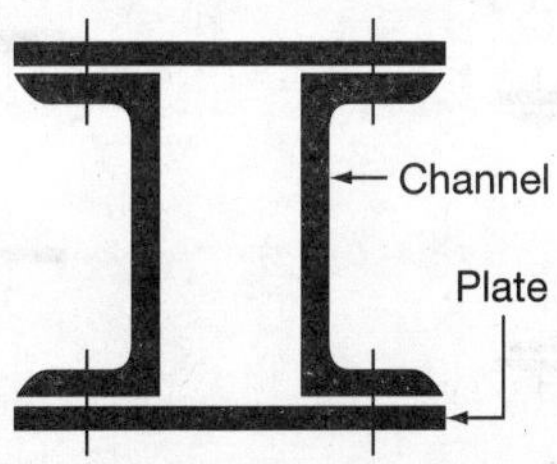

Figure 24.10 Girder

When very large loads are to be handled plate girders are used. A plate girder is one which is a built-up section consisting of top and bottom flanges made up of angles and plates. The web consists of one or more steel plates. The beam is formed by placing the individual pieces in positions and riveting or welding is done to keep then together (Fig. 24.11). As the plate girders are of considerable depth they have to be stiffened laterally to prevent buckling. Angles or T-sections are used as stiffeners and are riveted to the plate girder at suitable spacing (Fig. 24.12).

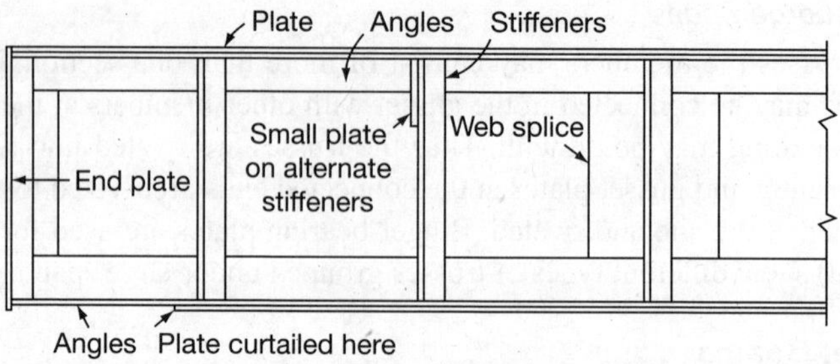

Figure 24.11 Part elevation of plate girder

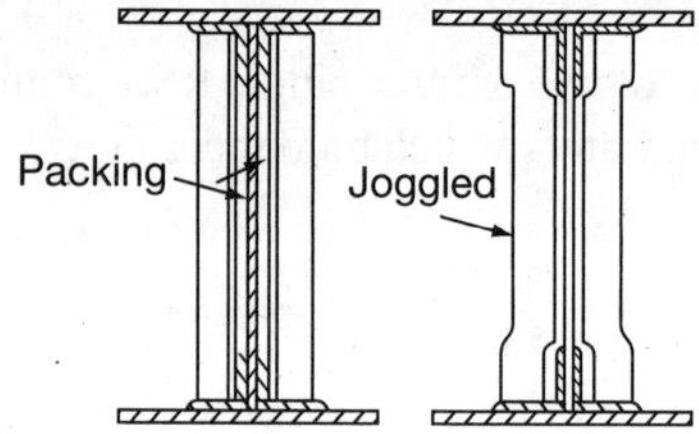

Figure 24.12 Stiffeners

24.5.3 Steel Columns

Steel columns of simple type comprises of a single RSJ or a RSJ with two flange plates (Fig. 24.13). Compound columns may be made by riveting together two or three RSJs (Fig. 24.14). Other forms of compound columns commonly used are a combination of RSJs and channels or two angles, channels with flange plates or four angles and a web plate or four angles joined together with lattice braces (Fig. 24.15).

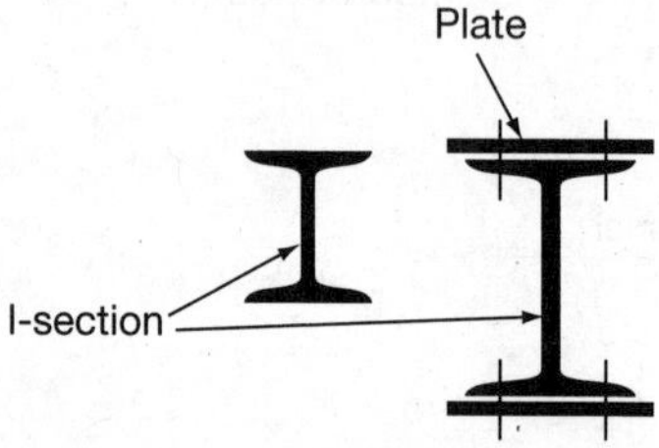

Figure 24.13 Simple type of column

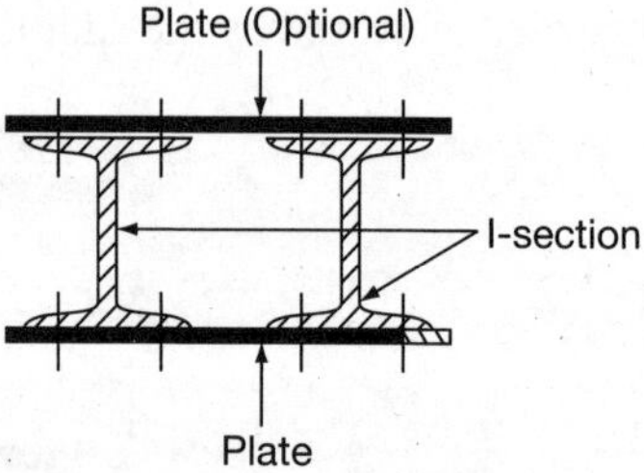

Figure 24.14 Compound column

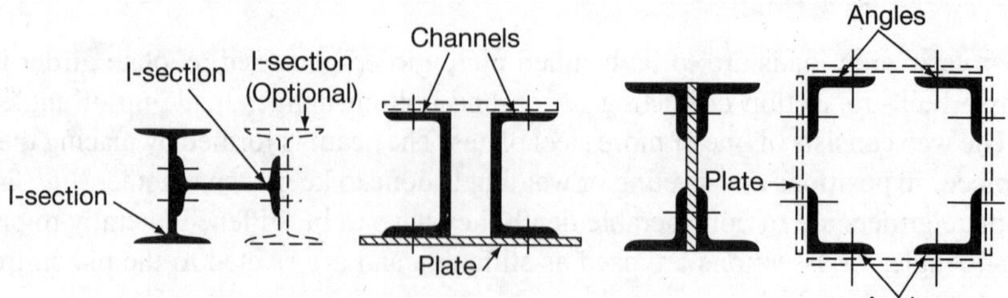

Figure 24.15 Other types of compound column

Whenever more than one member is joined it is necessary to interconnect them with the help of lacings and batten plates. This arrangement of lacings makes the column to act as a combined unit (Fig. 24.16).

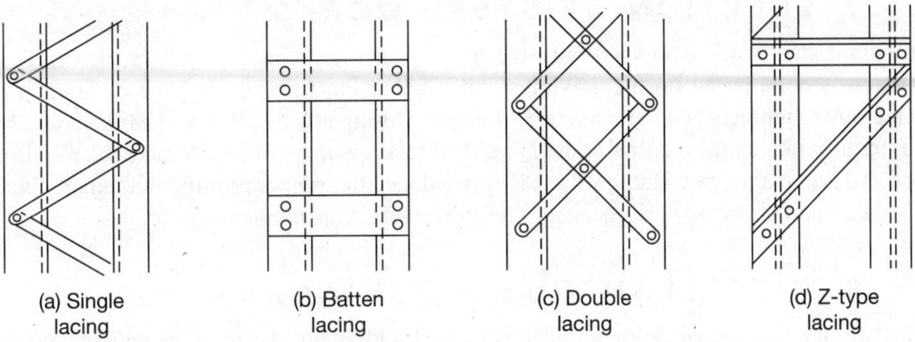

Figure 24.16 Types of lacings for columns

24.6 FABRICATION OF STEEL WORKS

Preparation of steel work for erection is called as fabrication. It includes all works necessary to layout, cut, drill, rivet or weld the steel sections. The fabrication has to be properly planned such that the work at the site of erection should be reduced as much as possible. Thus most of the work is carried out in the fabrication shop itself.

All materials procured from the mill or the markets are straightened if necessary. Cutting is effected by shearing, cropping or sewing. For mild steel gas cutting by mechanically controlled arcs is also used. For high tensile steel gas cutting is permitted under special care. Plates and angles are cut by shearing. Beams and channels are usually cut to the desired lengths in factories.

Templates according to the shape of the final job are prepared. Templates may be made of wooden strips showing location of all holes and cuts. Cardboard templates may be used for gusset plates.

All the materials are laid out. The centres of the holes are identified with a punch. Holes are drilled, punched or bored. Drilling is generally preferred as it gives the exact hole without spoiling the surroundings. For thick sections drilling is preferred.

When the components of a member are ready, they are held in position temporarily by shop bolts. Shop bolts are longer in size than the normal ones. There should be at least two bolts put in one part of a member. All assembled parts should be in close contact. Further all bearing stiffeners should bear tightly at top and bottom. In general, no drifting should be permitted.

Assembled parts are then riveted adopting hydraulic or pneumatic pressure. Rivets of diameter less than 10 mm are driven cold. Bigger rivets are heated by a steady flame produced by burning oil or using electric heaters. It is the practice that the rivet fills the hole fully and forms a head of standard size. While riveting, if loose, burnt or badly formed, rivets should be cut out and replaced.

After finishing the required structure formation, it is cleared thoroughly and one coat of red lead paint is given. All machined faces should be coated with a mixture of white lead and tallow. All inaccessible portions of the structure should be painted with two coats of read lead paint.

The completed work should be temporarily shop erected such that accuracy of fitness may be checked before despatch.

24.7 ERECTION OF STEEL WORKS

Structural steel work is erected with the use of derricks, slings, guys, cranes, etc. The erection is done based on the drawings and keeping the verticality of the columns and fixing other members and parts of steel structure. During erection the steel work should be temporarily braced till the final stage is reached to allow the structure to take the required load. All the riveting or welding works, if any, is done after proper position is attained. Final alignment and verticality are checked. The steel work is finally painted.

24.8 MISCELLANEOUS STEEL WORKS

Apart from various application of steel work for building construction, they are also used in buildings for minor works which are explained below.

24.8.1 Fixing of Steel in Wooden Frames

Round or square bars with flats or grills with certain design are fixed to the windows or ventilators for safety. These materials are basically of mild steel.

In simple design round or square bars are used along with flats of 4 mm thick. Fixing of bars through window frames apart from aesthetic value is economical also. Steel bars are fixed in wooden window frames through holes made on either side of frame and in between flats may be used to prevent bending. Instead well designed and factory made steel grill may be fixed while making the wooden frames. If grills are to be installed in frames that are already attached to walls, they are fixed with counters work wood screws with heads painted to conceal its position.

24.8.2 Fixing of Rolling Shutters

Rolling shutter consists of mild steel laths, 1.21 mm thick and 80 mm wide or as specified. The laths are machine rolled continuous strip with interlocking facility through the entire length. Such combined laths form a shutter.

The shutter is supported by means of spring barrel which in turn is supported by cast iron or steel brackets. The shutter slats shall coil round the spring board. Guide channels are provided on either side. The gap between the two legs of the guide channel shall be sufficient to allow the free movement of the shutter and at the same time close enough to prevent rattling of the shutter due to wind. The shutter shall operate easily and smoothly under all conditions. When rolled up open position is made. Locking facility is also provided.

SALIENT POINTS

1. Materials generally used for structural purposes in buildings are cast iron, wrought iron and mild steel. As mild steel can withstand tension and compression, it is largely used.
2. For buildings various types of sections and shapes are used, viz., angles, plates, flats, bars, channels, I-section, T-section, expanded metal and ribbed TOR steel bars.

3. Steel members are connected by rivetted connections, bolt connections and welded connections.
4. In order to make steel structures (e.g., beam, columns, etc.,), the standard sections are combined in a suitable way. Such a combined section is called built-up sections.
5. Different sections are used to form trusses, beams and columns. For this individual steel sections or built-up sections are used.
6. Steel roof trusses are designed in such a way that the members are either in compression or in tension only. Members of the truss which form the roof base depends upon the span, roof slope, covering materials, centre to centre of trusses, etc., T-section is the most suitable section for principal rafter.
7. For small spans up to 15 m, the following trusses are used: King post, Simple fink, Compound fink, Howe and Compound Howe trusses are used.
8. For large spans over 15 m, the following trusses are used: North-light, Bow-string and Arch trusses.
9. Simplest form of steel beam would be a single RSJ or an angle section for carrying small loads.
10. Compound beams comprise of two or more single RSJs connected together through bolts and separated suitably with separators which hold them in required position. Separators are placed 1½–2 m apart.
11. Girders formed by two channels placed back to back and their flange area increased by the addition of one or more plates at top and bottom are used for heavier loads.
12. For very heavy loads, plate girders are used. Plate girder is one which is a built-up section consisting of top and bottom flanges made up of angles and plates. The web consists of one or more steel plates. The plate girders are stiffened laterally to prevent buckling.
13. Steel columns of simple type comprise of a single RSJ or a RSJ with two flange plates. Compound columns may be made by riveting together two or three RSJs.
14. Preparation of steel work for erection is called as fabrication. It includes all work necessary to layout, cut, drill, rivet or weld the steel sections.
15. Fabrication has to be properly planned such that the work at the site of erection should be reduced as much as possible. Thus most of the work is carried out in the fabrication shop itself.

REVIEW QUESTIONS

1. How steel is put into use in building construction?
2. What are the various steel sections used?
3. What are steel connections? Explain.
4. Explain the methods adopted for connecting the members of steel work.
5. Explain the functions of rivets in a steel structure.
6. Describe the possible types of failures of riveted joints.

7. What are the different types of welding used in connecting steel members?
8. What is a built-up section?
9. What are the advantages of adopting built-up sections? Draw typical sketches of built-up sections.
10. Name the types of trusses used for short and long span.
11. Write a short note on the fabrication and erection of steel work.
12. As asbestos cement sheet roof has to be provided for a workshop building of 48 m × 12 m. Suggest a suitable steel truss to be supported on a brick wall. Sketch the truss and the junctions.

25 ROOFS AND ROOF COVERINGS

25.1 INTRODUCTION

Roof is the upper most part of a building which is constructed on structural members and provided with a covering material. The sole aim of roof is to enclose the space securely and protect the building and the occupants from atmospheric effects like sun, wind, rain, snow, etc. A roof is as important as that of a foundation.

Roofs are constructed keeping in view the weather conditions of the location, the material availability, architectural considerations, heat insulation, ventilation, strength and stability, etc.

The olden and conventional type of roof is sloping roofs which were used based on the availability of material. Such roofs have been improved with invention of new materials like steel, asbestos, etc.

The next improvement is flat roofs which have improved with the advent of reinforced cement concrete (RCC).

Modern construction techniques have yielded ways to have new type of roofs, viz., shells, folded plates, etc.

This chapter deals with different types of roofs and roof covering materials.

25.2 REQUIREMENTS OF A ROOF

A roof is provided to satisfy the following basic requirements:

1. It should give a protective covering against the adverse effects of atmosphere.
2. It should have adequate slope to drain the rain and snow.
3. It should be strong, stable and durable to take the external loads.
4. It should provide sufficient insulation against heat and sound.
5. It should meet the different climatic conditions and accordingly the covering material has to be chosen.
6. It should also provide good architectural appearance.

25.3 TYPES OF ROOFS

Roofs may be grouped under two major categories, viz., sloping or pitched roofs and flat roofs.

Slope or pitched roofs are those which have the surfaces with considerable slope for covering the building structure. The modern construction like shells may also be grouped under this category.

A roof which is nearly flat is known as a flat roof. It is the convention if the slope is less than 10°, it is considered as a flat roof. As a matter of fact no roof is laid flat.

Further discussion is made on different types of roofs in each category.

25.4 SLOPING ROOFS

Sloping roofs are comparatively light in weight and can be used in large to very large spans. They are quite suitable in areas where there are heavy rain and snow. Shapes of sloping roofs are dependent on the area to be covered, material used, light, ventilation needed, etc.

25.4.1 Terminologies

Following are the terminologies which are relevant to sloping roofs (Fig. 25.1).

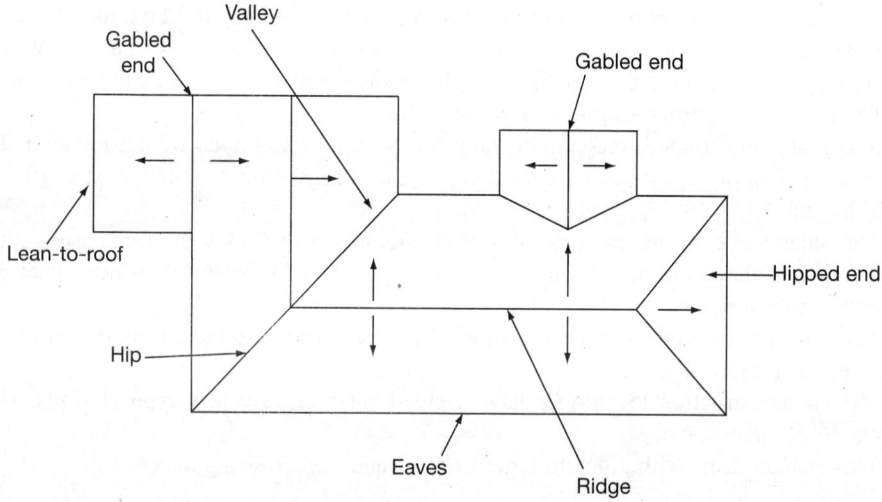

Figure 25.1 Parts of a sloping roof

1. Span

Clear span is the horizontal distance between the walls or the supports. Effective span is the horizontal distance between the centres of bearings of the supports.

2. Pitch

Pitch is the slope of a roof which may be expressed as degrees or inclination to the horizontal or the rise to the span.

3. Eaves

The lowest course of the roof is called eaves. Eaves are generally projected beyond the external face of the wall. This is made to protect the walls from the rains, for shading the walls from the sun and to provide a pleasing appearance.

4. Hipped End

Hipped end of the roof is the sloping end of a roof and is generally triangular in shape. The hipped end has the same slope as that of the main roof.

5. Hip

Hip of a roof is the line of intersection of two roof planes.

6. Gable

Gabled roof is a roof with slopes in two directions.

7. Ridge

It is the top most line formed by the intersection of roof surfaces.

8. Valley

The line of intersection of two roof planes containing an angle of less than 180°.

25.4.2 Types of Timber Sloping Roofs

Different types of timber sloping roofs are discussed below.

1. Lean to Roof

This is a roof which is usually used to cover verandah. This is used for spans up to 2.5 m. The roof covering is fixed to battens running across the rafters. It is usually constructed as a mono-pitched roof (Fig. 25.2).

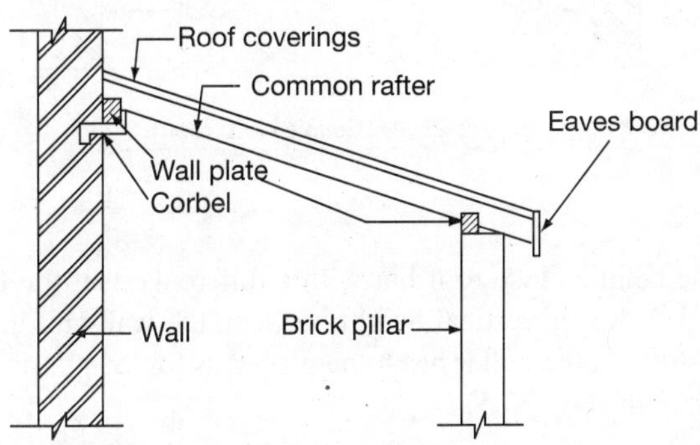

Figure 25.2 Lean to roof

2. Couple Roof

This is a roof which has two rafters with a central ridge piece at the top. The rafters are fixed to the wall plates. The roof covering is fixed to battens running across the rafters. This can be used only up to spans of 3.5 m (Fig. 25.3).

3. Couple Close Roof

This is similar to that of couple roof and the rafters are connected by wooden tie. The tie has two functions: (i) prevents the outward spreading of the walls and (ii) acts as a support for any ceiling which might be fixed on it. This is suitable for spans up to 4.5 m (Fig. 25.4).

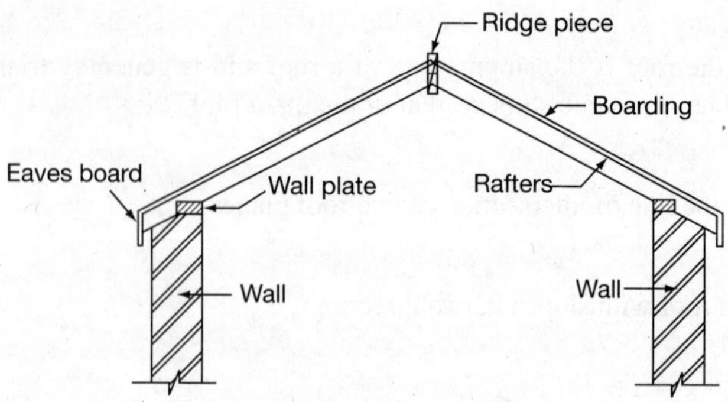

Figure 25.3 Couple roof

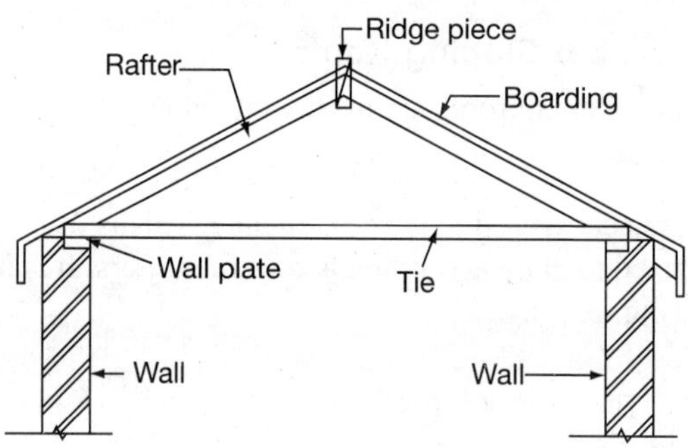

Figure 25.4 Couple close roof

4. Collar Tie Roof

This is similar to the couple close roof but with a difference that the tie (called collar) is fixed at a height of 1/2–1/3 of vertical height between the wall and the ridge. In order to gain the maximum strength the collar has to be placed as low as possible. This roof is suitable for spans of 4–5.5 m (Fig. 25.5).

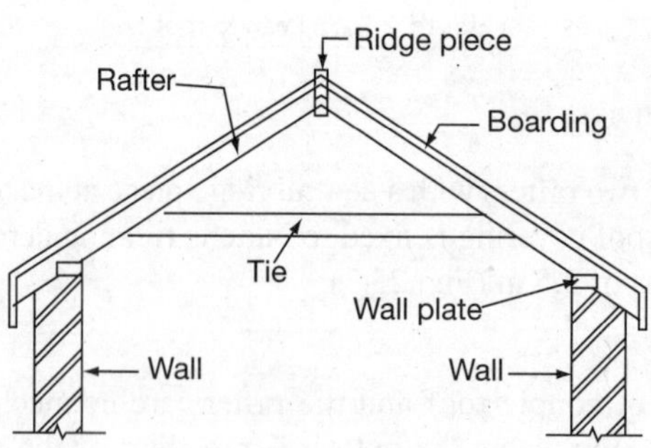

Figure 25.5 Collar tie roof

5. *King-Post Truss*

For large spans, King- and Queen-post trusses are used. In king-post truss, the total weight of the roof is transferred to the walls vertically. The frame work of the truss is built such that the shape is not altered when loaded. Members in this truss are subjected either to compression or tension. The shape of the truss being triangular in shape gives a greater rigidity. Different load bearing members are parts of the King-post truss which are shown in Fig. 25.6. This type is suitable for spans of 6–9 m.

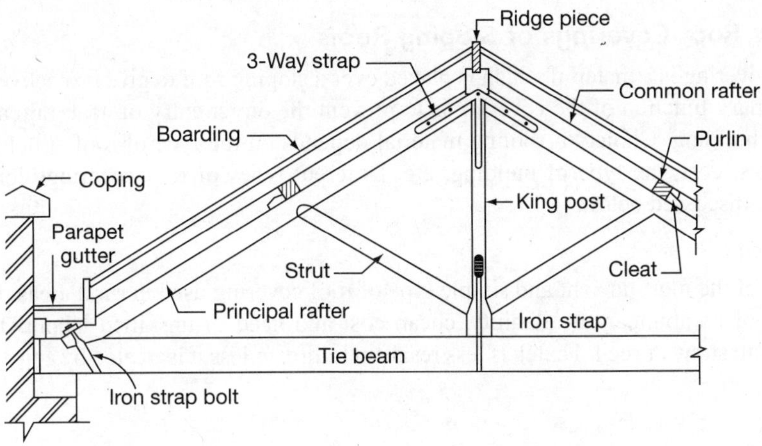

Figure 25. 6 King-post truss

6. *Queen-Post Truss*

This consists of two Queen posts instead of one as in King-post truss and also has one straining beam. The members which are subjected to compression are straining beam and struts whereas the Queen-posts and the tie beam will be under tension. This type of truss is suitable for spans up to 13.5 m. Different load bearing members and parts of the Queen-post truss are shown in Fig. 25.7.

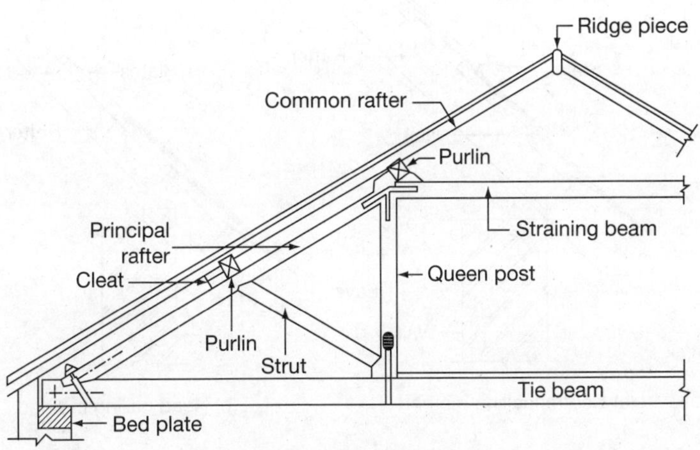

Figure 25.7 Queen-post truss

25.4.3 Types of Steel Sloping Roofs

Shape and positioning of members in a steel roof truss are designed in such a way that the members are subjected to either compression or tension. Trusses do not have any bending stress in them. The compression members of the truss are based on the span, slope, the covering material, and centre to centre of trusses. The various types of roof trusses are discussed in Section 24.5. For large spans steel trusses are generally preferred. This also can be used for small spans.

25.4.4 Roof Coverings of Sloping Roofs

A roof covering is a material which is placed over a sloping roof to give protective surface. The primary function of the covering is to prevent the direct entry of heat, rain and snow into the building. Choice of roofing material depends on the type of roof structure, local conditions, cost, the type of building, etc. Different types of roof coverings for sloping roofs are discussed below.

1. Thatch

It is one of the most ancient and simple type of roof covering used in rural areas. It is used because of its abundant availability, cheap cost and need of unskilled labour. Thatch is either from straw or reed. Thatch is less resistant to fire, unless it is treated by a fire proofing material.

2. Slate

It is basically a sedimentary rock which could be split into thin sheets. Slates are available in different natural colours. Most common colour is various shades of grey. A good quality slate should have (i) even colour, (ii) be hard, (iii) impervious and (iv) should give a ringing sound when struck. It should not be tough and have white patches.

The sizes of slates vary from 60 cm × 35 cm to 25 cm × 12.5 cm and the thickness varies from 1.5 to 4.5 mm for smaller sizes and 10 to 12 mm for large sizes. There are various methods of fixing slates (Fig. 25.8).

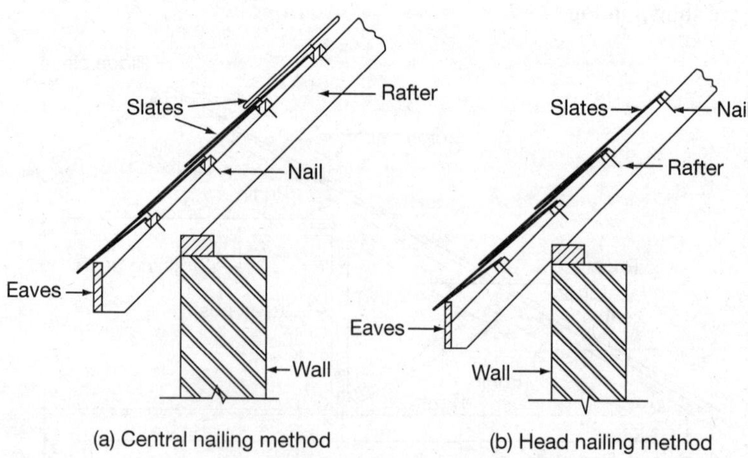

Figure 25.8 Methods of fixing slates

3. Tile

Like thatch, tile has been in use from olden days. The sizes and shapes have improved from time to time. Clay tiles are made similar to bricks. Concrete tiles are used to a limited extent.

Plain tiles are of sizes varying from 25 cm × 15 cm to 28 cm × 18 cm with thickness varying from 9 to 15 mm. These tiles are used along with special tiles such as hip and valley tiles.

Pan tiles are made of clay and it is the oldest type. The roof is covered by the use of bent tiles. This is comparatively less weight. The tiles are of sizes 30–35 cm in length and 20–2.5 cm width. There is little chance of leakage at joints.

Mangalore tile is one of the most popular type of patented tiles. These tiles are machine-pressed clay tiles which have better interlocking, enabling better protection against heavy rains.

Asbestos cement tiles of square or rectangular in shape are also in use. These tiles can be laid on boards of the roof or directly on the battens which are nailed to the rafters. Figure 25.9 shows the fixing details of asbestos cement (AC) tile.

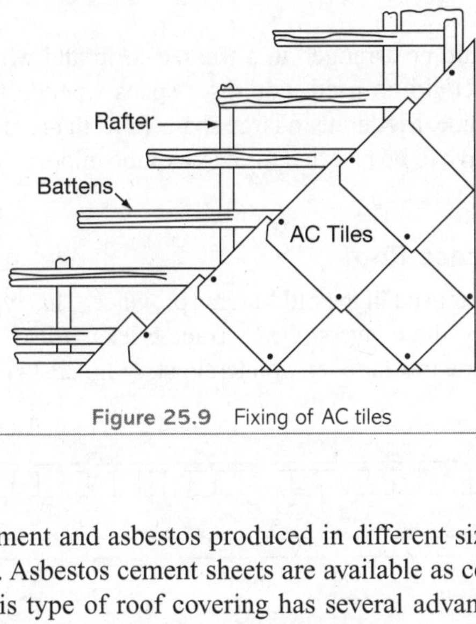

Figure 25.9 Fixing of AC tiles

4. AC Sheet

The combination of cement and asbestos produced in different sizes and shapes form the asbestos cement sheets. Asbestos cement sheets are available as corrugated roofing sheets and as well boards. This type of roof covering has several advantages, viz., durable, not heavy, impervious, fire and vermin proof, less labour cost, etc. These sheets are fixed on wood or steel purlins (Fig. 25.10).

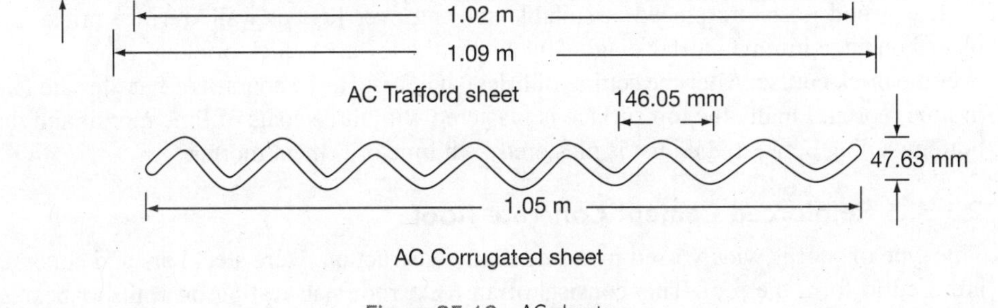

Figure 25.10 AC sheets

5. CGI Sheets

Corrugated Galvanised Iron (CGI) sheets is yet another type of roof covering widely used although the appearance is not that good. Like AC sheets, it has high durability, light in weight and fire proof. This consists of thin metal sheets which are galvanised and given the corrugation of bent shape for additional strength. These sheets are fixed on purlins with special screws.

6. Aluminium or PVC Sheets

Now-a-days Aluminium or PVC sheets similar to corrugated sheets are available which are light in weight, easy to handle and available in different colours.

25.5 FLAT ROOFS

Flat roofs are used in the areas which have less rainfall and no-snow fall during the year. Drainage in a flat roof is a major problem unlike a sloping roof which has an excellent drainage facility.

A material which could be arranged in a trussed form and which satisfies the general requirements of a roof could be used. For short spans wooden flat roofs are used with adequate weathering course. Moderate to large spans RCC flat roofs are preferred. For very large spans steel roofs have to be resorted to. Three of the important flat roofs are discussed below.

25.5.1 Madras Terrace Roof

It is an old type of flat roof used in the old Madras province. This type of roofing is not used now-a-days. It consists of three layers, viz., terrace bricks, brickbat concrete and flat tiles placed in this order in lime mortar over wooden joists (Fig. 25.11).

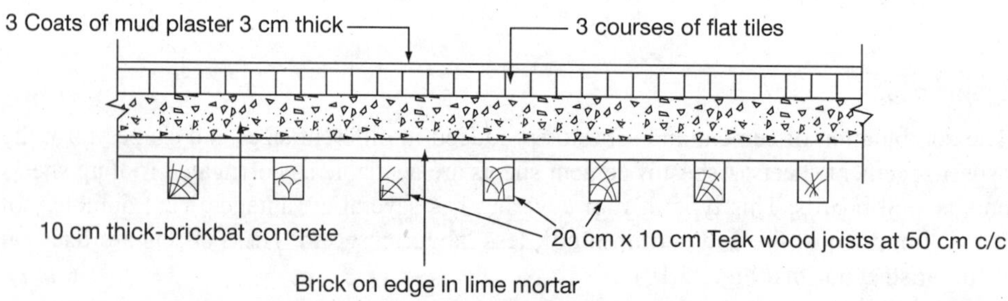

Figure 25.11 Madras terrace roof

Teak wood joists are placed at suitable spacing over RSJ or wall. Terrace bricks are placed on edge in lime mortar diagonally across the joists. A thick brickbat course is laid over the brick course. After the setting of brickbat course, three courses of flat tiles are laid in lime mortar. Finally the top surface is plastered with three coats of lime mortar and the bottom ceiling between the joist is plastered with lime or cement mortar.

25.5.2 Reinforced Cement Concrete Roof

This type of roof is widely used in the modern construction. Here steel bars and concrete are used to form the roof. This consists of an RCC roof slab resting on walls or beams.

For short spans (length to width ratio greater than 1.5) the slab is treated as one-way slab otherwise it is treated as a two-way slab and accordingly steel reinforcements are provided. The thickness and steel reinforcement required are based on the span and loading conditions. In case the slab is required to be supported by RCC beams, the size and the steel reinforcement of the beams are determined and provided accordingly (Fig. 23.8).

At the top level of the wall and beam, centring sheets are placed horizontally by means of grid over the centring sheets. A clear cover of 15 mm at the bottom and above the centring sheet is provided. Cement concrete mix (usually a mix of 1 : 2 : 4 is used) is placed on the centring sheets after adequate mixing with water. The concrete is rammed well by tampers or a vibrator is used to attain the maximum density and the required thickness. After about 12 hours of laying, the exposed surface is cured with water for 21 days.

In case of large spans, they are divided into bays, and ribs are provided to stiffen the slabs. These ribs are monolithically casted with the slab and function as beams in the shape of T, usually referred to as T-beams. The slab portion is termed the flange of T-beam. The width of the rib should be adequate to house the tensile reinforcement of beam. All modern constructions employ T-beam slabs for public buildings (Fig. 23.9).

25.5.3 Bengal Terrace Roof

This type of roof is used especially used in Bengal to cover verandahs and so get the name as Bengal Terrace Roof. This type of roof consists of rafters, battens, flat tiles and final finish.

At first the rafters are placed in position, sloping outwards from main wall at a spacing of 30–50 cm centre-to-centre, Fig. 25.12. One end of the rafters is inserted into the wall for length of 30 cm.

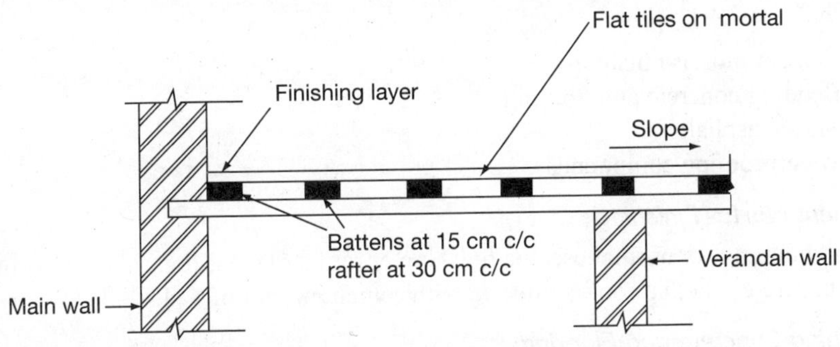

Figure 25.12 Bengal terrace roof

Battens or repairs of 5 cm × 1 cm are fixed to the upper surface of the rafters perpendicularly with a spacing of 15 cm. On the battens, a course of flat tiles well soaked in white-wash is placed in lime or cement mortar.

Finally the roof is finished by adopting any one of the following methods:

(i) Two courses of flat tiles are laid on mortar and finished finally with two or three coats of plaster out of which the last coat is rubbed smooth and polished.

(ii) Over the first course of tiles, a layer of fine jelly concrete of 4–5 cm thick is laid. Over this concrete layer one more course of flat tiles is laid and the surface is finally finished with two or three coats of plaster. The outer most surface is rubbed smooth and polished.

25.5.4 Jack Arch Flat Roof

The construction of Jack arch flat roof is similar to the construction of Jack arch floors. The covering or terracing may be a 10 cm thick layer of lime concrete to make the roof water tight along with some waterproofing material. Alternatively the covering may consist of a 10 cm thick earth filling which is finally plastered at the top. Some water proofing compounds, additives, may also be used.

25.5.5 Mud Terrace Roof

This type of roof, although oldest, is cheap and fairly water-tight. It is still used in villages where the rainfall is light. This mud terrace is constructed out of good white earth containing a large percentage of sodium salts.

25.5.6 Weather-Proof Course for Flat Roofs

Flat roofs are provided with a weather-proof course to sustain the effects of atmospheric agencies. The construction of waterproof course consists of providing one course of surki-concrete and two courses of flat tiles. The surki concrete (broken brick aggregate and lime with no sand) in the mix ratio of 1:2:5 is laid to a compacted thickness of 75 mm. The concrete layer is provided with a minimum slope of 1 in 50 slope directed towards the rain water drain pipes. After curing for 6 days, two courses of flat tiles (or one course of pressed tiles) are laid in a cement mortar of 1 : 3 with crude oil.

25.5.7 Waterproofing of Flat Roofs

The flat roof should be made waterproof by a surface covering. The following methods are employed:

(i) Cement mortar finishing
(ii) Bedding concrete and flooring
(iii) Mastic asphalt
(iv) Waterproofing compounds

1. Cement Mortar Finishing

For small buildings or normal use, the finishing of roof surface is done at the time of laying cement concrete. The flat roof is finished with cement mortar of 1 : 4.

2. Bedding Concrete and Flooring

The concrete surface is kept rough over which a brick bats lime concrete of 1:2:4 or brick bats cement concrete of 1 : 18 : 24 is laid for a thickness of about 10 cm. The bedding concrete is finished with tiles, stone slabs or terrazzo, etc. A convex joint is provided at the junction of parapet wall and the roof.

3. Mastic Asphalt Finish

A layer of hot mastic asphalt is laid on the roof surface and a jute cloth is spread over this surface. One more layer of mastic asphalt is laid over the jute cloth such that the jute cloth is sandwiched between to asphalt layers. Then sand is sprinkled over the entire surface of the roof. A led sheet is inserted at the junction of the parapet and the roof so as to have a better grip and easy draining.

25.5.8 Drainage of Flat Roofs

It is one of the basic requirements of flat roofs. The rain water should be drained off quickly so as to avoid leakage of roof. In order to achieve this satisfactorily watertight roof surface, adequate slope and drain outlets have to be provided.

The slopes in flat roofs vary from 1 in 20 to 1 in 40. The rain water outlets should be evenly spaced round the building some possible arrangements of drain outlets are shown in Fig. 25.13.

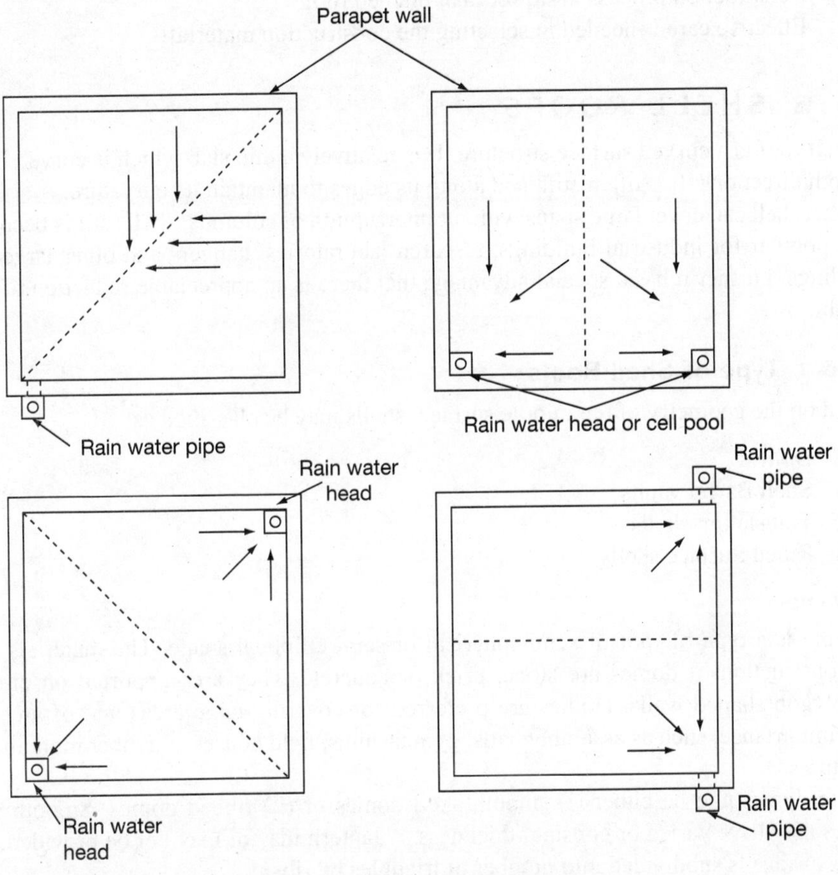

Figure 25.13 Different arrangements of drain-outlets

25.5.9 Merits and Demerits of Flat Roofs

1. Merits

(i) Flat roof tops may be used for several purposes, such as roof gardens, drying yard, conducting functions, etc.
(ii) Construction and maintenance easy.
(iii) Construction of additional floor can be taken.
(iv) Possess good heat and sound insulation.

(v) Easy to make fire-resistant than pitched roof.
(vi) Require less roofing material than pitched roof.
(vii) Economical.

2. Demerits

(i) Not suitable in places of heavy rainfall or snowfall.
(ii) Susceptible for cracks due to variations of temperature.
(iii) Heavy construction and involves high initial cost.
(iv) Construction process is slower than pitched roof.
(v) Effective care is needed in selecting the construction material.

25.6 SHELL ROOFS

A shell roof is a curved surface structure. It is relatively a thin slab which is curved in one or both directions. It is often stiffened along its edges to maintain its curvature.

As a shell can cover large spans without interruption of columns, shell roof is becoming very popular for industrial buildings, research laboratories, hangers and other large span buildings. Further it has a special advantage that there is an appreciable reduction of dead weight.

25.6.1 Type of Shell Roofs

Based on the geometry of the middle surface, shells may be classified as:

(i) Domes
(ii) Shell Barrel vaults
(iii) Translation shells
(iv) Ruled Surface shells

1. Domes

A dome is a type of roof of semi-spherical or semi-elliptical shape. The materials used for construction of domes are stone, brick or concrete. They are supported on circular or polygon shaped walls. Domes are preferred for covering large areas and of architectural importance, such as assembly halls, gymnasiums, field houses and other monumental structures.

The domes can be either (i) smooth shell domes or (ii) ribbed domes. Smooth shell domes may have varied or constant thickness. A lantern may or may not be provided. The dome surface is subdivided into number of triangles by ribs.

Here a tension ring constructed at the perimeter of the structure is a thrust resisting member. The ring is usually supported on columns spaced around the perimeter and braced to provide lateral stability for the structure. In order to support bearing walls are also constructed. In order to span the space between the ribs and to support the roof deck purlins are provided. Figure 25.14 shows three types of domes (Arora and Bindra, 2005).

2. Shell Barrel Vault

Elements of a shell barrel vaults consists of edge beams, end frame and curved membrane. Different parts are shown in Fig. 25.15.

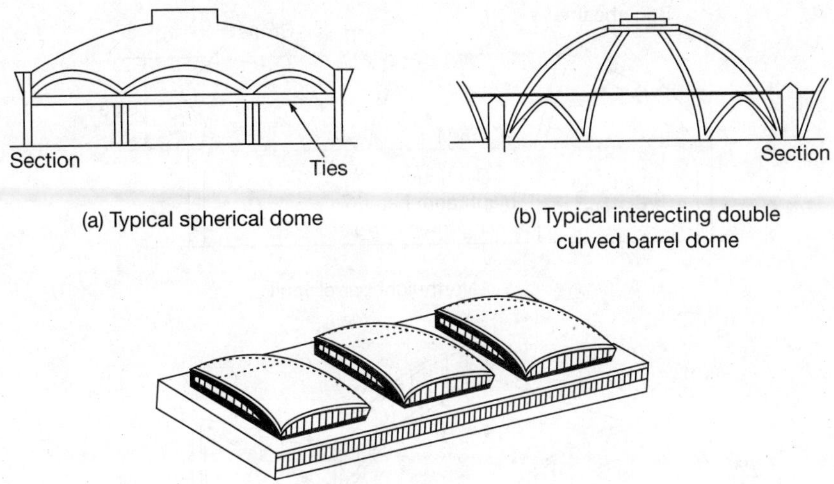

(a) Typical spherical dome

(b) Typical interecting double curved barrel dome

(c) Rectangular dome

Figure 25.14 Three types of domes

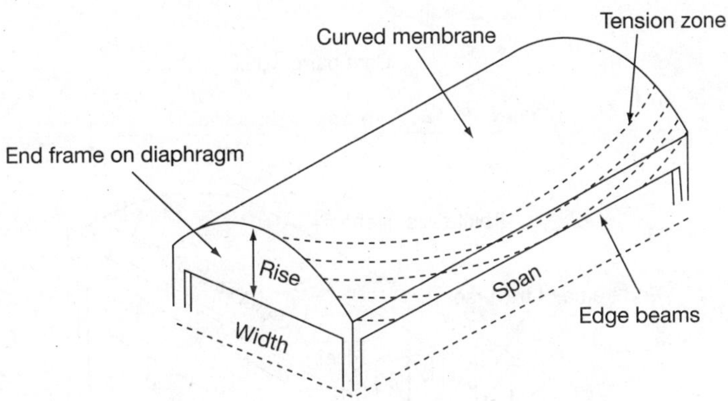

Figure 25.15 Elements of a shell barrel vault

Different types of shell barrel vaults are, short-span barrel vault, Multi-barrel double cantilever barrel vault, North-light barrel vault, Cant barrel vault, Cantilever barrel vault and Multiple barrel vault. Two of the barrel vaults are shown in Fig. 25.16.

3. Ruled Surface Shells

This type of shell is developed by moving a straight line so that it ends lie on two fixed vertical curves. These vertical curves may be different types or of similar type.

If two of these curves are of similar type, then the resulting shell is a cylindrical shell. Instead, if one of the curves is circular, elliptical, etc., then the resulting shell is a conoid.

Such shells also have two opposite curvatures, and have saddle surfaces. A cone is a special form of conoid in which the horizontal line is a point and the other curve is circular. Hyperbolic paraboloid is a ruled surface. Elements of a cylindrical shell are shown in Fig. 25.17.

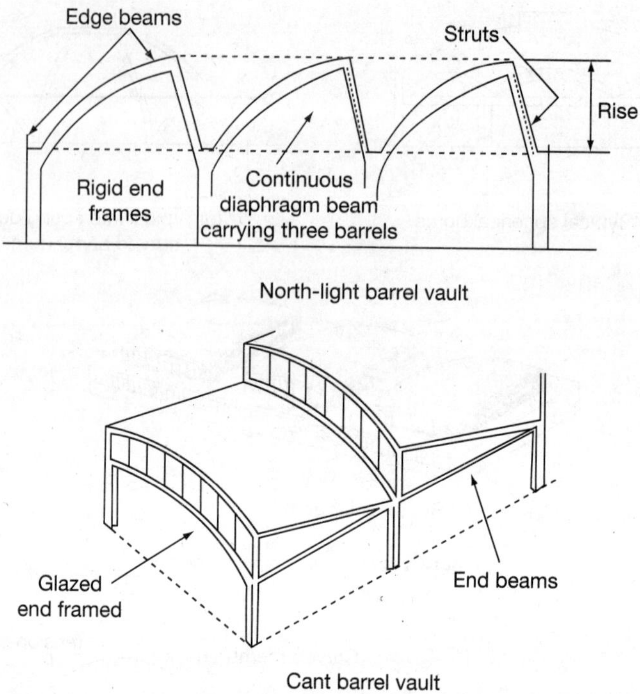

Figure 25.16 Two types of barrel vault

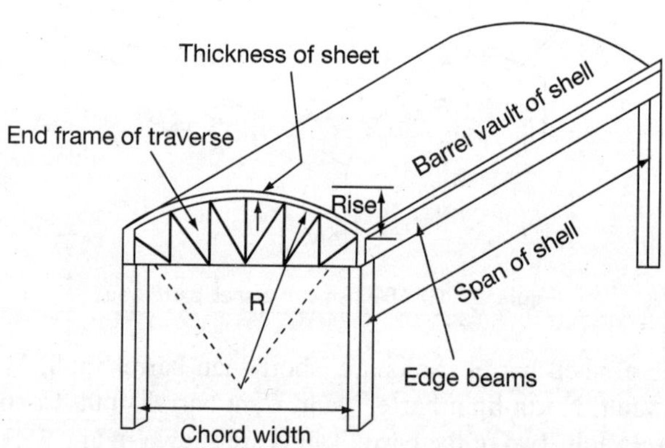

Figure 25.17 Elements of a cylindrical shell

25.6.2 Merits and Demerits of Shell Roofs

1. Merits

Following are the merits of shell roofs:

(i) Structural shells are capable of transmitting loads in more than two directions to support.

(ii) Shells are structurally highly efficient when they are so shaped, proportioned and supported.
(iii) Shells transmit the loads without bending or twisting.
(iv) Shells have small thickness compared to other dimensions.
(v) Shells provide uninterrupted space without columns.
(vi) Shells have less dead weight.
(vii) Shells are highly economical.
(viii) Shells provide better ventilation.
(ix) Shells provide better appearance and provide good reflecting surfaces.
(x) Shells require comparatively less time for their construction.

2. Demerits

Following are the demerits of shell roofs:

(i) Shells can sustain only direct stresses and no bending is permissible.
(ii) Shells can take only a negligible amount of shear stresses.
(iii) Damages caused to shells can not be repaired easily.
(iv) Colour washing the shell roof needs special ladder or temporary scaffolding.

25.6.3 Erection of Prefabricated Concrete Shell Elements

Prefabrication of shell units has been discussed in Chapter 9. Concrete shell elements, viz., doubly curved shells, cylindrical shells, etc. which are resting on load bearing structures of buildings are erected using cranes. Figures 25.18(a), (b) and 25.19 show the erection of short and long cylindrical shells and double curvature shells. Re-course is made to a ground level pre-assembly and load-gripping arrangement which eliminates erection stresses. Elements are pre-assembled in panels of 3 m × 6 m size or special mobile jig beds on the ground which are within the reach of an erection crane. After welding cover plates, two erection tie bars are fitted in each element.

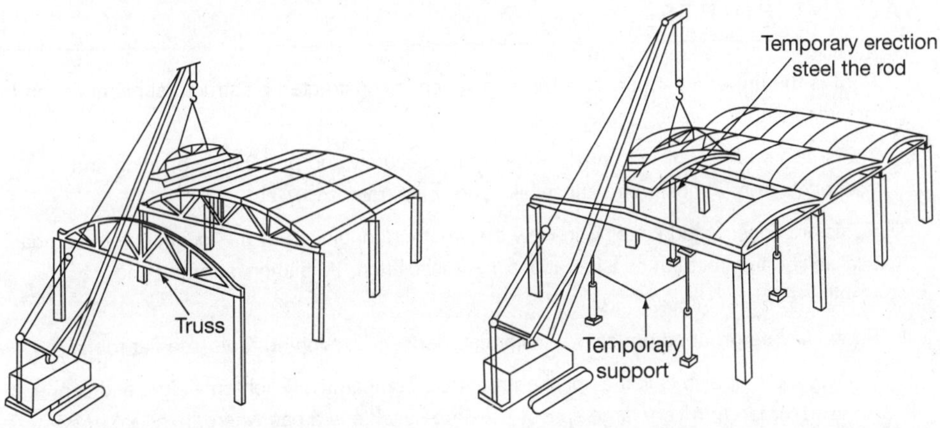

(a) Roof of short cylindrical shells

(b) Roof of long cylindrical shells

Figure 25.18 Erection of short and long cylindrical shells

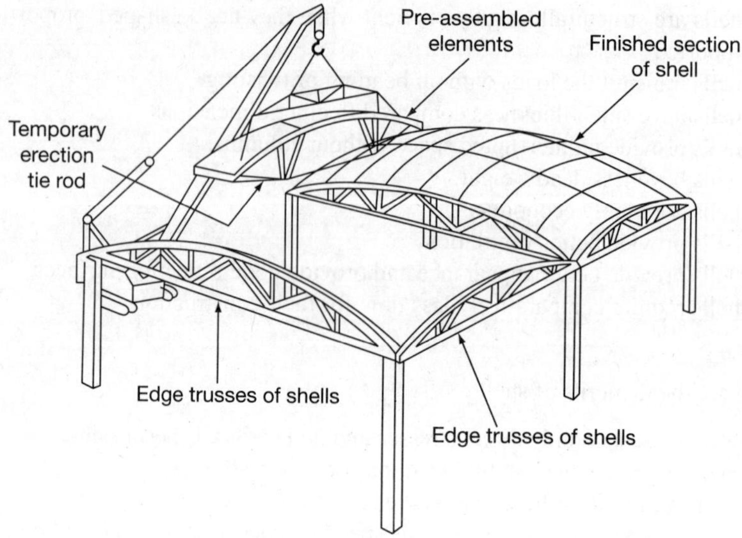

Figure 25.19 Erection of double curvature shell assembly

Following are the steps of operation (Atalev, 1985), in general:

(i) Pre-assembly of units of shells.
(ii) Trueing and securing of edge trusses within the reach of an erecting crane.
(iii) Welding of insets of elements of shells and of trusses.
(iv) Removal of temporary erection fastening devices and grouting with rich concrete in places where the elements adjoin trusses and shell joints.

SALIENT POINTS

1. Roof is the upper most part of a building which is constructed on structural members and provided with a covering material.
2. The sole aim of a roof is to enclose the space securely and protect the building and occupants from atmospheric effects like sun, wind, rain, snow, etc.
3. Roofs are constructed keeping in view the weather conditions of the location, the material availability, architectural considerations, heat insulation, ventilation, strength and stability, etc.
4. Roofs may be grouped under two major categories, viz., sloping roofs and flat roofs.
5. Sloping roofs or also called as pitched roofs are comparatively light in weight and can be used for large to very large spans. They are suitable in areas where there are heavy rains. Slabs of sloping roofs are dependent on the area to be covered, material used, light, ventilation needed, etc.
6. Lean-to-roof is a sloping roof which is usually used to cover verandah and suitable for spans up to 2.5 m.

7. Couple roof is a sloping roof which has two rafters with a central ridge piece at the top. The rafters are fixed to the wall plates.
8. Couple-close-roof is similar to that of couple roof and the rafters are connected by wooden tie. The tie has two functions: (i) prevents the outward spreading of the walls and (ii) acts as a support for any ceiling which might be fixed on it.
9. Collar-tie-roof is similar to the couple close roof but with a difference that the tie (called collar) is fixed at a height of 1/2–1/3 of vertical height between the wall and the ridge.
10. For large spans King-post truss is used. In this truss the total weight of the roof is transferred to the walls vertically. Members in the truss are subjected either to compression or tension.
11. Queen-post truss is also used for large spans which has two Queen posts instead of one as in King-post truss. The members which are subjected to compression are straining beam and struts whereas the Queen-posts and the tie-beam will be under tension.
12. Steel sloping roofs are also used which are designed in such a way that the members are subjected to either compression or tension. Trusses do not have any bending stress in them.
13. Roof covering is a material which is placed over a sloping roof to give protective surface. The primary function of the roof covering is to present the direct entry of heat, rain and snow into the building. Choice of roofing material depends on the type of roof structure, local conditions, cost, the type of building, etc. The roof coverings may be thatch, slate, tile, AC sheet, CGI sheet, Aluminium sheet or PVC sheets.
14. A roof which is nearly flat is known as a flat roof which as per convention has slope less than 10°. Madras terrace roof, RCC roof, Bengal terrace roof, Jack arch flat roof, and mud terrace roof are the types of flat roofs.
15. Waterproofing of flat roofs may be done by cement mortar finishing, Bedding concrete flooring finish, mastic asphalt finishing or water proofing compound finishing.
16. Drainage of flat roofs is one of the basic requirements of flat roofs. The rain water should be drained off quickly so as to avoid leakage of roof.
17. A shell roof is a curved surface structure. It is relatively a thin slab which is curved in one or both the directions. It is often stiffened along its edge to maintain its curvature.
18. Based on the geometry of the middle surface, shells may be classified as (i) domes, (ii) shell barrel vaults, (iii) translational shells and (iv) ruled surface shells.

REVIEW QUESTIONS

1. What are the requirements of a well planned roof?
2. Explain briefly the various types of sloping roofs indicating the span for which they are applicable.
3. Distinguish between a King-post truss and a Queen-post truss.

4. When does it become necessary to use a "Queen – Post Truss" for the construction of a sloping roof?
5. Explain the types of steel sloping roofs.
6. What are the methods employed to secure pitched roofs against uplift due to wind pressure?
7. A room of size 10 m × 21 m is to be covered by a steel roof. Suggest a suitable steel roof truss and show the details.
8. In automobile workshop shed of size 45 m × 10 m is to be constructed on a North-light truss. Suggest a suitable roofing material and give the reasons for your choice.
9. Discuss the use of different roof coverings and bring out application of each.
10. What is asbestos cement? What are the disadvantages of asbestos cement covering?
11. How slates are fixed to the rafters?
12. Distinguish between Madras terrace roof and Bengal terrace roof.
13. Explain RCC T-beam slab roof.
14. How weatherproofing of flat roofs is done?
15. Enumerate the methods of waterproofing of flat roofs.
16. How drainage of flat roofs is made?
17. Bring out the advantages and disadvantages of flat roofs.
18. What are the different types of domes? Explain.
19. Explain shell barrel vault.
20. Give the merits and demerits of shell roofs.
21. Explain the erection procedure of prefabricated concrete shell elements.

26

BUILDING FINISHES

26.1 INTRODUCTION

Building finishes comprise of plastering, pointing, painting, varnishing, white/colour washing or distempering. These finishes are performed to satisfy the following functions:

(i) These finishes primarily form protective coating to the exposed surface. This implies that the finishes preserve and protect the materials and thereby increase the life span of the materials.
(ii) These finishes add to the aesthetic view of the building and form a sort of decorative item.

Plastering is performed in majority of the cases and pointing is preferred in stone masonry and brick-masonry built by machine-moulded bricks.

Painting, varnishing, white/colour washing or distempering are applied to different type of materials which give a protection to the adverse weather and at the same time adds to the beauty of the building. These aspects are discussed in this chapter.

26.2 PLASTERING

Plastering is the method of covering rough and uneven surfaces of the various components of a building with a plastic material (i.e., mortar which is called the plaster).

26.2.1 Objects of Plastering

Object of plastering are as follows:

1. To provide an even and durable finished surface.
2. To enable the external surfaces to resist the atmospheric influence particularly the penetration of water.
3. To provide the internal surfaces a smooth surface such that dust and dirt can not lodge.
4. To protect the surface against vermin.
5. To form a good surface for taking colour wash, distemper or a paint.
6. To hide a poor workmanship or use of inferior quality materials.

26.2.2 Requirements of a Good Plaster

1. It should adhere to the surface and remain during all seasons withstanding the effects of atmospheric agencies.
2. It should not shrink with time and yield for formation of cracks.
3. It should provide an even surface with non-absorbent and washable qualities.
4. It should be hard, durable and high resistance against fire.
5. It should provide a surface as to form a decorative finish.

26.2.3 Preparation of Surface for Plastering

Durability of plastering depends on its adhesion with the background. A good-background is one which has enough strength and plain enough for suitable application of plaster. Types of background are (i) Solid Background, (ii) Lathing and (iii) Boards and Slabs.

1. Solid Background

The main characteristic of the solid background is that the nature of surface presented. Whatever may be the condition of solid background, preparation of surface is of prime importance. Following steps are undertaken:

(i) The background surface has to be first prepared by cleaning the surface and roughening the mortar joints.
(ii) Any projection more than 12 mm or the background surface should be removed and cavities and holes are fitted properly so as to obtain an even surface.
(iii) In order to get a good key for the plaster with the wall surface, all the joints in the masonry are raked out for a depth of at least 12 mm.
(iv) Dust and loose mortar are brushed out of the raked joints.
(v) The surface is rendered free from oily, greasy spots and efflorescence.
(vi) In case of re-plastering of old works, the smoothness of the surface is destroyed by hacking it with some tool.
(vii) Whenever concrete surfaces are to be plastered, rough sawn boards should be used. Mechanical keys may be applied to the inside shuttering or metal mesh may be left so as to help in the bonding of the plaster later on.

2. Lath Background

Laths are provided to form a foundation to the plaster work. When lath is used, the number of coats of plaster is to be three. Laths may be of two types, viz., metal laths and wooden laths.

(i) Metal Laths

In the market variety of laths are available. The most commonly used one is the plain expanded metal laths which is prepared from sheets of mild steel. The hyrib which is also a steel lath formed of sheet and perforated with projecting ribs.

Metal laths are supported and fixed with 9–12 mm diameter steel rods or steel channels. The plastering is carried out on both sides. They are fire-resistant.

(ii) Wooden Laths

Strips of well-seasoned wood are used for this purpose. The strips are generally 25 mm wide and 1–1.20 m long. Based on the spacing of timber joints the length of strips is adjusted. The thickness varies from 3 to 12 mm. The thicker strips are used for ceilings whereas thinner strips are used for walls and partitions.

3. *Board Background*

Boards and slabs have the advantage that a uniform thinner coat of plaster can be given. As the boards have relatively plain surface, they give good results with two coats or even with a single coat plastering. Special treatment may have to be needed for soft type of boards, so as to make better adherence of the plaster.

26.2.4 External Finishes

Based on certain factors such as desired appearance of the surface, cost, degree of maintenance, etc., the external walls are finished with suitable additional material than mortar. These are various forms of external finishes and some of them discussed below.

1. *Sand-faced Finish*

Sand-faced finish plaster comprises of two coats. The first coat is carried out in cement mortar of proportion 1 : 4 (i.e., 1 part of cement and 4 parts of clean coarse and angular river sand by volume). The thickness of the first coat is 12 mm and the plaster is watered for at least 7 days. Zigzag lines are made on the first coat so as to give adequate-adherence surface.

The second coat is of 1 : 1 plaster and the thickness is 8 mm. Sand used for the second coat should be of uniform size. When the coat is wet the sponge is worked such that the density of sand grains appearing on the surface is equal and uniform. After completion of the second coat the surface is kept well-wetted at least for 15 days.

2. *Pebble-dash Finish*

This type of finish is also called as Dry-dash finish. In this type of finish, the finishing coat is made of 12 mm thick. Over this plaster, small pebbles or crushed stones of suitable sizes (from 10 to 20 mm) are thrown and left exposed. The pebbles are sometimes lightly pressed or tapped into the mortar after throwing. It can resist good amount of aesthetic determination.

3. *Rough-Cast Finish*

This type of finish is also called as Spatter-dash finish. The plaster is of 1 : 3 proportion with one part of cement and three parts of coarse-grained sand by volume. Using a large trowel the mortar is dashed against the surface and the surface is roughly finished by the light movements of a wooden float.

4. *Scrapped Finish*

In this type of finish, the final coat of mortar is allowed to become stiff for a few hours. Then the surface is scrapped with a steel straight edge, or a board studded with nails or old saw blade in a particular pattern. The finish is a rough finish and is less liable to the cracks.

5. *Smooth-Cast Finish*

This finish is just similar to rough-cast finish and gives a smooth surface because of use of fine-grained sand. Generally a wooden float is used.

6. *Textured Finish*

In this type of finish, ornamental patterns or textured surfaces, such as ribbed stucco or fan textures, are made on the final coat of plaster with the help of using suitable tools.

7. Depter Finish

This is another form of rough-cast finish. As in the case of pebble-dash finish a rendering coat of 12 mm thick is prepared. While the coat is wet, pieces of gravel or flint are pressed with hand on the surface. It is possible to have the desired patterns selecting materials of different kind.

8. Machine Applied Finish

Machines are used to throw or spatter the material on the wall in a variety of ways which gives variety of finishes. The machines may be manually operated or mechanically operated. In order to through the material a sort of gun is used.

Any external finish, unless, it is painted should not require any maintenance over a long period of time. Light-coloured finish is required to be maintained only by periodical painting. An external finish is often required to be protected against rain penetration through the walls. Cracks also should be avoided.

26.2.5 Tools for Plastering

Following are the tools used for plastering:

1. Laying Trowel

Laying trowel is the ordinary trowel and is useful for applying or spreading or laying material and for trowelling so as to attain the desired finish. The trowel has a pointed or bull-nosed end. There are two types of trowels one can use to get a good work. The first one is with a stiff blade which is used for applying the rendering. The second one has got a thin blade which has a slightly springing action and is used exclusively for finishing coat (Fig. 26.1).

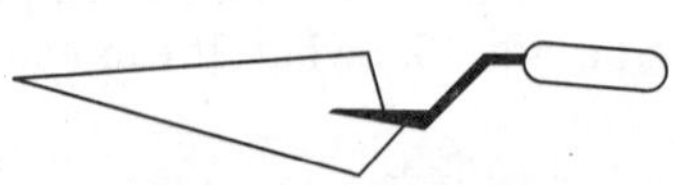

Figure 26.1 Laying trowel

2. Plum Bob

This is used to check the verticality of a surface (e.g., forming screeds in the same vertical plane) (Figs. 26.2 and 26.3).

Figure 26.2 Plumb bob

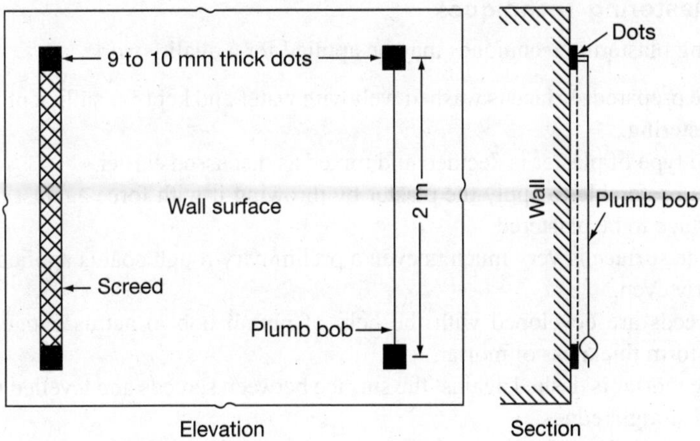

Figure 26.3 Dots and screeds

3. Gauging Trowel

This is another type of ordinary trowel and is useful for applying mortar to mouldings, corners, etc. It has a pointed as bull-nosed end.

4. Float

Floats are used to spread the mortar on the surface. The wooden float is known as the skimming float and is used for final or finishing coat of plaster. A float which is provided with nails projecting by about 3 mm from the surface is known as the devil float. It is used to make zigzag lines on the plastered surface so as to form a key for the next coat (Fig. 26.4).

Figure 26.4 Types of floats

5. Floating Rule

This tool is used to check the level of the plastered surface between the successive screeds (Fig. 26.5).

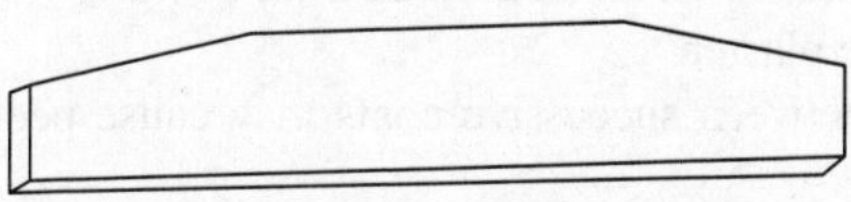

Figure 26.5 Floating rule

26.2.6 Plastering Techniques

The following plastering techniques may be applied for a quality work:

1. The prepared surface is washed well with water and kept for sufficient time before plastering.
2. The type of plaster is decided and mixed as discussed earlier.
3. It is preferable to apply the plaster by throwing it with force against the place or surface to be plastered.
4. If the surface is very much uneven a preliminary rough coat is applied to make it fairly even.
5. Screeds are developed with the help of plumb bob to act as gauges to obtain uniform thickness of mortar.
6. The mortar is dashed against the surface between screeds and levelled using floats and straight edges.
7. The final finish is obtained with trowels and the surface is watered for about 7 days.
8. For surfaces which need two coats, the first coat is made as discussed in steps up to 6 and then the surface is roughened.
9. The roughened surface forms a good seating for the second coat and the surface is watered for 2 days.
10. After this, the second coat is applied with a thickness of about 3 mm and the surface finished with trowel and is watered for about 7 days.

26.2.7 Defects and Remedial Measures in Plastering

1. Defects

Usual defects which are noticed in plastering are discussed below.

(i) Cracks may appear on the plastered surface which may be easily seen (called hair cracks).
(ii) Sometimes the soluble salts are brought to the surface. Such materials appear in the form of a whitish crystalline substance which is called efflorescence. Efflorescence affects the adhesion of paint with wall surface.
(iii) Small patches sometimes swell out beyond the plane of the plastered surface. These are called as blistering.
(iv) Because of inadequate bond available on brick work or concrete, the adhesion of plaster may not be complete.
(v) Presence of soluble salts on the surface adhesion might have been lost by their crystallization and expansion.
(vi) Failure of bond of successive layers may cause formation of loose mass which is called flaking.
(vii) Failure of plaster may also occur if the suction of the background is not uniform. Further-background which are absorbing a lot of water from the plaster may affect the strength of the plaster.
(viii) Failure of bond between successive coats may cause peeling of some portion of plaster.

(ix) Plaster sometimes contains particles which expand on being set. Because of this a conical hole is formed in the plastered surface. The conical hole is termed as the blow or pop.
(x) Because of poor workmanship, uneven surfaces may be formed.
(xi) Presence of excessive dampness at certain parts of the plastered surface makes that portion soft. Further thin coats may also cause soft points.
(xii) Plastering done on metal lath may show sometimes rust-stains.

2. Remedies

(i) Primarily the workmanship towards brickwork and plastering should be the best.
(ii) The bond of the brickwork should be properly maintained so as to give the plaster a proper bond.
(iii) The surfaces which are affected by efflorescence should be removed by rubbing with brushes. Then a solution of one part of hydrochloric acid or sulphuric acid with five parts of clean water is prepared. The solution is applied on the affected surface. Then the surface is washed with water. The efflorescence should be prevented by selecting a suitable material for mortar.
(iv) Good quality bricks only should be used.
(v) For brickwork and plastering work, water free from salts should be used.
(vi) In order to avoid the absorb of water in plaster by the background, the background should be well watered.
(vii) Excessive trowelling leads to cracking which should be avoided.
(viii) Prevention of water on to the wall should be avoided.
(ix) Damp proof courses should be provided whenever needed.
(x) The plastered surface should not be exposed to excessive water by natural agencies or otherwise.

26.3 POINTING

Finishing of mortar joints to have a good appearance is called pointing. In general, constructions made with good quality bricks and stone masonry are finished with pointing to have a pleasing appearance.

26.3.1 Mortar for Pointing

Pointing may be carried out either in lime mortar or cement mortar.

In lime mortar, lime and sand are mixed in equal volumes. These two materials are ground carefully in a mortar mill. Cement mortar also is of equal proportion of cement and sand. In both the cases the sand used should be clean, fine and free from any organic impurities. The mixed mortar should be used as early as possible and in no case later than 30 minutes.

26.3.2 Methods of Pointing

Following procedure is followed to perform the pointing work:

1. The existing mortar joint is raked out at least to a depth of 20 mm.
2. The joint is perfectly cleaned by using a brush.

3. The mortar joint is made wet by washing with clean water and kept wet for a few hours.
4. Over the prepared joint the mortar is carefully placed in the desired shape. For this purpose a small trowel is used and the mortar is pressed and brought into close contact with the old interior mortar of the joint.
5. For at least three days, the finished surface is well-watered if lime mortar is used and ten days if cement mortar is used.

26.3.3 Types of Pointing

Different types of pointing are in use. The choice of a particular type of pointing depends on the type of masonry and the effect required. Usual types of pointing are explained below.

1. Flush Pointing

This type of pointing is formed by removing the excess mortar from the joint and made flush with the surface of the wall (Fig. 26.6). Although it does not give a pleasant appearance, it is durable and does not give any room for accumulation of dust. Thus it has large usage.

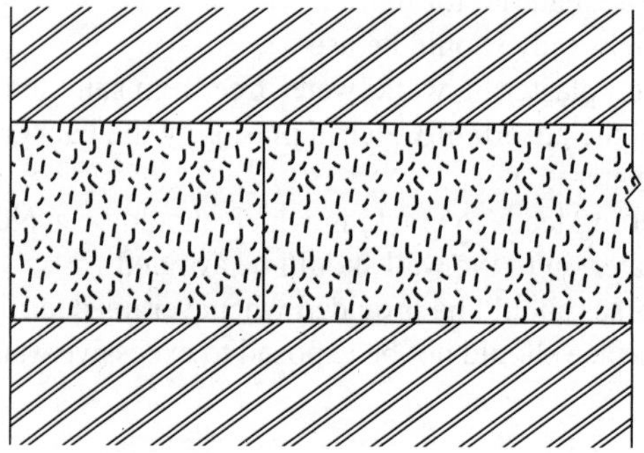

Figure 26.6 Flush pointing

2. Recessed Pointing

This joint is also has a vertical face as that of flush pointing, but pressed inside the wall surface using a special tool to a length of about 5 mm or more. This gives a very good appearance (Fig. 26.7).

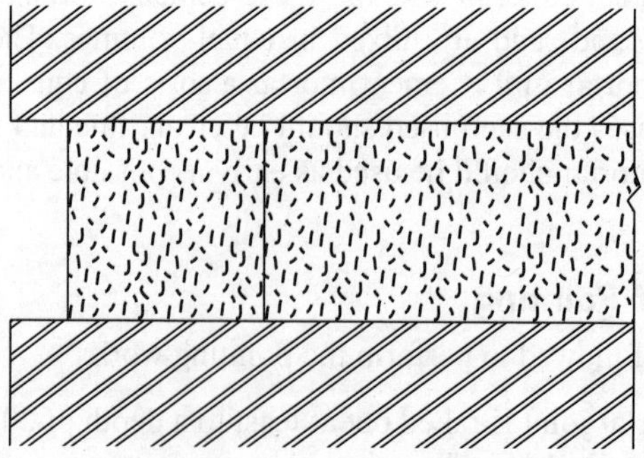

Figure 26.7 Recessed pointing

3. *Beaded Pointing*

This is a special type of pointing having a good appearance because of the beading point. This type is made using a steel or iron rod with a concave edge. This type of pointing is difficult to maintain (Fig. 26.8).

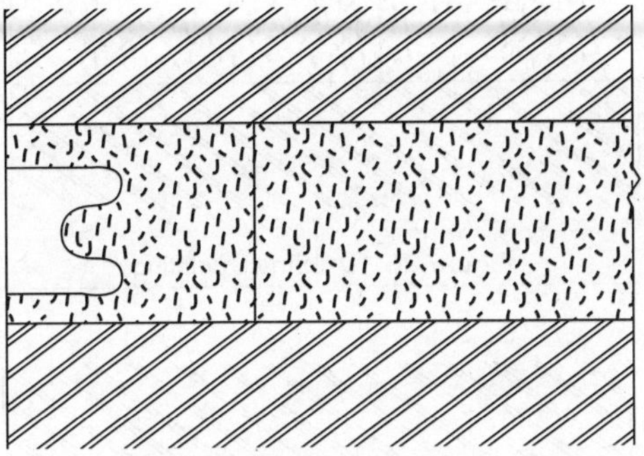

Figure 26.8 Beaded pointing

4. *Rubbed Pointing*

This type of pointing is also called as keyed or grooved pointing. Here a groove is made at the centre of height by a pointer. This type of pointing is generally adopted as it is easy to make and gives good appearance (Fig. 26.9).

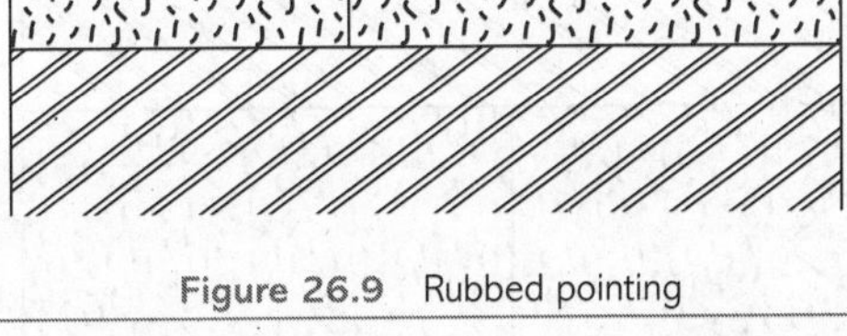

Figure 26.9 Rubbed pointing

5. *Struck Pointing*

In this type the face of the pointing is made inclined (Fig. 26.10) downwards. The upper edge of joint is at about 10 mm inside the wall surface and the lower edge is kept just inside the face of masonry. It is not a common type as water can get collected.

6. *Tuck Pointing*

In this type of pointing, a groove is cut at the centre of the mortar joint of 5 mm width and 3 mm depth. This groove is then filled with white cement putty with a projection of 3 mm (Fig. 26.11). If the projection is done in the mortar, is called half tuck pointing.

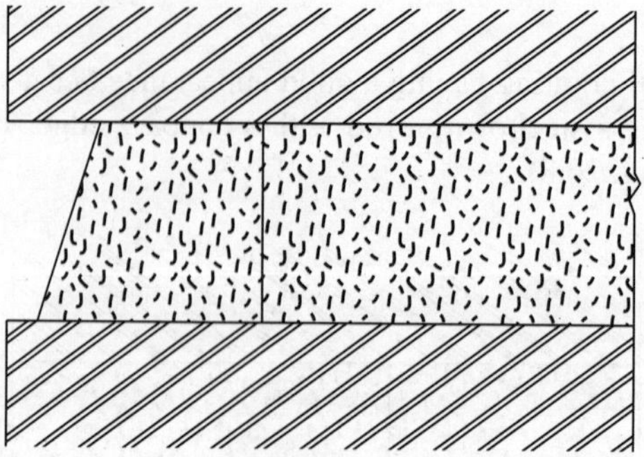

Figure 26.10 Struck pointing

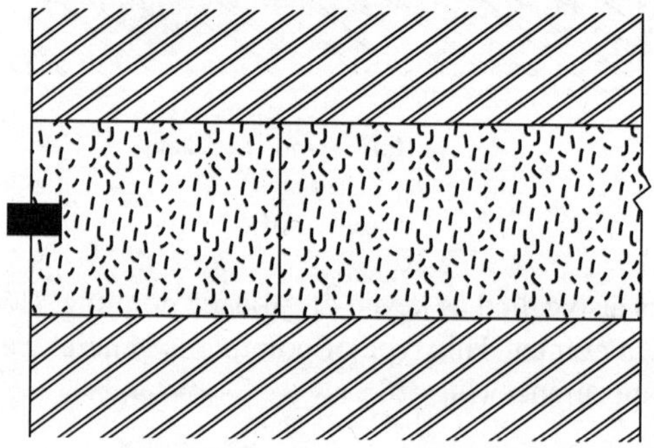

Figure 26.11 Tuck pointing

7. Vee-pointing

In this type of pointing a V-shaped groove is formed in the mortar joint (Fig. 26.12).

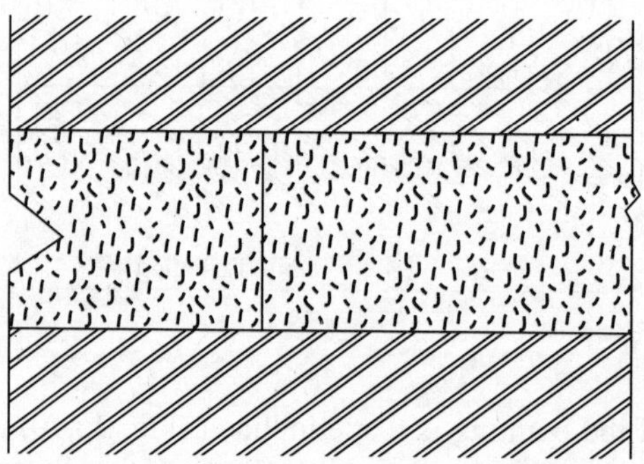

Figure 26.12 Vee-pointing

8. Weathered Pointing

In this type of pointing a V-shaped projection is made as shown in Fig. 26.13.

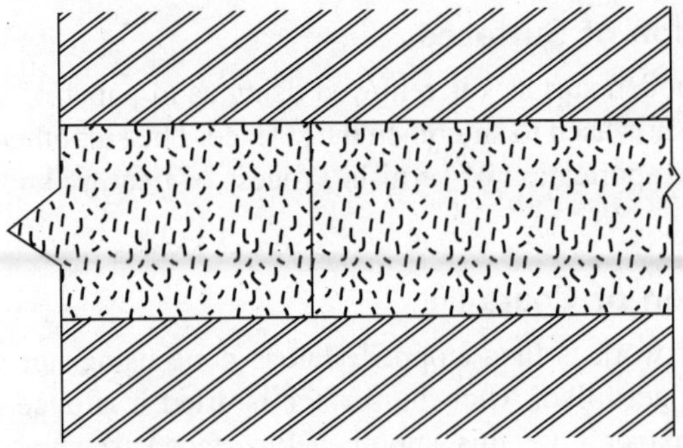

Figure 26.13 Weathered pointing

26.4 PAINTING

Paints are applied as a final finish to all surfaces such as walls, ceilings, wood work, metal work, etc.

Paints are classified based on different norms, viz., based on media, based on utilization, etc. Some of the common paints in use are: oil paints, cement paints, synthetic rubber paints and emulsion paints.

26.4.1 Preparation of Surfaces

It is the process by which the surface is cleaned to take an optimum paint. In wood work knots should be avoided as there is a possibility of resinous material coming out of them. If knots are inevitable they are property smoothened by paint knoting or lime knoting application or hot read lead.

In case of iron and steel works, the surfaces are freed from rust, greeze and seals before application of primary coat.

In the case of preparing for plastered surfaces, the cracks, nail holes, etc., are first filled with a pore filling solution. After drying, this results to an impervious surface. The surface is washed with a solution of size, soft soap and water in the proportion of 2 : 1 : 18.

26.4.2 Application of Coats

In wood work a primary coating is applied to the entire surface to be treated. This gives a hard and opaque covering. After drying the surface is smoothened by rubbing with pumice stone or sand paper. Successive coats with the chosen paint is given one after the other, after the previous coat has dried. Generally two such coats over the primary coating is enough.

In steel and iron work, the primary coat consists of read lead oxide in linseed oil. The successive coats with the decided paint are applied after the pervious coat has dried.

In the case of plastered surfaces, the first two coats consist of white lead and boiled linseed oil. The third coat is a combination of white lead, linseed oil and desired pigment. The finishing coat is similar to the third coat with more of thinner agent.

26.5 VARNISHING

Varnish is primarily used only for wood work. For decoration purposes it is also used on stone facings.

26.5.1 Preparation of Surfaces

The roughness of wooden surfaces is removed using sand paper of suitable grades. Then fine grained sand paper is used to smoothen the surface. The smoothened surface is cleared thoroughly before the application of varnish. Process of treating knots is done as done in painting.

26.5.2 Application of Coats

The surface of wood work is then stopped, that is done using hot weak glue size. This fill enables the pores get filled. After the surface is dried it is once again be rubbed and smoothened by sand paper. Over this smoothened surface varnish is applied in thin coats. Additional required coats are given once the previous one has thoroughly dried.

26.6 WHITE/COLOUR WASHING

The building units other than wood, aluminium and iron, other structurally finished surfaces are given a final finish to improve the appearance by using white/colour washing or distempering.

Pure lime or fat lime is slaked first and the required quantity of water is added to make the solution. The solution should have a milky consistency and in order to make the solution stick to the surface glue and salt are added.

The surface is first prepared by making good the nail holes and other surface defects. Then the surface is cleaned by rubbing with a hard brush. The water is sprinkled on the surface before application of the first coat of white wash. The first coat is applied by moving the brush from upper to lower and the second coat is applied by moving the brush from lower to upper after the drying of first coat. Generally three coats are applied on new surfaces.

The colour wash solution is prepared with adding a suitable pigment powder or liquid concentrate and 5% of gum. The colour wash application is similar to that of white wash. One or two coats of colour wash is given over the first white wash.

26.7 DISTEMPERING

Distempers are available both in powder (dry distempers) or paste forms (oil bound distempers). Dry distempers consist of 90–95% of whiting and 5–10% of glue. Oil bound distempers are made by grinding pigments with a medium composed of an emulsion of a drying oil in water. Linseed or tung oil is used. Vehicle for distempers is water.

Dry distemper is one of the cheapest forms of decoration. These give a pleasing coloured appearance and can stand a certain amount of dry rubbing and can be removed easily by washing.

Oil-bound distempers are more valuable for early decoration. This finish is comparatively better than dry distempers and is retained for a longer time. Cracking and peeling is not prevalent. This type is not suitable for places where there is possibility of wetting and drying for prolonged times. Distempers in general are suitable for interior surfaces as they are easily affected due to atmospheric agent.

The surface of distempering is prepared by applying the surface with 10% zinc sulphate solution. After drying the surface is smoothened with sand paper. Two coats are generally enough. The second coat is applied after the previous one has dried up.

SALIENT POINTS

1. Plastering is the method of covering rough and uneven surfaces of the various components of a building with a plastic material, i.e., mortar which is called the plaster.
2. Types of background are: Solid background, Lath background and Board background.
3. External finish is provided keeping in view certain factors such as desired appearance of the surface, cost, degree of maintenance, etc. Suitable additional material other than mortar is used.
4. Some of the external finishes are Sand-faced finish, Pebble-dash finish, Rough-cast finish, Scrapped finish, Smooth-cast finish, Textured finish, Depter finish and Machine-applied finish.
5. Tools used for plastering are laying trowel, plumb bob, gauging trowel, float, floating rule.
6. Finishing of mortar joints to have good appearance is called pointing.
7. Type of pointing are flush pointing, recessed pointing, beaded pointing, rubbed pointing, struck pointing, tuck pointing, vee-pointing and weathered pointing.
8. Paints are applied as a final finish to all surfaces such as walls, ceilings, wood work, metal work, etc.
9. Varnish is primarily used only for wood work. For decorative purposes it is also used on stone facings.
10. Building units other than wood, aluminium and iron, other structurally finished surfaces are given a final finish to improve the appearance by using white/colour washing or distempering.

REVIEW QUESTIONS

1. Enumerate the objects of plastering.
2. Explain the requirements of a good plaster
3. What is called a background with reference to plastering? Name different types.
4. With the help of sketches, describe the tools used for plastering.
5. Discuss various defects in plastering and how they are rectified.
6. Mention the necessity to have more than one coat.
7. Explain the special materials which are applied on the plastered surfaces to make them attractive and durable.
8. What factors will you keep in mind while selecting a suitable external finish?
9. Explain the defects in plastering. What are the precautions to be taken to avoid failures of plasters?

10. Why lath is used in plaster? Explain the use of metal laths and wooden laths in plastering.
11. On what situations you recommend pointing?
12. Explain with sketches different types of pointing.
13. How do you prepare the surfaces for
    (i) Plastering
    (ii) Painting
    (iii) Varnishing
    (iv) Distempering
14. Enumerate the objects of varnishing and point out the characteristics of an ideal varnish.
15. Discuss the usual defects which are found in the painting work.
16. What factors will you keep in mind while selecting a suitable external finish?

27 ESSENTIAL SERVICES IN BUILDINGS

27.1 INTRODUCTION

Essential elements for the human beings to live in a house with a minimum comfort are water supply, drainage arrangements and electrical power supply.

Biologically, water is required for the human beings for the satisfactory performance of physiological organisms, as a circulatory fluid, as a carrier of nourishing food and for the removal of products of wastes. Man uses water for variety of purposes. At the domestic end, a minimum requirement of water is needed for drinking, cooking, bathing and cleaning. Hence, a certain amount of water per head per day needed is being recommended. Accordingly the water requirement has to be made.

Disposal of wastewater, that is, used water from kitchen, toilet, cleaning, etc., has to be separately dealt for hygienic conditions.

At present, electricity has become an essential commodity as it is being used in the house for variety of works.

These essential services in a building are dealt in this chapter.

27.2 PLUMBING SERVICES

Plumbing services is the entire system of piping, fixtures, applications, etc., for providing water supply or drainage to a building or premises. A plumbing water supply system consists of water supply and distribution with the use of pipes, control valves, pipe layout, storage tanks, etc. A plumbing drainage system consists of collection and disposal of wastewater with the use of water closets, urinals, traps, vents, anti-siphonage pipes, house drains, house sewers, etc.

27.2.1 Plumbing Terms

Following are the terms to which a civil engineer should be familiar in connection with plumbing services.

1. Plumbing or Plumbing System

It is a general term encompassing the entire design practice, material and fixtures used in the installation and maintenance of all main fixtures and other appurtenances used in connection with water supply systems and wastewater collection and disposal systems.

2. *Available Head*

It is the pressure water available from a water main at the ground floor level to the premises.

3. *Water Main*

It is a main water supply pipeline intended for general use and maintained by local authority.

4. *Service Pipe*

It is a pipe branching out of the water main which is used for feeding water from the main pipe to the building.

5. *Storage Tank*

It is a tank for storing water from the water main by means of a supply pipe.

6. *Stop Cock or Tap*

It is a control valve used for regulating the supply of water.

7. *Stack*

It is a general term used for any line of drainage particularly vertical lines.

8. *Back-Siphonage*

It is the flow back of used or polluted water from a plumbing fixture or vessel into a water supply pipe due to negative pressure.

9. *Trap*

It is a fitting provided with a water seal so that when placed in a drainage pipe it prevents the passage of foul air or gas.

10. *Soil Pipe and Waste Pipe*

Soil pipe is used for liquid wastes carrying human excreta, whereas waste pipe carries only the liquid waste.

11. *Vent Pipe*

It is a pipe to allow air to or from a drainage system so as to prevent the water seal of traps against siphonage and back pressure.

12. *Flushing Storage Tank*

It is provided to supply water to all flushing cistern through downtake pipes.

13. *Supply Pipe*

It is the pipe which extends from the stop cock to the entrance of the storage tank.

14. *Distributing Pipe*

It is the pipe connecting the storage tank to various sanitary fixtures, taps, etc., for purposes of distribution of water supply inside the building.

15. *Sullage and Sewage*

Sullage is the spent water from bathrooms, kitchens, washbasins, sinks, etc. This does not include foul discharge (i.e., human or animal excreta as from water closets, urinals, hospitals, stables, etc.). Sewage includes both the sullage as well as the foul discharge explained above.

16. Night Soil

Night soil is a composite of human excreta and urine.

17. Sewer

It is a closed drain intended to be used for carrying sewage (i.e., night soil and other water-borne wastes).

18. Drain

It is a channel or pipe which carries wastewater or water-borne wastes in a building draining system.

27.2.2 Principles of Plumbing Services in Buildings

The following factors are to be considered while designing plumbing services to buildings:

1. All foul and waste matter should be removed as quickly as possible.
2. All possible passage of gases, odour or vermin from pipes and sewers into the building should be prevented.
3. All drainage pipes should be airtight, gas-tight and strong enough and durable to withstand the corrosive action of liquid wastes.
4. All pipe joints (both water supply and drainage) should be leak-proof and strong and durable.
5. The entire network of pipes should have ample means of cleaning and removing obstructions.
6. Every care has to be taken to check and rectify all possible air-locks, siphonage, under deposits and obstructions.

27.3 WATER SUPPLY

In villages and in some towns where municipal water supply is not available, water is obtained from private wells. In such places, usually the water is pumped up to an elevated storage tank from where the distribution to the building is carried out.

In large towns and cities, municipal water, after due treatment, is available through street mains. If adequate pressure is available, the municipal water can be directly connected to an overhead tank. If the pressure is less, a sump may be provided to collect the water from the municipal supply lime. The capacity of the low-level tank (sump) should be adequate enough to hold water depending on the supply system of the municipal authorities.

Water is pumped from this sump at regular intervals to an overhead tank from where the water is distributed to various parts of the building. In addition to the overhead tank, provision may be made to collect the municipal water directly for drinking and cooking purposes at a convenient place (kitchen) in the building. The water in the overhead tank will take care of other demands.

The capacity of the overhead tank is based on the number of persons using it. It is usually of 200–500 *l*. If it is more than 5000 *l*, it is better to average it in a series of interconnected tanks such that they can be easily cleaned. The overhead tanks may be of brick, concrete or readymade *PVC* tanks. An inlet pipe, on outlet pipe and a drainpipe are connected to each tank.

The underground (or sump) and overhead tanks should be cleaned at regular intervals. They should also be disinfected after cleaning by using chemicals. In some cases separate flushing storage overhead tank is provided exclusively for flushing water closets, etc., as the quantity of water used for this purpose is large and be of low quality. For example, bore well water may be used separately for this purpose. Flushing storage tanks may have the capacities as given in Table 27.1.

Table 27.1 Flushing storage capacities

Sl. No.	Classification of buildings	Storage capacity	
		WC seat	Urinal seat
1.	Residential houses	270 / for one seat and 180 / for every additional WC seat	
2.	Tenements having common convenience	900 / per seat	
3.	Factories and workshops	900 / per seat	180 / per seat
4.	Cinemas and assembly halls	900 / per seat	180 / per seat

27.3.1 Design of Water Distribution System

Water distribution system needed for a conventional residential building depends on the following aspects:

1. Estimation of water requirement
2. Determination of pipe sizes
3. Pipe layout principles

1. Estimation of Water Requirement

As per Indian code recommendation (IS: 1172–1983 and SP 57–1993) for residences, 135 *l* of water per head per day to be used for the design of water supply system. The details of the daily consumption is given Table 27.2.

Table 27.2 Domestic consumption of water

Sl. No.	Use	Litres	Percent
(A)	Drinking	5*l*	3.7%
(B)	Cooking	10*l*	7.4%
(C)	Washing Clothes	15*l*	11.0%
(D)	Cleaning of house	15*l*	11.1%
(E)	Bath Wash	35*l*	26.0%
(F)	Flushing Toilets, etc	55*l*	44.8%
	Total Use	135 Litres	100%

Source: IS: 1172–1983.

The daily consumption will vary depending on the activities in a building. For example, in a hospital with beds, the consumption may be of the order of 340 *l* per bed per day and for a day school, the consumption will be as low as 45 *l* per head per day.

2. Determination of Pipe Sizes

In general for a middle class house the average number of pipes needed shall be about 8 and pipes of minimum nominal size of 20 mm diameter are enough. For WCs which are directly connected to overhead tanks, a 25 mm diameter is adequate to get sufficient flow. Minimum recommended size of pipes for different purposes in a building is shown in Table 27.3.

Table 27.3 Recommended sizes of water supply pipes

Sl. No.	Pipe details	Diameter of Pipe (mm)
(A)	Lavatory	10
(B)	Water closet (Tank type)	10
(C)	Drinking fountain	10
(D)	Bath tub and shower stand	12
(E)	Kitchen sink	12
(F)	Service pipe	20
(G)	Water closet (Valve type)	25

A thumb rule to assess the size of pipes which can supply two or more branch pipes is as follows (Varghese, 2012):

(i) Up to 3 numbers of 10 mm branches can be supplied by a 12 mm pipe.
(ii) Up to 3 numbers of 12 mm branches can be supplied by a 20 mm pipe.
(iii) Up to 3 numbers of 20 mm branches can be supplied by a 25 mm pipe.

In order to control the output from overhead tanks, a stop valve or gate valve must be fixed in an accessible place. Apart from control this will help to stop the water supply during repairs.

For multi-storey buildings with many flats, a more complex arrangement of water distribution system should be planned.

3. Pipe Layout Principles

Following rules may be adopted in laying of water supply lines (Varghese, 2012):

(i) There should not be any cross connection and backflow. Backflow can be eliminated by ensuring the under mentioned conditions.
   (a) Adequate air gap is to be provided between the inlet and final flood level in all places particularly washbasins.
   (b) While connecting cistern to water closet, the inlet and float valve should be suitably fixed.
   (c) Positive pressure at the outlet pipes should always be assured.

(ii) Drainage pipes and water supply pipes should not be laid very close to each other. Especially drainage stoneware pipe which are liable to leak if damaged should not be laid above the water supply line. Further, water supply pipes should not be laid alongside of the pipes carrying foul water.

(iii) Planning of the distribution line should be such that it is accessible for inspection, replacement and repairs. *GI* pipes should not be buried as they tend to corrode in contact with soils, but they can be laid on walls.

(iv) All pipes including *PVC* pipes laid inside the building as concealed pipes should be tested for leakage.

27.3.2 Water Supply Fittings and Fixtures

Most of the fittings which are to be used for the water supply are standardised. Some of the fittings which are routinely used in residential water supply are discussed below.

1. Stop Valves

These are fixed in an accessible place for each outlet such as washbasins, water closets, etc., to control and facilitate repair.

2. Gate Valves

These operate as a gate in the pipe. It regulates the flow and when fully open, it offers no resistance to flow. These are used for closing the main pipes from tanks.

3. Ball or Float Valves

These are provided in storage tanks, flushing cistern, etc., to control the level of storage water. The supply is cut off automatically when the water reaches a particular designed level.

4. Bib Taps or Bib Cocks

This is the conventional type of fittings used in washbasins.

5. Pillar Taps

It is the outlet provided from a vertical inlet and horizontal outlet in washbasins in kitchens, lavatories, etc.

6. Self-Closing Taps

They remain open as long as the valve is pressed down and closes when the pressure is released. Such taps are provided in railway stations, hotels, etc.

7. Magnetic Taps

These taps operate as soon as the hand is placed below the tap and closes automatically as soon as the hand is removed. This is a very hygienic tap as one need not to touch the tap to operate. Such taps are provided in airport, hotels and hospitals.

27.3.3 Materials of Water Pipes

Two main types of pipes are Galvanised-iron (*GI*) pipes and unplasticised Poly-vinyl chloride (*UPVC*) pipes.

If the quality of water running through the pipe is good, and the pipes are buried *GI* pipes are excellent. These pipes are available in three grades, viz., *A*, *B* and *C*. Class *C* is the pipe meant for heavy duty and is cosliest of all. Grade *B* pipes are generally used for water supply in buildings. In order to identify the type of pipes different colours are used, viz., yellow for grade *A*, blue for grade *B* and red for grade *C*.

UPVC pipes are now-a-days preferred for water supply in buildings. They do not function well when exposed to sunlight. Further the coefficient of expansion of *UPVC* pipes is about eight times greater than that of *GI* pipes. *UPVC* pipes require supports at closer intervals.

27.4 DRAINAGE OF WASTEWATER AND SEWAGE

Wastewaters coming from kitchen sink, washbasins and urinals are not containing any solid matter. But sewage a solid material has to be treated such that it also could be converted as a wastewater.

27.4.1 Need for Treatment of Sewage

Nearly 75% of the water supplied returns back in the form of wastewater. As there is a need for water supply system, there is an equal necessity to dispose the wastewater. The night-soil and urinary content of domestic sewage (originating from residential areas) consists of potentially most dangerous organisms. Such organisms are responsible for the occurrence and spread of water-borne diseases.

Sewage treatment is meant the various steps undertaken in the process of transferring sewage into a harmless liquid. That is to render sewage inoffensive without causing nuisance or odour and to reduce or eliminate any possible contaminations of water supplies, bathing areas, etc., by the untreated or un-disposed sewage.

27.4.2 Sanitary Fittings and Appliances

Sanitary fittings and appliances comprise of traps, water closets, flushing cisterns, etc., which are discussed below.

1. Traps

Traps are fittings which are installed in drainage pipes to prevent the passage of foul air or gases through drains, waste or soil pipes. This is possible because traps are equipped with water seals having a minimum depth of 25 mm. Higher the depth greater the effectiveness of trap. Depth varies in practice from 25 to 75 mm and 50 mm depth is quite common.

Qualities of good traps are:

- (i) They should have adequate water seal at all times which is possible only when they are non-absorbent.
- (ii) There should not be any projections which may obstruct or retard the flow of water. Further it should be smooth.
- (iii) They should retain minimum quantity of water consistent with providing deep water seal with large area.
- (iv) Apart from self-cleansing, provision should be there for cleaning manually or otherwise.

Traps are classified based on shape as R-trap, Q-trap and S-trap and based on use as floor trap, gully trap and intercepting trap.

Figure 27.1 of traps based on shape.

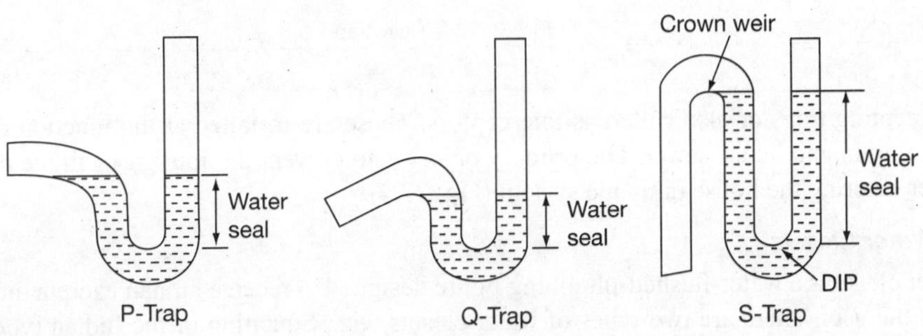

Figure 27.1 Traps based on shapes (P-Q-S)

Floor traps are intended to collect surface wash or wastewater from the floors of the bath and kitchen. These are provided with cast iron grating at top so as to retain coarse solid matter. Thus preventing the matter to go along with wastewater and thereby cause blockade (Fig. 27.2).

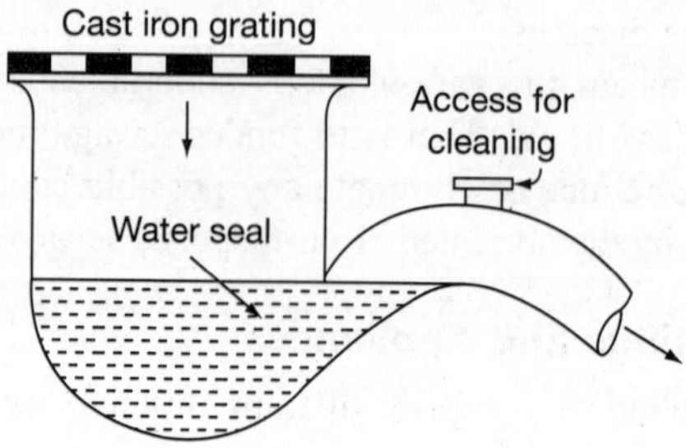

Figure 27.2 Floor trap

Gully traps are intended to receive sullage water from baths, sinks and washbasins. Further they are also used to receive rain or surface water from house-tops or back yards. The water seal is 50–75 mm. Waste pipes and rain-waste pipes are invariably connected to drains through the gully trap (Fig. 27.3).

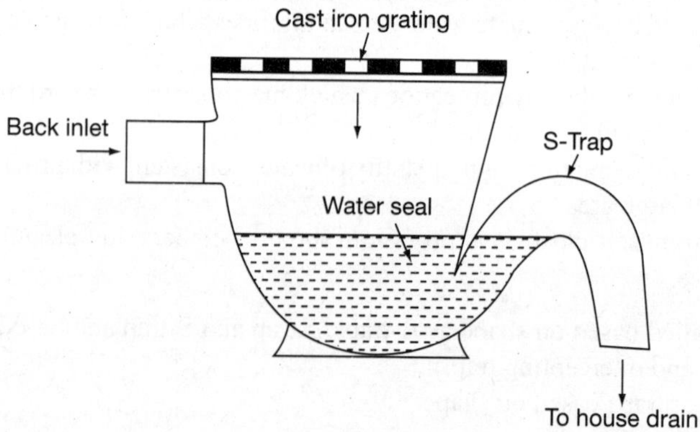

Figure 27.3 Gully trap

Intercepting traps or also called as interceptors. These are installed at the junction of the house drain and house sewer. The primary object is to prevent the foul gases in the public sewer entering the house drainage system (Fig. 27.4).

2. Water Closets

Water closet is a water-flushed-plumbing figure designed to receive human excreta directly from the user. There are two types of water closets, viz., Squatting or the Indian type and pedestal or the European Type. Figures 27.5(a) and (b) show the two different types.

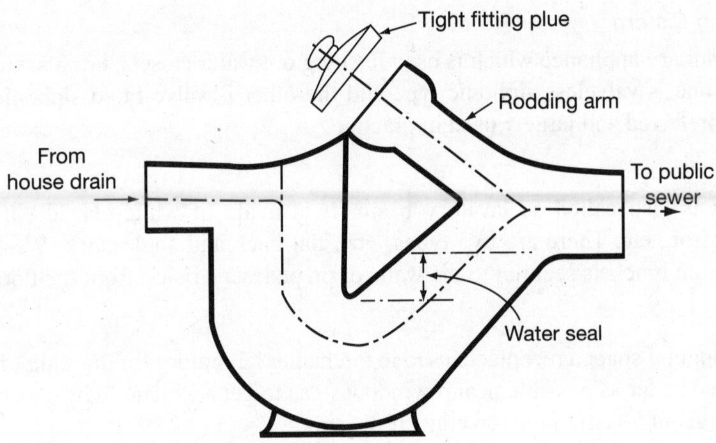

Figure 27.4 Intercepting drain

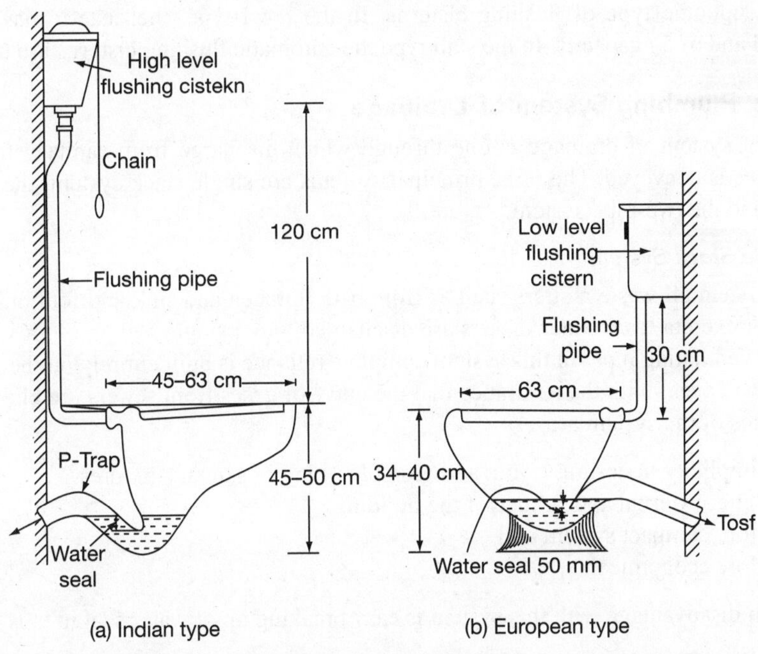

Figure 27.5 Water closets

(i) Indian Type

This has a squatting pan of 45–63 cm in overall length and 45–50 cm in height. The W.C. is connected to soil pipe through a P-trap or S-trap.

(ii) European Type

It is a wash down water closet provided with a seat and cover and fitted with P-trap and S-trap. It has a maximum overall height of 63 cm and a height of 34–40 cm.

Both the types of WCs are provided through flushing cisterns and flushing pipes.

3. Flushing Cistern

This is a sanitary appliance which is used flushing out water closets, urinals, etc. There are two types one is valveless siphonic type and the other is valve-fitted siphonic type. The former is preferred and largely used in practice.

4. Washbasins

Washbasin is also called as lavatory basin. It is made of white-glazed earthen wave, enamelled iron, etc. There are two types, viz., flat back and angle back. Washbasins are fixed either on brackets secured to the walls or on pedestals rising from the floor.

5. Sink

It is a rectangular shaped receptacle used in kitchen or laboratory for draining of water. It is to be located as far as possible near a window so as to get adequate light. The height from the floor is about 90 cm to the top edge of the sink.

6. Urinals

There are two types, viz., bowl type and the slab or stall type. Both the types are flushed through siphonic type of flushing cisterns. In the bowl type, the cistern may be hand operated and of 5 *l* capacity. In the stall type the automatic flushing cisterns are used.

27.4.3 Plumbing System of Drainage

Plumbing system of drainage is one through which discharge from sanitary fixtures or appliances is conveyed. The three principal systems are single stack system, the one-pipe system and the two-pipe system.

1. Single Stack System

In this system all waste matters such as from bath, kitchen and sink and foul matter from urinals or excreta from water closets are discharged into a single soil pipe. This pipe also acts as a ventilating pipe. In this system complete reliance is built entirely on the effectiveness of water seal with the assurance that the entry of gases from sewers are blocked. The advantages of the system are:

(i) Simplicity in design, layout and easy plumbing of sanitary fixtures.
(ii) Better external appearance of the building.
(iii) More compact system and
(iv) More economical.

The main disadvantage with this system is easy breaking up of water seal in taps.

2. One-Pipe System

In this system separate vent pipes are provided. All the traps of water closets, basins, baths etc., are adequately ventilated to preserve the water seal. In this system a caged dome is provided projecting above the roof top such that to allow a natural outlet to the foul gas. This system requires a difficult arrangement of pipe work and also costlier than the single-stack system.

3. Two-Pipe System

In this system all soil appliances such as water closets and urinals are connected to a vertical soil pipe. All wastewater appliances such as baths, washbasins, sinks, etc., are connected to separate waste pipe which is disconnected from the drain by means of a gully trap.

Thus in this system there are two sets of vertical pipes, viz., the soil pipe and the waste pipe, each one is provided with a vent pipe. This system is the best plumbing system which is most efficient in the conveyance of sanitary waste and also largely favoured for adoption. However, this is the costliest of all systems.

27.4.4 Layout of Building Drainage

The following aspects have to be considered in the design and construction of a drainage layout:

1. This layout should be simple.
2. All pipes should be laid in straight lines both in horizontal and vertical directions.
3. Abrupt changes in the direction of pipeline may affect the natural flow.
4. In the points of intersection of pipes and wherever bends occur, it is essential to provide inspection chambers or manholes.
5. The contained angle between the intersecting pipes should be less than 45° so as to ensure a gradual and smooth flow.
6. Sewers should not be ordinarily laid under a building. If it is inevitable a cast iron pipe should be laid in straight reach with a uniform gradient.

27.4.5 Pipes for Drainage

1. Pipes and Gradients

Pipe sizes for house drains and sewers are 100, 150 and 230 mm and in some cases 300 mm. In order to avoid deposition of solid matter, the pipes should be laid as per the recommendation of I.S. Code (IS: 1742–1972), Table 27.4.

Table 27.4 Recommended gradient of sewer discharge pipes

Sl. No.	Diameter (Pipe)	Gradient	Sewer Discharge
(A)	300 mm	1 in 200	1.66m³mt
(B)	230 mm	1 in 120	1.93m³mt
(C)	150 mm	1 in 65	0.42m³mt
(D)	100 mm	1 in 35	0.18m³mt

Source: IS: 1742–1972.

2. Positioning of Pipes

The following points shall be kept in view while deciding the positioning of pipes:

(i) All soil pipes, waste and ventilating pipes should be conveniently grouped in shafts or ducts with adequate capacity. This provision is needed to allow for inspection and repair work.

(ii) All the pipes needed for different purposes have to be adequately provided for during the construction stage itself at their appropriate positions.

(iii) When pipes are not embedded it should run clear of the wall with a minimum clearance of 5 cm.

(iv) Waste pipes are to be separated from the house drains by using gully traps. This arrangement is essential to prevent entry of foul air or gas, vermin, etc., into the building.

(v) The soil, waste and vent pipes should be vertically carried above the top of the building. They are to be covered by copper, plastic or galvanised iron wire domes. The arrangement is needed to prevent nesting of birds or inadvertent falling in of objects inside the pipe.

(vi) Diameter of pipes used for various purposes are given below (Duggal, 1988)

(a) Soil pipe	100 mm
(b) Waste pipe – horizontal	32–50 mm
(c) Waste type, vertical	75 mm
(d) Vent pipe	50 mm
(e) Anti-siphonage pipe:	
Connecting soil pipe	50 mm
Connecting waste pipe	40 mm

27.4.6 Sludge and Effluent Disposal

Sedimentation is carried out with the object of removing such suspended mineral and organic matter from sewage. Sedimentation tanks are units in which sedimentation is brought about. The lighter organic sewage solids which settle in the sedimentation tanks are termed as sludge. The sewage that has been partially classified by the settling out of the solids is known as the effluent.

In a general set up the disposal of sludge can be done by adopting any one of the following methods:

(i) Dumping into waste bodies
(ii) Shallow Burial
(iii) Lagooning
(iv) Mechanical dewatering and
(v) Drying in Beds

If such a general disposal system is not available one has to go in for a sedimentation tank. Generally sedimentation tank adopted is a septic tank which is a horizontal continuous flow sedimentation tank. A description of septic tank is explained in next section. The putrescible and highly odorous efficient from the septic tank requires to be properly treated and disposed of sub-surface irrigation field method or discharging into other soil absorption system such as soak pit and leading cell pool which is described later.

1. Septic Tank

A septic tank is a horizontal continuous flow sedimentation tank. Here the sewage is allowed to move very slowly so as to retain for a period sufficient to develop 60–70% of suspended matter to settle in the form of sludge. Lighter solids float to the surface and combine with grease and fat and form floating scum. The scum and sewage are allowed to stay for a period of 7 months during which period complete decomposition takes place through a process called sludge digestion. The volume of sludge also gets reduced for easy disposal. The effluent from septic tank has very bad colour. It is dark in colour with fine solid particles and has to be disposed with utmost case. A schematic layout of a septic tank is shown in Fig. 27.6.

A septic tank is constructed in such a way so as to prevent direct current between the inlet and the outlet. This provides a better sedimentation and is achieved by using T-pipes

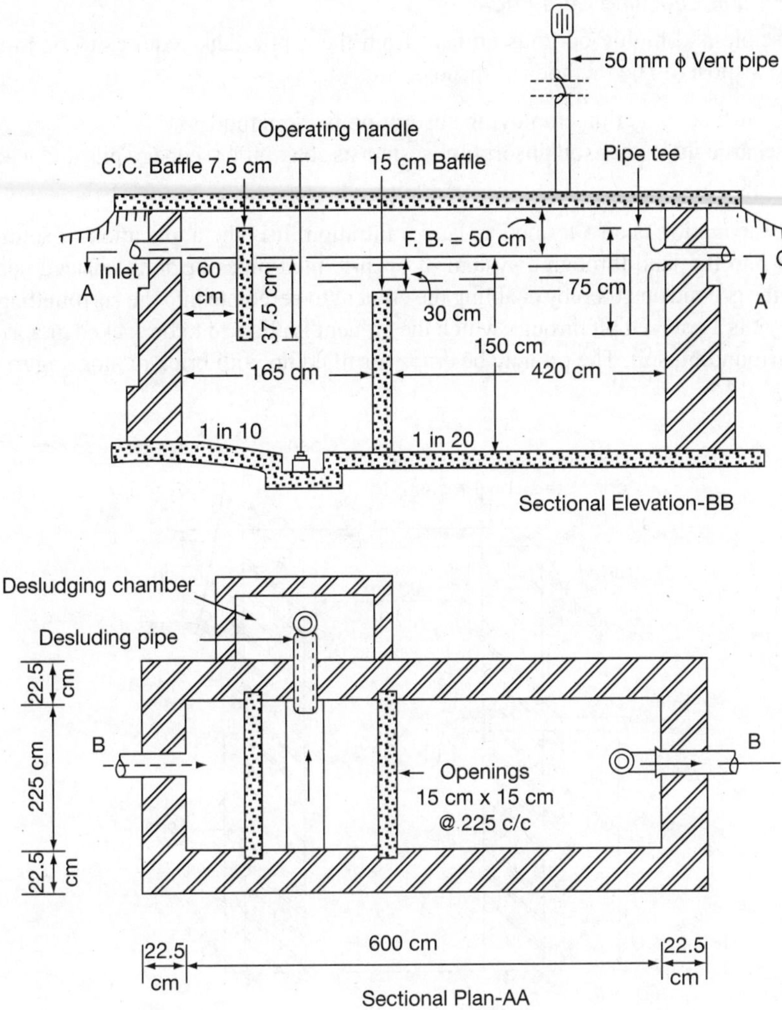

Figure 27.6 Septic tank

with submerged ends as inlet and outlet. As an alternative, baffle walls may be provided. The T-pipe or the baffle at the outlet also helps in retaining the scum in the tank. Certain quantity of scum is needed to hold-back odours and to create a form of heat insulations which in turn aids the bacterial action. Usually a manhole is fixed on the RCC cover slabs. The tank cover allows to keep the sewage warm, lessening odour, etc. Gases are separately removed through a vent pipe. Sludge is removed periodically.

Septic tanks are to be located at a place which is exposed to sky and accessible for cleaning. The sewage in a septic tank can be taken as only a primary treatment effect. Before disposal the effluent needs a secondary treatment. The septic tank has only a restricted use in practice, such as schools, hospitals, small residential colonies and other public institutions where sewers are not yet laid.

2. Soak Pit and Leaching Cess Pool

The putrescible and highly odorous effluent from the septic tank requires to be properly treated and disposed. The methods of disposal are:

(i) Sub-surface irrigation employing absorption field method.
(ii) Discharge into other soil absorption system as soak pit or seepage pits and leaching cess-pools.

Sub-surface irrigation is also termed as land infiltration. It is the application of sewage or its effluent into the land through a system of open-jointed pipes or drains placed near the surface of the ground and thereby enabling the effluent to percolate into the surrounding soil.

A soak pit is a covered pit through which the effluent is allowed to be soaked or absorbed into the surrounding soil. The pit may be empty or filled up with brick or stone aggregates (Fig. 27.7).

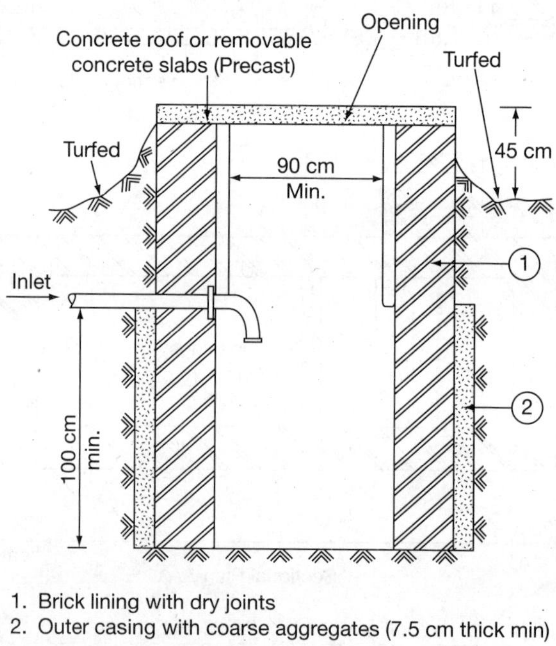

Figure 27.7 Soak pit

A leaching cesspool is a pool such that the top portion acts as an absorption field and the bottom as a septic tank. This is done by providing open-jointed lining at the upper portion which enables the sewage effluent to the easily dispersed to the surrounding soil. The bottom portion is of solid wall which functions as a septic tank (Fig. 27.8).

27.5 ELECTRICAL SUPPLY

Electrical power is provided to the building by the Government. Electricity is distributed in small towns and villages by overhead wires. But in large cities and towns the distribution is done through underground cables. The supply of electricity is including the cables up to

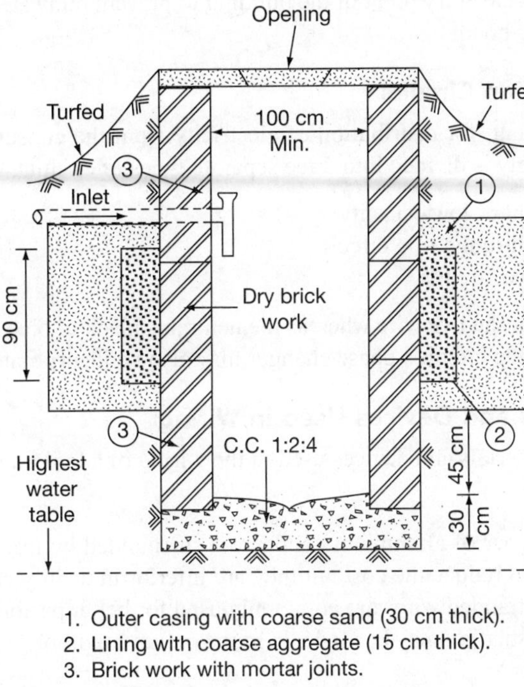

Figure 27.8 Leaching cesspool

the electric metre. In order to isolate the main supply from the building, a system of fuses is installed between the supply inlet and the electric metre board. From the fuses, the wires are led to the metre and from there to the consumer unit. From this point the electricity is distributed to the various parts of the building. It is the responsibility of the owner of the building to provide wiring and other fixtures from the metre to the distribution board and other parts of the building (Varghese, 2012).

27.5.1 Single and Three-Phase Supply

The electricity from the sub-stations are stepped down through local transformers and supplied to nearby buildings as single-phase or three-phase supply (50 cycles per second at 220 volts between the phase and the neutral). Then the supply is 230 volts between a phase and a neutral and it is $400\left(=\sqrt{3}\times 230\right)$ volts between the phases. In a single-phase supply, there shall be one live-phase wire and a neutral. In a three-phase supply, there shall be three live phases and a neutral.

When the electrical power load is small and the total current drawn is less than 30 amp, a single-phase may be sufficient. That is only for lighting and use of fans a single-phase may be enough. However, when the load is heavy, as when use of many equipment such as air conditioners, cooling ranges, etc., it is necessary to draw current from three live phases and a neutral. The electric current metre for a three-phase supply is different from that of single-phase. It is a condition that all equipment used in residences must be 220–230 volts.

Instead of underground cable, if supply is drawn from overhead insulated cable special care has to be taken. In this case it is to be first run down the walls with insulated cables and then turned up at the entry point of the building to prevent rainwater running along the cable into the electric board.

27.5.2 Distribution Circuits

In an ordinary building, for distribution of electricity from the consumer control unit to other points, the supply is divided into three separate types of circuits, viz.,

1. Lighting circuit of low capacity
2. Power circuit through ring circuit
3. Fixed appliance circuits

Red wires are used for the live wire whereas the neutral is black or blue and earth wires are green and yellow. In the circuit a phase changer may also be incorporated.

27.5.3 Materials and Devices Used in Wiring

Following are the materials and devices used in the wiring of a building.

1. Cables

Cables consist of copper of aluminium conductors surrounded by insulation. Aluminium conductors are used to reduce the cost but they are inferior in quality compared to copper cables. Now-a-days stranded wires are commonly used for better performance.

Rubber or *PVC* insulation may be made. In moist or wet conditions *PVC* insulation may fare better than rubber.

2. Fuses

Basically a fuse protects appliances and cables from damage by electrical faults. The current rating of fuses should be lower than that of the cable it has to protect. Two types of fuses are in use, viz., Rewirable or semi-enclosed fuses and cartridge fuses.

3. Miniature Circuit Breaker

Now-a-days miniature circuit breakers (*MCB*) are used instead of fuses. Miniature circuit breakers are single pole switches which automatically switches off when excess current flows. An MCB trips at 1.25 times its rated current capacity. It is only an overload device. But for the initial cost it has many advantages.

4. Earth Leakage Circuit Breaker (ELCB)

An electric appliance with a fault is connected to a line and if the metal parts are not earthed the metal may become live without the fuse blowing. Such a situation may cause a shock to the person touching it. On the other hand, if the metal body is earthed, then current will flow through the earth wire. In such a condition, it is advisable to use a tripping device called the earth leakage circuit breaker. It automatically trips even where a small current flows through the earth wire. This switch is connected to the neutral line of the electric supply. When it trips, the supply to the building will be completely cut off.

5. Residual Current Circuit Breaker (RCCB)

This type of breaker is much more positive protection device which can be provided instead of earth leakage circuit breaker to avoid electric shocks and current leakages. An *RCCB* is connected to all the three phases and the neutral. It is more expensive but provides excellent protection.

27.5.4 Wiring of Buildings

The following are three types of wires:

1. Sheathed wiring by surface fixing
2. Conduit installation on walls and ceilings
3. Concealed conduit wiring

Now-a-days in most of the buildings concealed *PVC* conduit wiring is used. The *PVC* conduits may be laid on the surface or concealed. In both the cases care has to be taken to see that no water enters the pipes. In the case of concealed pipe, it is safe to surround the conduit in chicken mesh reinforcement.

SALIENT POINTS

1. Plumbing services in the entire system of piping, fixtures, applications, etc., for providing water supply or drainage to a building or premises.
2. A plumbing water supply system consists of water supply and distribution with the use of pipes, control valves, pipe layout, storage tanks, etc.
3. A plumbing drainage system consists of collection and disposal of wastewater with the use of water closets, urinals, traps, vents, anti-siphonage pipes, house drains, house-sewers, etc.
4. Where municipal water supply is not available water is obtained from wells as in villages and in small towns. In large towns and cities, municipal water supply is available through street mains. In both the cases the water may be pumped to an overhead tank from which the entire water distribution can be taken care of. In case the pressure in mains is less a sump may be provided.
5. Capacity of an overhead tank is based as the number of persons using it. It is usually of 200–500 *l*.
6. Water distribution system needed for a conventional residential building depends on the following aspects.
   (i) Estimation of water requirement.
   (ii) Determination of pipe sizes.
   (iii) Pipe layout principles.
7. As per Indian Code recommendations for residences 135 *l* of water per head per day to be used for the design of water supply system.
8. In general for a middle class house the average number of pipes needed shall be about eight and pipes of minimum nominal size of 20 mm diameter are enough.
9. Two main types of pipes are Galvanised-iron pipes (*GI* pipes) and unplasticised polyvinyl chloride (*UPVC*) pipes.
10. Traps are fittings which are installed in drainage pipes to prevent the passage of foul air or gases through drains, waste or soil pipes.
11. Traps are classified based on shape as P-trap, Q-trap and S-trap and based on use as Floor trap, Gully trap and Intercepting trap.

12. Floor traps are intended to collect surface wash or wastewater from the floors of the bath and kitchen.
13. Gully traps are intended to receive sullage water from baths, sinks, and washbasins. They are also used to receive rain or surface water from the house tops or back yards.
14. Intercepting traps are also called as interceptors and are installed at the junction of the house drain and house sewer.
15. Water closet is a water-flushed-plumbing fixture designed to receive human excreta directly from the user. There are two types, viz., squatting or the Indian type and Pedestal or the European type.
16. Flushing cistern is a sanitary appliance which is used flushing out water closets, urinals, etc.
17. Plumbing system of drainage is one through which discharge from sanitary fixtures or appliances is conveyed. The three principal systems are single stack system, the one-pipe system and the two-pipe system.
18. In the single stack system all waste matters such as from bath, kitchen and sink and foul matter from urinals or excreta from water closets are discharged into a single soil pipe.
19. In the one-pipe system separate vent pipes are provided. All the traps of water closets, basins, baths, etc., are adequately ventilated to preserve the water seal. In this system a caged dome is provided projecting above the roof top such that to allow a natural outlet to the foul gas.
20. In the two-pipe system all soil appliances such as water closets and urinals are connected to a vertical soil pipe. All wastewater appliances such as baths, washbasins, sinks, etc., are connected to separate waste pipe which is disconnected from the drain by means of a gully trap. Thus in their system there are two sets of vertical pipes, viz., the soil pipe and the waste pipe, each one is provided with a vent pipe.
21. In a general set up the disposal of sludge can be done by adopting any one of the following methods:
    (i) Dumping into waste
    (ii) Shallow burial
    (iii) Lagooning
    (iv) Mechanical dewatering
    (v) Drying in beds.
22. A septic tank is a horizontal continuous flow sedimentation tank. Here the sewage is allowed to move very slowly so as to retain for a period sufficient to develop 60–70% of suspended matter to settle in the form of sludge.
23. The putrescible and highly odorous effluent from the septic tank requires to be properly treated and disposed. The methods of disposal are
    (i) Sub-surface irrigation employing absorption field method.
    (ii) Discharge into other soil absorption system as soak pit or seepage pits and leaching cell pool.
24. A soak pit is a covered pit through which the effluent is allowed to be soaked or absorbed into the surrounding soil. The pit may be empty or filled up with brick or stone aggregates.

25. A leaching cesspool is a pool such that the top portions acts as an absorption field and the bottom as a septic tank.
26. Electrical power is supplied as single phase or three-phase supply. In a single-phase supply, there shall be one line-phase wire and a neutral. In a three-phase supply there shall be three line phases and a neutral.
27. For distribution of electricity from the consumer control unit to other points, the supply is divided into three separate types of circuits, viz.,
    (i) Lighting circuit of low capacity
    (ii) Power circuit through ring circuit
    (iii) Fixed appliance circuits.

REVIEW QUESTIONS

1. How the water requirement of a building is assessed? What is the requirement for a residential building?
2. Why it is necessary to store water in case of residential building?
3. List the various appliances used for water supply in a residential building.
4. What is meant by a backflow? How it is prevented in the layout of water supply pipes in a residence.
5. How the sizes of pipes are selected for water distribution in a residential building?
6. What are traps? Why is it necessary that all sanitary fittings should be provided with individual traps?
7. Explain the one- and two-pipe systems of plumbing and state their specific use.
8. How effluents are discharged?
9. What is a septic tank? How is it effectively used in a residence?
10. Explain soak pit and leaching cesspool.
11. Distinguish between a one-phase and two-phase systems.
12. Explain the materials and devices used in wiring of a building.
13. Distinguish between Miniature circuit Breaker and Earth Leakage circuit breaker.
14. Explain Residual Current Circuit Breaker.
15. Bring out the advantages of concealed wiring.

28 SPECIAL SERVICES IN BUILDINGS

28.1 INTRODUCTION

As the atmospheric conditions vary, there is a need to condition the air to suit the need of the occupants. This is done by the process of air-conditioning the air so as to simultaneously control temperature, humidity and movement of the air with due control on distribution and cleanliness. This requirement of occupants of a building has to be satisfied both during summer and winter seasons.

Difference in temperature in various parts of the building and inside-outside temperatures result in a transfer of heat from the warm area to the cool area. Such a variation in temperature causes discomfort to the occupants. It is appropriate to maintain almost a constant temperature in all parts of the building by proper thermal insulation.

Apart from the above two special services needed in a building, it is also necessary to design and construct buildings of different units of buildings to get proper acoustical conditions also.

These three special services to be incorporated in buildings are dealt in this chapter.

28.2 AIR-CONDITIONING OF BUILDINGS

Air-conditioning primarily consists of refining air in order that it may be utilised to the best advantage of the user. The user may be human beings, materials, equipment, etc. Thus conditioned air would mean an atmosphere having controlled temperature, humidity and velocity condition with dust-free cleanliness to meet the requirement.

28.2.1 Purposes of Air-Conditioning

Air-conditioning of any building or space is adopted to satisfy one of the objects mentioned below.

1. Residential Buildings

For human beings living in residential buildings, the conditioned air is expected to give a better health or to preserve and maintain the health, comfort and convenience.

2. Industrial Premises

Industrial air-conditioning may be required to preserve the quality of industrial product and at the same time to create, control and maintain the working of different processes such as artificial silk, cotton cloth or preservation of fruits, etc.

3. Commercial Premises

Commercial premises such as theatres, offices, banks, restaurants, shops, stores, etc., are air-conditioned to improve their working conditions and to maintain comfort within the concerns.

4. Transport Vehicles

Air-conditioning plays an important role in providing facility and comfort in transport vehicles such as aeroplanes, railway coaches, road-cars, buses, ships, etc.

28.2.2 Terminologies

Following terminologies are used in air-conditioning:

1. Psychometrics

It is the branch of science that deals with the study of mixture of dry air and water vapour and is also the subject connected with the study of the behaviour of moist air.

2. Dry Air

It is a mechanical mixture of gases comprising of oxygen, nitrogen, carbon-dioxide, hydrogen, argon, neon, helium, ozone and xenon. In nature, a complete dry air does not exist but some quantity of water vapour is diffused in it.

3. Moist Air

It is the mixture of dry air and water vapour.

4. Dry Bulb Temperature

Temperature of air recorded by an ordinary thermometer is known as dry bulb temperature.

5. Wet Bulb Temperature

Temperature recorded by a mercury thermometer when its bulb is covered with a wet cloth and is exposed to a current of moving air at a velocity of 300 ml min is known is wet bulb temperature.

6. Due Point Temperature

Temperature at which liquid droplets just appear when the moist air is cooled continuously is known as dew point temperature.

7. Relative Humidity

It is the ratio of partial pressure of water vapour to the partial pressure of saturated water vapour at that temperature or the ratio of mass of water vapour to the mass of saturated water vapour.

8. Specific Humidity

It is the ratio of mass of water vapour to the mass of dry air in a given volume of the mixture.

9. Degree of Saturation

It is the ratio of prevailing specific humidity to the saturated specific humidity. Degree of saturation is a measure of the capacity of air to absorb moisture.

28.2.3 Classification of Air-Conditioning

Classification of air-conditioning may be made based on functional point of view and process point of view.

1. Functional Point of View

Following are the two types of air-conditioning based on functional point of view.

(i) Comfort Air-conditioning

In this type of air-conditioning, air inside the room or the stipulated space is created so as to give the maximum comfort.

(ii) Industrial Air-conditioning

In this type of air-conditioning, such an atmospheric condition is created, as would suit to the needs of the industry during material processing, manufacturing, storage, etc.

2. Process Point of View

Following are the types based on the viewpoint of air-conditioning process (Arora and Bindra, 2012):

(i) Summer Air-conditioning

During summer the outside temperature is high and above the inside temperature of the room. Therefore the cycle of operations needed are air-cooling, dehumidifying, air distribution and air-cleaning. The cycle of operation for summer air-conditioning is shown in Fig. 28.1.

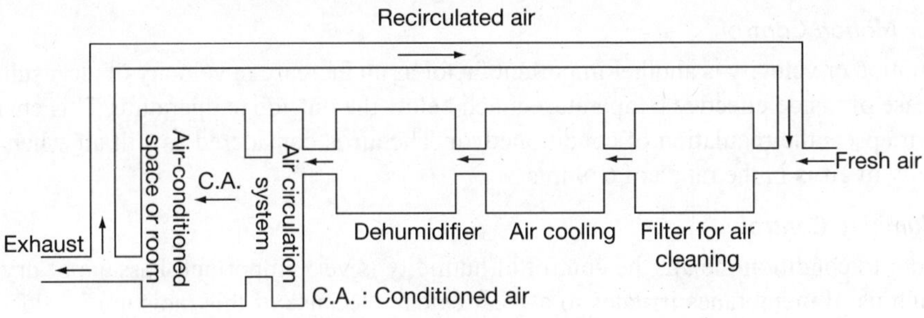

Figure 28.1 Cycle of operation for summer air-conditioning

(ii) Winter Air-conditioning

In winter, the outside temperature is low and below the inside temperature of the room and the cycle of operations needed are air-heating, humidification, air distribution and air-cleaning. This cycle of operation for winter air-conditioning is shown in Fig. 28.2.

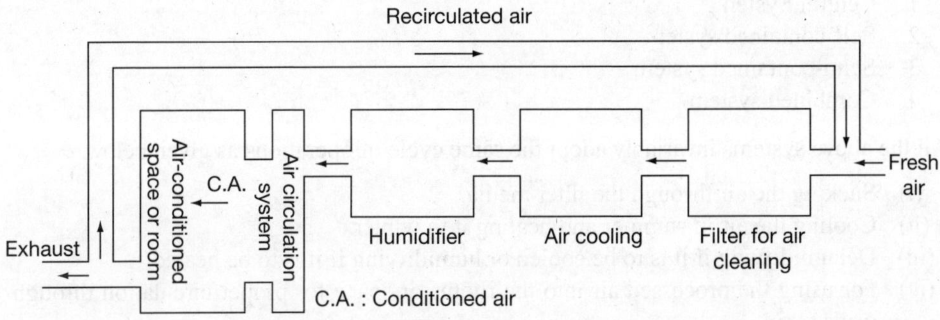

Figure 28.2 Cycle of operation for winter air-conditioning

(iii) Composite Air-conditioning

In this type, air-conditioning has to be done throughout the year irrespective of the variation of outside temperature. In this the cycle of operations involves humidification together with air-distribution and air-cleaning.

28.2.4 Principles of Comfort Air-Conditioning

Air-conditioning for comfort has to provide a healthier atmospheric conditions. As atmosphere is a function of temperature, air motion or velocity and humidity, the principles of air-conditioning should involve the proper control of these parameters.

1. Temperature Control

The range of temperature which is suitable to majority of people is called the *comfortable zone*. The comfortable zones are different for summer and winter seasons due to changes in body conditions and the clothing of human beings. The effective comfortable zones for summer and winter range between 20–23°C and 18–22°C, respectively. This effective temperature range is the measure of comfort of human body under most conditions of humidity and air motion. Irrespective of the outside temperature, a temperature range of 21–22.5°C is required for comfort conditions.

2. Air Motion Control

Air motion or velocity is another important factor as an increase in velocity of air results in decrease of inside effective temperature much below the outside temperature. This change in turn upsets the regulation of conditioned air. The air is considered as still air when the velocity of air is in the range of 6–9 m/s.

3. Humidity Control

In order to condition the air, the control of humidity is very important. Passage of dry air through nasal membranes irritates to a great extent. Because of this moisture is added to the heated air (i.e., humidification) or extracted (i.e., dehumidification) depending on the necessity. During summer a humidity of 40–50% and during winter 50–60% are the best. On an average, the value of relative humidity between 40 and 60% is considered desirable.

28.2.5 Systems of Air-Conditioning

Based on the location of air-conditioning equipment, the systems of air-conditioning are classified into the following four systems:

1. Central System
2. Self-contained system
3. Semi-contained system
4. Combined system

All the above systems invariably adopt the same cycle of operations as given below.

(i) Sucking the air through the filter media.
(ii) Cooling the air in summer and heating it in winter.
(iii) Dehumidifying if it is to be cooled or humidifying if it is to be heated.
(iv) For using the processed air into the rooms or space for proper circulation through grills.
(v) Collecting the used air through exhaust and mixing with the outside air and sucking again the air through the filter medium, which completes the cycle.

1. Central System

In this system all the equipment and accessories needed are installed one focal point or central point. From this point the conditioned air is distributed to all the room and intended spaces through concealed ducts.

This system requires less accommodation as all the units are placed at one place and for the same reason it is also economical. Maintenance is also simple. However, the ducts occupy more space at the ceiling level, Fig. 28.3 (Sharma, 1988).

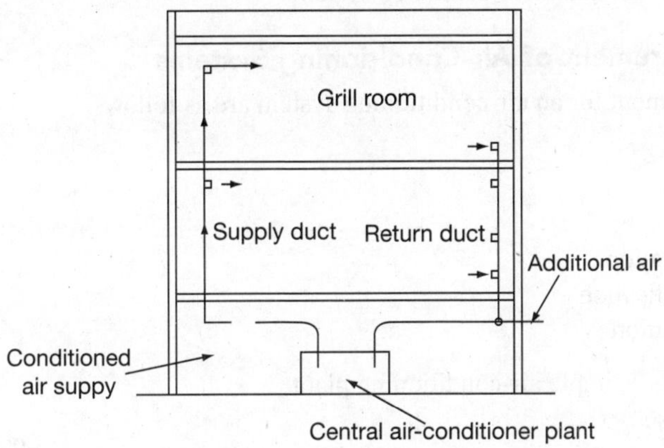

Figure 28.3 Layout of a central system of air-conditioning

2. Self-contained System

This is also called a unit system. In this system, the conditional air and distribution are kept in place that is why it is called self-contained system. Here no ducts are provided. Special, portable and attractive cabinets are provided for the distributing unit. Window air-conditioner is a typical self-contained system, Fig. 28.4.

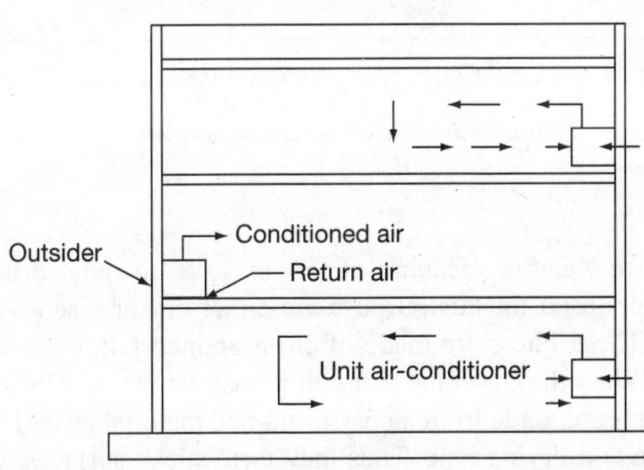

Figure 28.4 Layout of self-contained system of air-conditioning

3. Semi-contained System

In this type of system, every room is provided with an air-conditioning unit. This unit receives the supply from a central system. Because of this, lesser size of ducts are used.

In another set up of this type, the conditioned air may be supplied from a central unit but the heating or cooling may be done in the room itself.

4. Combined System

This combined system comprises of.

(i) Central and self-conditioned system,
(ii) Central and semi-contained system and
(iii) Self-contained and semi-contained system.

28.2.6 Requirement of Air-Conditioning Systems

Essential requirement for an air-conditioning system are as follows:

1. Filters
2. Heating
3. Cooling
4. Humidification
5. Dehumidification
6. Air circulation

Figure 28.5 shows a simple air-conditioning plant.

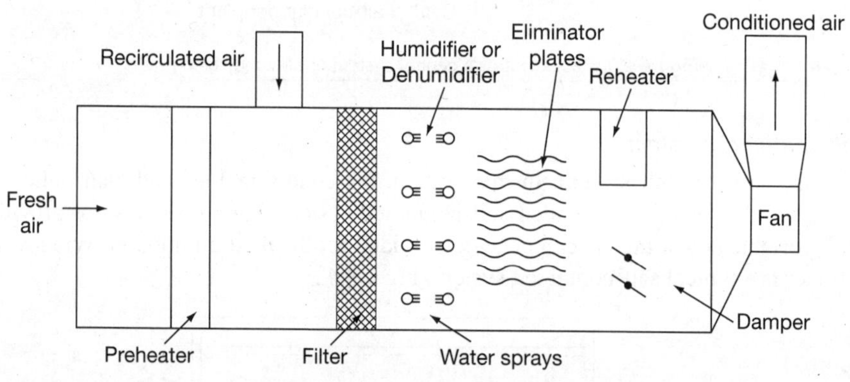

Figure 28.5 Typical air-conditioning plant

1. Filters

These are used to have effective cleaning of the air so as to remove dust or other suit particles from it. Filters used for this purpose are either dry or viscous type. Dry filters are commonly used. These filters are made of close-grained felt, cloth or paper through which air is forced. When they become dirty, they may be cleaned by vacuum-cleaning methods. Viscous filters are made from a coarse-grained material coated with an adhesive or a viscous oil of the non-drying type. Thus they form sticky surfaces which attract and remove the dust particles. With use they become dirty, which are washed, recoiled and used again.

When exceptionally dust-free air is required, electric precipitators are used. They consist of a high-intensity ionising field with a secondary field where the dust is precipitated. These are expensive.

2. Heating

During winter season, the air-conditioner has to provide slightly heated air to the inside of the building. This arrangement further compensates the heat loss from the room. The pre-heating incoming air may be carried out by warm-air furnaces or by coils around wherein the hot water or steam is circulating. Before entering the room the incoming air is allowed to pass over warm-air furnaces. As a matter of fact, the circulation of hot water or steam around the coils makes them hot and where the incoming air comes into contact with heated coils, it also gets heated.

3. Cooling

During the summer season, the air-conditioner has to provide cooled air inside the room, for which the incoming air is to be cooled before its entry into the room. It is further to state that the cooling is necessary to remove excessive heat from the air. The cooling can be done by surface cooling, spray cooling or evaporative cooling.

4. Humidification

In winter the outside air may have high relative humidity. The relative humidity becomes low when the air is heated. If too low air becomes too dry giving a parched feeling. However, if air is too much humidified, it also results in feeling of discomfort. For human comfort proper control of humidity is therefore necessary. Hence, humidification is an important part of the air-conditioning system and so it is also called as the heart of air-conditioning system.

Humidification may be achieved by passing air over pans of water with large surfaces or by passing air through spray nozzles which break the water into tiny drops. The sprayed water may be reused. Eliminator plates are used to remove the droplets from the air (Fig. 28.5). They are arranged in such a way that the air is made to follow a tortuous path between them.

5. Dehumidification

This is obtained by first lowering the temperature of the air below the dew point. This is followed by condensing out the necessary amount of moisture and then re-heating with dry heat to the desired temperature. Cooling of the air is done then. Two common methods of cooling, viz., mechanical refrigeration or cooling by water spray may be adopted. Mechanical method is similar to household refrigerator. The second method is one wherein warm air is passed through a water spray. By the action of evaporation the air gets cooled. This method is good for dry climates. In humid climates, mechanical refrigeration would be preferred. Re-heating is done to ensure no heaviness due to high humidity.

6. Air Circulation

Air circulation or air distribution is the act of circulation or distribution of conditioned air inside a room or enclosure. This is one of the most essential requirement of the air-conditioned system. The minimum fresh air requirement is based on the number of users and whether they are smokers or non-smokers. The conditioned air should be spread in the room at such height and location so as to mix thoroughly with the air present in the room. The air circulation is performed using one of the apparatus, viz., air pumps, air delivery system (consisting of supply and return ducts) and air distribution system (consisting of inlets and outlets).

28.3 THERMAL INSULATION OF BUILDINGS

Temperature difference exists between outset and inside of a building and at difference units of a building. Heat is transferred from warm to cool place. The transfer of heat may take place by conduction, convection or radiation. Heat transfer mainly happens due to conduction through the solid building materials. This heat transfer depends on the conductivity of the intervening materials, temperature difference between two surfaces, the thickness of the material, the area of exposed material and the time through which the heat flow takes place.

Convection and radiation also play their role in the process of heat transfer through the building materials. Convection may take place through large areas and the thickness of the building material by currents of air transmit. Similarly, polished surfaces on the exposed side of a building may retard the inward flow heat and prevent heat penetration (Sharma, 1988).

28.3.1 Purpose of Thermal Insulation

1. The main purpose of thermal insulation is to conserve a constant heat or temperature inside a building.
2. Thermal insulation provides comfortable living and working indoor conditions. That is, to keep the room cool in summer and warm in winter.
3. In order to reduce the demand of heating in winter and refrigeration in summer.
4. In order to prevent the condensation of moisture on interior walls, ceilings, windows, etc.
5. In order to reduce the risk of water-freezing in case of pipes and heat loss in case of hot water system.

28.3.2 Causes for Heat Transfer

Following are the factors which contribute for transfer of heat:

1. Value of thermal insulation of the exterior shell (e.g., external walls, roofs, etc.).
2. Difference in temperature between the outside and the inside. If the difference is greater; greater shall be the loss of heat.
3. Surface area of the building exposed at the external face. Evidently for greater area the heat transfer shall be more.
4. Surface areas of external windows and doors. For example, the heat transfer of a glazed window is about three times more than a typical wall construction.
5. Rate of air movement through any cavities in the enclosing structure. For example, air spaces through the roof or below a suspended floor may cause considerable heat transference.

28.3.3 Principles of Thermal Insulation

Based on the causes identified for heat transfer, the following general principles govern the thermal insulation:

1. Materials of low thermal conductivity have to be adopted. That is the material used should have adequate heat insulation value.
2. As the thermal resistance of a material is directly varies with the thickness and so the material to be provided should have adequate thickness.

3. Provision of open spaces like cavity walls provides excellent means of thermal insulation.
4. External exposure of windows, ventilators and doors play a major role and hence the orientation of the building and positioning of doors and windows play a vital role.
5. To some extent thermal insulation can be achieved by providing sunshades, increasing the height of parapet walls, etc., can reduce the heat transfer.

28.3.4 Heat Insulation Materials

Good insulating material should possess the properties such as adequate fire-resistant, termite resistant, moisture-resistant, adequate strength and stability and long life.

Some of the insulting materials which are in use are:

1. Blanket insulation
2. Bats insulting materials
3. Block or slab insulation
4. Cement concrete products
5. Insulating boards
6. Loose fills
7. Reflecting sheet materials

1. Blanket Insulation

Blanket insulators are flexible fibrous materials supplied in rolls or otherwise. They are made out of mineral wool, processed wood fibre, cotton and animal hair. They are available in thickness varying from 1 to 8 cm which can be spread directly on the surface of walls and ceilings.

2. Bat Insulating Materials

These materials are similar to the above type but are smaller in size and greater in thickness. Commonly available thicknesses are 5, 7 and 9 cm. For framing purposes they are made in smaller sizes also.

3. Block or Slab Insulators

Block or slab insulators are small rigid small units of 2.5 cm thickness and of sizes 60 cm × 120 cm are available. These boards made out of cork boards, cellular glass blocks, rubber blocks, mineral wood slabs or wood fibre boards together with cement. These small units may be fixed on the walls and roofs for lining purposes.

4. Cement Concrete Products

Cement concrete products in general have less insulation value. However, use of cement concrete with light weight aggregates such as blast furnace slag, burnt clay aggregate, vermiculate, etc., the resistance against can be improved.

5. Insulating Boards

These are the boards which are prepared by reducing wood cane or other materials to pulp and then re-assembling fibres into boards. In order to keep the fibres in the desired position, adhesives are used. These boards, available in different sizes and thickness, are used for interior lining of walls and thickness, are used for interior lining of walls and roofs and for partitions.

6. Loose Fills

These are fibrous materials like rock wool, slag wool, glass wool, cellulose or wood fibre wool. They are loosely placed into the required space for heat insulation purposes.

7. Reflecting Sheet Materials

These materials primarily depend on their surface characteristics for their heat resistant properties. They are used usually along with the air spaces so that the reflecting insulating surface is exposed. These reflective insulations consist of sheet or gypsum boards, steel sheet reflecting materials, aluminium foils, etc.

28.3.5 Methods of Thermal Insulation

Based on the causes of heat transfer, the areas to be concentrated for heat insulation accordingly the methods to be adopted are:

1. Thermal Insulation of roofs
2. Thermal Insulation of exposed walls
3. Thermal Insulation of external windows, and doors

1. Thermal Insulation of Roofs

The insulation can be done on outside or inside of a roof. Generally it is better to put a thermal insulation above the slab. Following methods may be adopted:

(i) Heat insulating materials may be installed over the roof but below a water proof course. If it is an internal application of the material, it may be fixed by adhesives or otherwise on the underside of the roofs.
(ii) False ceiling may be provided inside by insulating material with certain air gaps.
(iii) On the top of the roof reflecting insulating material may be laid.
(iv) Air spaces may be provided on the top of the flat roof by using asbestos sheets over the bricks.
(v) Spraying water regularly on the top of the roof may reduce the heat flow.
(vi) White-washing of roof before onset of each summer also prevents heat transfer.
(vii) Providing adequate shading on the exposed roof surface also reduces the heat flow.

2. Thermal Insulation of Exposed Walls

(i) Increase in thickness of walls to some extent forms a thermal insulator. But it is expensive.
(ii) Providing cavity walls effectively acts as a thermal insulator.
(iii) Heat insulating materials may be used along with other building materials during the construction of walls.
(iv) Heat insulators explained earlier may be installed or fixed on the inside and outside of the exposed wall in order to reduce the thermal transmittance.
(v) For partition walls air spacing may be created by fixing sheathing of hard boards or battens on either side of the wall.
(vi) Exposed surfaces may be white-washed or light colour distemper may be used on the exposed surface.

3. Thermal Insulations of External Windows and Doors

Heat is considerably transmitted through windows and doors which are exposed. This can be reduced by adopting the following method:

The incidence of solar heat on exposed doors and windows have to be reduced. This can be done by use of external shading such as louvered shutters, sun breakers, chajjas, etc. Internal shading such as curtains in heavy folds and venition blinds may be used. In the glazed windows, the heat reduction may be achieved by insulating glass or double glass with air space may be used.

28.4 ACOUSTICS AND SOUND INSULATION

Sound is produced when part of the atmosphere is compressed suddenly. It is transmitted in the form of waves. The waves are a series of compressions and rare factions created in the air medium. The average sound travels at a speed of 340 m/s at ordinary temperature. It depends on the medium through which it travels.

28.4.1 Effects of Audible Sound

Sound can travel through some medium like air. It can not travel in vacuum. Thus for the sound to be audible to the ears, the sound source and ear must be connected by an elastic medium like air. The characteristics of audible sound are discussed below.

1. Frequency of Sound

Frequency of pitch of sound is defined as the number of cycles or vibrations per second. The highest audible sound (e.g., whisle) has a frequency of 20000 cps (cycles or vibrations per second). The lowest audible sound has a frequency of 20 cps (e.g., whispering). The frequency is a measure of the quality of sound.

2. Intensity of Sound

The intensity of sound is defined as the flow of sound energy per second through unit area. The intensity of sound is the strength of the sensation received by the human ear. Intensity of sound is a purely physical quantity. But loudness of sound depends on the characteristics of ear.

3. Measurement of Sound

The range of intensity of sound is very large. The loudest sound is about 10^{13} times the sound which is just audible by the human ear. There is a wide range of sound levels, it is realised that a scale has to be adopted as a guidance. The intensity of sound is measured on a logarithmic scale due to wide range of variation of the intensity of sound. Bel is the measure of intensity of sound named after Graham Bell, the inverter of telephone.

As the unit of bel is comparatively large, hence a shorter practical unit decibel (*db*) equal to 1/10th of a bel unit. The range of audible sound to painful noise varies from 1 to 10^{13} which is covered on logarithmic scale between 1 and 130 *db* units. Thus one *db* unit is approximately the smallest change of sound intensity which the human ear can hear. Table 28.1 (Sharma, 1988) shows the rating to sound generally experienced.

28.4.2 Principles of Acoustics

The behaviour of sound plays an effective role in the acoustical design of different types of buildings and in the sound insulation process.

A sound originating from a source, such as music or operation of machine, is transmitted through the medium in all directions. The transmitted sound strikes on some surface, like wall, ceiling, floor or any other barrier. Depending on the type of surface part of it is

Table 28.1 Intensity of sound and comfort

Average intensity levels (db)	Feeling levels	Types of sound
10	Very Faint	Rustle of leaves, whisper, sound proof room, threshold of Audibility
30	Faint	Quiet house or private office, average auditorium, quiet conversation, etc.
50	Moderate	Noisy house, average office or conversation, quiet radio, etc.
70	Loud	Noisy office, average street noise or radio, average factory noise, etc.
90	Very Loud	Noisy factory area, loud street noise, police whistle, etc.
110	Deafening	Thunder, artillery, boiler factory, pneumatic hammers, etc.
180	Pain and Discomfort	Loudest sound due to operation of machines like pneumatic drill or aeroplane at a distance of 4 m, etc.

reflected back, and a part being absorbed by the surface. If the sound is not absorbed by the material, it will be transmitted in part to another side of barrier.

If the ultimately reflected sound is not properly controlled the reflection may result in acoustical defects, viz., echoes and reverberations. This reflected sound is important in the acoustical design of buildings. The part of sound absorbed by the surface is represented by an absorption coefficient. This coefficient is the ratio of the energy absorbed by the area of the surface to the energy striking the area. This coefficient is a function of the frequency of sound. The values of the absorption coefficient for different surfaces for a normal frequency of 500 cps are presented in Table 28.2. It is assumed that the sound wave will completely pass through a window, then it is taken as 1.

Table 28.2 Absorption coefficient for different surfaces

Sl. No.	Surface	Absorption coefficient per m²
1.	Marble, metal	0.01
2.	Plaster on wall surface, stage curtain	0.02
3.	Unplastered brick wall (40 cm thick), Glass against solid surface, Linoleum or concrete floor	0.03
4.	Solid wooden floor	0.09
5.	Framed wooden floor	0.13
6.	Window glazed	0.18
7.	Plywood on battens	0.17–0.26
8.	Ventilators	0.10–0.50
9.	Sound absorbent or acoustical materials	0.50
10.	Curtains in heavy folds	0.40–0.75
11.	Open window	1.00

The reduction of intensity of sound of a transmitted sound through a barrier is called as *transmission loss*. This transmission loss is a measure of the effectiveness of a surface as an insulating material.

Thus transmitted and absorbed sounds have important bearing on the acoustical condition of a building. However, both transmitted and absorbed sounds are inter related and influence the acoustic and sound insulation.

Highly porous materials have the quality to dissipate considerable energy and the absorption will be relatively high. An ideal material for sound absorption should possess the following qualities:

(i) It should have a high coefficient of sound absorption of economy.
(ii) It should be fire resistant.
(iii) It should have high structural strength.
(iv) It should be unaffected by decay, attack by insects and vermin.
(v) It should be easily workable to fit in any location.

28.4.3 Acoustical Defects

The acoustical design of an enclosed space is basically depend on the behaviour of the reflected sound. Due to the reflection of sound two main defects are developed, viz., echoes and reverberation.

1. Echoes

Echo is said to be produced when a reflected sound wave reaches the ear just when the original sound from the same source has already been reached. The sensation of sound persists for one-tenth of a second after the source has ceased. Thus an echo forms when the time lag between the two sounds is about $1/17^{th}$ of the second.

Further considering the velocity of sound in the atmosphere air as 34.3 m/s, it is shown that when the distance of the reflecting surface is between 8 and 17 m, echoes are formed.

The defect of echoes also occurs when the shape of the reflected surface is curved with smooth character. Echoes are unpleasant to hear and cause disturbance to hearing.

2. Reverberation

When the surfaces of an indoor place are hard and smooth, very small energy is lost at each impact of sound and many reflections take place before the sound dies down. This repeated reflection of sound is called prolongation or reverberation. Then if sound exists too long, then successive words of a speech will overlap and confuse.

The remedy for this defect is to select a correct time of reverberation known as optimum time of reverberation. This is achieved by suitably selecting a proper absorbent or acoustical materials for different reflecting surfaces.

Acoustical level of a room and reverberation times are shown in Table 28.3, Table 28.4 shows the recommended values of the optimum reverberation time in India for halls after considering the indicated audience strengths.

28.4.4 Acoustical Materials

Common building materials are absorbents of different levels. Such materials are called as absorbent materials.

Table 28.3 Acoustics level and reverberation time

Acoustics level	Reverberation time
Excellent	0.50-1.50
Good	1.50-2.00
Fairly good	2.00-3.00
Bad	3.00-5.00
Very bad	Above 5.00

Table 28.4 Optimum reverberation time for closed area

Sl. No.	Type of hall	Audience factor	Optimum reverberation time (sec)
(A)	Big halls	Full	2.00–3.00
(B)	Music concert halls	Full	1.60–2.00
(C)	Churches	Two-thirds	1.8–3.00
(D)	Theatre	Two-thirds	1.30
(E)	Public-lecture halls	One-thirds	1.50–2.00
(F)	Law courts, Committee rooms, Conference hall	One-thirds	1.00–1.50
(G)	Parliament house, Assembly halls, Council chambers	Quorum	1.00–1.50

Qualities of a good acoustical materials are as follows:

(i) An absorbent material should be economical for construction and should be water-proof, fireproof, adequate strength and heat resistant.
(ii) It should be durable and easy for maintenance.
(iii) It should have high absorption capacity.
(iv) It should be easy to fix and should be self-supporting.

Some of the acoustical materials are discussed below.

1. Acoustic Plaster

It is a granulated insulation material mixed with cement. The quantity of cement used in such materials is to provide enough strength and have sufficient pores. The coefficient of absorption varies from 0.15 to 0.30.

2. Acoustic Tiles

These are commercially available materials with adequate absorbent qualities. The absorption of sound is uniform and most suitable for small rooms. But it is costly compared to other acoustical materials.

3. Porous Boards

Straw boards, pulp boards and compressed fibre boards have varied coefficient of absorption. These boards can be fixed by ordinary panelling. The thickness is about 17 cm and coefficient of absorption varies from 0.17 to 0.50.

4. Perforated Boards

The material can be formed by combination of mineral wool and cement asbestos or with mineral wool and hard board. The coefficient of absorption for the first combination is about 0.95 and for the latter is 0.20. Such boards are generally suspended from trusses.

5. Quilts and Mats

These materials are made out of mineral wool or glass wool. The coefficient of absorption varies which depend on the thickness of the board or mat, density, perforations, mode of fixing and frequency of sound.

SALIENT POINTS

1. Air-conditioning primarily consists of refining air in order that it may be utilised to the best advantage of the user.
2. Classification of air-conditioning based on functional point of view are: Comfort air-conditioning and Industrial air-conditioning
3. Classifications of air-conditioning based on process point of view are: Summer air-conditioning, Winter air-conditioning and Composite air-conditioning.
4. Principles of air-conditioning are: Temperature control, air motion control and humidity control.
5. Based on the location of air-conditioning equipment, the systems of air-conditioning are classified into four systems are: central system, self-contained system, semi-contained system and combined system.
6. Essential requirements for an air-conditioning system are filters, heating, cooling, humidification, dehumidification and air circulation.
7. In general, transfer of heat may take place by conduction, convection or radiation.
8. Heat transfer mainly happens due to conduction through the solid building materials. This heat transfer depends on the conductivity of the intervening materials temperature difference between two surfaces, the thickness of the material, the area of exposed material and the time through which the heat flow takes place.
9. Heat insulating materials are blanket insulation, bats insulating materials, block or slab insulators, cement concrete products, insulating boards, loose fills and reflecting sheet materials.
10. Methods of thermal insulation adopted are thermal insulation of roof, thermal insulation of exposed walls and thermal insulation of external windows and doors.
11. The characteristic of audible sound is based on frequency of sound and intensity of sound.
12. Intensity of sound is defined as the flow of sound energy per second through unit area. The intensity of sound is the strength of the sensation received by the human ear.

13. *Bel* is the measure of intensity of sound named after Graham Bell, the inventor of telephone.
14. As the unit of *bel* is comparatively large, hence a shorter practical unit decibel (*db*) equal to 1/10th of a *bel* unit. Thus one *db* unit approximately the smallest change of sound intensity which the human ear can hear.
15. The part of sound absorbed by the surface is represented by an absorption coefficient. This coefficient is the ratio of the energy absorbed by the area of the surface to the energy striking the area. This coefficient is a function of the frequency sound.
16. The reduction of intensity of sound of a transmitted sound through a barrier is called as transmission loss. This transmission loss is a measure of the effectiveness of a surface as an insulating material.
17. Transmitted and absorbed sounds have important bearing on the acoustical condition of a building. However, both transmitted and absorbed sounds are inter related and influence the acoustic and sound insulation.
18. Echo is said to be produced when a reflected sound wave reaches the ear just when the original sound from the same source has already been reached.
19. When the surfaces of an indoor place are hard and smooth, very small energy is lost at each impact of sound and many reflections take place before the sound dies down. This repeated reflection of sound is called prolongation or reverberation. Thus if sound exists too long, then successive words of a speech will overlap and confuse.

REVIEW QUESTIONS

1. What are the purposes for which air-conditioning is used?
2. Differentiate between
   (i) Dry air and moist air
   (ii) Dry bulb temperature and wet bulb temperature.
3. Define relative humidity.
4. How do you classify based on (i) functional point of view and (ii) process point of view?
5. Discuss the principles of comfort air-conditioning.
6. What are the requirements of a successful system of air-conditioning?
7. Explain the purpose of thermal insulation.
8. What are the principals of thermal insulation?
9. Discuss different heat insulating materials.
10. Briefly explain the methods of thermal insulation.
11. Discuss the effects of audible sound.
12. Explain the principles of acoustics.
13. What are acoustical defect?
14. Briefly discuss the acoustical materials used in practice.

15. Explain the usual defects of an auditorium, bring out the reasons for such defects and the necessary remedies.
16. List the precautions to be taken in the acoustic design of a studio.
17. What are filters? Explain their desirable properties.
18. How conditioned air is distributed in a room.
19. What are the factors responsible for heat transference?
20. Explain the various aspects of thermal treatment in tropical and sub-tropical regions.

29 PROTECTION OF BUILDINGS

29.1 INTRODUCTION

Although buildings are constructed based on the space requirement and safety needs, they are also to be protected from natural calamities such as action of termites, dampness, fire, lightening, earthquake, etc. Steps taken during construction itself to protect the buildings from the above calamities the durability and life of the building will be increased.

Termites are commonly found in tropical and sub-tropical countries. These termites are nothing but white ants which live in a colony and destroy textiles, woodwork, paper products, etc. As doors and windows and other cubboards and furnitures are of wood, it is essential to safeguard these from termite. Necessary steps should be taken during construction of the building.

One of the essential requirements of a building is that it should be dry. Dampness in building may occur due to several reasons such as improper design, faulty construction, use of poor quality materials, ground water position, etc. Dampness not only affects the life span of the building but also creates unhygienic conditions to the occupants. Damp prevention is considered to be one of the important items of work in the construction of a building.

Fire protection is another important requirement to ensure the safety of the building. Protection of a building against fire can be attained by the use of special materials and construction techniques which aim at giving adequate resistance to the spread of fire within the building. The building components should be fire resistant, at least for a few hours, so that the occupants can live safely. Also, there should be sufficient appliances available to stop fire from spreading.

Tall buildings and buildings constructed in open areas are to be protected from lightening failing which the building will be subjected to large quantity of electricity. This may cause electrification of occupants. Hence, necessary lightening-protection system should be provided in the building.

Buildings are also to be protected from the dangerous natural calamity, the earthquake. Protection from the above calamities is discussed in the following sections.

29.2 ANTI-TERMITE TREATMENT

29.2.1 Termites

Termites, popularly known as white ants, are found in groups in tropical and sub-tropical countries. They are very fast in eating wood and other cellulosic materials as food. They also damage non-cellulosic materials like plastics, leather, etc. The term termite-proofing is used to indicate the treatment which is given to a building. This is done to prevent or control the growth of termite in a building. Termites are of two types, viz., dry wood termites and subterranean termites.

29.2.2 Principles of Termite-Proofing

While making a building termite-proof, the following points have to be observed:

(i) It should be observed that no part of a building is bridged to untreated soil.
(ii) If any fill material is found to contain termite colonies, it should be eradicated.
(iii) Joint fillers or metal strips may be used to make floor-joints as termite-proof.
(iv) Termites can not enter through dense concrete. Hence, foundations should be carried out with good quality concrete or superior quality materials with good workmanship.
(v) The site should be cleared before taking up the construction from all old tree stumps, dead wood, etc.
(vi) Superstructure materials should be treated with suitable preservatives so as to prevent termite.
(vii) Selection of the method to be adopted for termite-proofing should be carefully selected.
(viii) Pre-treatment is cheaper compared to post-treatment and should be resorted to.

29.2.3 Methods of Termite-Proofing

Termite-proofing methods may be grouped under the following two categories:

1. Soil treatment with chemicals
2. Structural barriers

1. Soil Treatment with Chemicals

(i) Chemicals for Treatment

Following are the emulsifiable chemicals recommended in BIS code:

(a) Chloropyrifos concentrate 1% by weight
(b) Heptachlor concentrate 0.5% by weight
(c) Chloride concentrate 1.0% by weight

Chloropyrifos has been widely used. They are available with agrochemical agencies which are used to control termites in crops. This chemical with 1% concentration is mixed in water for soil treatment and in kerosene oil for treatment of wood. As the solution is toxic, it should be washed immediately by soap and water, if it comes into contact with part of the body.

(ii) Pre-Construction Treatment

Treatment should be started when foundation trenches and pits are excavated to the size and level. Treatment should not be done when the soil is wet due to rain or sub-soil water. Further, the treated area should not be disturbed. The treatment is performed in soil in five stages as discussed below.

In the first stage, treatment of wall trenches and basement excavation are done. If termite hills are noticed at the site during site clearance chemicals should be sprayed first. All side surfaces of trenches/pits and bottom of wall trenches/pits and basement excavations should be treated to a height of 30 cm from the bottom. The usage of solution should be at the rate of 5 l/m^2 of surface area (Fig. 29.1).

In the second stage, treatment of refill in contact with foundation is attended. All the refill earth in the excavation immediately in contact with both sides of the wall footing should be treated (Fig. 29.1) for a distance of 30 cm. Similarly all the four sides of a column should be treated and treatment should be at the rate of 3–5 l per linear metre vertical surface of the wall.

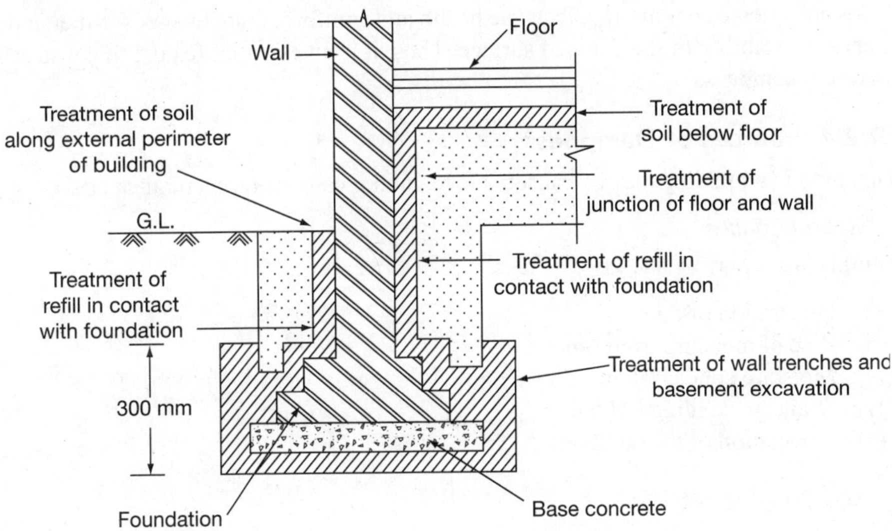

Figure 29.1 Anti-termite treatment of load bearing wall foundation

The third stage is treatment of soil below floors. The earth fill below the floors up to the plinth level has to be treated after the fill has been made. This is performed by making holes 5–7.5 cm deep at 15 cm centres in a grid pattern. The holes are filled with the solution at the rate of 5 l/m^2 of treated surface (Fig. 29.1).

The fourth stage consists of treatment of junction of floor and wall. Before laying the sub-grade, channels of 3 cm wide and 3 cm deep are dug along the junctions of floor and wall. Along the channel, holes at 15 cm apart are made and treated at 15 l/m^2 of the wall surface and the chemical is allowed to seep through the bottom. The soil is tampered back in position after the operation (Fig. 29.1).

The fifth stage is treatment of soil along external perimeter of the building. After the completion of building, holes are made along the external perimeter at intervals of 15 cm and to a depth of 30 cm. These holes are filled with the chemical emulsion at the rate of 5 *l*/m length of wall (Fig. 29.1).

Apart from treating the soil, it is also important to treat the expansion joints, soil around the pipes and conduits and wooden surfaces for a complete protection from termite.

2. Structural Barriers

In order to prevent the termites' entry through walls, impenetratable physical structural barriers may be provided at the plinth level continuously. Such a structural barrier is generally of concrete or sometimes metal. The cement concrete may be from 50 to 75 mm thick. It is recommended to project about 50–75 mm internally and externally. The metallic materials may be sheets of non-corrodible one, such as copper or galvanised iron with a thickness of 0.80 mm.

29.3 DAMP PREVENTION

Presence of hydroscopic moisture on a surface is called dampness. In general dampness causes unhygienic conditions, affects the health and comfort of inhabitants. Further, it deteriorates the stability of the damped surface. Thus, it is an essential feature in construction to prevent dampness.

29.3.1 Sources of Dampness

Dampness in a building may be caused due to natural causes or structural causes.

1. Natural Causes

Natural causes may be due to:

(i) Penetration of rain
(ii) Rise of moisture from ground
(iii) Moisture condensation
(iv) Drainage condition of the site
(v) Orientation of the building

(i) Penetration of Rain

Rain may penetrate a building through the top of the walls, through the surface of the walls and through the roof. Rain may penetrate the unprotected parapet walls of a building during heavy rains.

(ii) Rising of Moisture from the Ground

The ground or sub-soil on which a building is constructed may give an access to water to enter the structure and cause dampness. Gravel and sandy soil are highly permeable and allow water to pass through them easily. But clay and clayey silt or clayey sand are less permeable but cause dampness due to heavy capillary rise.

(iii) Moisture Condensation

Whenever the warm air in the atmosphere is cooled, the process of condensation takes place. Because of this condensation, moisture is deposited on the areas of walls, floors and ceilings.

(iv) Drainage Conditions of the Site

The elevation of a ground on which a building is constructed has significance on dampness condition. Buildings built on a higher ground can be drained easily and are, hence, less liable for dampness. Low-lying sites can not be drained easily and may cause water logging. The condition will get still worse if impervious soil is present beneath the structure. Hence, it is an essential condition to be looked into before starting any construction.

(v) Orientation of the Building

Depending on the climatic conditions and monsoon walls of buildings may be subjected to constant splash of rain water. Hence, the construction of a building should be accordingly planned.

2. Structural Causes

Structural causes may be due to faulty design of the building, faulty construction, poor workmanship or use of inferior quality material in construction.

(i) Faulty Design of Structure

Improper mix design of concrete not only has less strength but of high permeability causing dampness. Improper mortar proportions may also cause water penetration. Non-provision may also cause water penetration. Non-provision of a damp-proof course in the design and improper plinth-protection measures may cause dampness in the building.

(ii) Faulty Construction of Structure

Improper construction of various parts, for example, fixtures in a building, joints in the roofs, throating of sills and copings, joining of walls, etc., may cause dampness by entry of water.

(iii) Poor Workmanship and Materials

Skilled workers should be used for construction of walls, roofs, floorings, etc., and electrical and plumbing works. Further, good quality materials should be used to get effective construction.

29.3.2 Effects of Dampness

In general, presence of dampness results in poor functional performance, shabby appearance and structural weakness. Following are the effects of dampness:

(i) Presence of damp conditions causes efflorescence on the surfaces leading to disintegration of bricks, stones, tiles, etc., and in reduction of strength.
(ii) Plastering may get softened or crumbled due to dampness.
(iii) Paints on surfaces may get blistered, bleached and disfigured.
(iv) Dampness may cause corrosion of metals used in the construction.
(v) Materials used as floor coverings, such as tiles, marble stones, etc., may be damaged due to loss of adhesion with the floor bases.
(vi) Doors and windows and other timber works may get warped due to dampness.
(vii) All electrical fittings get deteriorated and liable to cause short circuits.
(viii) Dampness promotes the growth of termites and hence deteriorates materials and causes unhygienic conditions.
(ix) Dampness added with warmth and darkness leads to breeding of germs which may cause some disease.
(x) Dampness creates an unhygienic working condition for the occupants.

29.3.3 Methods of Damp-proofing

Damp-proofing courses (*DPC*) of suitable materials are provided at appropriate locations for their effective use (Sharma, 1988). *DPC* prevents basically the entry of the water from ground in buildings. The best position for *DPC* is the plinth level in buildings.

Following general principles should be adopted while providing *DPC* in buildings:

(i) *DPC* should cover the full thickness of the wall.
(ii) Mortar bed on which the *DPC* is laid should be level and there should not be any projection.
(iii) In places where a vertical *DPC* is provided, it is to be laid continuously with a horizontal *DPC* and a fillet.
(iv) *DPC* course should be continuous and should form as a bearer from the entry of moisture.
(v) *DPC* should not be exposed in total.

29.3.4 DPC Treatment in Buildings

Provision of damp proof course at plinth level is shown in Fig. 29.2, whereas Fig. 29.3 shows the *DPC* in basement.

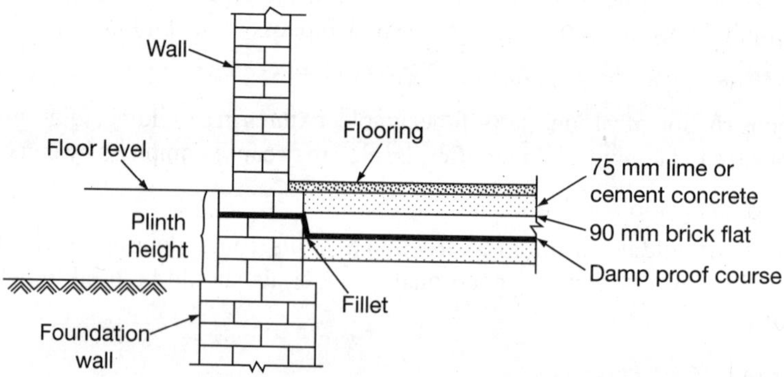

Figure 29.2 DPC above ground level

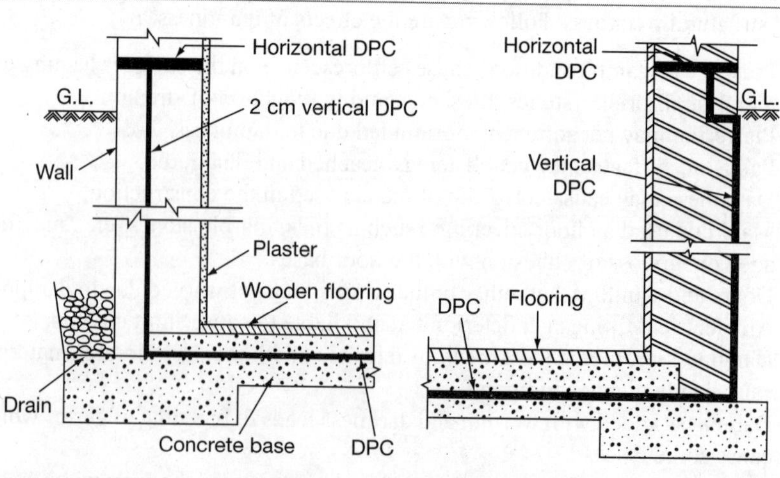

Figure 29.3 DPC in basement

Prevention of damp along parapet walls and flat roof are shown in Figs. 29.4 and 29.5.

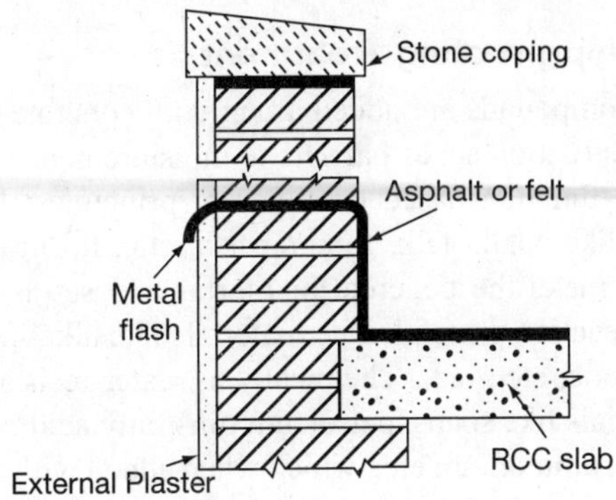

Figure 29.4 DPC in parapet wall

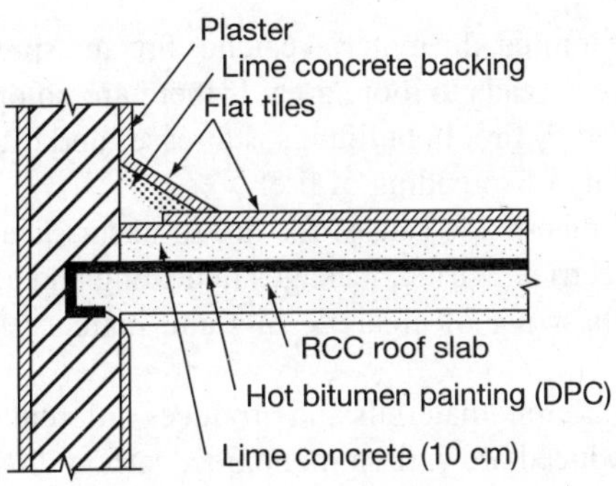

Figure 29.5 DPC in flat roof

29.3.5 Materials used for DPC

Materials generally used for *DPC* are flexible materials like, hot bitumen, bituminous felts, bituminous sheets, polythene sheets, metal sheets of lead, copper, etc.; semi-rigid materials like mastic asphalt or combination of materials or layers and rigid materials like first-class bricks, stones, slates in courses and cement-concrete stones, slates in courses and cement-concrete or mortar layers, etc.

29.3.6 Damp-proof Surface Treatment

In this method the surface exposed to moisture is treated by providing a thin film of water-repellent material over the surface. Such a surface treatment may be external or internal. Generally the external treatment is more effective in damp prevention when compared to internal treatment.

Surface treatments include pointing, plastering, painting, distempering, etc., Lime-cement plaster mix (1 cement:1 lime:6 sand proportion) is more effective.

Materials used for surface treatment are sodium or potassium silicates, aluminium or zinc sulphates, barium hydroxide and magnesium sulphate, soft soap, linseed oil, coal tar, soap, bitumen, remix and gums, etc., applied in alternate layers with suitable combination depending on the climatic conditions.

29.3.7 Integral Damp-proofing Treatment

In this process certain compounds are added along with concrete or mortar while mixing which when used in construction act as barriers to moisture penetration.

The added materials function based on different principles. Based on the mechanical principle, materials like chalk, talk, fullers earth, etc., fill in the pores present in the concrete or mortar and make the concrete or mortar and set as a waterproofing agent. Based on the chemical reaction principle, the materials like alkalines, silicates, aluminium sulphates, calcium chloride, etc., react chemically as water-resistant. Based on the repulsion principle, the materials like soaps, petroleum oils, fatty acid compounds such as stearates of calcium, sodium, ammonium etc., which when added with concrete or mortar react with it and become water repellent.

29.4 FIRE PROTECTION

When some materials get ignited, the material catches fire and spreads. If there are opening in walls and floors the fire spreads to more area. If there are no openings, the temperature of the structure is increased by fire. In buildings, staircases and lift shaft act as flues for fire and increases the possibility of spreading of fire.

There are natural and man-made causes for fire to occur. They may be caused due to faulty workmanship in electrical wiring, leakages in heating and cooking equipment, flammable liquids, careless throwing of cigarette bits and matches, lightening, spontaneous combustion, etc.

The fire spreads over different materials and produces different gases of which some are poisonous. The gases produced are carbon monoxide, carbon dioxide, hydrogen sulphide, nitrogen dioxide, etc. Thus to protect the goods and activities within a building or structure and of adjacent buildings fire-protection has to be resorted to. Fire resistance of a material is the time during which a structure fulfils its function with reference to safety when a fire prevents with a particular intensity.

29.4.1 Fire-Resisting Properties of Building Materials

With reference to fire, materials may be classified as combustible materials and non-combustible materials.

Combustible materials are the materials which combine exothermally with oxygen and give rise to flame at a particular range of temperature. Examples of such materials are wood, wooden products, animal products, and man-made products like fibreboard, strawboard, etc.

Non-combustible materials are those which when decomposed by heat will do so endothermically. These materials do not catch fire by or decomposed at a particular range of temperature. Examples of such materials are metal, stone, glass, concrete, clay products, gypsum products, asbestos products, etc.

The building materials have varying fire-resistant properties which are discussed below.

1. Bricks

Bricks in general have good fire-resistant property. Particularly first class bricks are fireproof and can withstand heat for a considerable length of time. As bricks are made out of clay, which is a poor conductor, can withstand heat as high as 1300°C. Special type of bricks called fire-bricks are best for use in fire-resistance constructions. In total, brick masonry is most suitable to withstand fire hazards.

2. *Terracotta*

Terracotta is also a clay product which has better fire-resisting properties than bricks. As the cost is high, it is used only in restricted places.

3. *Stone*

Although stone may resist high temperature but deteriorate due to sudden cooling. Thus stone should be used only for a limited use in buildings with reference to fire-resistance. Granite although very strong crumbles or cracks when subjected to heat. Compact sand stone has better fire-resistant capacity. But lime stone is not all desirable as fire-resistant material.

4. *Concrete*

As concrete is a bad conductor of heat it has high fire-resistance capability. The extent of fire resistance depends on the aggregate, density and position of reinforcement in RCC. Use of foamed slag, blast furnace slag, crushed brick, cinder, crushed limestone, etc., form the best aggregate for fire-resisting concrete.

5. *Mortar*

Mortar is a cheap and best incombustible material. Cement mortar is better fire-resistant than lime mortar as lime plaster is susceptible by calcination. In order to increase the fire-resistant property, the thickness may be increased. Cement mortar with surki or pozzolana shows very high fire-resistance capability.

6. *Asbestos Cement*

Asbestos cement is formed by combing fibrous mineral with cement. This material shows a very high fire-resistant property. The products of asbestos cement are widely used for the construction of fire-resistive partitions, roofs, etc. Any structural member formed by combing asbestos cement offer great resistance to fire, less susceptible for cracking, or disintegration at high temperatures.

7. *Steel*

Steel is incombustible at moderate temperatures but shows very low fire resistance at high temperatures. At high temperatures the yield stress reduces and deforms when quenched with water in the process of extinguishing a fire. That is, all the exposed steel should be protected against fire by covering them with materials like bricks, terracotta, etc.

8. *Wrought-Iron and Cast-Iron*

These materials have the same behaviour as that of steel. The only difference being that it has less elasticity and retains compression and tension compared to steel. Cast-iron should not be used as a fire-resistant material as it is susceptible for deterioration when subjected to cooling.

9. *Aluminium*

It is a poor performance material as a fire-resistant and recommended only in places where low fire risks are expected. It is a very good conductor of heat but possesses adequate tensile strength.

10. Glass

Glass is a poor conductor of heat and undergoes a very small compression or expansion. Thus it is a good fire-resistant material. But sudden change in temperature leads to cracks or fracture. But a reinforced glass possesses high melting point and thus recommended for fire-resistant doors, skylight, windows, etc.

11. Timber

In general timber is a combustible material. But it has a special property of self insulation by forming a charred face when exposed to fire which forms a protective cover. Use of timber in large sizes offers a better fire-resistance. Timber may be made fire-resistant by impregnating it with fire-retarding chemicals such as ammonium phosphate and sulphate, boric acid, zinc chloride, etc.

29.4.2 Fire-Load

The amount of heat liberated in combustion of any content or part of the building of a floor area is referred to as fire-load. It is represented in kilojoules per square metre (kJ/m^2).

The fire-load is the ratio of the weight of all combustible materials (by their respective calorific values) to the floor area under consideration. For example, let a floor area of 120 m^2 contain 18×10^3 N of combustible material having calorific value of 1.5×10^3 J/N, then the

$$\text{Fire} - \text{load} = \frac{18 \times 10^3 \times 1.5 \times 10^3}{120}$$

$$= 225 \times 10^3 \, \text{J/m}^2$$

The fire-load is used as a measure of grading of occupancies by BIS (BIS 1641–1968). Accordingly the classifications are as follows:

(i) Low fire-load
(ii) Moderate fire-load
(iii) High fire-load

Table 29.1 shows the classification of occupancies.

Table 29.1 Grading of occupancies by fire-load

Sl. No.	Class of fire-load	Limit in kJ/m^2 Min	Max	Occupancies
(A)	Low		Not to exceeding 1.15×10^6	Domestic buildings, hotels, boarding houses, restaurants, schools, hospitals, temples, mosques, offices, factories where NH materials are used, etc.
(B)	Moderate	1.15×10^6	2.30×10^6	Retail shops, emporium, markets, factories, workshops, etc.
(C)	High	2.23×10^6	4.60×10^6	Godowns and similar structures used for bulk storage of NH materials and goods.

29.4.3 BIS Grading

Bureau of Indian Standards (BIS 1641–1968) has graded the structural elements into five grades with respect to 'time in hours for resisting standard fire', as shown in Table 29.2.

Table 29.2 Fire resistant grades

Sl. No.	Time in hours for resisting standard fire	Fire-load in kJ/m²		Class of fire
		Min	Max	
(A)	6 hrs		4.60×10^6 and over	Very high
(B)	4 hrs	2.10×10^6	4.60×10^6	High
(C)	2 hrs	1.15×10^6	2.10×10^6	Medium
(D)	1 hrs	Less than 1.15×10^6		Low
(E)	0.5 hrs	–	–	Very low

National Building Code graded type of construction into four categories as Type 1 to 4 as given in Table 29.3.

Based on the availability of firefighting equipment in the premises or the public fire brigade availability, the duration of fire-load of 2.10×10^6 to 4.60×10^6 is usually considered as less than 3 hours. Hence, all the normal buildings are considered to come under Type 1 construction. Further care should be taken for ventilation and escape of gases.

Table 29.3 Types of construction and hours of resistance

Sl. No.	Type of construction	No. of Hours of fire resistance for all structural components
(A)	1	4
(B)	2	3
(C)	3	2
(D)	4	1

29.4.4 General Safety Requirements Against Fire

All building should satisfy certain safety requirements against fire, smoke and fumes.

1. Maximum Height

The height of a building is restricted depending on the number of storeys, the number of occupancy and the type of construction. Furthermore, all the above factors in turn depend on the width of the road in front of the building, floor area ratio and the local firefighting facility available.

2. Open Space

In general, every room for use by human beings should abutt on an interior or exterior open space or on an open verandah. The open spaces inside or outside should be able to provide sufficient lighting and ventilation. Further, the open space adjoining a road should be well inside giving scope for widening of the road.

3. Mixed Occupancy

When a building is used for more than one type of occupancy, for example, residential, godown, shops, etc., it should conform to the requirements for the most hazardous of the occupancies. Such mixed occupancy should be avoided as there is more risk for life of occupants. If mixed occupancy is separated by walls of 4-hour fire resistance, then the occupancy can be treated individually and safety measures can be taken.

4. Openings in Separating Walls and Floors

The openings in separating walls and floors should be designed in such a way that necessary protection is guaranteed to all such factors which may spread fire. For types 1–3 construction a door way or opening in a separating wall may be limited to about 6 m^2 (i.e., height 2.75 m and width 2.1 m). Such wall openings should be provided with fire-resisting doors or steel rolling shutters. All openings in the floors shall be protected by vertical enclosures. In Type 4 construction, openings in the separating walls or floors should be fitted with 2-hour fire-resisting assemblies.

5. Enclosure on all openings

Wherever openings are permitted, they should not exceed three-fourths the area of the wall in the case of external wall and should be protected with fire-resisting assembles or enclosures. Such assembles and enclosures shall also be capable of preventing the spread of human or smokes.

6. Power Installations

Electrical power installations and gas connections for kitchen, if any, should be done as per norms and requirements from the point of view of fire safety.

7. Materials of Construction

The structural elements of the building such as floors, partitions, roofs, walls, etc., should be invariably constructed with fire-resisting materials. In general non-combustible materials like stones, bricks, concrete, metal, glass, clay products, etc., should be used in construction. Combustible materials such as wood and wood products, fibreboards, strawboards, etc., should be avoided or used only for the most essential places.

29.4.5 Emergency Fire Safety Measures

Apart from the steps taken in construction of buildings the following general measures of fire safety have to be adopted.

1. Alarm Systems

Alarm systems are installed with a view to give an alarm and to call for assistance from neighbours in case of fire. As per the saying 'prevention is better than cure', the first five minutes of fire should be stopped instead of fighting to extinguish the fire for five hours. Further, safety alarm also gives enough time and warning for the occupants to save important materials and to reach to a safe place.

The alarm system may be manual or automatic. The manual alarm system may consist of a horn bell or siren by which the occupants can be alerted. The automatic alarm system is usually installed in large industrial building which is unoccupied at night. The automatic fire alarm, apart from sending information to the nearest control point, also alerts the nearest fire brigade station.

2. Fire-extinguishing Arrangements

Various types of extinguishing arrangements are provided to extinguish the fire depending on the importance of the building.

(i) Portable Fire Extinguishers

The purpose of portable fire extinguishers is intended for immediate use in case of an outbreak of fire. The portable extinguishers in common use are carbon dioxide type, foam machines, large foam generators, etc. Carbon dioxide type extinguishers are the most common for small fires. Sometimes small fires can be extinguished by keeping buckets of water, sand and asbestos blankets.

(ii) Fire Hydrants

Fire hydrants may be installed inside or outside the building. But they should be located in a suitable position such that water is made available easily. For large and close buildings the fire hydrants should be located 90–120 m apart. For open areas the distance may be 300 m or more. One hydrant for an area of 4000–10000 m^2 is provided depending on the population and importance of the region. Generally, hydrants are installed at all street corners.

(iii) Automatic Sprinkler System

This consists of pipes and sprinkles. They are installed in such a way as to operate automatically by the heat of fire and sprinkles water on the fire. This arrangement is suitable for the internal protection of premises. This arrangement is provided in industries which produce combustible materials like textile mills, paper mills, gas industries, etc.

(iv) Escape Routes

Adequate passages to escape in times of emergency have to be made in the building. This is more important in public buildings like theatres, town halls, schools, restaurants, etc. In case of buildings more than 25 m, it is recommended to provide at least one fire tower as the escape route. All escape routes over roofs and strairs should be protected with railings.

29.5 PROTECTION FROM LIGHTNING

Lightning protection should be provided in the following areas:

(i) In areas where lightning can occur often.
(ii) Buildings located in exposed areas.
(iii) Height of building is more compared to the surrounding buildings and places.

The lightning-protection system consists of an unbroken chain of conductors from the roof of a building to the ground. This provides an easy path for the heavy electrical power released by the lightning to discharge to the earth in the shortest time possible.

The conductor should be pure copper. The conductors should be of shortest length without sharp bends, kinks, etc. The area of influence of a lightning conductor is assumed to be a cone with the top most point of the conductor as the apex and a radius related to the height of the apex. This radius may be taken as equal to the height of the conductor on a safe side.

29.6 EARTHQUAKE-RESISTANT BUILDINGS

29.6.1 Causes of Earthquakes

Earthquakes may be caused by natural reasons or due to man-made activities. Natural causes are tectonic forces or volcanic eruption and man-made activities such as reservoir-associated forces.

1. Tectonic Earthquakes

Earthquakes are mainly caused due to sudden movement along faults which in turn due to tectonic origin. Such earthquakes generally result from sudden yielding to strain produced on the rocks by accumulation of stresses. Because of this the rock break along the weakest plane or otherwise and produces relative displacement of the rocks. Along the fault-planes the movement occurs after overcoming the frictional resistance along the fault-plane. Earthquakes due to fault line failure is an established fact (Parbin Singh, 2012).

2. Volcanic Earthquakes

Earthquakes associated with volcanoes are more localised. Compared to failure along faulting planes, the extent of damage and the intensity of wave produced are generally less. Volcanic earthquakes may be caused due to one of the following mechanisms:

(i) Explosion of volcano may take place due to the relax and expansion of gases and lavas.
(ii) Faulting may also occur within a volcano and thereby causing high pressures in the chamber of molten rock.
(iii) Centre of volcano may collapse and thereby extrusion of gases and molten matter.

3. Reservoir-associated Earthquakes

Only during the second half of the twentieth century, a new class of earthquake associated with reservoir has been recognised. It is believed to have caused due to impounding of water in artificially created reservoirs. Areas which were region of seismic activity (discussed elsewhere) have shown sign of disaster due to earthquake. Seismic shocks associated with filling of water in reservoirs have also been recorded in different parts of the world.

Reasons for such earthquakes have been identified due to (i) Sagging effect of the load and (ii) Increased pore pressures (Parbin Singh, 2012).

29.6.2 Magnitude of Earthquake

Magnitude of an earthquake is a measure of the amount of ground shaking based on the amplitude of elastic wave it generates. Richter's magnitude scale, named after Prof. Charles Richter, a geologist is most often used. The Richter scale starts from 2, and there is no upper limit. Table 29.4 gives the description of an earthquake in relation to its magnitude on the Richter scale.

Table 29.4 Magnitude of an earthquake

Magnitude	Description
> 8.0	• Great earthquake
7.0 – 7.9	• Major earthquake
6.0 – 6.9	• Strong earthquake
5.0 – 5.9	• Moderate earthquake
3.0 – 5.0	• Minor earthquake

The Richter scale is a logarithmic one; that is, an earthquake of magnitude 4 causes 10 times as much ground movement as one of magnitude 3, 100 times as much as one of magnitude 2, and so on. The Richter scale is widely used throughout the world.

Seismograph is an instrument designed to record earth motion set up by seismic waves. The actual record of motion produced by a seismograph is called a seismogram. Seismograph is designed to record both the horizontal and vertical component of ground motion.

29.6.3 Seismic Zones of India

Varying geological conditions at different locations of the country may have at any time damaging earthquakes to occur. Thus there is a need for seismic zone map of the country so as to design structures taking into effect the magnitude of earthquake likely to occur at a particular location.

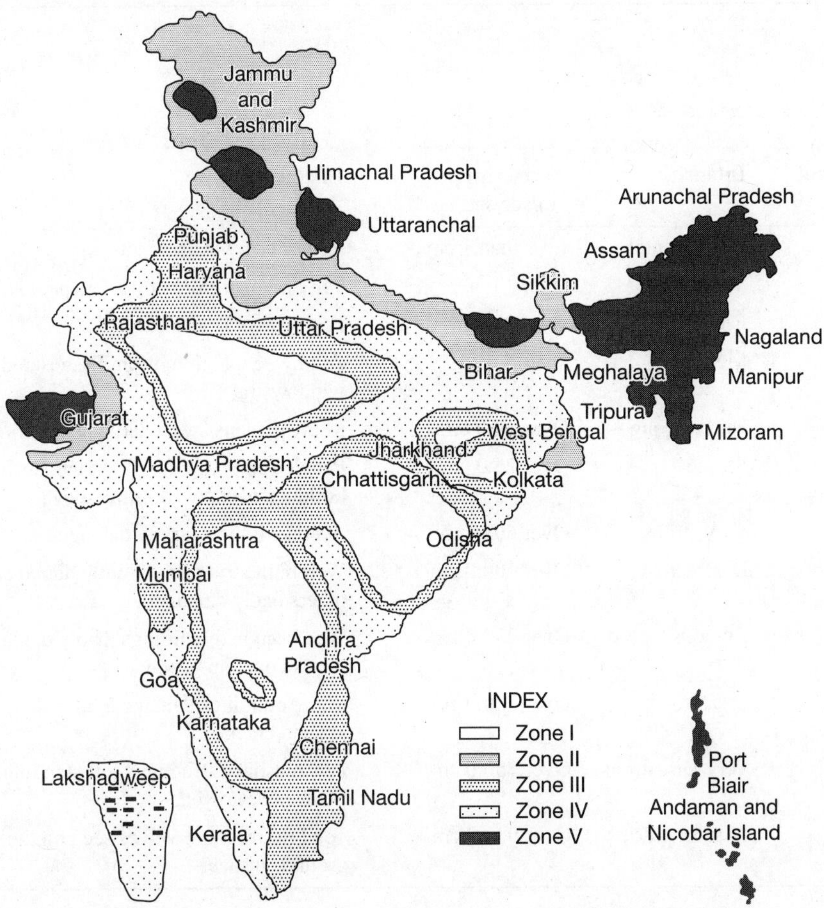

Figure 29.6 Location of epicentre from travel-time records (*Source:* IS: 1893–1984)

The zone map (IS: 1893–1984) sub-divides India into five zones, I, II, III, IV and V (Fig. 29.6). The corresponding intensity and acceleration are shown in Table 29.5 which is based on Mercali scale. Mercali scale is shown in Table 29.6.

Table 29.5 Intensity of earthquake

Symbol	Intensity	Acceleration cm/s/s	Effects
I	Instrumental	Less than 1	Recorded by instruments only.
II	Very feeble	Over 1	Felt on upper floors by a few people.
III	Feeble	Over 2.5	Felt by people at rest.
IV	Moderate	Over 5.0	Felt by people in motion. Dishes and windows rattle.
V	Fairly strong	Over 10.0	Many persons awakened. Dishes broken.

Table 29.6 Mercali scale

Symbol	Intensity	Acceleration cm/sec/sec	Effects
I	Instrumental	Less than 1 cm	Recorded by instruments only.
II	Very feeble	Over 1 cm	Felt on upper floors by a few people.
III	Feeble	Over 2.5 cm	Felt by people at rest.
IV	Moderate	Over 5.0 cm	Felt by people in motion. Dishes and windows rattle.
V	Fairly strong	Over 10.0 cm	Many persons awakened. Dishes borken.
VI	Strong	Over 25.0 cm	People run outdoors. Slight damage to buildings. Heavy furniture moved.
VII	Very strong	Over 50.0 cm	Average homes lightly damaged.
VIII	Destructive	Over 100.0 cm	Well built structures slightly damaged, others badly damaged.
IX	Ruinous	Over 250.0 cm	Well designed buildings, badly damaged. Cracks open in ground.
X	Disastrous	Over 500.0 cm	Many buildings damaged, rails bent, ground cracked.
XI	Very Disastrous	Over 750.0 cm	Few masonry structures left standing. Bridges destroyed.
XII	Catastrophic	Over 980.0 cm	Total destruction. Waves seen on the ground surface.

Seismic zone maps are to be revised periodically with the better understanding gained with time. For instance, the Koyna earthquake classified under Zone I in 1966 was changed to Zone IV in 1970.

Epicentre is the point on the earth's surface vertically above the focus of an earthquake. Shaking is highest at the epicentre and gradually decreases outwards. The difference in primary waves (P – waves) and secondary waves (S – waves) may be used to determine the epicentre.

29.6.4 Effects of Earthquake on Structures

During an earthquake, ground motions occur in a random fashion, both horizontally and vertically, in all directions, radiating from the epicentre. The ground accelerations cause structures to vibrate and induce inertial forces on them. Hence, structures to be constructed in earthquake-prone areas need to be suitably designed and detailed to ensure stability, strength and serviceability.

The magnitude of the forces induced in a structure due to a given ground acceleration will depend, amongst other things, on the mass of the structure, the material and type of construction, the damping, the ductility and energy dissipation. Ductility of a material is the ability of a structure or member to undergo inelastic deformations beyond the initial yield deformation with no decrease in the load resistance.

Thus by enhancing ductility and energy dissipation capacity in structure, the induced seismic forces can be reduced and also the probability of collapse reduced (Pillai and Menon, 2012). Further, it is desirable to avoid discontinuities in mass or stiffness in plan or elevation. Torsional effects should particularly be accounted for in buildings with asymmetry in plan or elevation.

29.6.5 General Requirement

Bureau of Indian Standard have specified the minimum design requirements for earthquake–resistance design in IS codes: 1893 (Part I): 2002, IS: 4326: 1993 and IS: 13920: 1993. These requirements have been stipulated after considering the characteristics and probability of occurrence of earthquakes, the characteristics of the structure and the foundation and the amount of damage that is considered tolerable. Codal provisions from other countries are also available.

The criteria adopted by codes for fixing the level of the design seismic loading are generally as follows:

1. Structures should be able to resist minor earthquakes without damages.
2. Structures should be able to resist moderate earthquakes without significant structural damage, and
3. Structures should be able to resist major earthquakes without collapse, but with some structural and non-structural damages.

29.6.6 Major Design Considerations

1. Objective

Bureau of Indian Standards IS: 13920: 1993 recommends for special design to ensure adequate toughness and ductility (with ability to undergo large inelastic reversible deformation) for individual members such as beams, columns and walls and their connections and to prevent other non-ductile types of failure.

As a general rule, to maintain overall ductile behaviour of structure with minimal damage, it is necessary to provide the following combinations (Pillai and Menon, 2012):

1. Strong foundations and weak superstructure.
2. Members stronger in shear than in flexure.
3. Strong columns, and beams with little over-strength.

2. Means of Providing Ductility

Some of the main design considerations in providing ductility include:

1. Using a low tensile steel ratio (with relatively low grade steel) and/or using compression steel.
2. Providing adequate stirrups to ensure that shear failure does not precede flexural failure.
3. Confining concrete and compressions steel by closely spaced hoops or spirals, and
4. Proper detailing with regard to connections, anchorage, splicing, minimum reinforcement, etc.

3. Requirements of Stability and Stiffness

Under a severe earthquake, large lateral deformation and oscillations are induced resulting in formation of reversible plastic hinges at various locations. Thus a structural system should be designed to ensure that the formation of plastic hinges at suitable locations may, at worst condition, result in the failure of the individual element rather than progressive collapse.

Apart from the stability, the structure should have sufficient stiffness to limit the lateral deflection or drift. As per code the inter-storey drift is to be limited to 0.004 times the storey height to account for stiffness.

4. Requirements of Materials

As mentioned earlier use of relatively low grade steel is recommended. Further, lower the grade of steel, higher is the ratio of the ultimate tensile strength (f_u) to the yield strength (f_y). A high ratio of f_u/f_y is desirable, as it results in an increased length of plastic hinge and thereby an increased plastic rotation capacity.

For all buildings, which are more than three storeys, in height, have to use M20 as a minimum grade of concrete. Low density concrete lead to poor performance under reversed cyclic loading, whereas very high strength concrete is associated with lower ultimate compressive strain which adversely affects ductility.

5. Foundation

It is most important in the design to ensure that the foundation of a structure does not fail before the possible failure of superstructure. The maximum seismic forces transmitted to the foundation shall be governed by the later loads at which actual yielding takes place in the structural elements transforming the later loads to the foundation. Thus to ensure a safe foundation, it has to be ensured the foundation is stronger than the superstructure. Such a design concept is necessary to provide for ductile behaviour of the superstructure without serious damage to the foundation.

6. Flexural Yielding in Frames and Walls

As reinforced concrete is less ductile in compression and shear, dissipation of seismic energy is best achieved by flexural yielding. Thus it is necessary to avoid weakness of structure in compression and shear in relation to flexure.

In a structure composed of ductile movement-resisting frames and/or shear (flexural) walls, the desired inelastic (ductile) response is developed by formation of plastic hinges (flexural yielding) in the members, as shown in Fig. 29.7.

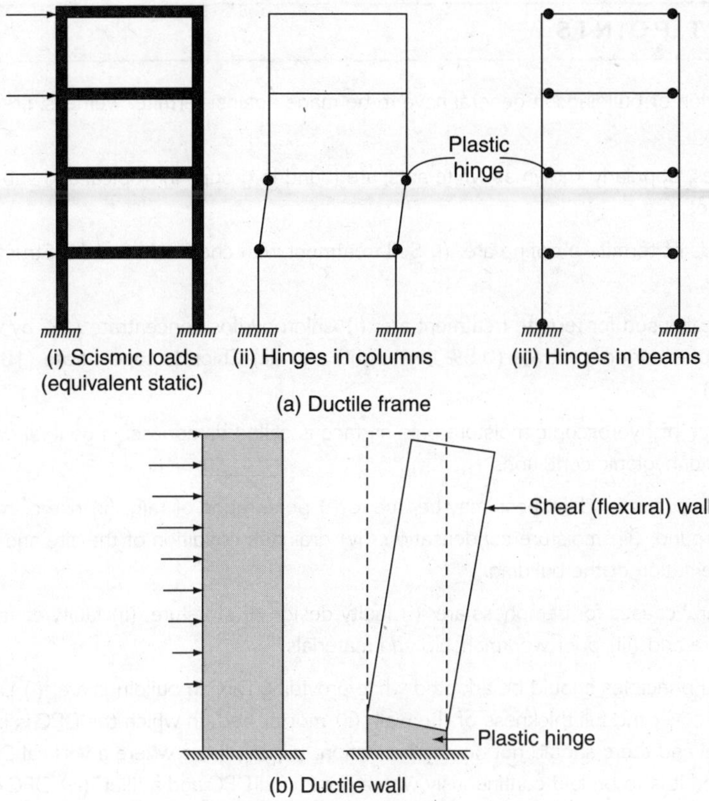

Figure 29.7 Formation of plastic hinges in a ductile structure

In ductile frames, plastic hinges may form in the beams or in the columns (Fig. 29.7a). It is always desirable to design the frame such that the plastic hinges form only in the beams rather in columns. The reasons for such a condition are as follows:

1. Plastic hinges in beams have larger rotation capacities than in columns.
2. Mechanisms involving beam hinges have larger capacity – absorptive capacity on account of the larger number of beam hinges (with large rotation capacities) possible.
3. Eventual collapse of a beam generally results in a localised failure, whereas collapse of a column may lead to a 'global' failure, and
4. Columns are more difficult to straighten and repair than beams, in the event of residual deformation and damage.

Ductility and strength assessment of an entire structure requires non-linear analysis, considering material and geometric non-linearities.

SALIENT POINTS

1. Protection of buildings in general have to be made against termite, wetness, fire and lightning.
2. Termites, popularly known as white ants, are found in groups in tropical and sub-tropical countries.
3. Methods of termite-proofing are: (i) Soil treatment with chemicals and (ii) Structural barriers.
4. Chemicals used for termite treatment are: (i) Chloropyrifos concentrate (1% by weight), (ii) Heptachlor concentrate (0.5% by weight) and (iii) Chloride concentrate (10% by weight).
5. Presence of hydroscopic moisture on a surface is called dampness. In general dampness causes unhygienic conditions.
6. Natural causes for dampness may be due to (i) penetration of rain, (ii) rise of moisture from ground, (iii) moisture condensation, (iv) drainage condition of the site and (v) orientation of the building.
7. Structural causes for dampness are: (i) faulty design of structure, (ii) faulty construction of structure and (iii) poor workmanship and materials.
8. General principles should be adopted while providing DPC in buildings are: (i) DPC should cover the full thickness of the wall, (ii) mortar bed on which the DPC is laid should be level and there should not be any projection, (iii) In places where a vertical DPC is provided it is to be laid continuously with a horizontal DPC and a fillet, (iv) DPC course should be continuous and should form as a bearer from the entry of moisture and (v) DPC should not be exposed in total.
9. Materials used for DPC are flexible materials like hot bitumen, bituminous felts, bituminous sheets, polythene sheets, metal sheets of lead, copper, etc.; semi-rigid materials like mastic asphalt or combination of materials or layers; and rigid materials like first-class bricks, stones, slates in courses and cement-concrete or mortar layers, etc.
10. Combustible materials are the materials which combine exothermally with oxygen and give rise to flame at a particular range of temperature.
11. Non-combustible materials are those which when decomposed by heat will do so endothermically.
12. The amount of heat liberated in combustion of any content or part of a building of a floor area is referred to as fire-load.
13. Fire-load is the ratio of the weight of all combustible materials (by their respective calorific values) to the floor area under consideration.
14. General measure of fire safety to be adopted are: (i) Alarm systems and (ii) Fire-extinguishing arrangements.
15. Lightning protection should be provided (i) in areas where lightning can occur often, (ii) if building is located in exposed areas and (iii) if height of building is more compared to the surrounding buildings and places.

16. Earthquakes may be caused by natural reasons or due to man-made activities. Natural causes are tectonic forces or volcanic eruption and man-made activities such as reservoir-associated forces.
17. Earthquakes are mainly caused due to sudden movement along faults which is due to tectonic origin.
18. Earthquakes associated with volcanoes are more localised.
19. Reservoir-associated earthquakes have been identified due to (i) sagging effect of the load and (ii) increased pore pressures.
20. Magnitude of an earthquake is a measure of the amount of ground shaking based on the amplitude of elastic wave it generates.
21. Richter's magnitude scale starts from two and there is no upper limit.
22. Epicentre is the point on the earth's surface.
23. Seismic map of India is based on Mercali's scale.

REVIEW QUESTIONS

1. Give the name of chemicals used for anti-termite treatment.
2. Describe the general principles of termite treatment.
3. What are the methods adopted for termite-proofing?
4. What is dampness in buildings?
5. Discuss the natural causes for dampness. What are their effects?
6. What are the general principles to be observed in all the methods of damp-proofing?
7. Explain the requirements of an ideal damp-proofing material.
8. What are the materials used for damp prevention?
9. Prepare a list of materials which are commonly used for a damp-proofing course. Briefly explain each of them.
10. Briefly discuss the methods generally adopted in preventing the defects of dampness.
11. What is integral damp-proofing treatment?
12. It is intended to construct a basement of 4 metre height and of 5 m × 8 m in size. There is a likelihood of dampness occurring in the inside of the basement. Explain briefly the various methods of damp prevention in this situation if the walls are to be brick masonry and concrete floor.
13. What do you understand by fireproof construction?
14. Discuss fire-resisting properties of different materials.
15. Define fire-load.
16. Discuss grading of occupancies by fire-load.
17. What is fire-grading? Explain.

18. Bring out the special measures to be adopted for safety against fire in case of theatres.
19. Enumerate general safety requirements against fire.
20. Briefly explain the emergency fire safety measures to be adopted.
21. What steps do you take to protect a building from lightning?
22. How earthquakes are caused?
23. Explain reservoir-associated earthquake.
24. What is Richter scale? How is it measured?
25. Explain the earthquake zones of India.
26. Explain the major design consideration to protect buildings from earthquake.

30 PREVENTIVE MEASURES AND MAINTENANCE OF BUILDINGS

30.1 INTRODUCTION

Primarily it is needed to take preventive measures wherever feasible before facing any failure in small or large scale. A small investment or preventive measure taken at the initial stage not only prevents failure but also gives a long life of the building.

Then maintenance plays a major role to make every component of a building to function properly and at the same time increases the lifespan of the building.

Maintenance of floors, doors and windows, sanitary appliances, water supply, drainage system and electrical system are all important.

Further, checking of leakages, identifying the causes of cracks and repairing them or make some preventive measures to avoid such fractures in future. Special repairs, if any, are to be attended to then and there. Apart from the routine maintenance annual general maintenance of a building should be made and implemented.

30.2 MAINTENANCE OF FLOORINGS

As there are many types of floors constructed, maintenance of some of the important floors are dealt below.

30.2.1 Terrazzo Floors

Cleaning of terrazzo floors with water may be taken up after 3 months of laying. After this period, it should be swabbed daily using clean water and a clean rough swab (floor cloth) which should be rinsed in water frequently so as to avoid dirt being back on to the floor. The floor is then allowed to dry in air. After this initial cleaning, ordinary swabbing shall keep the floor shining.

Sometimes acid polishing of terrazzo floors is resorted to after machine polishing during the first laying of the floor. If need be, this can be repeated. Oxalic acid is used for acid cleaning. After acid polishing, the following day the floor is wiped with a moist rag and washed clean with water.

30.2.2 Marble Floors

For marble floors daily maintenance is to be made. This is carried out by mopping with mild detergent in water. Marble gets easily stained and scarred. If there is a spill on the floor, it should be cleaned immediately.

Stubborn stains should be removed based on the type. If the stain is from grease, a paste of chalk dust or whiting with acetone. This mixture is applied on the stain and allowed to stand overnight. The mixture is sponged off and the treated area is buffed. If the stain is from any organic source like tea, fruit juice, etc., chalk powder with hydrogen peroxide mixture is used. This paste can also remove stains due to rust. Such stains can also be removed by dry borax and damp cloth followed by warm water rinsing.

30.2.3 Linoleum Floors

For this floor mopping with kerosene oil can pick up all the dirt. After the removal of dirt, the floor may be wiped with water and mopped well to set a polished surface.

30.2.4 Concrete Floors

A properly constructed concrete floor should be dust free under normal use. Dusts in floors can be reduced by the application of a hardener, or by waxing, or by painting it with a solution of oil spirit. For such application, raw or boiled linseed oil mixed with an equal amount of turpentine has been found to be effective. Painting with epoxy, acrylic or chlorinated rubber base is also used.

Wax polishing with proprietary polish is also popularly used. This polish may be made at site by mixing a 25% solution of melted bees wax in turpentine together with pigment at the rate of 60% by weight of bees wax. The wax is applied on the floor which hardens after 12 hours and the surface is sprinkled with French chalk and polished with dry cloth. About 200 gm of bees wax shall cover an area of 10 m^2.

If the floor has to be roughened, it is usually done by washing the floor with dilute hydrochloric acid. The dilute acid is applied at the rate of 0.3 l/m^2 of floor area. The acid is left for 15 minutes and then brushed. Then the floor is thoroughly cleaned with water. If need be three successive applications may be given.

30.2.5 Other Floors

All stone floors may be maintained by mopping with mild detergent and water at regular intervals.

Ceramic tile floors are easy to maintain. Soap and water can be used to clean the floors and special care has to be taken to clean the joints. Stains, if any, may be removed as explained in marble floors.

Wooden floor is to be cleaned and kept bright by polishing with bees wax or ready-made wax polish.

30.3 MAINTENANCE OF DOORS AND WINDOWS

In the maintenance of doors and windows both wood work and glass work should be taken care of.

30.3.1 Maintenance of Woodwork

Woods in general get deteriorated with time due to water and direct exposure to sun. However, teak wood may need less maintenance. Other woods should be periodically painted and kept clean. With moist condition both dry and wet rots grow which will be worst in toilet room. This can be prevented by providing adequate ventilation and painting.

Bottom of bath room doors have to be protected by painting with waterproof paints. White putty glazing is used for ordinary wood and in teakwood glazing with beads is usually carried out.

30.3.2 Maintenance of Glass

Glasses fitted to doors and windows are kept clean using clean tepid water. In order to remove dirt and stains cleaning liquids available in market may be used. Instead a cleaning can be made by mixing two tablespoons of household ammonia or white vinegar with 1.25 *l* of water. This is sprayed using a sprayer. Cleaning has to be taken up from top to bottom and the glass is dried by a cotton swab followed by a crumbled newspaper. One side has to be wiped horizontally and the other side vertically.

30.4 MAINTENANCE OF WATER SUPPLY SYSTEM

Sumps and overhead tanks are to be in closed conditions and should be cleaned periodically. Exposure of sump and overhead tanks lead to mosquito breeding and formation of algae.

Taps and other fixtures should be kept in working condition so as to avoid wastage of water. They should be repaired wherever they become defective. Taps, chrome fittings, basins, etc., can be cleaned with liquid ammonia in water. Material for polishing of metals are also available in market.

30.5 MAINTENANCE OF SANITARY APPLIANCES

The main problems with sanitary applications like water closet, washbasins, etc., are the growth of fungus and algae. A solution of chloride of lime (bleaching powder) in hot water will remove surface stains from ceramic ware and tiles. Simple application of moistened cloth in hot water and a little paraffin will effectively restore the lustre of porcelain and glazed surfaces. There are ready-made cleaning materials which are also available in market. If the sewage is to be treated by a septic tank, chemicals and detergents should not be used as they may kill the bacteria in the septic tank.

30.6 MAINTENANCE OF DRAINAGE SYSTEM

All the soil and waste-disposal system must be kept clean and should be in working condition. Otherwise nuisance and unhealthy conditions prevail. Main item of the drainage system to be taken care are discussed below (Varghese, 2012).

30.6.1 Fittings Inside the Buildings

Water taps, wastewater pipes, traps in fittings, grating above floor trap, etc., should be periodically checked. This is done by pouring water on them and examining their performance. Otherwise they are liable to be clogged if the passage of water is not with adequate force. Gratings at bathrooms have to be periodically checked for clogging due to hair, oil, etc. Blocking if any may be removed using drain cleaning compounds or manually wherever feasible.

30.6.2 Waste and Soil Pipes

Waste pipes and soil pipes which are fixed on the walls are to be checked periodically. Wherever feasible *PVC* pipes may be used with balloons on the top should be checked and replaced if necessary.

30.6.3 Gulley and Manholes

Before the onset of monsoon, gulleys, manholes, manhole covers and drain pipes are to be examined. In order to prevent backflow from street sewer, the outlet to the sewer should be checked. By chemical sprays, cockroaches are to be destroyed and stagnation of water should be avoided.

30.6.4 Roof Drainage System

Before the rainy season, the roof drainage system should be inspected. Horizontal and vertical drainage systems are to be cleaned and leakages repaired. Rectification should be attended to, particularly horizontal drain pipes, prevent blockage of wastewater.

30.6.5 Septic Tanks

Grease, slow decomposing matter, etc., should be avoided in the drains for the efficient working of a septic tank. Further drain-cleaning chemicals should be avoided which can kill the bacteria in the septic tank.

When the depth of the sludge and scum exceed half the depth of the tank, digested sludge should be removed. A portion of the sludge should be left in the tank to act as a seed to the fresh sewage. If the action in septic tank is dull ¼ kg of brewer's yeast is added to the tank to hasten the action. Signs of a defective system is to be noticed when foul colours appear from the drains or the tank and growth of lush vegetation over the tank. If water backs up in the drain, it is to be realised that clogging by sludge and scum has happened.

30.7 MAINTENANCE OF PAINTWORK

Washable paints can be cleaned with clean water. Most of the paints, viz., soda, soft soap and other alkaline substances are injurious. In order to remove stains, they should be highly diluted and removed off the paint with clean water as soon as the dirt is removed.

Re-paint of surfaces have to be done before the old paint disintegrates. Quality of a paint contributes for the life of the work. Frequent painting of external surfaces may be prevented by use of special exterior paints. Also special paints may be used to paint the underside of the wet area, such as bath rooms and lavatories if wall tiles are not provided.

All steel works are to be carefully observed. They should be regularly cleaned with clean water. Rusted parts, if any, should be removed and re-painted. Special attention are required if steel windows are used in buildings.

30.8 MAINTENANCE OF BRICKWORK

Cracks are frequently found in brick masonry wall. Causes for such cracks have been discussed in Section 17.2. It has been discussed that prevention of cracks in brick masonry may be avoided by proper foundation design and providing different joints such as expansion joints, isolation joints, sliding joints, etc.

Maintenance of brickwork involves in cleaning, removal of efflorescence, repairing of cracks, re-painting brickwork and re-painting old brick joints.

30.8.1 Cleaning Brickwork

Brickwork has to be cleaned with steam or steam and hot water jets. This treatment is suitable for fine-textured and hard-burnt bricks. Sand blasting can also be done. However, this disturbs the texture of the bricks and absorbs the painting.

30.8.2 Removal of Efflorescence in Brickwork

Soluble salts after drying get deposited on the surface of the walls as white layer. These salts may be present in the material of the walls and come out. Sources of these salts are mortars and plasters. As moisture is the cause for the creation of efflorescence, this can be checked by proper damp prevention.

Efflorescence can be removed by scrubbing those portions on the wall with water and a tough brush. For effective use instead of water, 10% of solution of muriatic acid may be used. After this treatment, the wall has to be rinsed with clean water or with a weak ammonia solution.

30.8.3 Repairing of Cracks in Brickwork

Cracks in brickwork are rectified by adopting the following procedure:

1. If the cracks are less than 1.5 mm in width, the cracks may be filled with commercially available putty.
2. Wider cracks are repaired by releasing them out to about 25 mm and re-filling with 1 cement, 2 lime, 9 sand mortar or any crack-filling material available in the market.
3. Still wider cracks may have to be stitched by using reinforced mortar or concrete stitching blocks.

30.8.4 Re-pointing Old Brickwork

In the case of brickwork which are pointed may be needed to re-point in order to improve the appearance and to make it water tight. Generally mortar at the pointed portions gets lost due to rain, wind, freezing, etc. Following procedure may be followed for re-pointing:

1. The joint of the old mortar is loosened and scrapped out and cleaned for a depth of at least 3 mm and wetted with water.
2. New mortar is applied to the joints with special trowel and pointed as per practice.

30.8.5 Re-painting Brickwork

Re-painting is necessary for walls which have been previously painted. The type of paint and colour should be decided and accordingly the surface has to be prepared.

30.9 MAINTENANCE OF STONEWORK

Maintenance of stonework involves in removing efflorescence and stains, repair of cracks and waterproofing.

30.9.1 Removal of Efflorescence in Stonework

Appearance of efflorescence occurs commonly due to use of certain type of mortars. Removal of efflorescence has to be done as done in brickwork. Stones are to be kept saturated with water to remove the action of acids.

30.9.2 Removal of Stains in Stonework

Removal of stains can be done easily if its origin is known. Iron stains show rusty appearance which can be removed by rinsing the area with a solution of ½ kg oxalic acid in 5 *l* of water. After 2–3 hours, the surface should then be scrubbed with brush and clean water.

Darker stains may be removed with six parts of water, one part of sodium citrate and one part of glycerine, made into a paste with adequate whiting. This paste is applied on the surface and then kept for a few days.

For very deep and dark stains, one part of sodium citrate is mixed with six parts of water. This solution is applied on the surface of stains. This is followed by covering the surface with a thin layer of sodium hydro-sulphite crystals and removed after an hour. This may be repeated if needed.

30.9.3 Repair of Cracks in Stonework

Repairing of cracks has to be attended by ensuring that the settlement causing the crack has ceased. This generally takes about one year for the complete settlement.

Small cracks are cleaned with a wire brush and a thick paste of cement mix is forced into the crack. In order to get a firm key for the mortar, large cracks should be raked. An inverted V groove is made with a minimum of 1 cm depth. A cement mortar of 1 : 2 ratio is applied with less water. The mixture is forced into the crack to get a firm bond. Along with this mixture expansive agents are added to improve the bond.

30.9.4 Waterproofing of Stonework

In order to keep the stone masonry free from damp, efflorescence, frost action, etc., colourless waterproofing materials is applied waterproofing materials obtained from heavy petroleum distillates, oils or in soluble soaps are best. It is difficult to waterproof very closely textured stones. Waterproofing materials substances containing resins are not suitable for fine-textured stonework. Waterproofing causes some dislocation which goes from the stones in course of time.

30.10 CRACKS AND REMEDIAL MEASURES IN CONCRETE

Causes of cracks, remedial measures and repair of cracks in concrete are discussed.

30.10.1 Causes of Cracks in Concrete

Cracks in concrete may be plastic shrinkage cracks, settlement cracks, cracks due to bleeding, cracks due to delayed concrete casting and cracks due to construction effects are discussed in Section 8.11.

30.10.2 Preventive Measures for Crack in Concrete

Preventive measures for cracks in concrete are discussed in Section 8.11. Apart from them, joints in concrete structures are provided to continue a specific work or prevent cracks due to change in temperature.

The following two types of joints are usually provided in concrete structures:

(i) Construction joints
(ii) Expansion and contraction joints

1. Construction Joints

This type of joints are provided at the location where the construction is stopped at the end of day's work or for any other reason so as to bridge the old work and the new work by a proper bond. Such a situation generally occurs when large concrete work has to be executed which cannot be done within a day.

If the work is well planned such that the day's work is to be stopped at an expansion or contraction joint, in such a case there is no need for a construction joints.

The construction joints may be vertical, horizontal or inclined depending on the type of structure. In the case of inclined or curved members the joint should be perpendicular to the axis of the structural member. The position of the construction should be well planned and constructed keeping in view the stability of the structure. Construction joints may be provided for structures which are vertical, horizontal or inclined (Fig. 30.1).

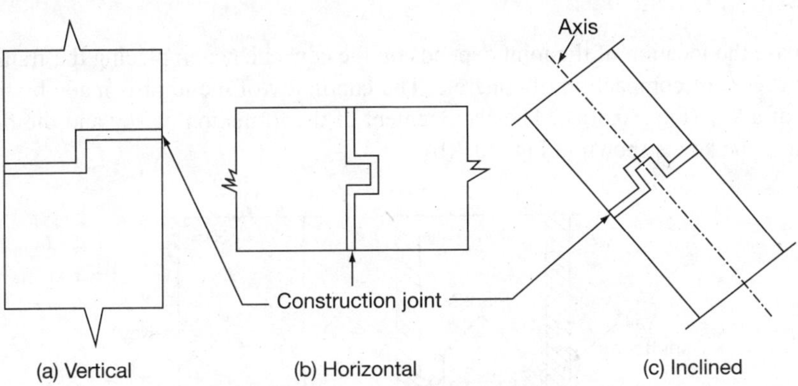

Figure 30.1 Construction joint in different positions

Depending on the type of concrete structures the following factors should be considered.

(i) Columns

Columns should be concreted to a few centimetres below its junction with the lowest soffit of the beam. Construction above joint should be at least 4 hours after the completion of the joint. Care should be taken such that the construction joint is at the location of least bending moment.

(ii) T or L Beams

The ribs of T or L beams are first concreted and then slabs forming the flanges are concreted up to the centre of the rib (Fig. 30.2). In certain situations, if a construction joint has to be provided between the slab and beam, the rib of the beam is concreted up to 25 mm below the level of soffit of slab and the joint should be located at that level.

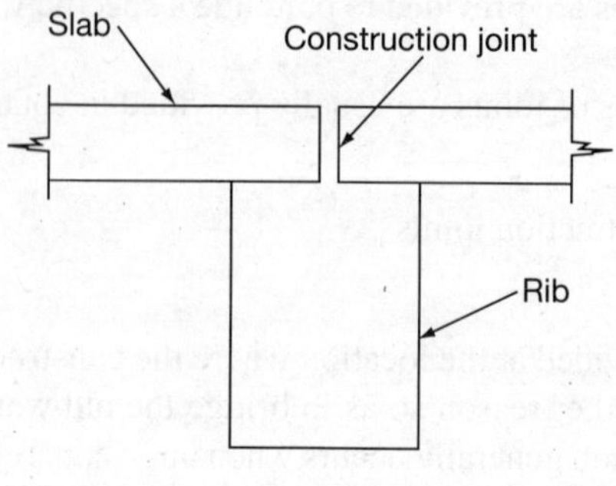

Figure 30.2 Construction joint on T-beam

(iii) Simply Supported Slabs

In slabs supported on two sides the construction joint should be vertical and parallel to the main reinforcement. Alternatively the joint may be provided at the middle of the span perpendicular to the main reinforcement. For two-way slabs the construction joint is provided near the middle of either span.

(iv) RCC Wall

In this case the location of the joint depends on the convenience in placing the framework and the access of compaction of concrete. The continuity of the joint is made by the formation of a key (Fig. 30.3(a)). The arrangement of the formation of key and the finished position of the key is shown in Fig. 30.3(b).

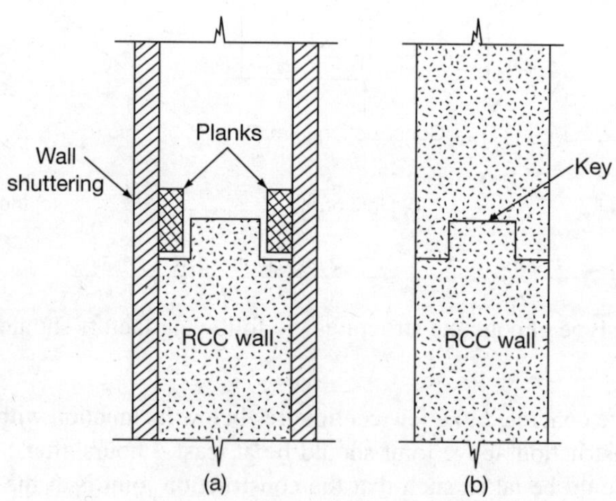

Figure 30.3 Construction joint in RCC Walls

2. Expansion and Contraction Joints

These joints are provided in all concrete structures when the length exceeds 12 m. These joints are made to satisfy the following requirements:

(i) To accommodate for changes in volume of concrete due to temperature.
(ii) To retain the appearance by maintaining the same shape of the concrete structures.

The joints are filled with some elastic materials like filter or dowels of keys. The quality of the filter should be in a position to withstand cold weathers. It should be compressible, cellular and not brittle. The Conventional materials used as filter are strips of metal, bitumen-treated felt, cane fibre-board, cork, soft wood, etc.

Provision of dowels or keys is to transfer the load. The movement due to shrinkage is controlled by the contraction joint. The contraction joint may be complete or partial. In the complete contraction joint there is complete discontinuity of both concrete and steel (Fig. 30.4). In the partial contraction joint, there is a discontinuity of concrete but the reinforcement bars are continued across the joint (Fig. 30.5).

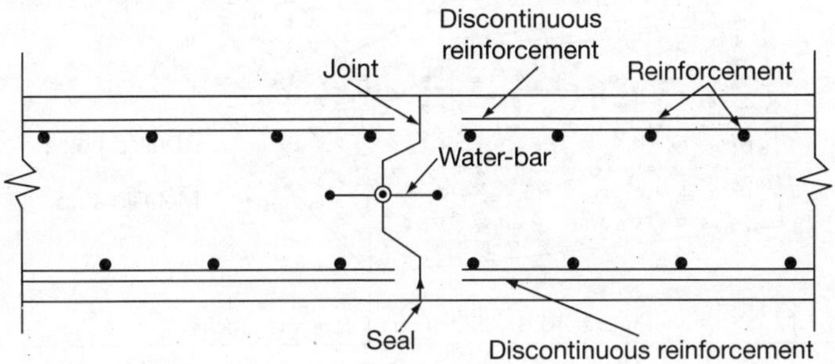

Figure 30.4 Complete contraction joint

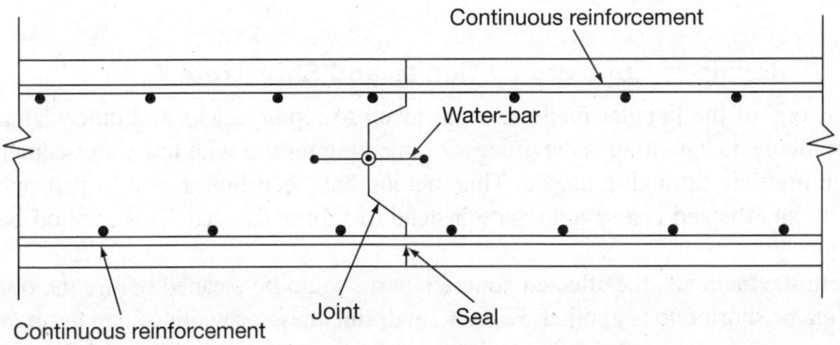

Figure 30.5 Partial contraction joint

30.10.3 Stitching of Cracks in Concrete Members

At times cracks do develop in concrete members due to several reasons such as design of construction deficiency, problems with concrete, disturbance at early stage, etc. If such cracks are localized in nature they can be repaired either by epoxy grouting or by stitching. The choice of either of them depends on the discretion of the concerned engineer and the intensity of the cracking (Gambhir, 2010).

Stitching consist of drilling holes on both sides of the crack, and inserting U-shape metallic strips (stitching dogs) spanning across the cracks. The bonding of the metallic strips with parent concrete can either by with non-shrink grout or epoxy-based bonding system. It is preferable if the metallic strips are variable in length and orientation with closer spacing at the ends of cracks (Fig. 30.6).

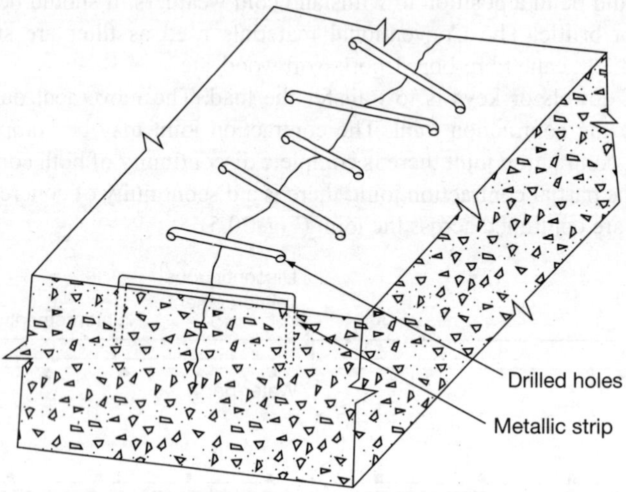

Figure 30.6 Stitching of concrete cracks

However, stitching does not close a crack, but only prevents it from further widening. Such stitches, when encased with an overlay have been reported to be performing well in several cases.

30.10.4 Repair of Concrete by Gunite and Shortcrete

Gunite is one of the popular methods undertaken to repair cracks and other damages in concrete members. Guniting is the process of injecting mortar with low water-cement ratio at a high pressure through a nozzle. This method has been improvised in that only concrete with small-sized coarse aggregate instead of mortar is used. This method is called shortcrete.

In both the methods, the affected concrete part should be cleaned before the operation and gunite or shortcrete is applied. Former small thickness or width of gap gunite is used whereas shortcrete is recommended for larger thickness. The materials used for gunite are cement, sand and water and for shortcrete are cement, sand and small aggregate and water. For quick setting of gunite, accelerators are also added.

30.11 CHECKING BUILDING LEAKAGE

Although adequate care has been taken by providing weatherproofing of roofs, leakage may occur with time due to lack of proper maintenance. Leakage of roof due to rain water may spoil the materials stored in the room, may damage the electrical circuits, may wet the walls, etc. Leakages may occur in different types and different parts of a roof which are discussed below.

30.11.1 Leakage in Sloping Roofs

All parts of sloping roof are to be checked for leakage. It is difficult to repair after construction, particularly tiled roofs. It is generally recommended to apply waterproofing coating at the construction stage itself. Tiled roofs leak during heavy rain due to lack of this precaution.

Other sloping roofs like folded plates, shells, etc., are made out of concrete. These roofs would not have been compacted properly due to sloping surface. During normal rains there is no problem of leakage. During heavy rains, water may seep through the concrete and tend to leak. A cement coating may fill the porous concrete followed by a waterproofing coating may stop the further leakage.

30.11.2 Leakage of Flat Roofs

Following preventive measures and repairs may be undertaken to prevent leakage.

1. During construction stage itself, flat concrete roofs are to be cured perfectly. Any leakage noticed at this stage can be rectified easily. All surfaces can be given a coat of hot blown asphalt after curing.
2. Adequate slope may be given to the drain pipe joints such that stagnation of water is prevented.
3. Wherever the horizontal surfaces meet the vertical or inclined faces, fillets are to be provided. The gap formed with time has to be closed periodically.
4. All the entry points of the rain water should be checked and should not be blocked by leaves, wastes, etc.
5. Weathering courses, like flat tiles, laid on the roof should not crack. If cracks are found they should be closed immediate by grouting or otherwise.
6. Very porous concrete slabs may have to be grouted.

30.11.3 Leakage Through Junction of Walls and Sunshades

This is very common place where the leakage occurs which are to be checked and rectified as discussed below.

1. At the construction stage itself, proper slope and drainage arrangements are to be checked. Any architectural features made over sunshade should not give room for stagnation of water. A throating at the bottom end of sunshades shall make the water to drip properly.
2. In long sunshades there is a possibility of cracking at the ends. Such cracks make room to flow of water on the walls. These cracks should be filled with bituminous or elastomeric compounds. Top of sunshade should be kept clean always, particularly before the onset of monsoon.

30.11.4 Leakage Through Wall Cracks

All the external cracks found on walls should be deepened to 5 mm and filled with paintable sealant. In order to keep the external walls crack-free, waterproof cement or external paint may be used. Walls containing air-conditioning or other electrical connections may be painted with silicon paint to prevent wetting of walls during rains.

30.12 MAINTENANCE OF ELECTRICAL SYSTEM

It is better to have an ELCB trip switch in the electrical system of the buildings. This is helpful to show the earth leakage. All electrical switches should be kept clean, particularly ones at the kitchen which are liable to be clogged up by grits accumulated. In order to avoid menace of ants and white ants, it is better to keep anti-ant chemicals in the switch boxes. If the performance of bearings, condensers, regulators, etc., is not good, they have to be attended to. High amperage gadgets like heaters, air-conditioners, etc., are liable to get heated up and burnt. They should be plugged tightly or should be connected through fused outlets. It is advisable to have phase-changing devices in the electric supply system and thereby changing phases wherever there is a need. Savings in electricity can be made by the use of fluorescent lamps, LCD lamps for lighting and capacitance type fans.

SALIENT POINTS

1. Cleaning of terrazzo floors with water may be taken up after 3 months of laying. After this period it should be swabbed daily using clean water and a clean rough swab (floor cloth) which should be rinsed in water frequently so as to avoid dirt being back on to the floor. The floor is allowed dry in air. After this, initial cleaning by ordinary swabbing should keep the floor shining.
2. For marble floors daily maintenance is to be made. This is carried out by mopping with mild detergent in water. Stubborn stain should be removed based on the type.
3. For linoleum floor mopping with kerosene oil can pick up all the dirt. After the removal of dirt the floor may be wiped with water and mopped well to get a polished surface.
4. A properly constructed concrete floor should be slush-free under normal use. Dust in floors can be reduced by the application of a hardener, or by waxing, or by painting it with a solution of oil spirit.
5. Woods in general get deteriorated with time due to water and direct exposure to sun. With moist condition both dry and wet rots grow which will be worst in toilet room. This can be prevented by providing adequate ventilation and painting.
6. Glasses fitted to doors and windows are kept clean using clean tepid water. In order to remove dirt and stains cleaning liquids available in market may be used.
7. Sumps and overhead tanks are to be in closed conditions and should be cleaned periodically. Exposure of sump and overhead tanks lead to mosquito breeding and formation of algae.
8. Main problems with sanitary application like water closet, washbasins, etc., are the growth of fungus and algae. A solution of chloride of lime (bleaching powder) in hot water will remove surface stains from ceramic ware and tiles. Simple application of moist cloth in hot water and a little paraffin will effectively restore the lustre of porcelain and glazed surface.
9. All the soil and waste-disposal system must be kept clean and should be in working condition. Otherwise, nuisance and unhealthy conditions prevail. Main items of the drain systems to be taken care of are fittings inside the building, waste and soil pipes, gulley and manholes, roof drainage system and septic tank.
10. Washable paints can be cleared with clean water.

11. All steel works are to be carefully observed and rested parts should be removed and re-painted.
12. Cracks are frequently found in brick masonry walls. Prevention of cracks in brick masonry may be attained by proper foundation design and providing different joints.
13. Maintenance of brickwork is carried on by cleaning brickwork, removing efflorescence and repairing of cracks and re-painting.
14. Maintenance of stonework is carried out by removing efflorescence, removal of stains, repair of cracks, and waterproofing.
15. Cracks in concrete may be plastic shrinkage cracks, settlement cracks, cracks due to bleeding, cracks due to delayed concrete casting and cracks due to construction effects.
16. Preventive measures for cracks in concrete are by providing construction joints, expansion and contraction joints wherever needed.
17. Repairing of concrete work can be done by stitching or by gunite or shortctete.
18. Building leakages are checking (i) leakage in sloping roof, (ii) leakage of flat roofs and (iii) leakage through junction of walls and sunshade and leakage through wall cracks.
19. It is better to have an ECCB trip switch in the electrical system of the buildings. This is helpful to show the earth leakage. In order to avoid menace of ants and white ants, it is better to keep anti-ant chemicals in the switch boxes.

REVIEW QUESTIONS

1. What steps would you take to maintain terrazzo floors and marble floors?
2. How are concrete floors maintained?
3. State briefly the basic principles of maintenance governing the water supply in buildings.
4. Indicate the method you would recommend for maintenance of water closets, sinks, washbasins, etc., of a hospital building.
5. Enumerate the steps to be taken in the maintenance of electrical system.
6. What are the causes of cracks in the brick masonry? Explain.
7. Explain the prevention of cracks in brick masonry
8. Explain different types of joints.
9. How would you maintain brickwork?
10. How are efflorescence and stains in stonework removed?
11. How damages caused to concrete are rectified? Explain.
12. How joints provided in concrete help in the prevention of cracks and failures?
13. Mention the method of preventing leakages in sloping roofs.
14. Explain the prevention of leakages in flat roofs.
15. How leakages through junctions of walls and sunshades are prevented?
16. What are the steps to be taken to maintain electrical systems.

31 CONSTRUCTION PLANNING AND SCHEDULING

31.1 INTRODUCTION

Planning is the process of devising of a scheme for doing, making or arranging a project or programme before the commencement of a work. Thus, construction planning is a process requiring the use of intellectual faculties, imagination, foresight and sound judgement to decide in advance as to what is to be done, how and where it is to be done, who will do it and how the results are to be evaluated.

In any project the sequence through which the project is to develop should be decided. The requirement is met in a systematic manner by scheduling or programming. Scheduling considers the main construction activity and also supporting elements such as labour, material, equipment and expenditure. All the activities consume resources of three kinds, viz., time, labour and material.

Here, the construction project may be housing-complex project, multi-storeyed building projects, factory-building projects, which involves in land acquisition, planning of the project, method of construction, material procurement, construction stages, etc.

31.2 OBJECTIVES OF PLANNING

The main objective of any planning, particularly construction planning, is to execute the project most economically both in terms of money and time. Effective planning includes the following factors:

1. Design of each element of the project.
2. Selection of equipment and machinery depending on the magnitude of the project.
3. Arrangement of repair of equipment and machinery near the site of project to keep them ready to work at all times.
4. Procurement of material well in advance.
5. Employment of trained and experienced staff on the project.
6. Arrangement of welfare schemes for the staff members and workers such as medical and recreational facilities.
7. Arranging constant flow of funds for the completion of the project.
8. Providing safety measures such as ventilation, arrangement of light, water and first aid.
9. Arrangements or means of communication and feedback, etc.

31.3 STAGES OF PLANNNING

During the planning of a construction project, the entire project is phased out identifying the sequence of construction. Secondly each phase is divided for operation into a number of jobs or units of construction. Different stages of planning are as follows:

1. Job planning
2. Technical planning
3. Contract planning

31.3.1 Job Planning

Each job or unit of construction has to be further planned with respect to the following:

1. Manner of Execution of the Job

The jobs may be executed departmentally or through contractor. In case the work is done through the contractor, type of contract is to be finalised.

2. Duration of the Job

The factors which contribute for the duration of job or the proposed period of its completion are:

(i) Urgency of the work
(ii) Availability of resources
(iii) Position of the construction with reference to network

3. Planning of Resources

Resources of a construction project comprises of the following:

(i) Plant, equipment and machinery
(ii) Construction stores
(iii) Both technical and non-technical staff and skilled and unskilled labour
(iv) Construction Material

31.3.2 Technical Planning

Technical planning is done by engineer or concerned authority for economical execution of the construction project. It starts after the administrative approval of the work is obtained for technical sanction. Technical planning is carried out in order to satisfy the following objects:

(i) Finalisation of design, provide detailed drawing and specification to be adopted.
(ii) Preparation of a detailed estimate and modification of estimate if necessary.
(iii) Deciding the executive authority, like departmentally or though contract.
(iv) Planning resources and initiating procurement action.
(v) Foresee the obstacles in the completion of the project and take necessary steps for fruitful completion.

31.3.3 Contract Planning

Contract planning is divided into the following two categories:

1. Pre-tender planning
2. Post-tender planning

1. Pre-tender Planning

Planning required for the time of inviting of tenders up to the receipt of the same is termed as pre-tender planning. It consists of the following works:

(i) Finalisation and Acquisition of Site

Before issue of order to contractors the acquisition of the project site has to be finalised and legal issues if any should be finalised.

(ii) Planning of Resources

Resource planning includes the following: (a) availability of site, (b) availability of stores, (c) availability of labour and (d) availability of equipment and plants.

(iii) Planning Time Limit

The requirement of resources and time limit are inter-related and both these aspects are to be taken together. In some cases the time needed for completion is decided based on the available resources. Such is the case when the work has to be completed departmentally. Even when the work has to be executed by the contractor, circumstances of the project has to be decided and modified before inviting tenders. As a matter of fact the contractor submits tender considering the following aspects: (a) site survey, (b) availability of materials, (c) equipment and plants to be hired and purchased, (d) fuel, (e) labour, (f) facilities for camp and (g) study of drawing/design/specifications.

2. Post-tender Planning

In post-tender planning the following aspects are considered irrespective of the work done by contractor or departmentally: (a) setting camp site office, (b) welfare of staff/labour amenities, (c) materials required, (d) labour required, (e) equipment, (f) safety measures to avoid accidents and theft, (g) follow-up of drawing/specifications and (h) billing, to record progress, to calculate materials, labour, etc.

In the works executed by contractors, responsibility is fixed for recording the work in diary the day-to-day details of work done. The planned progress in the form of bar charts or progress reports should be known to the contractor as well as to the engineer's representative. It is normally the practice for all civil engineering projects to make controlled construction stores available to the contractor.

31.3.4 Advantages and Limitations of Planning

Following are the advantages and limitations of planning:

1. Advantages

(i) Decision of network or bar chart makes the contractor to know more about the job.
(ii) With proper planning, the financial burden will be lessened for the contractor.
(iii) By proper programming, weekly supply of labour can be attended.
(iv) Programme provides a standard against which actual work can be measured.

2. Limitations

(i) Effectiveness depends on correct assumptions.
(ii) Planning is expensive.
(iii) Planning delays action.
(iv) Planning encourages of false sense of activity.

31.4 SCHEDULING PROCEDURES

Various scheduling procedures are discussed below.

31.4.1 Construction Schedule

This schedule consists of duration of construction of various components of the work. Based on the availability of the following data, the construction schedule can be made:

(i) Number of activities
(ii) Quantity of various items of works
(iii) Labour contribution and number of field workers
(iv) Machinery required
(v) Stages of activities to be executed

31.4.2 Labour Schedule

Under this schedule one can know in advance the type of labour and their numbers required from time to time. This is needed to recruit the persons (both skilled and unskilled labour) needed so as to have a continuous flow of work without any disruption.

31.4.3 Material Schedule

This is an important schedule by which one will be in a position to know the requirement of various construction materials well ahead. This helps to start the work and continue without any dearth for materials at any stage. If adequate fund and storage facility are available, materials which can not be spoiled with time may be stored to some extent.

31.4.4 Equipment Schedule

The success of any project depends to a large extent on the proper use of construction equipment. Properly maintained equipment and application in construction can bring down construction time and cost. Further proper deployment of equipment judiciously yields economical result without heavy financial burden in the project. Thus project management should properly plan the judicious deployment of equipment.

31.4.5 Expenditure Schedule

The project engineer and his team have to primarily judge to give the most economical construction. Thus it is essential to review the financial position of the project at regular intervals so as to ensure the following:

(i) The annual expenditure, if any, of the project has to be kept within the allotment.
(ii) A properly balanced expenditure will result in an early completion of work.

31.5 SCHEDULING TECHNIQUES

In any project the sequence through which the project is to develop should be decided. The requirement is met in a systematic manner by scheduling techniques. The project scheduling techniques are concerned with the resource time. One of the objectives of project management is to optimise the use of resources. Scheduling techniques offer solutions to optimisation of project time.

Technical scheduling can be done by different methods depending on the size of the project. The methods used are as follows:

1. Bar charts
2. Milestone charts
3. Network analysis

31.5.1 Bar Charts

1. Principle of Bar Chart

In dealing with complex projects, a pictorial representation showing the various jobs to be done and the time and money they involve is generally helpful. One such pictorial chart, also known as the bar chart, was developed by Henry Gantt around 1900. Bar chart is also referred to as Gantt chart.

A bar chart consists of two coordinate axes, viz., one representing the time elapsed and the other representing the job or activities to be performed. The jobs are represented in the form of bars as shown in Fig. 31.1.

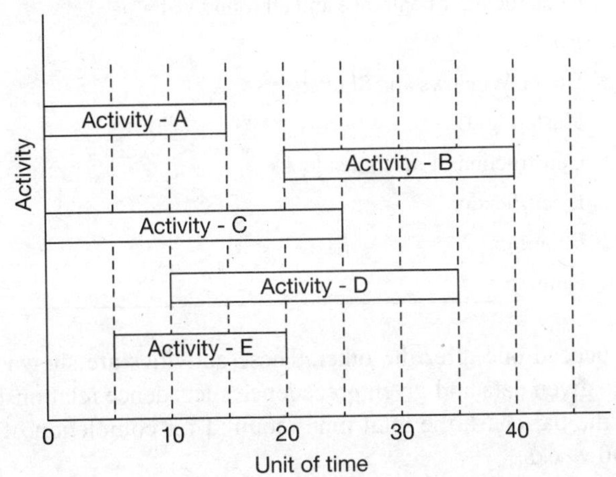

Figure 31.1 Bar chart

The beginning and end of each bar represent the time of start and time of finish of that activity. The length of the bar represents the time required for the completion of the job or activity.

In any project there may be some activities which are to be taken up simultaneously but may take different lengths of time for their completion (e.g., Activities *A* and *C*) and some activities have to be taken up only after the completion of some other activity (e.g., Activities *A* and *B*). There may also be some activities which are independent of all other activities (e.g., Activities *D* and *E*).

2. Example

Let us consider the steps involved in the construction of a factory shed. Major activities identified in the construction of a factory shed where machines are to be erected are presented in Table 31.1. The activities have been identified and arranged in logical sequence. At the factory site no water is available and is not feasible to get water from surrounding. Hence, a bore well has to be sunk before starting the actual construction. Hence, excavation of foundation and sinking of bore well may be started simultaneously. That is, at the end of one week, water will be available.

Table 31.1 Activities of construction of factory shed

Activity no.	Activities	Time required (weeks)
1.	Excavation for foundation	4
2.	Sinking of tube well	1
3.	Base concrete and placing of column reinforcement	2
4.	Shuttering and construction of columns	4
5.	Fabrication of truss at site	1
6.	Erection of trusses	4
7.	Construction of entrance and all round walls	4
8.	Roofing	2
9.	Doors, Windows and Shutters	2
10.	Plastering	1
11.	Construction of machine blocks	2
12.	Electrification	1
13.	Flooring	1
14.	Painting	2

Other activities succeed one after the other. These activities are shown in a bar chart in Fig. 31.2, with the given data and given precedence/succedence relationships.

As seen from the bar chart, the total time required for completion of the factory shed project is about 30 weeks.

3. Limitations of Bar Chart

Following are the limitations of a bar chart:

1. A bar chart can be successfully used only for small projects.
2. Bar charts can not be updated when there are many changes.
3. The progress of the project can not be assessed at different stages scientifically.
4. Bar charts do not equate time with cost. That is time-cost relationship can not be derived.

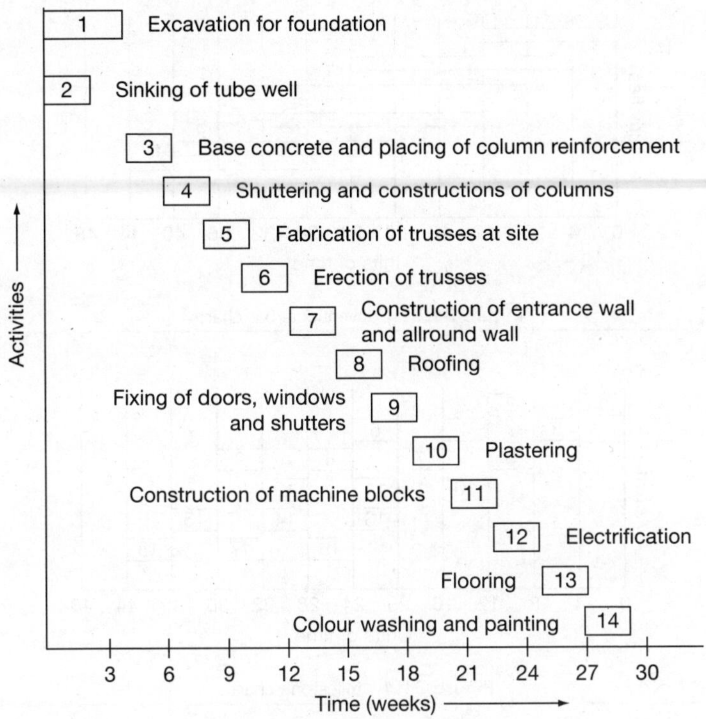

Figure 31.2 Bar chart for factory shed

5. Delay in the work can not be deducted.
6. It does not clearly show the interdependencies between the various activities in the project.
7. It does not reflect the critical activities of the project.
8. Bar charts do not provide methods for optimising allocation of resources.

31.5.2 Milestone Charts

1. Principle of Milestone Chart

The shortcomings or the inadequacies of the bar chart have been modified to some extent in milestone chart. In every activity, there are certain key events which are to be carried out for the completion of the activity. Such key events are called milestones and they are represented by a square or circle. These events are those which can be easily identified over the main bar representing the activity.

It has been observed in a long time activity the details will be lacking. If the activity is broken into a number of sub-activities or key events each one can be recognised during the progress of the project. In such cases controlling can be done easily and some inter-relationships may be established between the activities. In a milestone chart, the events are in chronological, but in a logical sequence. Figures 31.3 and 31.4 show the conventional bar chart and milestone chart.

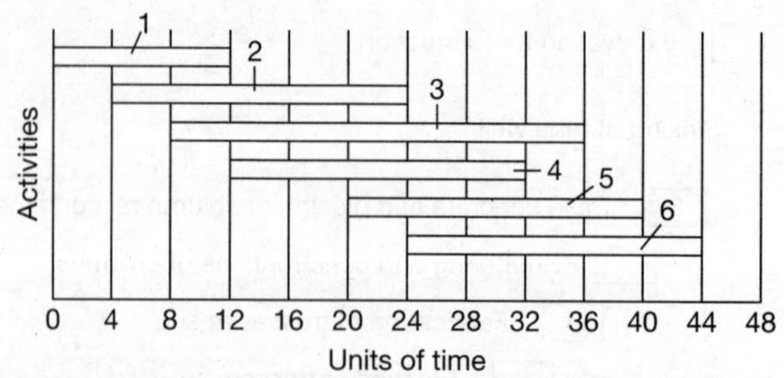

Figure 31.3 Conventional bar chart

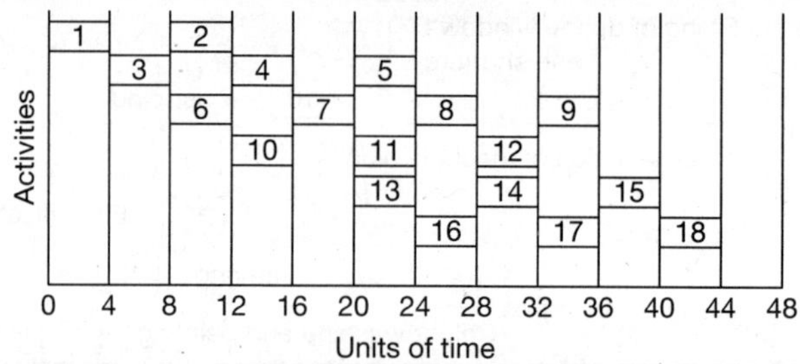

Figure 31.4 Milestone chart

2. Limitations of Milestone Chart

Following are the limitations of a milestone chart:

(i) This method has one great deficiency, that is, it does not clearly show the interdependencies between events.
(ii) Here the events are in chronological order but not in a logical and sequential order.

31.5.3 Network Analysis

1. Principle of Network Analysis

The network analysis techniques, developed between 1950s and 1960s, have now come to be used as an effective management tool for planning, scheduling and controlling of complex projects.

The term project network analysis is a general term covers all the network techniques used for planning, scheduling and controlling of projects. The three techniques commonly used are as follows:

(i) Critical Path Method (CPM)
(ii) Program Evaluation and Review Technique (PERT) and
(iii) Precedence Diagramming Method (PDM) or Precedence Network Analysis (PNA).

These network techniques produce time-oriented diagrams having activities organised into a logical order.

2. Network Diagram

Network Diagram is a graphical flow plan of the activities that are to be accomplished for completing the project. The precedence and succedence relationships between activities are identified first.

Sequencing activities refers to finding out the logical relationships among the activities of a project and arranging them accordingly. By studying the project features the different activities involved in a project the concerned officer-in-change of the project establishes the precedence-succedence relationships between the activities.

While deciding the logical relationships between the different activities, activities which are mandatory to be operated before some of the activities to be taken up have to be identified. In addition there may be certain activities which may not be mandatory dependencies (such activities are called discretionary dependencies) are to be identified. After finalising all the activities and finding their characteristics the network diagram for the project can be drawn.

Activities of construction of a building are given in Table 31.2 and shown as network diagram in Fig. 31.5.

Table 31.2 Activities of construction of a building

Activity	Description	Immediate predecessors
A	Plan approval	–
B	Site preparation	–
C	Arranging foundation materials	A
D	Excavation for foundation	B
E	Carpentry work for Doors and Windows for main supporting frames	A
F	Laying foundation	C, D
G	Raising wall from foundation to sill	F
H	Raising wall from sill to lintel level	E, G
I	Casting of lintels	H
J	Raising wall above lintel up to roof base	I
K	Electrical wiring	I, J
L	Casting of roof	J
M	Fixing frames of doors and windows	L
N	Making of shutters	J
O	Plumbing works	L
P	Plastering	O
Q	Making ready shutters to Doors and Windows	P
R	White washing and painting	Q
S	Fitting Electrical and plumbing parts	R
T	Clearing the site before handing over	S

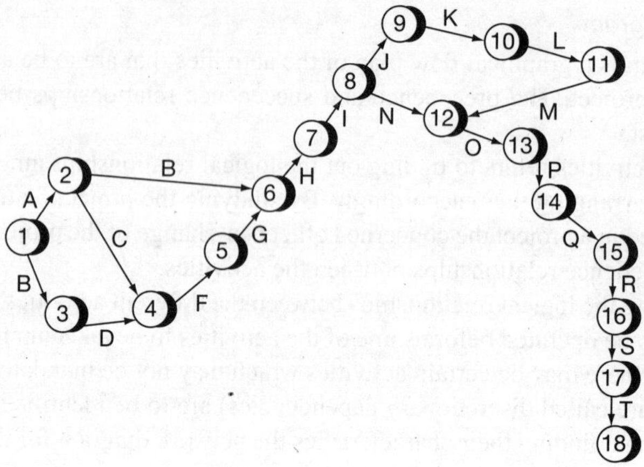

Figure 31.5 Network diagram for construction activity

3. *Definitions of Network Techniques*

(i) Critical Path Method (CPM)

In a network the sequence of activities arranged in each path will have different duration. The path that has the longest duration is called critical paths.

(ii) Program Evaluation and Review Technique (PERT)

It uses three times, viz., optimistic time estimate, pessimistic time estimate and most likely time estimate.

(iii) Precedence Network

Here each activity is represented by a rectangular or square box. The time duration of the activity is incorporated inside the modal box.

31.6 SITE CLEARANCE

Clearing of vegetation and trees is usually necessary before undertaking any construction operation. It is very difficult to handle equipment and materials in the presence of trees, bushes, grass and weeds. Presence of these materials along with earth materials may lead to settlement at a later stage. Crawler dozers, equipped with special cleaning blades, are the best equipment needed for site clearance.

Type of vegetation, condition of soil and topography, quantity of clearance needed, and the purpose for which cleaning is done are the factors to be considered in planning the sequence of operation.

The factors are:

(i) Removing trees and stumps and roots
(ii) Removing surface vegetation
(iii) Disposing of the vegetation by stacking and ultimately burning.
(iv) Knocking out all vegetation stumps and roots, chopping or crushing them and later burning.
(v) Killing or retarding the growth of bush by cutting the roots below the surface of the ground.

31.7 JOB LAYOUT

Overall economy of a project can be achieved by a thorough coordination in planning of services and the construction plant. Such coordinated work will frequently lead to modification in plant facilities and in layout of services.

Thus layout of a construction project camp is governed by the following factors:

(i) Type of project.
(ii) Location where it is situated.
(iii) Method of execution of the project.
(iv) Resource availability.
(v) Availability of water supply and electrical power.
(vi) Waste and sewage disposal works.
(vii) Availability of medical facilities.
(viii) Availability of welfare facilities such as postal service, market, canteen, etc.
(ix) Religious facilities.
(x) Newspaper and magazines TV and other recreation facilities.

31.8 TEMPORARY SHEDS

As explained in job layout different facilities have to be provided. Since the project has to be completed in a short period facilities can not be made permanently. Thus only temporary sheds made out of sheets are provided with adequate safety.

Facilities like drinking water, sleeping areas, dwelling areas, office, store and all are to be provided in temporary sheds. However, depending on the period of construction the sheds are made big or small with additional strength and durability.

The sheds should withstand all weather conditions and unexpected storm on heavy rain. No water logging should be allowed.

31.9 SETTING OUT WORKS

Any civil engineering work has to be set out before starting of construction. Primarily excavation has to be started for which marking has to be done. Marking for excavation of a typical building area is explained below.

In order to carry out construction exactly according to plan, the outline of excavation and centre line of walls are marked on the ground (Fig. 31.6).

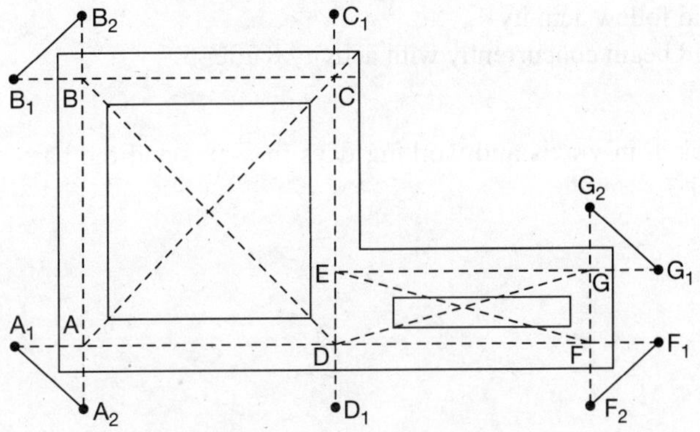

Figure 31.6 Setting out for a building

The following procedure is followed:

1. Based on the plan the centre lines of the walls are calculated. The centre lines of the room are set out by setting perpendiculars in the ratio 3:4:5. The corner points are identified as *A*, *B*, *C*, *D*, *E*, *F* and *G*. These points are marked by driving pegs.
2. The setting of corner points are checked based on the diagonal distances *AC*, *BD*, *DG* and *EF*.
3. At the time of excavation the pegs at *A*, *B*, *C*, … may be removed. The centre lines are extended and the centre points are marked 2 m away from the outer edge of excavation. Accordingly the points A_1, A_2, B_1, B_2 … are marked with stout pegs.
4. Then the width of excavation is set around the centre line and marked by thread with pegs at appropriate positions.
5. Further the excavation line is then marked by lime or by making a ferrow with a spade.
6. Based on field conditions more pegs are driven.

Example 31.1

The following project is to be represented by a bar chart. The duration of each activity is in working days. Draw the bar chart with working days in the horizontal scale.

Activities	Working days
1	7
2	5
3	6
4	10
5	2
6	4
7	5
8	10

Activities 1 and 2 can occur concurrently.
Activity 3 can start 5 days after the commencement of activity 2.
Activities 4 and 6 can start concurrently after activity 3 is completed.
Activity 5 should follow activity 4.
Activity 7 should begin concurrently with activity 8 after 5.

Solution:

Activities are taken in y-axis and working days in x-axis and the bar chart is drawn as shown in Fig. 31.7.

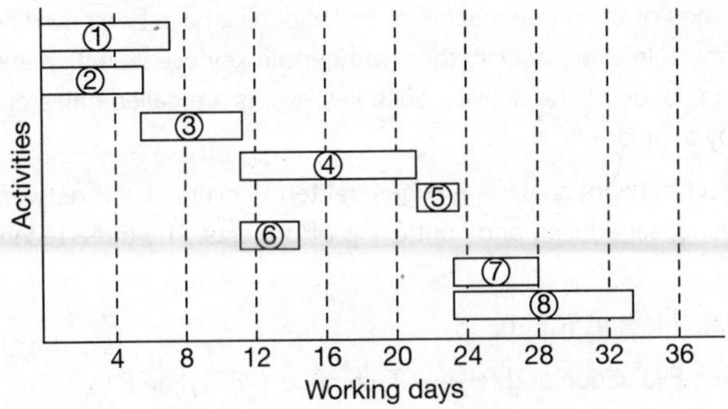

Figure 31.7 Bar chart

SALIENT POINTS

1. Planning is the process of devising of a scheme for doing, making or arranging a project or programme before the commencement of work.
2. During the planning of a construction project, the entire project is phased out identifying the sequence of construction.
3. Different stages of planning are:

   (i) Job Planning, (ii) Technical Planning and (iii) Contract Planning.
4. Each job or unit of construction has further planned considering:
   (i) Manner of execution of Job
   (ii) Duration of the job
   (iii) Planning of resources
5. Technical planning is done for economical execution of the construction project. It starts after the administrative approval of the work is obtained for technical sanction.
6. Contract planning comprises of (i) pre-tender planning and (ii) post-tender planning.
7. Construction schedule consists of duration of construction of various components of the work.
8. Under labour schedule, one can know in advance the type of labour and their numbers required from time to time.
9. Material schedule is an important schedule by which one will be in a position to know the requirement of various construction materials well ahead.
10. The success of any project depends to a large extent on the proper use of construction equipment.
11. Technical scheduling methods are (i) bar charts, (ii) milestone charts and (iii) network analysis.
12. In dealing with complex projects, a pictorial representation showing the various jobs to be done and the time and money they involve is generally helpful. One such pictorial chart is called bar chart.

13. The shortcomings or the inadequacies of the bar chart have been modified to some extent in milestone chart. In every activity there are certain key events which are to be carried out for the completion of the activity. Such key events are called milestones and they are represented by a square or circle.
14. The term project network analysis is a general term covers all the network techniques used for planning, scheduling and controlling of projects. The three techniques commonly used are:
    (i) Critical Path Method (CPM)
    (ii) Programme Evaluation and Review Technique (PERT), and
    (iii) Precedence Diagramming Method (PDM) or Precedence Network Analysis (PNA).
15. Network diagram is a graphical flow plan of the activities that are to be accomplished for completing the project.
16. In a network the sequence of activities arranged in each path will have different duration. The path that has the longest duration is called critical path.
17. Program Evaluation and Review Technique (PERT) uses three times, viz., optimistic time estimate, pessimistic time estimate and most likely time estimate.
18. In precedence network, each activity is represented by a rectangular or square box. The time duration of the activity is incorporated inside the modal box.

REVIEW QUESTIONS

1. Discuss the objectives of construction planning.
2. Explain different stages of planning.
3. What is job planning?
4. Explain what is meant by Technical planning
5. Distinguish between pre-tender planning and post-tender planning.
6. What are the aspects considered by a contractor in submitting a tender?
7. What is a bar chart?
8. How bar chart is different from milestone chart?
9. What is a network diagram?
10. Define critical path.
11. Explain the principle of network analysis.
12. Bring out the importance of site clearance.
13. Explain the factors to be considered in job layout.
14. Explain setting out of works with a suitable example.
15. Bring out the advantages of CPM.
16. What is PERT? In which situation this can be used.

32 CONSTRUCTION EQUIPMENT

32.1 INTRODUCTION

The type of equipment to be used for a specific project depends on the scope of the project involved. Every project is executed not by department but by contracting the work. So it is the choice of contractor to use suitable equipment. If a particular equipment is required to be purchased, he has to think whether he could get the amount of investment for the equipment during the project itself.

32.2 COST OF EQUIPMENT

32.2.1 Procurement Cost

Procurement cost includes (i) the cost price of the equipment, (ii) interest on money invested on the purchase, (iii) taxes paid and (iv) insurance cost.

Evidently the owner of the equipment pays cash for the procurement of the equipment. It is to be noted that if the same amount is invested in some other scheme what would be the turnover. One should get a higher amount annually by investing in the equipment. Each year the equipment earns for the owner and in the process its value gets depreciated. Thus the investment in the equipment is therefore reduced by the amount of depreciation. A realistic rate of interest would be the one based on the average value of the equipment during its useful life. The average value of the equipment has to be determined.

On the depreciated value of the equipment the insurance and taxes are paid. The expenses also should be worked out on the average value of equipment. These aspects are to be considered before investing on the procurement of equipment.

32.2.2 Operating Cost

Operating cost of an equipment is based on the following factors:

(i) Cost of investment
(ii) Depreciation cost
(iii) Cost of major repair
(iv) Cost of fuel and lubricants
(v) Cost of labour
(vi) Servicing and field repairs
(vii) Overheads

1. Cost of Investment

By investment made on the equipment the owner is losing an amount equal to the interest or return one would have otherwise received from bank or other sources of investment. For the purpose of calculating the total cost of investment, the interest, the taxes, insurance, etc., may be taken as 10–15% of the cost of equipment.

2. Depreciation Cost

Because of wear and tear the cost of equipment gets depreciated. An amount of earning has to be set aside, to balance the depreciation, so that the accumulated amount can be used at the end. This amount is recovered from the earnings of the equipment and forms a charge on the operating cost.

3. Cost of Major Repair

Major repairs are those which are incurred when the equipment is taken to a workshop. This repair involves replacement of major parts of the equipment, overhauling and servicing. This incurred expense is added to the capital cost so that the expenditure is distributed over the years and included in the operating cost of the machine.

4. Cost of Fuel and Lubricant

While calculating the fuel consumption of the equipment, it is the general practice to consider the full load condition and constant speed under favourable conditions. As the equipment would not be working always at the ideal condition and so a reduction factor of 60% is considered.

5. Cost of Labour

The cost of labour includes (i) the wages of the working crew and (ii) wages of supervisory staff.

6. Servicing and Field Repair

All expenses involved in the maintenance normally met during the operation of equipment except major repairs are included under this. This demands periodical service of the equipment.

7. Overhead

In a large project the whole establishment is work charged. Any unforeseen expenses which are not work charged or non-productive are distributed to the various work producing elements for purpose of cost control.

32.3 TYPES OF EQUIPMENT

For a building project the selection of equipment depends on:

1. The magnitude of the building project, viz., a single unit, multi-storeyed building, housing complex, etc.
2. The type of equipment, its size and other particulars.
3. Whether the equipment are to be purchased, rented or contracted for the work.

Major equipment needed for a building project are the following:

1. Earthwork equipment
2. De-watering equipment

3. Compaction equipment
4. Pile driving equipment
5. Movement and hoisting equipment
6. Hauling equipment

32.3.1 Earthwork Equipment

Earthwork and moving equipment are available to suit different field conditions which are discussed below.

1. Excavators

Excavators are digging machines. These machines consist of the following components (Fig. 32.1):

(i) An undercarriage – this gives mobility to the excavator. This may be mounted with crawler track or wheel.
(ii) A superstructure with operator's cabin – this could traverse through 360° or fitted on a rigid frame.
(iii) Hydraulically articulated booms and dipper arms with bucket.

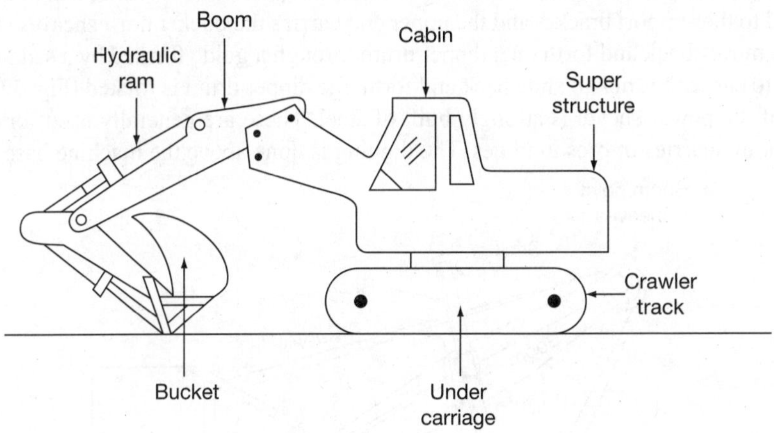

Figure 32.1 Excavator

Excavators may be of mechanical (or cable operated) or hydraulic types. Hydraulic excavators have more advantages. They can be fitted with various multi-purpose attachments for various earth moving works. Most efficient action of the machine is that the hydraulic unit of action of the excavator which directs the bucket teeth at their most efficient angle during dipping operation.

Mechanical or cable operated excavators are of either hoe or shovel type. In hoe type the bucket opening fences towards the machine whereas in shovel the bucket opening faces away from the machine.

Excavators are of four types, viz., crawler-mounted excavator, truck-mounted excavator, self-propelled excavator, and excavator mounted on barge or rail.

Excavators include dipper or power shovel, dragline, clamshell and drag shovel. All these machines basically fall under shovel family in that the front end only is changed. Shovel based machines have a common type of operating and tracking mechanism.

2. Shovels

Shovel is often used for a specific type of excavating machine fitted with a short length boom and working with forward strokes. The following are the four types of machines which have the same basic structure but the front-end attachments are different:

(i) Dipper shovel
(ii) Drag shovel or hoe
(iii) Dragline
(iv) Clamshell

The frontline attachments are as follows:

(a) Shovel boom and dipper stick for a dipper shovel.
(b) Shovel boom, jack boom and stick for a drag shovel.
(c) Crane boom with a failed and a loosely attached bucket for dragline.
(d) Crane boom with a special type of bucket for a clamshell.

(i) Dipper Shovel

The dipper shovel or power shovel is the most popular of the shovel excavator. It consists of a boom, dipper stick, the bucket and the mechanism of operation. The lower end of the boom is hinged to the support bracket and the upper end carries the bucket hoist sheaves. The dipper stick moves back and forth on a dipper drum through a guide formed by a saddle block. In order to cause the stick to slide back and forth, the dipper drum is rotated (Fig. 32.2). The bucket of the power shovel is strongly built of steel. These are generally used for digging hard rock in quarries or ores in mines. The digging is done above the machine base level

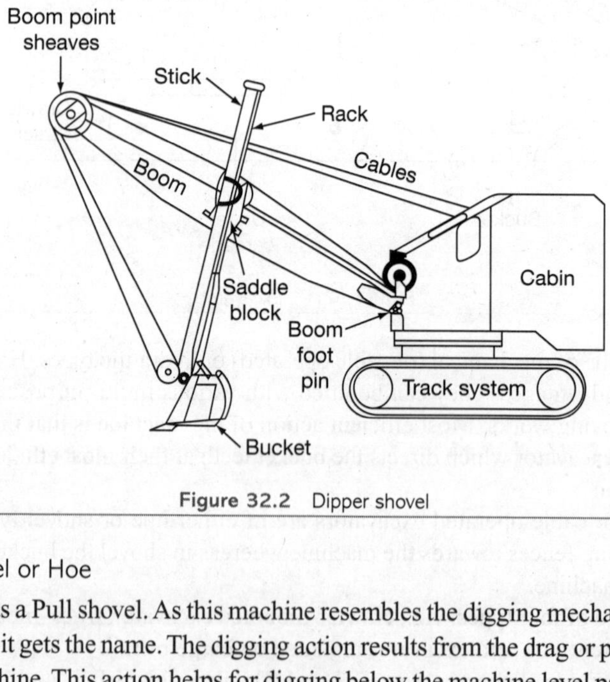

Figure 32.2 Dipper shovel

(ii) Drag Shovel or Hoe

It is also called as a Pull shovel. As this machine resembles the digging mechanism of an ordinary garden hoe it gets the name. The digging action results from the drag or pull of the bucket towards the machine. This action helps for digging below the machine level particularly while digging for trenches, footings, basements, etc., Fig. 32.3 (Mahesh Varma, 1979).

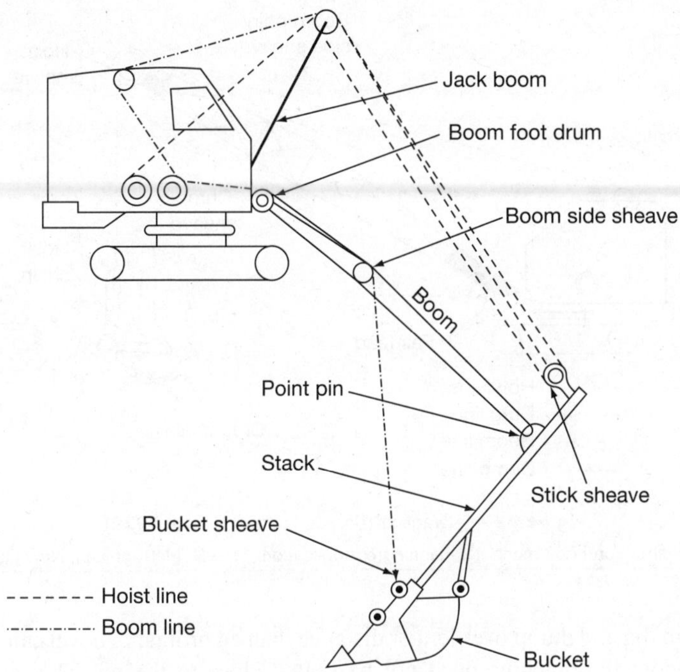

Figure 32.3 Drag shovel (*Source*: Mahesh Varma 1979)
(Construction Equipment: It's Planning & Applications, 1979, Metropolitan, New Delhi)

This shovel consists of a boom, a stick, a jack boom and a bucket which is attached at the end of the stick. The stick is hinged by a pin with a boom. This enables to turn over the pin to take any desired direction best suited for digging or dumping operations. The lower end of the stick carries the bucket and the upper end carries a sheave having the hoist cable supported on a jack bottom at the other end. This arrangement enables to hoist the cable when pulled and the reaction at the stick enables to move the boom with stick up or down. Thus the boom is capable of taking any position in vertical plane.

Drag shovels have application in

(a) Digging trenches, footings or basements
(b) Digging for hard materials
(c) Providing close trimming whenever needed, and
(d) Where excavated materials need to be dumped at a short distance.

(iii) Dragline

As this machine drags the bucket against the material to be dug it attains this name. This consists of a boom, a bucket, a fairlead and hoist, dump and drag cables, Fig. 32.4 (Mahesh Varma, 1979). The boom is similar to that of a crane boom. The top and bottom ends of the boom are known as the point and the foot respectively. The boom is attached to the deck with pin at foot and supported at the point through the boom hoist cable. This cable passes over a pair of sheaves placed one on each side. They have three types of buckets, viz., light, medium and heavy, which are used depending on the nature of material.

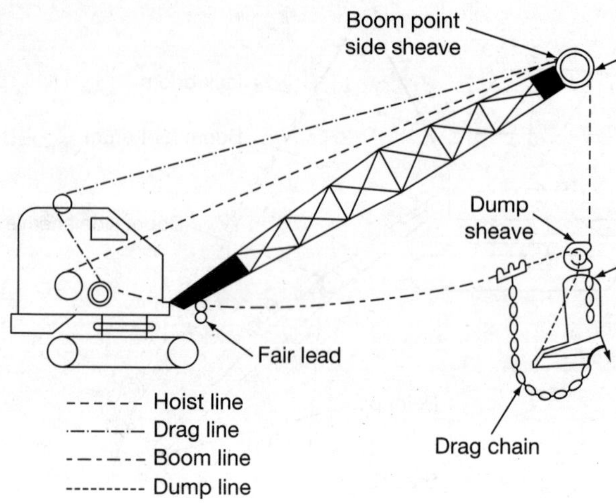

Figure 32.4 Dragline (*Source:* Mahesh Varma, 1979)
(Construction Equipment: It's Planning & Applications, 1979, Metropolitan, New Delhi)

A dragline can dig and dump over longer distance than an ordinary shovel can do. Because of its long boom the machine need not be seated close to the pit. Thus it can handle digging of underwater excavation while stationing the machine on a firm soil away from the location of the pit. This can dig below the track level and can handle soft materials comfortably.

(iv) Clamshell

It gets the name as its bucket resembles to a clam which is a shellfish with a hinged double shell. It is having most of the characteristics of dragline and crane. Digging is done as in a dragline and soon after the bucket is filled, it works like a crane. It has a crane boom with a specially designed bucket attached to it at the upper end through cables. The bucket is hinged at top and has either sharp edge or the teeth at the bottom (Fig. 32.5).

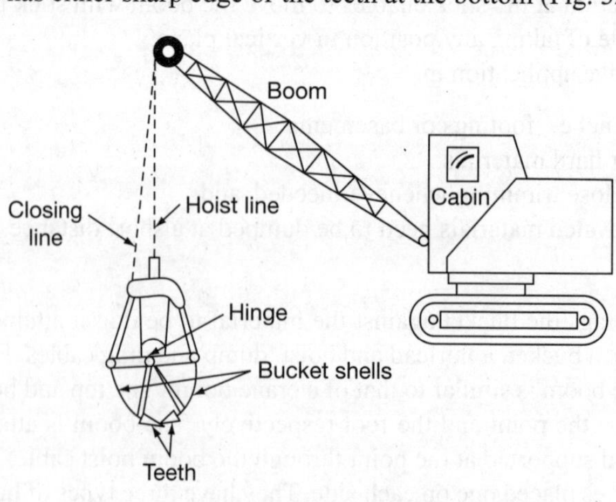

Figure 32.5 Clamshell

During digging process the bucket is lowered with shells open. When reaching the surface to be dug it makes a good contact with it. The weight of the bucket helps the sharp-edge or the teeth at the bottom of the bucket to dig into the surface or material, thereby filling the bucket. Once the bucket is filled the shells close.

The bucket is then hoisted and swung to the position where dumping has to be done and contents are dropped. Buckets may be light or heavy weight and used depending on the type.

Its applications are widely used where

(i) The digging or dumping vertically is needed.
(ii) The material is relatively soft or medium hard.
(iii) Digging of trenches.
(iv) Loading materials in a bin or a stock pile.
(v) Accurate dumping of materials.

3. Bulldozers

It is an important equipment on a construction project. It is basically a scraping and pushing unit. However, it is a multi-purpose equipment which can be used for different purposes with some modifications. Accordingly they are called as angle dozer, tilt dozer, tree dozer, and push dozer. Up to 100 m distance these can be used to haul.

Angle dozer pushes its load at an angle (nearly 25°) to the direction of travel. This is helpful (i) when the material has to be pushed down the slope on hill work and (ii) where a long windrow has to be made during the travel of the dozer.

Tilt dozer is used to start excavating a ditch or a trench or for excavation in hard ground. For this the blade is required to be tilted by raising one end up to 25 cm above the other. Push dozer is used to push the scraper unit after digging and also during loading operation using the pusher plate.

Tree dozer is used to uproot and remove trees.

4. Tractors

A tractor is a multi-purpose machine. This comes in varied type of light model to heavy model. The light model is used for agricultural or small haulage purposes. Heavy model equipped with several special rings are used for earth moving work. This is an important piece of equipment which is indispensable on all important projects.

Two principal applications of tractors are as follows:

(i) Clearing and excavating machinery.
(ii) Hauling and conveying machinery.

There are two types, viz., wheeled tractors and crawler tractors. The wheeled type is used for light and speedy jobs. As regards to its applicability it falls between the crawler tractor and the truck. The crawler tractors are rugged machines which are used for heavy duty work. It is used particularly when there is a demand for more tractor power and speed of movement. Now-a-days wheeled units have been made to work on the jobs which are intended for the crawler type. Wheel tractors are now available for all practical earth moving jobs including nipping and dozing.

5. Motor Graders

A grader is primarily a device for loading or finishing earth work. Sometimes it is also used for mixing gravel, making windrows and trimming slopes. There are two types of graders, viz., towed and motorised.

The towed type is of small size with a tractor. The controls are in general manual and sometimes a small petrol engine is fitted on the framework of the grader to operate the controls. It is operated by separate operator.

Attachments to the grader include the blade (also called as moldboard), the scarifier, the bulldozer or the snowplow, the elevator and the roller. The versatility of the machine is increased by the addition of these tools. The blade is thick and like a blade of dozer and can swivel through 180°. The scarifier is the tool for loosening hard soil and may be mounted in the front or in the rear of the machine. The bulldozer and the snowplow (a V-shaped/ blade) are front-cut attachments enables the grader to pick up the material cut by the blade and drop it over an inclined belt conveyor which transmits it into carrier unit or discharges it aside the windrow. In order to compact or smoothen the ground surface, a roller may be attached to the rear of the grader.

6. Scrapers

Scrapers are the devices to scrap the ground to load the material, to transport to the required distance, to dump at the intended place, to spread the dumped material over the required area, to attain the desired thickness and to return back to do the next cycle. In simple terms scrapers are designed to dig, load, haul, dump and spread. As a scraper does a multiple works it is also called a *carry all.*

A scraper is provided with a bowl, apron, ejector and a hydraulic system which are explained below (Fig. 32.6).

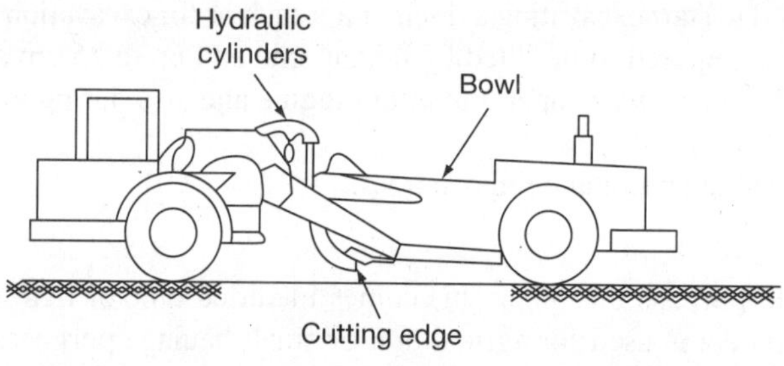

Figure 32.6 Scraper

(i) Bowl

It is a pan which is to hold the scraped material. It is hinged at the rear corners to the rear axle inside the wheels. It can tilt down for digging or ejecting. The size of a scraper is specified by the size of the bowl. At the bottom of the bowl a cutting edge is attached. In order to make a shallow cut, the cutting edge is lowered into material or dirt.

(ii) Apron

It is a wall located in front of the bowl to open or close in order to regulate the flow of the material in and out of the bowl. Further it is capable to open or to close during the carrying position also.

(iii) Ejector

It is also called as a tail gate which is the rear of the pan. It is provided with forward and backward movement inside the bowl. During loading it remains at its rear wall and moves in the forward direction to help in the ejection of the load during dumping.

(iv) Hydraulic System

All these operations are controlled by hydraulic cylinders.

The operation of a conventional scraper comprises of (i) digging or loading, (ii) transporting and (iii) unloading.

(i) Digging or Loading

Keeping the ejector at the rear and the apron raised (approximately to 40 cm) the operator moves to the cut. The bowl is then lowered to the required depth of cut, and the engine speed is increased and moved forward keeping the optimum depth of cut. After filling the bowl the apron is closed and the bowl is then raised.

(ii) Transporting

The bowl is transported in raised position so as to provide sufficient clearance from the ground. In order to prevent the loss of the collected material the apron is fully closed.

(iii) Unloading

The process of unloading by the scraper is called as dumping and spreading operation. In this operation the bowl is positioned so as to spread the material. So as to have an even spreading, a partial opening of the apron in the stage of unloading will be helpful. In the case of wet and sticky material, the apron should be raised and lowered separately. This helps the material behind it is get loosened and material drops out of the bowl by moving the ejector forward, the remaining material is pushed out at a uniform rate. After the dumping is complete, the tail gate is fully retarted, the apron is dropped and the bowl is raised to the transporting position. Using the cutting edge final finishing is done.

7. Loaders

Loader is a machine in which a bucket is attached to the arms and capable of being raised, lowered and dumped through mechanical or hydraulic controls. A very common type is one in which the loader is having the bucket in the front which is known as *front-end loader*.

Loaders are self-propelled and versatile equipment which are mounted either on crawler or wheel-type running gear. These loaders are fitted with front mounted general purpose bucket.

These loaders are operated through hydraulic rams with which they dig, scoop, lift, transport, carry, dump or load into hauling units, bins, hoppers, conveyors and stockpiles. With the additional parts added to the front and rear mounted attachments these can doze, scrap, grab, forklift, trench, grade, ditch, rip, clamp and winch.

Loaders are used for handling coal, sugar, sand, salt, stone, etc., to earth moving and digging works. Two types of loaders are crawler loaders and wheeled loader.

8. Trenchers

Trenchers are used to make trenches. Trenchers are used for excavating trenches or ditches of variable width and depth. The present day trenchers can be used to make trenches of width up to 12 m and depth up to 3 m can be excavated in one pass. Basically a trencher consists of:

(i) Self-propelled tractor (Fig. 32.7) or carrier mounting fitted with crawler or wheel-type running gear (Fig. 32.8).
(ii) An excavating device consisting of several buckets on the periphery of the wheel and alternatively several cutters are provided for digging and discharging continuously.
(iii) A provision conveying the excavated material.

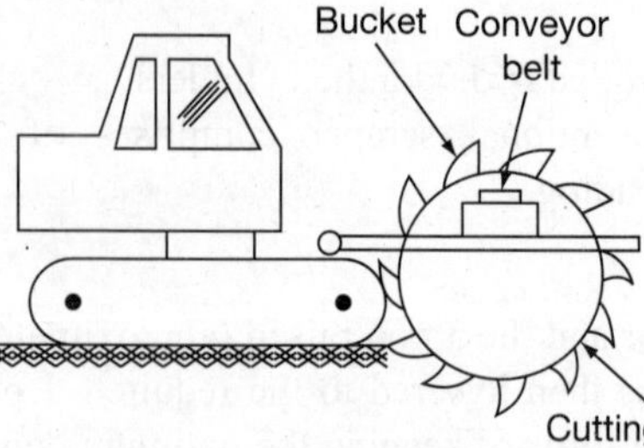

Figure 32.7 Trencher with tractor

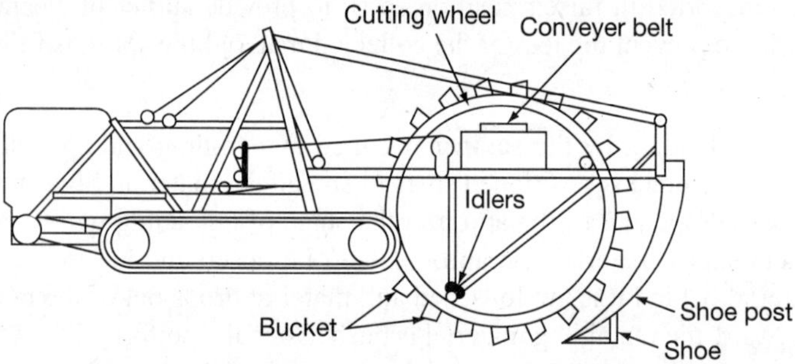

Figure 32.8 Wheel-type trencher or ditcher

The common type of trench excavator is the wheel trenchers. It consists of a pair of circular rims whose outside diameters are connected by V-shaped buckets or cutters. The wheel is turned by a chain drive which connects to the power source. The wheel moves to the top position when discharging the material and at the bottom position while digging. The buckets or cutters on the wheel perform the work of excavator while travelling upward. When the wheel reaches the top position the excavated material drops. The dropped material is carried on conveyors and discharged alongside of the trench.

The selection of a trenching equipment depends on various factors, viz., depth and width of trench, type of soil, disposal of excavated earth, ground water position, and the nature of the job.

32.3.2 De-watering Equipment

Groundwater and seepage control needs a most significant consideration in the construction of buildings. Groundwater can be controlled depending on the geological conditions and characteristics of the soil. In such a situation, groundwater can be controlled by adopting one or more types of de-watering systems or drain appropriate to the size and depth of excavation.

Following are some of the de-watering methods:

1. Pumping Methods
   (i) From open sumps and ditches
   (ii) From well-point systems
   (iii) From deep-well drainage
   (iv) From vacuum de-watering.

2. Electro-osmosis
3. Elimination or reduction of ground water by
   (i) Cement grouting
   (ii) Displacement grouting
   (iii) Chemical consolidation
   (iv) Freezing

Pumping and electro-osmosis methods are discussed below. For other methods, reference may be made to Purushothama Raj (2013).

1. Pumping Methods

(i) Open Sumps and Ditches

The essential feature of the method is a sump below the ground level of the excavation at one or more corners or sides (Fig. 32.9).

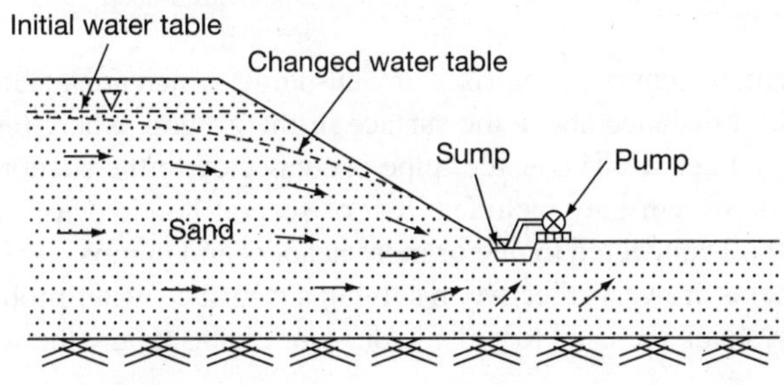

Figure 32.9 De-watering through sumps

In fairly permeable soils the head is low and flow does not emerge through the excavation slopes. In such cases pumping out can be done by collecting the seepage in the sump and pumping out.

In case of excavation in clays the groundwater seeps from fissures and can readily pumped out from sumps. Salty soils generally pose problems. In rocks ground water discharges in the form of sprays or weeps from fissures. Therefore no instability problem in rocks except the water seeps through a weak or shattered rock. The discharges can be collected in an open sump and pumped out.

In order to provide the required number and capacities of pumps the quantity of water to be pumped should be known. This is calculated from Dupit's formula.

(ii) Well-Point Systems

Filter wells or well-points are small well screens of sizes 50–80 mm in diameter and 0.3–1 m length. Well-points are made of brass or stainless steel screens and of closed ends or self-jetting types. When well-points are required to remain in the ground for a long period disposable plastic well-points are used. The plastic well-points are of nylon mesh screens surrounded by flexible plastic riser pipes. Water drawn through the screen enters the space between the gauze and the outside of the riser pipe through holes drilled in the bottom of the pipe and then reaches the surface. The well-points are installed by jetting them into the ground. A typical layout of a single stage well-point is shown in Fig. 32.10.

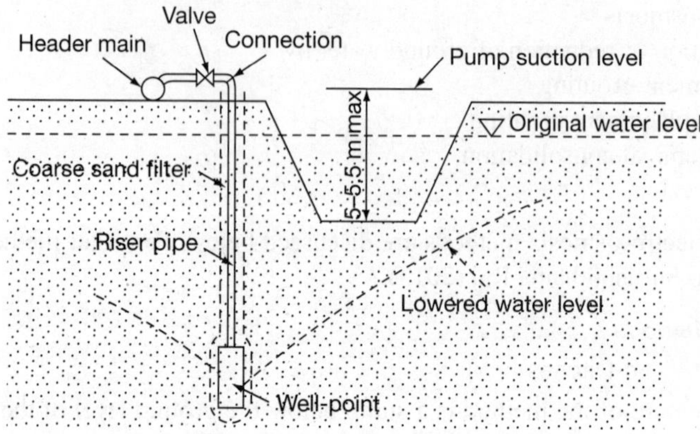

Figure 32.10 Single stage well-point installation

A well-point system, in general, comprises of well-points which are attached to riser pipes, which extend a short distance above the surface of the ground, where they are connected to a large pipe called header. The header pipe is connected to the suction of a centrifugal pump. A well-point system may include a few or several hundred well-points (generally 50–60 well points) all connected to one or more headers and pumps.

Well-point systems are very effective in solving subsurface water problem on construction sites. These systems are used to provide dry work areas below the water table for the following works:

(a) Foundation work including buildings, bridges, dams and dry dock.
(b) Trench work including buildings, bridges, dams and dry dock.
(c) Tunnel work such as subway construction.

(iii) Deep-well Drainage

If the soil formation is such that it is pervious with depth, large diameter deep-wells are suitable for lowering the ground water table. An advantage of the system is that it can be installed outside the zone of construction operation and drainage is effected to the depth of excavation. Deep-wells may be combined with the well-point system on certain occasions for lowering the ground water table (Fig. 32.11).

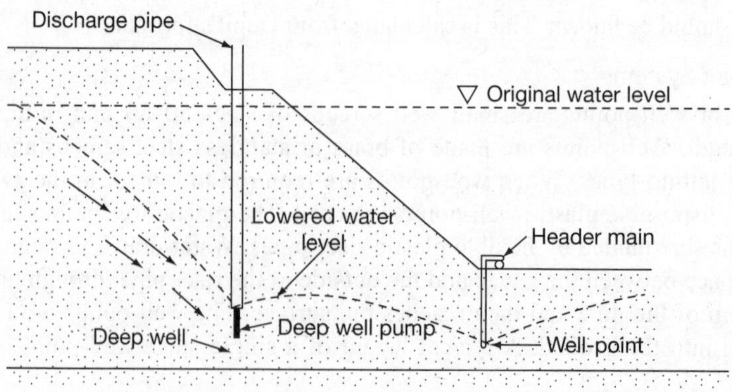

Figure 32.11 Deep-well de-watering

Presence of artesian pressure in some field conditions may be relieved by deep-wells. The cost of installation of deep-well is high. Thus it is preferred in jobs which have a long construction period.

The deep wells are installed by sinking a cased borehole having a diameter of about 300 mm larger than the well casing. The diameter of the latter depends on the size of the submersible pumps. The inner well casing is inserted after the completion of the bore hole. A perforated screen is provided over the lengths where de-watering of the soil is required. Graded filters of gravel are provided between the well casing and the outer borehole casing over the length to be de-watered. The outer casing is withdrawn in stages as the filter material is placed. The space above the screen is backfilled with any available material. The details of such a deep well are shown in Fig. 32.12. The spacing of deep well vary from 10 to 100 m depending on the field condition.

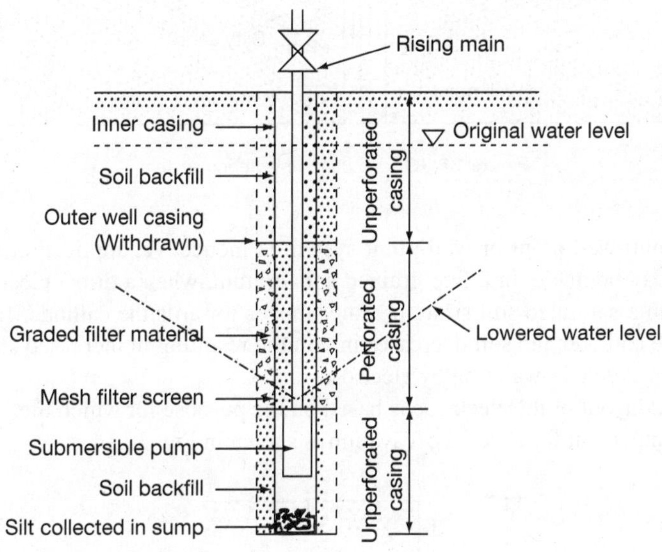

Figure 32.12 Details of deep-well installation

(iv) Vacuum De-watering System

Gravity methods explained so far are not very effective in fine-grained soils. Such soils can be stabilised by means of a vacuum well or well-point system. A vacuum de-watering system primarily consists of well or well-points with the screen and riser pipes.

A stabilising fine soil such as bentonite or impervious soil seal is provided at the remaining portion of the hole. By creating and maintaining a vacuum in the well screen and the sand filter, the flow towards the well is increased. A closer spacing is required (Fig. 32.13) for proper de-watering.

2. De-watering by Electro-osmosis

This is also a method applicable for fine-grained soils. This is not a general pumping method but collecting the water through some process to a well and pumping out.

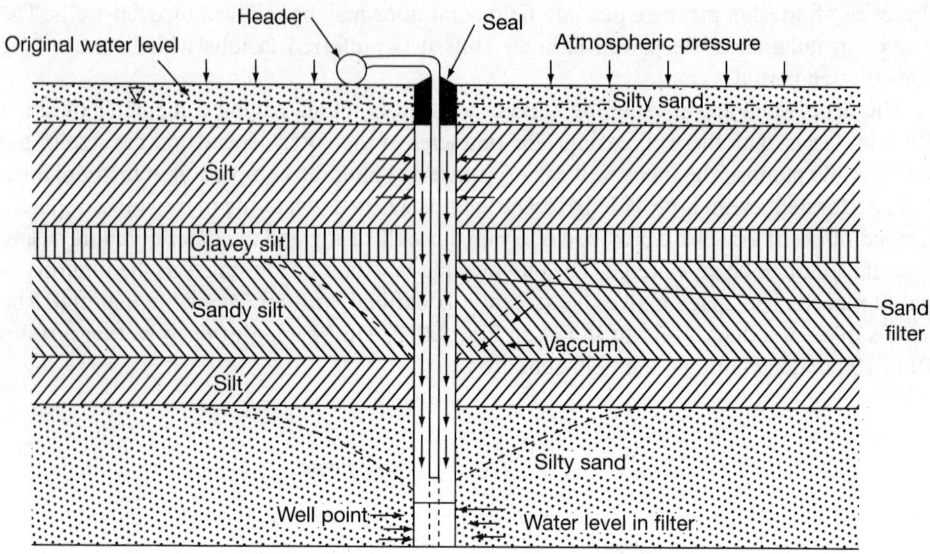

Figure 32.13 Vacuum de-watering system

If the vacuum well-point or well-point system is ineffective, application of an electrical gradient may be made. In a fine-grained soil stratum, when a direct electric current is passed through a saturated soil stratum, water moves towards the cathode. If the water is removed at the cathode, the soil decreases in volume resulting in increased shear strength. This process is called de-watering by electro-osmosis.

The general layout of the electrode is based on the purpose for which they are intended. Electrode arrangement for sheeted excavation is shown in Fig. 32.14.

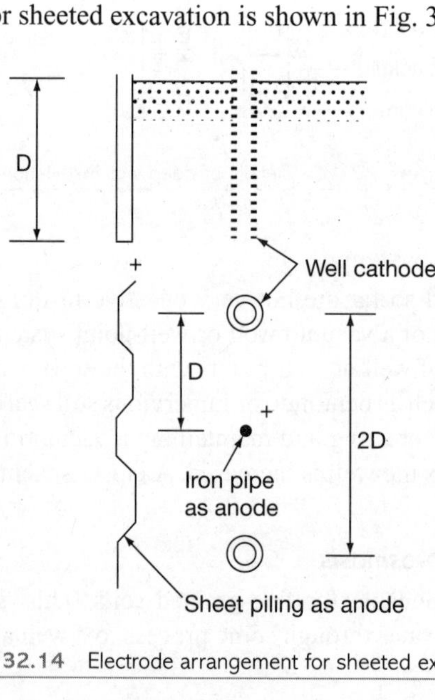

Figure 32.14 Electrode arrangement for sheeted excavation

Sheet piles of any shape and old pipes of 25–50 mm diameter can be made as anodes. Since the anodes corrode considerably in the course of a few weeks of electro-osmotic treatment, they should be replaced as soon as the current drops to less than 30% of the initial consumption. For cathode perforated tubes are used and the cathode wells are connected to a pumping system.

Electro-osmotic method is used only when other methods fail as the cost of installation and maintenance are very high.

32.3.3 Compaction Equipment

Compaction is the process by which particles of materials are packed closely together by reducing air void content. In general the compaction is achieved by mechanical means.

1. Mechanical Forces

Mechanical forces may be applied by the following ways, viz., by rolling, by kneading, by vibration and by ramming.

(i) Rolling

In the rolling process heavy weights in the form of rollers are used to press the soil particles together. Smooth-wheeled rollers are the typical example which falls under this category.

(ii) Kneading

In the kneading process the soil while at kneading will be subjected to some pressure. Sheepsfoot roller is a typical example of this type of compaction.

(iii) Vibrations

In the vibrations process the soil particles are shaken together to form a compact mass. Vibrating rollers are typical example of compaction equipment which operate on this principle.

(iv) Ramming

In this process the soil particles are forced to move closer together by pounding action. Hand-tampers and mechanical tampers are the typical examples of compaction equipment which operate under this principle.

2. Rollers

(i) Smooth-Wheeled Rollers

These rollers are used with or without ballast and may be provided with three wheels or two wheels of equal width called tandom type. These rollers are generally used for most of the works. But these rollers are not effective in uniformly graded sand, gravel or silt and on cohesive soil with high moisture content due to poor traction.

These static rollers, also called as dead weight rollers, are diesel powered. These rollers rely on the weight only to compact the materials by passing over them. Units of 8–10 tonnes can impact a pressure of 20–40 kg per linear cm are generally in use. Rollers with weight up to 1 tonnes are used for light work.

(ii) Sheepsfoot Rollers

Sheepsfoot or Padfoot rollers are suitable for cohesive soils. These may be self-driven or tractor driven and are especially useful when the water content is on the higher side.

The mass of the drum can be varied by adding ballast. For effective rolling, the lift thickness should be small and the contact pressure under the projection very high. These rollers are specially recommended for water-retaining earth works.

(iii) Pneumatic-tyred Rollers

In pneumatic-tyred rollers wheels are placed close together on two axles and placed such that the rear set of wheels overlap the lines of the front set to ensure complete coverage of the soil surface. In order to avoid the lateral displacement of soil, wide tyres with flat threads are provided. The compaction produced by these types is better than that of the smooth wheel rollers.

(iv) Tandom Compactors

Tandom compactors have two equal sized rollers and are centred in line-tandon. These rollers have smooth surface. Improvements have been made on these type of compactors as tandom vibratory compactors. Large size tandom vibratory compactors are generally preferred now-a-days as they can be used either as static compactor or as a vibratory compactor as per the requirement.

32.3.4 Pile Driving Equipment

As discussed earlier piles are generally classified as bored piles or driven piles depending on the method of installation. Only driven piles are to be installed by driving using an equipment. Bored piles are installed after making a hole in the ground and inserting a cage of reinforcement followed by concreting. In some cases reinforcement is not provided.

Driven piles are made out of some materials like wood, concrete, steel and sometimes with composite section of wood and steel, wood and concrete or steel and concrete. They are manufactured in special lengths.

Sheet piles, as discussed earlier, may be wood, concrete or steel. These piles are also driven.

1. Process of Pile Driving

The process of pile driving consists of lifting the pile into position, holding it and driving it to refusal or a desired depth. In this exercise, long piles need to be cut to the required length and the short piles need to be extended for safe and economical driving operations, proper judgement, experience and the combined skill of the crew. Driving of sheet piles greatly affect the seepage of the ground water and thereby pose special problem. As a guide to decide the depth of penetration pile driving formula is used.

2. Components of Equipment

Pile driving equipment comprises of the following components (Chellis, 1951):

(i) Driving Rigs
(ii) Guiding leaders
(iii) Pile hammer with Accessories
(iv) Additional aids for pre-boring and jetting
(v) Boiler for steam raising or air compressor

(i) Driving Rigs

Driving rig provides basic operations of lifting the pile, holding the pile in position, hammering it into the ground or of pulling it out of the ground and guiding the pile in the desired direction of movement. The rig supports the boom and winch mechanism, driving hammer, the guiding leaders, and a platform for mounting of auxiliary equipment such as a jet pump, drilling auger steam boiler or air compressor.

(ii) Guiding Leaders

The leaders guide the pile and the hammer during operation which extends to the entire height of the rig. In case of piles to be driven below the level of the rig into excavations, trencher or water, telescopic or extensible leaders can be used. The leader should enable the hammer to deliver blows axially to the pile.

During the process of driving, the driving rig should be strong and stable. In case a boom is used, adequate space should be available between the pile top and the point for the hammer to work.

There are two types of rigs, viz., skid-mounted and crane-mounted. The skin-mounted rigs are provided with rail wheels or with long steel rollers for movement. The crane mounted rigs are mounted with a crawler or a truck chassis with a swinging deck. In the case of floating pile driving, both the rigs can be mounted on a barge.

(iii) Driving Hammers

Pile driving hammers impart energy required to drive the pile into the soil. The routinely used pile hammers work by hitting the pile on its head. The vibratory and sonic type of hammers are the two new types.

Hammers are classified as follows:

(a) Drop hammers
(b) Single-acting hammers
(c) Double-acting hammers
(d) Differential-acting hammers
(e) Diesel hammers
(f) Hydraulic hammers
(g) Vibratory hammers
(h) Sonic hammers.

Out of these hammers drop hammer and single-acting hammer are generally used which are explained below. For details of other hammers reference may be made to Purushothama Raj (2015).

(a) **Drop Hammer:** This is the simplest form of hammer which does not use any external sources of power. The only mechanism needed is to lift the hammer through a cable. Although the process is slower, it is more efficient as it uses only the gravity. The drop hammer is basically a block of suitably shaped cast-iron with its centre of gravity centred near the base in order to facilitate smoothness of fall.

(b) **Singe Acting Hammer:** The functioning of single acting hammer differs from drop hammer only in the manner of lifting of the ram after each blow. A conventional single acting hammer employs a piston connected to a ram at its bottom end and moving inside a cylinder. The hammer may be of an open type or closed type. Steam power or compressed air is used in the single acting hammer wilthout any adjustment or alterations and the pressure remains unchanged. This pressure ranges from 5.6 to 10.5 kg per cm^2 and used depending on the size of the hammer and its weight. The operation of the single-acting hammer costs less compared to a double acting hammer but its speed is slower.

32.3.5 Movement and Hosting Equipment

Materials needed for construction are to be shifted from one place to another which may involve horizontal movement, vertical movement or other type. In such cases the following movement devices, are used:

1. Vertical Movement Devices
2. Horizontal Movement Devices
3. Combined Movement Devices

1. Vertical Movement Devices

The vertical movement devices are:

(i) Block and Tackle
(ii) Winch
(iii) Hoist
(iv) Elevators

(i) Block and Tackle

This is a vertical distance movement device which is the oldest and the simplest device. It depends on mechanical power and gives only mechanical advantage. It is the most expensive device but waste of manpower (Fig. 32.15(a)).

(ii) Winch

By winding the rope of cable on the drum vertical movement is attained. Manpower or other power can be used to wind and a greater mechanical advantage than that of block and tackle. It is after used to load heavy equipment into ships, construction equipment, etc. (Fig. 32.15(b)).

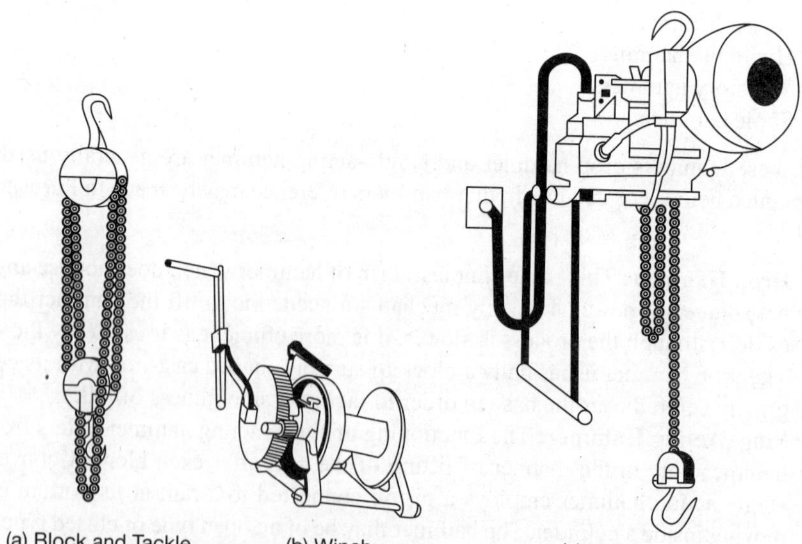

(a) Block and Tackle (b) Winch (c) Hoist

Figure 32.15 Vertical movement devices

(iii) Hoist

Hoist is operated between fixed guide rails for vertical lifting of things. Hoists are operated by hand, compressed air or by electric power. Variety of hoists is available to suit a specific purpose. The simplest is the chain hoist. Hoists are similar to elevators except that operator does not go up but operated from one point to the other (Fig. 32.15(c)).

(iv) Elevators

This is similar to hoist but with a difference that the operator can ride with the load. Among the different types of elevators, the electrical one is often used. In places where electric power can not be used hydraulic elevators are used.

2. Horizontal Movement Devices

The horizontal movement devices are:

(i) Hand Trucks
(ii) Narrow-gauge Rail Road
(iii) Tractors and Trailers
(iv) Skids

(i) Hand Trucks

Wheel barrows and hand trucks are the simplest transporting devices which are still in use. These devices involve large amount of manpower to move a small load. The advantages of these devices are small cost, flexibility, easy transportation from one place to another (Fig. 32.16(a)).

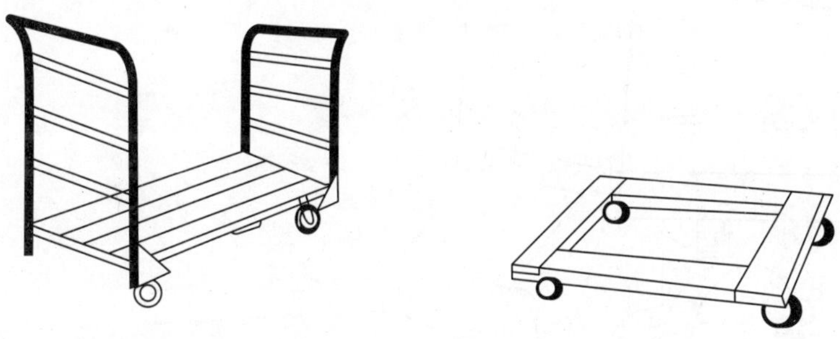

(a) Hand Truck (b) Skid

Figure 32.16 Horizontal movement devices

(ii) Narrow-gauge Rail Road

As it is very expensive, this is adopted only in very large projects. It is used to industries like blast furnace, copper refineries and steel-rolling operations.

(iii) Tractors and Trailers

These are the commonest modes of horizontal transportation. Trailers can be lift-loaded and can be towed to tractors. Different types of trailers can be picked up by tractors. This is one of the mostly used methods of handling materials from one place to another. This is less costly compared to narrow-gauge rail.

(iv) Skids

Skids are wheeled or plain trollies over which materials can be loaded and then picked up with lift trucks. Skids can be used to load and transfer from position to position without subsequent loading and unloading. The most common skids, which can be picked up in either direction, consist of a wooden platform and four steel legs (Fig. 32.16(b)).

3. *Combined Devices*

Combined devices comprise the actions of lifting, lowering or transportation. Combined devices are as follows:

(i) Chute
(ii) Lift truck
(iii) Forklift truck
(iv) Cranes
(v) Conveyers

(i) Chute

Chutes are the devices which are adopted for horizontal and vertical movements. The chutes may be straight or of spiral form.

(ii) Lift Truck

These are similar to roller skids but provide provision for a large platform to lift and place the materials and move them horizontally through power to another location (Fig. 32.17).

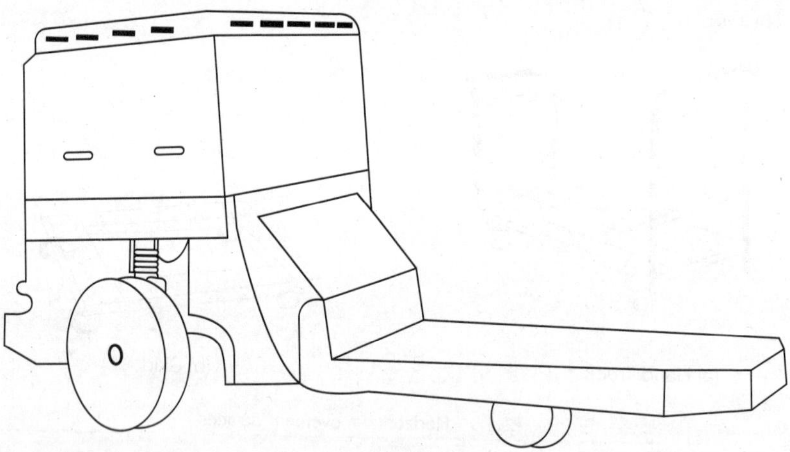

Figure 32.17 Lift truck

(iii) Fork Lift Truck

It is provided with fork which receives the load at ground level and elevates it hydraulically to the desired height. There is no need for manual lifting. Self-loading or unloading can be carried out by providing a fork at the front end of the truck. Fork lift trucks are used in construction industry (Fig. 32.18).

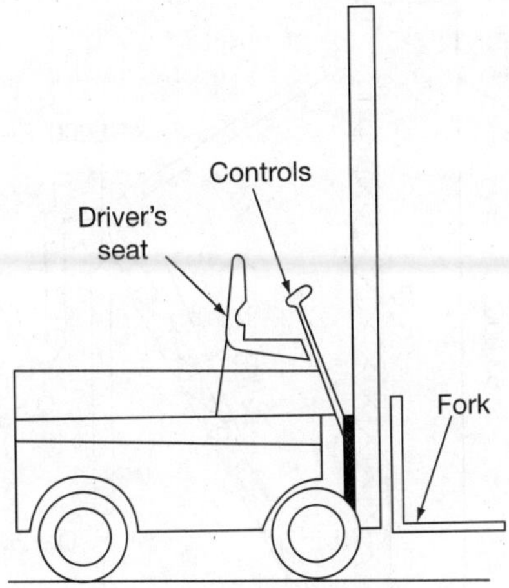

Figure 32.18 Fork-lift truck

(iv) Cranes

Cranes are electrically or diesel operated equipment used to lift and move heavy materials and machinery. Cranes have wide application in construction projects, industries and in shipping yard. Cranes have three motions, viz., hoisting, derricking and slewing.

Cranes are classified as given below:

(a) Derrick cranes
(b) Tower cranes
(c) Pillar cranes
(d) Overhead or gantry cranes
(e) Self-propelling boom cranes
(f) Crane trucks

(a) **Derrick Cranes:** They consist of a mast, a boom and a bull wheel on which it rotates about a vertical axis and with supporting members (also referred to as guys).

Most of the derrick cranes are supported by a number of guys. The boom can revolve through 360° and passes below these guys. A bull wheel is attached to the mast and rotates it. These cranes are operated by diesel engine or by an electric motor.

Guy derricks are available in lifting capacities of 5–200 tonnes and used mainly for erecting heavy structures, Fig. 32.19 (Ataev, 1985).

These cranes are used in construction projects like industries or multi-storeyed buildings, loading and unloading of cargoes at ports, in ship building, etc.

(b) **Tower Cranes:** These are used mainly in erection of apartment and high-rise industrial buildings.

Main advantages of tower rail-mounted cranes are their stability and a large reach of hook. Main construction is served by tower cranes of capabilities varying from 8 to 10 tonnes lifting capacities (Fig. 32.20).

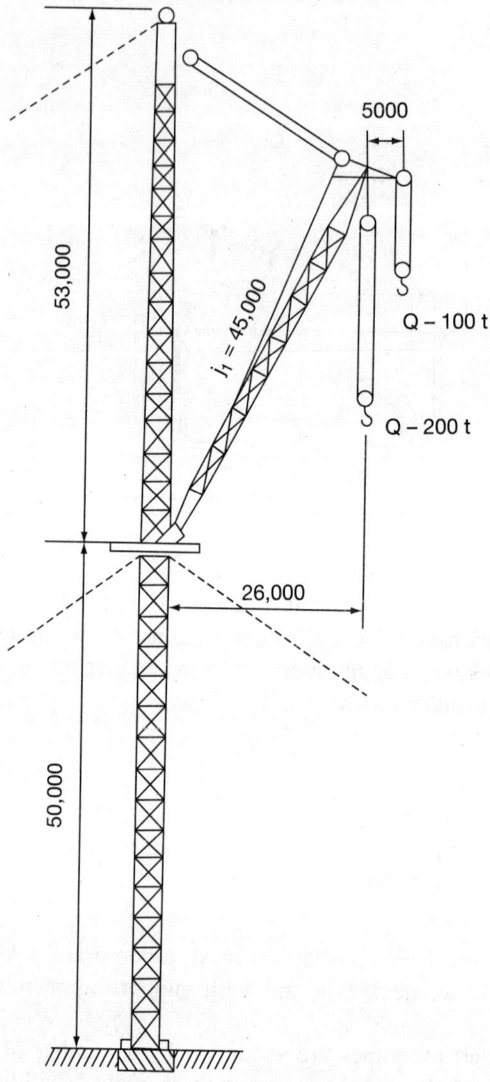

Figure 32.19 200-tonne guy derrick

(c) **Pillar Cranes:** Pillar cranes may be a stationary or mobile type. It is used for light load (up to 20 tonnes). In order to lift up or lower down the load a job or inclined boom is fixed to the lift and the lifting is done with the help of rope and pulley arrangement. All the movements to the crane required for a particular situation are provided by gearing and electric power drive (Fig. 32.21).

(d) **Overhead Cranes:** Overhead cranes are fixed in one location and can not be moved to place to place. These are provided in big workshops, boundaries, powerhouses, chemical plants, research stations, etc. These are operated by electric power. These cranes have the advantage of providing large service area. All the operations can be done by an operator seated in the cabin (Fig. 32.22).

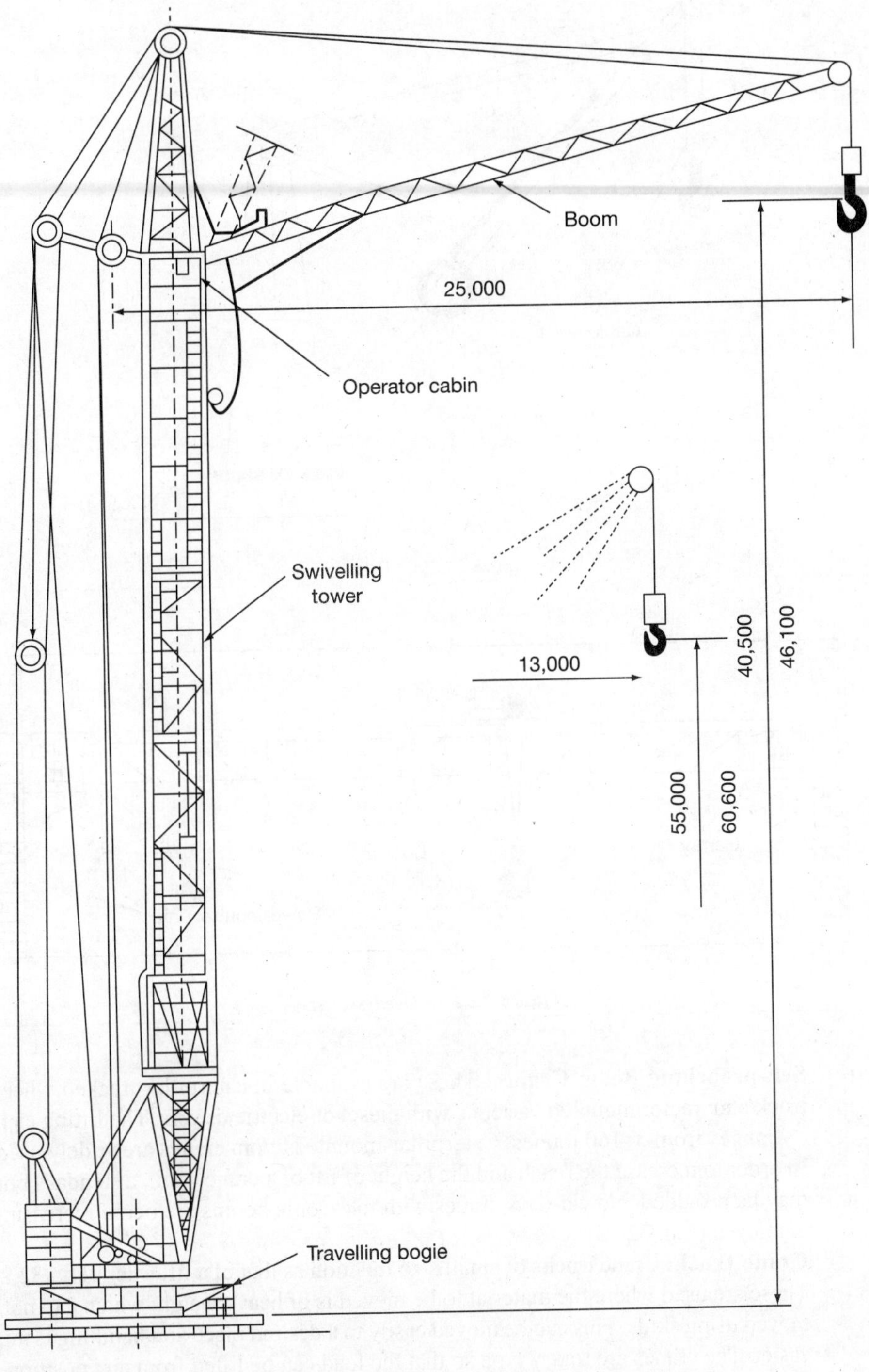

Figure 32.20 Tower crane for high-rise house construction

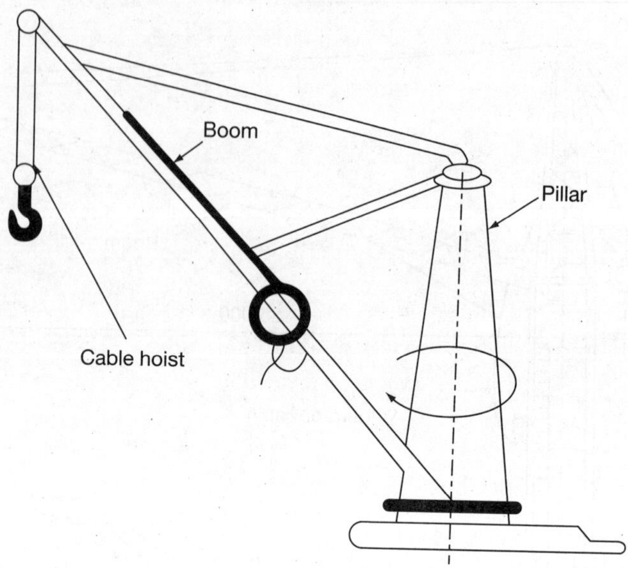

Figure 32.21 Pillar crane

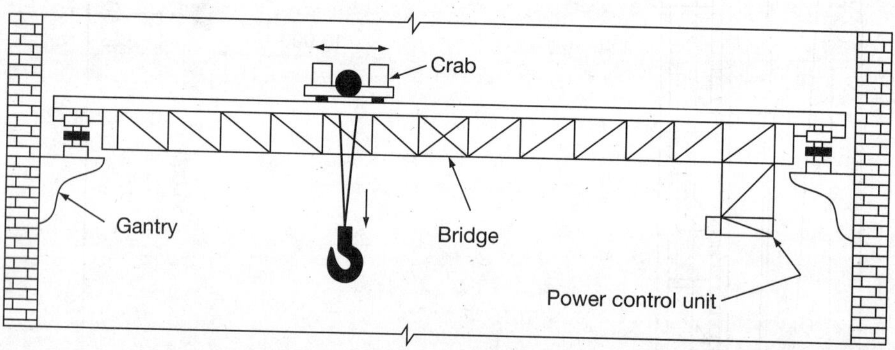

Figure 32.22 Overhead crane

(e) **Self-propelling Boom Cranes:** These are available in caterpillar-tracked wheeler, truck and tractor mounted varieties with diesel or electric drivers. The lifting capacity ranges from 3–160 tonnes. Caterpillar-mounted boom cranes are widely used. In order to increase the reach and the height of lift of a crane hook, a standard boom may be provided. Now-a-days cranes with telescopic booms are being increasingly employed (Fig. 32.23).

(f) **Crane Truck:** Crane trucks of small size function as that of a lift truck (Fig. 32.24). These are used where the material to be moved is of heavy weight which can not be moved using skids. This can be moved easily to a desired place and handling is much easier. The cranes are rotary type so that the load can be lifted from any position.

(g) **Conveyors:** These are the material transportation devices used when the parts of flow of material is fixed. Because of this desired fixity, lifting and lowering of

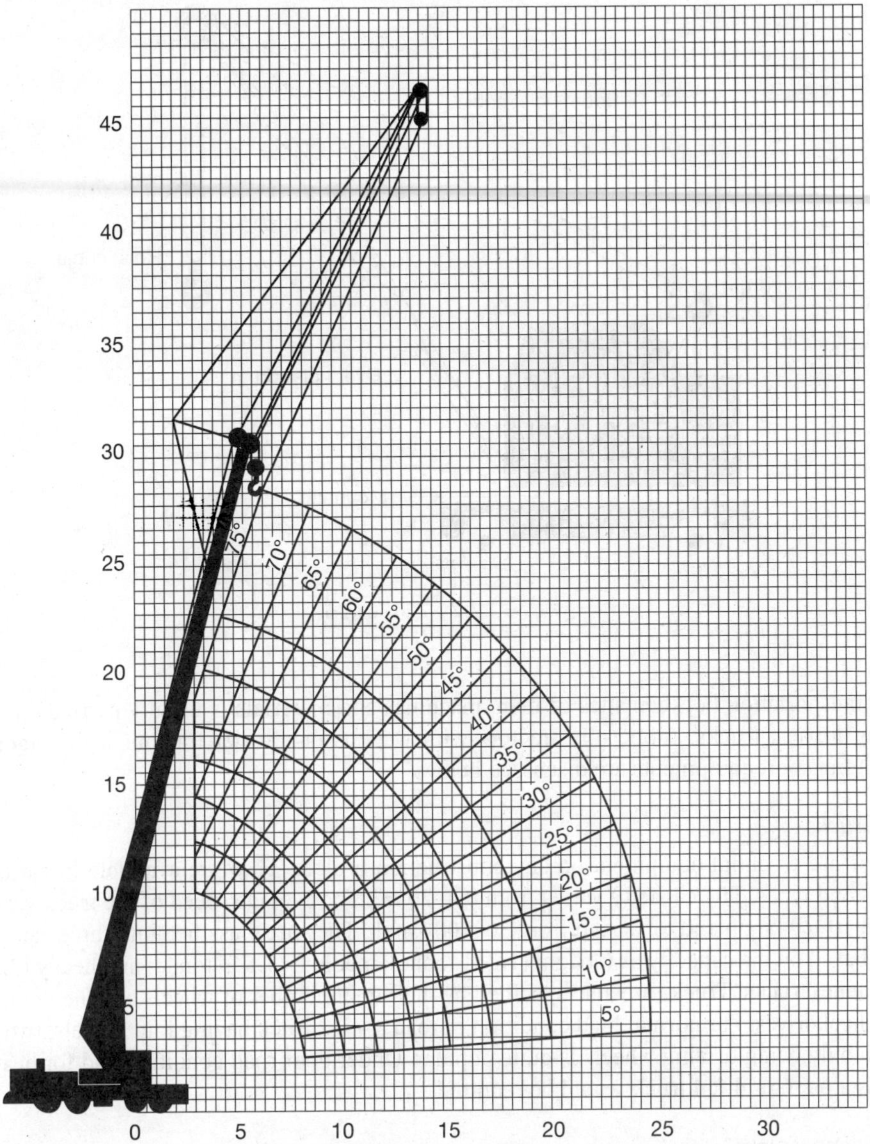

Figure 32.23 Crane with telescopic boom

materials are done automatically. Conveyors require no stopping or starting but the operation is continuous. The transportation is effected by friction between materials being transported and the belt or roller.

32.3.6 Hauling Equipment

Hauling of materials, machines and/or personnel around and between building sites can be very time consuming, expensive and unproductive. Movement of materials within a short distance can be done by using any one of the movement-equipment explained in the previous section.

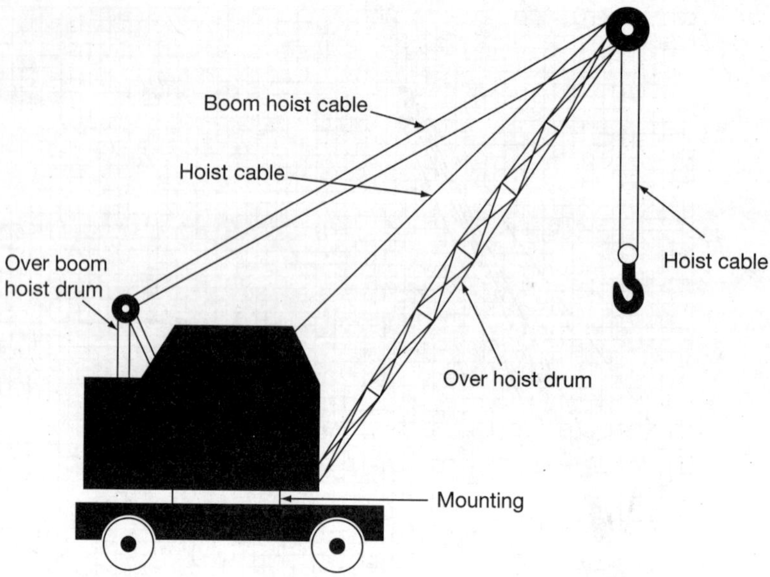

Figure 32.24 Crane truck

Transportation between sites of men, machines and materials is usually carried out by using lorries, trucks, rubber-tire tractors with wagons or crawler tractors with wagons. Such transport equipments are discussed below.

1. Trucks

These are essential equipment on any construction project. They are available in various sizes and types. The capacity varies from 0.4–20 m^3. The average speed of the truck varies from 10–20 kmph and top speed from 30–100 kmph. They are classified into three categories based on the capacity as (i) light (1/2–1 tonne), medium (1.5–3 tonne) and heavy (3.5–10 tonne) trucks. Trucks are also specified by their total number of wheels and the number of wheel drives. For normal load condition of road a 4×2 truck having four wheels, two of them only being driving wheels is quite popular. Under poor road condition and for heavy loads a 6×6 or 4×4 truck would be required.

2. Dump Trucks

These trucks are fitted with automatic unloading devices. The loading is done either by loading shovels or loaders. These trucks may have capacities as high as 50 tonnes. These trucks can be rear dump trucks, bottom dump trucks and side dump trucks.

3. Dumpers

Trucks with special pneumatic wheels and capable of moving at high speeds having short chassis and strong dumping bodies are known as dumpers. The material can be dumped in front or at the back. The loading, transporting and dumping can be done in quick succession. They are available in capacities up to 4.5 m^3.

4. Trailers

Trailers are carriages hauled by power units such as tractors or trucks. Trailers are of two types, viz., full trailer or wagons and semi-trailers. Full trailers are provided with power unit with swivelling axle and drawbar. The semi-trailers are supported in the front by the power unit and at the rear on their own wheels. These trailers can also be operated in a smaller space. Both these trailers are provided with hydraulically-operated breaking systems.

SALIENT POINTS

1. Procurement cost includes (i) the cost price of the equipment, (ii) interest on money invested on the purchase, (iii) taxes and (iv) the insurance cost.
2. Operation cost of an equipment is based on the following factors: (i) cost of investment, (ii) depreciation cost, (iii) cost of major repair, (iv) cost of fuel and lubricants, (v) cost of labour, (vi) servicing and field repairs and (vii) overheads.
3. Major equipment needed for a building project are (i) earthwork equipment, (ii) de-watering equipment, (iii) compaction equipment, (iv) pile-driving equipment and (v) movement equipment.
4. Earthwork equipment comprises of excavators, shovels, bulldozers, tractors, motor graders, scrapers and loaders.
5. Shovels are classified as (i) dipper shovel, (ii) drag shovel or hoe (iii) dragline and (iv) clamshell.
6. Bulldozers are classified as angle dozer, tilt dozer and tree dozer.
7. Scrapers are provided with bowl, apron, ejector and hydraulic system.
8. Operations of a conventional scraper are digging or loading, transporting and unloading.
9. Loaders are of two types, viz., crawler loader and wheeled loader.
10. De-watering methods are: (i) pumping methods, (ii) electro-osmosis and (iii) elimination of groundwater by cement grouting, chemical consolidation, displacement grouting and freezing.
11. Pumping methods comprise of open sumps and ditches, well-point systems, deep-well drainage and vacuum de-watering.
12. Compaction of material is obtained by mechanical means such as rolling, kneading vibrations and ramming.
13. Rollers are classified as smooth-wheeled rollers, sheepsfoot rollers, pneumatic-tired rollers and tandom compactors.
14. Pile driving equipment comprises of the following components: (i) driving rigs, (ii) guiding loaders, (iii) pile hammer with accessories, (iv) additional aids for pre-boring and jetting and (b) boiler for steam raising or air compressor.
15. Moving equipment are (i) vertical movement devices, (ii) horizontal movement devices and (iii) combined movement devices.

16. Vertical movement devices are (i) block and tackle, (ii) winch, (iii) hoist and (iv) elevators.
17. Horizontal movement devices are (i) hand trucks, (ii) narrow-gauge rail load, (iii) tractors and trailers and (iv) skids.
18. Combined devices are (i) chute, (ii) lift truck, (iii) fork-lift truck, (iv) cranes and (v) conveyers.
19. Hauling equipment are trucks, dump trucks, dumpers and Trailors.

REVIEW QUESTIONS

1. Explain the factors to be considered in the selection of construction equipment.
2. How do you select equipment for earthwork?
3. What are the different types of excavators? Explain any one in detail.
4. Distinguish between dipper shovel and drag shovel.
5. Explain a clamshell along with its applications.
6. How a dragline works? In what situation can they be used?
7. Suggest the method of dewatering the foundation trenches under the following condition.
   (i) Excavation is to be carried out at a location where the area is water-logged.
   (ii) If the permeability of a site very much less than 10^{-4} cm per second.
   (iii) Relatively homogeneous soft day.
8. Briefly explain the process of electro-osmosis for dewatering the foundation trenches.
9. What are the different types of tractors? Explain.
10. Explain the different parts of a scraper.
11. Discuss the operations of a scraper.
12. What is a front-end loader?
13. Explain the well-point system of de-watering.
14. What is deep-well drainage?
15. Explain vacuum de-watering system.
16. Explain the methods of de-watering foundation excavation.
17. What are the different types of rollers? Explain.
18. Distinguish between drop hammer and single-acting hammer.
19. Name the different components of a pile-driving equipment.
20. Explain briefly different horizontal movement devices.
21. What are different types of handling devices? Explain.
22. What are the different types of cranes? Explain the types in detail.
23. Explain different types of hauling equipment.

OBJECTIVE TYPE QUESTIONS AND ANSWERS

CHAPTER 1

Properties of Construction Materials

1. Mass per unit volume of a homogeneous material is termed as
   (a) Unit weight
   (b) Specific gravity
   (c) Density
   (d) Porosity
   (e) Density index

2. Spalling resistance of a material depends mainly on
   (a) Coefficient of linear expansion of the material
   (b) Coefficient of compression of the material
   (c) Contraction property of the material
   (d) Non-homogeneous property of the material
   (e) Isotropic property of the material

3. Hygroscopicity of a material is the property of a material
   (a) To absorb water through the pores
   (b) To absorb water vapour from air
   (c) To absorb pure air from the atmosphere
   (d) To collect the dust from air
   (e) To collect all the impurities in water

4. Various mechanical properties of a material are given below. Identify the wrong property.
   (a) Elasticity
   (b) Plasticity
   (c) Conductivity
   (d) Hardness
   (e) Toughness

5. The hardest material is
   (a) Quartz
   (b) Calcide
   (c) Talcum
   (d) Rock-salt
   (e) Diamond

6. Dielectric strength is the insulating capacity of a material against
   (a) High temperature
   (b) High voltage
   (c) Low current
   (d) Low resistance
   (e) High density

8. Coercive force is the __________ which is necessary to neutralize completely the magnetism in an electromagnetic field.
   (a) Tensile force
   (b) Compressive force
   (c) Magnetising force
   (d) Shear force
   (e) Neutralising force

7. Electrical conductivity is the reciprocal of
   (a) Current
   (b) Voltage
   (c) Superconductivity
   (d) Electrical resistivity
   (e) Potential difference

9. Some of the chemical properties are given below. Choose the wrong one.
   (a) Corrosion resistance
   (b) Chemical composition
   (c) Acidity
   (d) Alkalinity
   (e) Permeability

CHAPTER 2

Building Stones

1. Cooling and hardening of mortar magma resulted in the formation of
   (a) Magna rock
   (b) Metamorphic rock
   (c) Intrusive rock
   (d) Sedimentary rock
   (e) Igneous rock

2. Sediments when subjected to heavy pressure undergo compaction and cementation, resulting in
   (a) Metamorphic rocks
   (b) Igneous rocks
   (c) Extrusive rock
   (d) Sedimentary rock
   (e) Intrusive rocks

3. Metamorphic rock are given below. Choose the wrong one.
   (a) Gneiss
   (b) Schist
   (c) Slate
   (d) Sandstone
   (e) Marble

4. Stratified structure is formed due to the splitting of a series of paralled layers of
   (a) Metamorphic rocks
   (b) Extrusive igneous rocks
   (c) Sedimentary rocks
   (d) Intrusive igneous rocks
   (e) None of the above

5. Calcium carbonate or lime is the main constituent in this type of rock.
   (a) Argillaceous rocks
   (b) Calcareous rocks
   (c) Siliceous rocks
   (d) Foliated rocks
   (e) Un-stratified rocks

6. Basalt is of __________ origin.
   (a) Igneous rock
   (b) Sedimentary rock
   (c) Metamorphic rock
   (d) Plutonic rock
   (e) Stratified rock

7. Sandstone is
   (a) Igneous rock
   (b) Volcanic rock
   (c) Sedimentary rock
   (d) Metamorphic rock
   (e) Plutonic rock

8. Marble is classified as
   (a) Foliated rock
   (b) Metamorphic rock
   (c) Igneous rock
   (d) Sedimentary rock
   (e) Volcanic rock

9. Methods of quarrying are given below. Choose the wrong-one.
   (a) Digging
   (b) Heating
   (c) Wedging
   (d) Dressing
   (e) Blasting

10. Agencies which are responsible for the deterioration of stones are rain, wind, chemicals, vegetable growth and __________
    (a) Living organism
    (b) Moon light
    (c) Colour of the stone
    (d) Grains of the rock
    (e) Origin of rock

11. Acid test on stones is conducted to find the
    (a) Hardness
    (b) Toughness
    (c) Strength
    (d) Presence of alkaline or lime content
    (e) Presence of impurities

12. Smith's test is performed on stones to find the presence of
    (a) Earthly matter
    (b) Iron
    (c) Dissolved salt
    (d) Cracks
    (e) Foliation

13. Impact test determines the _________ of a stone
    (a) Abrasion quality
    (b) Toughness
    (c) Mineral constituents
    (d) Texture
    (e) Fissures

14. A good building stone should have
    (a) Strength
    (b) Hardness and toughness
    (c) Resistance to fire
    (d) Good appearance and colour
    (e) All the above

CHAPTER 3

Bricks

1. The main constituents of Brick earth are given below. Choose the wrong one.
   (a) Alumina and magnesia
   (b) Lime
   (c) Mica
   (d) Silica
   (e) Iron Oxide

2. Which one of the following impurity is not desirable in the soil used for brick formation?
   (a) Alkali
   (b) Kankar
   (c) Iron oxide
   (d) (a) and (b) above
   (e) (b) and (c) above

3. For good quality bricks, the percentage of alumina by weight should be
   (a) 40 to 50
   (b) 50 to 65
   (c) 20 to 30
   (d) 85 to 50
   (e) 10 to 20

4. Sometimes additional soil such as sandy or calcareous clays are added in suitable proportion to the natural soil to improve the quality. This process is known as
   (a) Mixing
   (b) Blending
   (c) Tempering
   (d) Weathering
   (e) Digging

5. The process of kneading the soil under feet of men or cattle after adding necessary quantity of water in order to make the soil stiff and homogeneous is called
   (a) Blending
   (b) Weathering
   (c) Spreading
   (d) Tempering
   (e) None of the above

6. Bricks attain red colour due to the presence of
   (a) Iron oxide
   (b) Lime
   (c) Silica
   (d) Magnesia
   (e) Alumina

7. Nominal size of modular bricks is
   (a) 18 cm × 8 cm × 8 cm
   (b) 18 cm × 9 cm × 4.5 cm
   (c) 19 cm × 9 cm × 9 cm
   (d) 18 cm × 9 cm × 9 cm
   (e) 20 cm × 10 cm × 10 cm

8. Heating the brick earth beyond 1300°C the material gets
   (a) Vitrified
   (b) Expanded
   (c) Red hot
   (d) Boiled
   (e) Liquified

9. As per Indian Standard bricks are classified based on _________ strength.
   (a) Tensile strength
   (b) Compressive strength
   (c) Shear strength
   (d) (a) and (c)
   (e) None of the above

10. Bricks containing a relatively large proportion of soluble salts are liable to become discoloured by the formation of a whitish deposit, known as
    (a) Chuffs
    (b) Bloating
    (c) Efflorescence
    (d) Lamination
    (e) Nodules

11. Brick moulded with a rounded angle is termed
(a) Circular brick
(b) Plinth brick
(c) Squint brick
(d) Bull-nose brick
(e) Copying brick

12. Refractory bricks are prepared from __________ clay in the same manner as the ordinary bricks.
(a) Fire
(b) Soft
(c) Stiff
(d) Silty
(e) Sandy

CHAPTER 4

Tiles and Ceramic Materials

1. Terracotta is a hard brownish unglazed vitrified __________ material used for ornamental purposes.
(a) Clayey sand
(b) Refractory clay
(c) Ceramic
(d) Ground stone
(e) Soft clay

2. Important characteristics of good tile are
(a) Should be uniformly burnt with uniform colour
(b) Should have even and compact surface
(c) Should not have cracks, flows or bends
(d) Should be hard and durable
(e) All the above

3. Common clay tiles are classified as clay Roof tiles, Clay Floor Tiles, Clay ceiling Tiles and
(a) Mangalore clay tiles
(b) Clay terracing tiles
(c) Clay pot tiles
(d) Country tiles
(e) Clay flat tiles

4. Manufacturing of common clay tiles involve the following operations.
(a) Preparation of clay
(b) Moulding of tiles
(c) Drying of tiles
(d) Burning of tiles
(e) All the above

5. Burning of tiles in kilns are done by firing initially with 800°C, slackened for six hours and increased to __________ for about 3 hours.
(a) 1000°C
(b) 1200°C
(c) 1300°C
(d) 1400°C
(e) 1600°C

6. The clay selected for terra-cotta for tiles should contain __________ of iron oxide and about 1% of lime.
(a) 5 to 8%
(b) 6 to 10%
(c) 3 to 5%
(d) 9 to 12%
(e) 7 to 15%

7. For terra-cotta tile making the temperature of the kiln is raised to about
(a) 1500°C
(b) 1400°C
(c) 1300°C
(d) 1200°C
(e) 1100°C

CHAPTER 5

Lime

1. Limestone is obtained by burning one of the naturally available materials such as
(a) Limestones found in limestone hills
(b) Limestone builders found in the bed of old rivers
(c) Kankar found below ground
(d) Shells of sea animals
(e) One of the above

2. In order to get the best quality lime it is recommended to have __________ of clay.
(a) 8 to 30%
(b) 5 to 15%
(c) 25 to 40%
(d) 40 to 60%
(e) 35 to 80%

3. Hydraulicity is returned to lime when the magnesium carbonate content is about __________.
   (a) 10%
   (b) 20%
   (c) 30%
   (d) 40%
   (e) 50%

4. It is undesirable to have thin material in limestone as it is harmful and liable to produce poor quality of lime. The material is
   (a) Alumina
   (b) Iron oxide
   (c) Silica
   (d) Magnesia
   (e) Pyrites

5. Increase in clay content in lime makes the slaking difficult and increases the __________ properties.
   (a) Hydraulic
   (b) Chemical
   (c) Physical
   (d) Fatigue
   (e) Shrinking

6. For manufacturing of fat lime, the percentage of impurities in limestone should not exceed
   (a) 30%
   (b) 25%
   (c) 20%
   (d) 10%
   (e) 5%

7. In this kiln for making lime, the fuel is not allowed to come in contact with limestone. Name the kiln.
   (a) Intermittent – flare kiln
   (b) Intermittent – flame kiln
   (c) Clamp burning
   (d) Continuous kiln
   (e) None of the above

8. The process of adding water to lime to convert it to hydrated lime is known as
   (a) Calcination
   (b) Watering
   (c) Quenching
   (d) Slaking
   (e) Hydration

9. Physical tests on limestone are based on the following properties
   (a) Texture
   (b) Appearance
   (c) Colour
   (d) Odour
   (e) All the above

10. Limestone loses its weight due to heating because of removal of
    (a) Carbon dioxide
    (b) Sulphur dioxide
    (c) Oxygen
    (d) Hydration
    (e) None of the above

11. In acid test, abundant efflorescence indicates high percentage of
    (a) Sodium carbonate
    (b) Calcium carbonate
    (c) Potassium carbonate
    (d) Sodium sulphate
    (e) Calcium sulphate

12. Use of pozzolanic materials provide the properties given below. Choose the correct one
    (a) Improves workability
    (b) Lower the heat of hydration
    (c) Reduces shrinkage
    (d) Improves hydraulic properties
    (e) All the above

CHAPTER 6

Cement

1. The main constituent which contributes for strength of cement is
   (a) Silica
   (b) Alumina
   (c) Lime
   (d) Iron Oxide
   (e) Magnesium oxide

2. When water is added to cement and mixed, it forms a paste which gradually becomes hard. The time taken for the entire process is called
   (a) Setting time
   (b) Elapsed time
   (c) Calculated time
   (d) Consolidated time
   (e) Strength time

3. Cement which has less expansive qualities when mixed with water is termed
   (a) Fitness
   (b) Solidification
   (c) Soundness
   (d) Calcinated
   (e) Hydrated

4. Unsoundness of cement may be controlled by
   (a) Limiting the MgO content to less than 0.5%
   (b) Fine grinding
   (c) Allowing the cement for aeration for several days and
   (d) Through mixing
   (e) All the above

5. The reaction of silicates and aluminates of cement with water forms a binding medium which solidifies into a hardened mass which is termed as
   (a) Hydration
   (b) Water tightening
   (c) Solidification
   (d) Liberation
   (e) Contraction

6. An inactive part of cement is called the insoluble residue. The maximum allowable value is
   (a) 0.50%
   (b) 0.65%
   (c) 0.70%
   (d) 0.80%
   (e) 0.85%

7. Clinker from the kiln is cooled and then ground in a ball mill with the addition of __________ of gypsum.
   (a) 1 to 2%
   (b) 2 to 3%
   (c) 3 to 4%
   (d) > 3%
   (e) > 5%

8. Out of the constituents of cement like Tri-calcium silicate, Di-Calcium silicate, Tri-calcium aluminate, the first to set and harden is
   (a) Tri-calcium silicate
   (b) Di-calcium silicate
   (c) Tri-calcium aluminate
   (d) (a) and (b) above
   (e) All set simultaneously

9. Portland slag cement is made by inter-grinding Portland Cement clinker and
   (a) Dicalcium silicate
   (b) Granulated blast furnace waste
   (c) 4% of Gypsum
   (d) Magnesium oxide
   (e) Free lime

10. Hydrophobic Portland cement is produced by inter-grinding Portland cement with 0.1 to 0.4%
   (a) Dilute Hydrochloric acid
   (b) Hydrogen per oxide
   (c) Stearic acid
   (d) Sulphur-di-oxide
   (e) Calcium chloride

11. In quick setting cement the compared added is
   (a) Aluminium Sulphate
   (b) Gypsum
   (c) Aluminium silicate
   (d) Calcium sulphate
   (e) Magnesium sulphate

12. Fineness of cement is represented by specific surface which is expressed as total surface area in
   (a) Square cm
   (b) Square cm/gm
   (c) Cubic cm/gm
   (d) Gm/square cm
   (e) Gm/cm

13. Le Chatelier's apparatus is used to find the __________ of cement.
   (a) Initial setting time
   (b) Final setting time
   (c) Soundness
   (d) Strength
   (e) Consistency

14. Strength of cement is found from conducting test on.
   (a) Cement mortar cube of 1:3 mix
   (b) Cement mortar cylinder of 1:3 mix
   (c) Cement mortar cube of 1:4 mix
   (d) Cement concrete cube of 1:2:4
   (e) Cement concrete of cylinder 1:1½:2

15. Brazilian test is conducted to find the __________ strength
    (a) Compress
    (b) Bending
    (c) Shear
    (d) Tensile
    (e) Bond

16. Ratio of percentage of alumina to that of iron oxide in Ordinary Portland Cement is
    (a) > 0.70
    (b) $= 0.80$
    (c) $\nless 0.66$
    (d) $\ngtr 0.66$
    (e) Between 0.80 and 0.90

17. Loss on ignition in cement should
    (a) be equal to 0.30%
    (b) $\ngtr 4\%$
    (c) $< 2\%$
    (d) Between 2 to 2.5%
    (e) None of the above

CHAPTER 7

Mortrar

1. Coarse sand is one which is passing through a screen with clear openings of
   (a) 3.18 mm
   (b) 2.18 mm
   (c) 1.38 mm
   (d) 8.31 mm
   (e) 1.81 mm

2. The fineness modulus of sand should be between
   (a) 1 and 2
   (b) 2 and 3
   (c) 3 and 4
   (d) 4 and 5
   (e) 5 and 7

3. Water is taken as free from organic matter if the *pH* value lies between
   (a) 8 and 10
   (b) 9 and 10
   (c) 6 and 8
   (d) 4 and 6
   (e) 3 and 5

4. Presence of salts in water should __________.
   (a) be 8000 ppm
   (b) not be greater than 15000 ppm
   (c) not be greater than 25000 ppm
   (d) not be greater than 10000 ppm
   (e) be between 34000 to 26000 ppm

5. Cement performs the following functions when used in cement mortar. Choose the correct one
   (a) Primarily fills the voids formed by fine aggregates
   (b) Binds the fine aggregates into a solid mass with time when added with water to the mix
   (c) Impacts strengths after setting
   (d) Only (b) and (c)
   (e) (a), (b) and (c)

6. Surkhi is another form of fine aggregate which is usually prepared by powdering
   (a) Limestones
   (b) Unburnt slag
   (c) Unburnt bricks
   (d) Furnace waste
   (e) Sandstone

7. For external walls without protection the grade of cement mortar to be used is
   (a) MM 2
   (b) > MM 3
   (c) MM7.5
   (d) MM 5 to MM 7
   (e) < MM 2

8. Fire-resistant mortar is obtained by adding aluminuous cement to the finely ground powder of
   (a) Half-burnt bricks
   (b) Slag
   (c) Fly-ash
   (d) Fire-bricks
   (e) Table-moulded bricks

9. Packing mortar is a special type of mortar possessing the property of
   (a) High homogeneity and water resistance
   (b) Pre-determined setting time
   (c) Ability to form solid and water proof plugs
   (d) (b) and (c) only
   (e) (a), (b) and (c)

10. For sound-absorbing mortar the type of crushed aggregate used is
    (a) Uniformly graded gravel
    (b) Light-weight porous material
    (c) Well-graded strong aggregate
    (d) Poorly-graded sandstone
    (e) River-pebbles

11. X-ray shielding mortar is obtained from cement, admixtures and
    (a) Light-weight porous pumice
    (b) Uniformly graded sand
    (c) Coarse sand
    (d) Crushed heavy rocks
    (e) Crushed cinders

12. The cement mortar used for pointing works is
    (a) 1.5 to 1.6
    (b) 1.3 to 1.4
    (c) 1:6
    (d) 1:3
    (e) 1:1 to 1:2

CHAPTER 8

Concrete

1. Strength of cement concrete depends on
   (a) Quality of water
   (b) Quality of aggregate
   (c) Quantity of aggregate
   (d) Quality of cement
   (e) Water-cement ratio

2. Super plasticizers permit reduction of water up to __________ without reduction in workability.
   (a) 10%
   (b) 20%
   (c) 30%
   (d) 40%
   (e) 50%

3. The commonly used retarder is
   (a) Calcium sulphate
   (b) Potassium sulphate
   (c) Barium sulphate
   (d) Sulpur chloride
   (e) Calcium chloride

4. Accelerators are used in situations as given below. Choose the correct one
   (a) Early removal of formwork
   (b) Reducing the period of curing
   (c) Accelerating the setting time in cold weather
   (d) Energy repair work
   (e) All the above

5. Different air entraining agents behave differently depending on the elasticity of the film of the bubble formed and the extent to which the __________ is reduced.
   (a) Compressive force
   (b) Surface Tension
   (c) Shear force
   (d) Torsion
   (e) Bond stress

6. Surface loss of water from concrete depends upon
   (a) Air temperature
   (b) Relative Humidity
   (c) Wind velocity
   (d) Fresh concrete temperature
   (e) All the above

7. Curing of concrete is done by
   (a) Spraying method
   (b) Bonding method
   (c) Covering with moist cloth
   (d) Covering with moist sand
   (e) Any of the above methods

8. The minimum water cement ratio required for a workable concrete is
   (a) 0.4
   (b) 0.5
   (c) 0.6
   (d) 0.7
   (e) 0.8

9. Cement concrete grading is based on the characteristic compressive strength of 150 mm cube at __________ days.
   (a) 7
   (b) 15
   (c) 21
   (d) 28
   (e) 30

10. M 45 Grade Designation is given to __________ concrete.

(a) Ordinary
(b) Low strength
(c) Standard
(d) Medium Strength
(e) High Strength

11. Separation of water on fresh concrete is called
(a) Segregation
(b) Honeycombing
(c) Hydration
(d) Bleeding
(e) Dilution

12. Bulking of sand, occurs due to
(a) Viscosity
(b) Capillary action
(c) Surface tension
(d) Moisture in voids
(e) Air in voids

13. Under-compaction makes the concrete
(a) Impermeable
(b) Segregated
(c) Lean
(d) Thin
(e) Tough

14. Exterior application of rendering made on concrete structure is denoted as
(a) Smooth finish
(b) Applied finish
(c) Rough finish
(d) Clean finish
(e) None of the above

15. Slump test is the most commonly used method of measuring __________ of concrete
(a) Consistency
(b) Compactness
(c) Toughness
(d) Rigidity
(e) Hardness

16. Normally strength of cylindrical specimens tested under compression is taken as __________ times the compressive strength of cubical specimens.
(a) 0.95
(b) 0.90
(c) 0.85
(d) 0.80
(e) 0.75

17. Schmidt's Rebound Hammer is a commonly adopted non-destructive equipment for measuring the
(a) Surface cracks
(b) Cracks at the bottom of the member
(c) Adequacy of reinforcement
(d) Surface hardness
(e) Mix proportion

18. Gamma-rays transmission method is particularly used to measure the __________ of concrete slabs of known density.
(a) Formation of cracks
(b) Thickness
(c) Compactness
(d) Rigidity
(e) Flexibility

CHAPTER 9

Precast Concrete Units

1. For maximum flexibility and convenience, the value of the basic module is
(a) 20 mm
(b) 150 mm
(c) 100 mm
(d) 90 mm
(e) 80 mm

2. Normal Length of precast slabs or other precast structural flooring units shall be in multiples of __________ M, where M is module
(a) 5 M
(b) 6 M
(c) 4 M
(d) 3 M
(e) 2 M

3. Normal width of precast beams shall be in multiples of
(a) M
(b) M/2
(c) M/4
(d) 2 M
(e) 1.5 M

4. Crucial factors which are to be considered for effective production and supply of precast concrete components are:

(a) Storage facilities
(b) Suitable transport facilities
(c) Erection equipment
(d) Availability of raw material
(e) All the above

5. Prefabricated RCC lintels are preferred for small spans up to _________
(a) 1 m
(b) 2 m
(c) 3 m
(d) 2.5 m
(e) 3.5 m

CHAPTER 10

Timber and Industrial Timber Products

1. Which one of the following is not an endogeneous tree?
(a) Bamboo
(b) Teak
(c) Palm
(d) Cane
(e) Coconut

2. Which one of the following is not an exogeneous tree?
(a) Teak
(b) Babul
(c) Sal
(d) Palm
(e) All of the above

3. Qualities of a softwood are the following. Choose the correct one.
(a) Light in weight and colour
(b) Has straight fibres and resinous
(c) Has distinct annular rings
(d) Weak and can split easily
(e) All the above

4. Qualities of a hardwood are the following. Choose the correct one
(a) Heavy in weight and dark in colour
(b) Close-grained and non-resinous
(c) No distinct annular rings
(d) Strong and durable
(e) All the above

5. Age of a tree may be estimated from
(a) Height of the tree
(b) Diameter of the bark
(c) Number of heart wood rings
(d) Cambrian layers
(e) Number of rings on the bark section

6. The central portion of a tree is called
(a) Cambrian layer
(b) Pith
(c) Sapwood
(d) Heart wood
(e) Inner Bark

7. In a tree the layer between the inner bark and sapwood is called
(a) Pith
(b) Cambium layer
(c) Heart wood
(d) Medullary rays
(e) Outer bark

8. Following are the type of woods which can be used for permanent structures. Choose the wrong one
(a) Hardwood like teak
(b) Softwood like deodar
(c) Hardwood like seal
(d) Softwood other than deodar
(e) (a) and (c)

9. When the diameter of the knot is _________ it is called small knot.
(a) < 6 mm diameter
(b) 6 to 20 mm diameter
(c) > 20 mm diameter
(d) > 40 mm diameter
(e) None of the above

10. The shakes which split or crack with more width at the centre and diminishing towards the circumference is called
(a) Heart shakes
(b) Star shakes
(c) Cup shakes
(d) Radial shakes
(e) Ring shakes

11. Sawing of a log along the diameter is called
(a) Half Sawing
(b) Quarter Sawing

(c) Ordinary Sawing
(d) Tangential Sawing
(e) Rift Sawing

12. Seasoning of timber is done to
(a) Heat the timber
(b) Make it water proof
(c) Make it free from organic matter
(d) Expel the moisture from timber
(e) Increase the strength

13. Superior quality of timber is obtained by
(a) Air seasoning
(b) Water seasoning
(c) Kiln seasoning
(d) Chemical seasoning
(e) Electrical seasoning

14. By this preservative treatment the timber is preserved against rot and attack by white ant.
(a) Tarring
(b) Charring
(c) Soaking
(d) Creosoting
(e) Hot and Cold

15. In plywood manufacturing, the veneers are placed so that the grains of adjacent veneers
(a) Run parallel
(b) Run at right angles
(c) Run at 60°
(d) Run at 45°
(e) Run at 30°

16. As per (CPWD) specification hardwood is said to be medium type when the density is in the range of
(a) 200 to 400 kg/m^3
(b) 480 to 800 kg/m^3
(c) 700 to 800 kg/m^3
(d) 800 to 1200 kg/m^3
(e) 1200 to 1400 kg/m^3

CHAPTER 11

Metals and Alloys

1. Pig iron is extracted from one or all of the following iron ores.
(a) Magnetite
(b) Haematite
(c) Limonite
(d) Siderite
(e) All the above

2. Percentage of iron content is least in the iron ore
(a) Haematite
(b) Limonite
(c) Magnetite
(d) Pyrite
(e) Siderite

3. Cast iron is an alloy of iron and carbon with carbon more than ________.
(a) 2%
(b) 3%
(c) 4%
(d) 5%
(e) 6%

4. Hard steel contains percentage of carbon as
(a) < 0.25%
(b) up to 0.25%
(c) 0.25 to 0.70
(d) 0.70 to 1.00%
(e) 0.70 to 1.50%

5. Addition of ________ provides high electrical resistance and magnetic permeability used in electrical machinery.
(a) Manganese
(b) Tungsten
(c) Nickel
(d) Chromium
(e) Silicon

6. Aluminium is an abundantly available through the ore called.
(a) Bauxite
(b) Galenite
(c) Zincite
(d) Dolomite
(e) Tinstone

7. Magnesium occurs in the following mineral
(a) Magnesite
(b) Dolomite
(c) Kieserite
(d) Carnalite
(e) All the above

8. An increase in percentage of tin increases the __________ strength of bronze.
   (a) Compressive
   (b) Shear
   (c) Bond
   (d) Tensile
   (e) All the above

9. The mechanical properties of brass and corrosion resistance of the brasses are further improved by addition of __________.
   (a) Ferrous metals
   (b) Non-ferrous metals
   (c) Aluminium alloys
   (d) Magnesium alloys
   (e) Nickel alloys

10. Principal elements which are alloyed with pure aluminium to improve its tensile strength and hardness are
   (a) Copper and silicon
   (b) Manganese and zinc
   (c) Magnesium and nickel
   (d) None of the above
   (e) All of (a), (b) and (c)

CHAPTER 12

Surface Finishing Materials

1. Surface finishing materials are
   (a) Paints and Varnishes
   (b) Distempers and white washing
   (c) Distempers and colour washing
   (d) Paints and distempers
   (e) All the above

2. The chief ingredient of a paint is
   (a) Inert filler
   (b) Vehicle
   (c) Base
   (d) Pigment
   (e) Thinner

3. A cheap pigment added to a paint to reduce its cost is
   (a) Inert filler
   (b) Base
   (c) Thinner
   (d) Pigment
   (e) Vehicle

4. Addition of this material increases the workability of the paint.
   (a) Base
   (b) Thinner
   (c) Drier
   (d) Vehicle
   (e) Inert filler

5. This type of pain is applied on the surfaces which are exposed to acidic gases and steam
   (a) Bituminous paint
   (b) Cellulose paint
   (c) Asbestos paint
   (d) Plastic paint
   (e) Luminous paint

6. The resinous substances are
   (a) Amber
   (b) Common resin
   (c) Copal
   (d) Lac
   (e) All the above

7. French Polish is prepared by dissolving black or light brown shellac in one litre of methylated spirit without heat.
   (a) 0.10 kg
   (b) 0.12 kg
   (c) 0.15 kg
   (d) 0.20 kg
   (e) 0.25 kg

8. Distemper comprises of the following material
   (a) a base
   (b) a binder
   (c) a thinner
   (d) a pigment
   (e) All the above

9. Casein is extracted from __________ and acts as a binder.
   (a) Resins
   (b) Slaked Lime
   (c) Milk curds
   (d) Glue
   (e) Plaster

10. Linseed oiling consists of a mixture obtained by heating 3 parts of double boiled linseed oil and 1 part of bees wax to which 1 part of __________ is added by weight.

(a) Resins
(b) Plaster of Pairs
(c) Varnish
(d) Turpentine
(e) Vinegar

CHAPTER 13

Other Building Materials

1. The original source of asbestos was the fibrous mineral
   (a) Actinolite
   (b) Chrysolite
   (c) Amonite
   (d) Crocidolite
   (e) None of the above

2. Asbestos boards can withstand temperatures up to
   (a) 200°C
   (b) 250°C
   (c) 300°C
   (d) 540°C
   (e) 600°C

3. Asbestos sheets are used for the following purposes
   (a) Wall-lining
   (b) Wall-panelling
   (c) False ceiling
   (d) Partitions side-cladding
   (e) All the above

4. Among the constituents of fly-ash given below which one has the lowest percentage.
   (a) Aluminium oxide
   (b) Silicon dioxide
   (c) Unburnt fuel
   (d) Calcium oxide
   (e) Sulphur trioxide

5. In order to use fly-ash as pozzolana, the minimum specific surface should be
   (a) 320 m^2/kg
   (b) 400 m^2/kg
   (c) 410 m^2/kg
   (d) 520 m^2/kg
   (e) 600 m^2/kg

6. Bitumen is found in
   (a) Gaseous form
   (b) Liquid form
   (c) Semi-solid form
   (d) Solid form
   (e) In all the above forms

7. When the viscosity of a bitumen is reduced by a volatile ________, the bitumen is called as cut back bitumen.
   (a) Gas
   (b) Mixture
   (c) Dilutant
   (d) Solid
   (e) None of the above

8. Tar is the viscous liquid produced when natural organic materials such as coal, petroleum or wood, etc are ________
   (a) Carbonised
   (b) Stabilised
   (c) Liquified
   (d) Solidified
   (e) Sulphodised

9. Glass is any substance or combination of substances which has solidified from the liquid state without
   (a) Any reaction
   (b) Forming semi-solid
   (c) Crystallization
   (d) Any physical change
   (e) None of the above

10. The old broken glass of the same type as that one is intended to be manufactured is called
    (a) Glasset
    (b) Pyrex
    (c) Flint
    (d) Cullet
    (e) Crown

11. Plastics which become soft when heated and become hard when cooked is called
    (a) Thermoset plastics
    (b) Thermo plastics
    (c) Perspex
    (d) PVC
    (e) Bakelite

12. Geomembrances are thin materials with very low ________
    (a) Conductivity
    (b) Durability
    (c) Flexibility
    (d) Permeability
    (e) Resistivity

CHAPTER 14

Planning of Buildings

1. Central Building Research Institute (CBRI) is a research body set up by the ________
   (a) Ministry of Education
   (b) Ministry of Industries
   (c) National Building Organisation
   (d) Council of Scientific and Industrial Research
   (e) National Building Construction Corporation

2. Hindustan Housing Factory is a Government of India Organisation involved in the manufacture of
   (a) Bricks
   (b) Building Tiles
   (c) Precast Components
   (d) Industrial timbers
   (e) Paints and Varnishes

3. Building are classified based on fire resistance offered as
   (a) Type – 1
   (b) Type – 2
   (c) Type – 3
   (d) Type – 4
   (e) All the above

4. Defining the maximum benefit from the minimum dimensions of a room is called
   (a) Aspect
   (b) Privacy
   (c) Prospect
   (d) Roominess
   (e) Elegance

5. Horizontal circulation in a building is provided by
   (a) Passages
   (b) Corridors
   (c) Halls
   (d) Lobbies
   (e) All the above

6. The usable area of a building is called
   (a) Covered area
   (b) Plinth area
   (c) Floor area
   (d) Building area
   (e) Carpet area

7. The portion of a structure immediately above the ground and between the surface of the floor and the surrounding ground and the road surface so as to provide adequate drainage of the site is called
   (a) Superstructure
   (b) Substructure
   (c) Damp Proof course
   (d) Plinth
   (e) Basement

8. Factors affecting orientation of building are
   (a) Solar radiation
   (b) Clouds and rainfall
   (c) Winds
   (d) Surrounding site conditions
   (e) All the above

9. Hot-humid zones are found in India
   (a) In the interior
   (b) Away from the coast
   (c) Along the coast
   (d) In the central part
   (e) In the northern part

CHAPTER 15

Foundations

1. That part of a structure which is in direct contact with the ground and transmits the load of the structure to the ground is called
   (a) Superstructure
   (b) Plinth
   (c) Foundation
   (d) Basement
   (e) Brickwork

2. Methods of exploration are
   (a) Indirect methods
   (b) Semi-direct methods
   (c) Direct methods
   (d) (b) and (c)
   (e) All methods (a), (b) and (c)

3. Direct methods are boring and sampling methods which continuously provide
   (a) Only representative samples
   (b) Only undisturbed samples
   (c) Only disturbed samples
   (d) Representative or undisturbed samples
   (e) None of the above

4. Black cotton soils are spread over
   (a) Central India
   (b) Parts of Tamil Nadu
   (c) Parts of Karnataka
   (d) Parts of Andhra Pradesh
   (e) All the above

5. The object of soil investigation is to get information regarding
   (a) Nature, thickness and variation of soil strata at a place
   (b) Physical properties of the soil strata encountered
   (c) Seasonal variation in ground water table and its effect on the soil strata
   (d) (b) and (c) only
   (e) All the above

6. The bearing capacity of a soil can not be increased by
   (a) Chemical treatment
   (b) Grouting
   (c) Compacting
   (d) Moistening the soil
   (e) Drawing the moisture of the soil

7. The process of strengthening soil by providing high strength thin horizontal members is called
   (a) Geotextile
   (b) Geosynthetics
   (c) Soil reinforcement
   (d) Soil consolidation
   (e) Chemical stabilization

8. Structures on rigid ground undergo
   (a) Tilt
   (b) Non-uniform settlement
   (c) Differential settlement
   (d) Uniform settlement
   (e) Immediate settlement

9. Types of shallow foundations are
   (a) Spread footings
   (b) Combined footings
   (c) Strip footings
   (d) Mat foundation
   (e) All the above

10. In case of pile if the load is supported by friction between the pile and the surround soil and also by resting the pile on a firm soil structure, such a pile is called as
   (a) Friction pile
   (b) End Bearing pile
   (c) Friction and End Bearing pile
   (d) Battered pile
   (e) Sheet pile

11. In order to resist horizontal and inclined forces in water and earth retaining structures __________ piles are used
   (a) Point-bearing
   (b) Friction
   (c) Uplift
   (d) Batter
   (e) Tension

12. In case of grillage foundations, the distance between the flanges of grillage beams should be equal to
   (a) Half the flange width
   (b) Flange width
   (c) Twice the flange width
   (d) Four times the flange width
   (e) None of the above

13. Foundations on weaker soils are
   (a) Grillage foundation
   (b) Pile foundation
   (c) Raft or mat foundation
   (d) Any of the above
   (e) None of the above

14. For heavy structures on sandy soils the type of foundation generally provided is
   (a) Pier foundation
   (b) Strap footing
   (c) Raft foundation
   (d) Combined footing
   (e) None of the above

15. Bored cast-in-situ and bored compaction concrete piles with enlarged base is called
   (a) Under reamed pile
   (b) Bored pile
   (c) Driven pile
   (d) Caisson
   (e) Well foundation

16. Excavation beyond depth of __________ is generally categorized as deep excavation.
   (a) 1.0 m
   (b) 1.5 m
   (c) 2.0 m
   (d) 2.5 m
   (e) 3.0 m

17. Types of sheet piles commonly used are
    (a) Flat web
    (b) Arch web
    (c) Trough web
    (d) Z-piling
    (e) Except (c)

18. Main causes of foundation settlement are
    (a) Elastic compression of the foundation
    (b) Inelastic compression of the underlying soil
    (c) Ground water lowering
    (d) (a) and (b)
    (e) (a), (b) and (c)

CHAPTER 16

Masonry Construction

1. Brick laid with its breadth or width parallel to the face or direction of a wall is called
    (a) Squint
    (b) Quoin
    (c) Header
    (d) Stretcher
    (e) Closer

2. A piece of thick straight wood ruler with a piece of string which is fixed in the wood at one end and a solid metal piece is called
    (a) Tri square
    (b) Bolster
    (c) Lin and Pins
    (d) Plumb bob
    (e) None of the above

3. A bond produced by laying alternate stretchers and headers in each course is called
    (a) English Bond
    (b) Flemish Bond
    (c) Stretcher Bond
    (d) Header Bond
    (e) None of the above

4. In stone masonry, a horizontal course of masonry continuously projecting from the face of the wall is called
    (a) Bed surface
    (b) Corbel
    (c) String course
    (d) Stretcher course
    (e) Header course

5. A stonework wherein blocks of stones are either undressed or roughly dressed and have wider joints is called
    (a) Ashlar masonry
    (b) Random masonry
    (c) Rubble masonry
    (d) Chamfered masonry
    (e) None of the above

6. Following are the qualities of Brick masonry. Choose the wrong one
    (a) Construction cost is less
    (b) More resistance to atmospheric effects
    (c) Massive appearance
    (d) High fire resistance
    (e) No special lifting devices are needed

7. In the construction of Load bearing walls using hallow blocks, joint reinforcement is used for
    (a) Crack control
    (b) Settlement control
    (c) Shear control
    (d) All the above
    (e) None of (a), (b) and (c)

8. In reinforced brick masonry walls iron bars or expanded metal mesh are provided at __________ course.
    (a) Every course
    (b) Alternate course
    (c) Third or fourth courses
    (d) First two courses only
    (e) First three courses only

CHAPTER 17

Walls

1. Wall consisting of two structural leaves separated by an uniform continuous space is called
    (a) Faced wall
    (b) Solid wall
    (c) Cavity wall
    (d) Solid wall with piers
    (e) Veneered wall

2. Some of the important load bearing walls are
   (a) Panel wall
   (b) Partition wall
   (c) Curtain wall
   (d) Free-standing wall
   (e) All the above

3. Following are the free standing walls the wrong one.
   (a) Parapet wall
   (b) Compound wall
   (c) Panel wall
   (d) Shear wall
   (e) Buttress wall

4. If both lateral and rotational restraints are offered by a support, then the wall is said to be
   (a) Partially Tree
   (b) Free
   (c) Partially restrained
   (d) Fully restrained
   (e) Restrained

5. The ratio of effective height or effective length to the effective thickness of a masonry unit is called
   (a) Flexible ratio
   (b) Stiffness ratio
   (c) Slenderness ratio
   (d) Stability ratio
   (e) Lengthwise ratio

6. Some of the factors which contribute for eccentricity on brick walls are
   (a) Long floor edges
   (b) Relative stiffness
   (c) Flexibility and Geometry of support
   (d) Unequal spans
   (e) All the above

7. In general expansion joints have to be provided for every
   (a) 5 m
   (b) 7.5 m
   (c) 10.0 m
   (d) 12.5 m
   (e) 15.0 m

8. Isolation joint is similar to joint
   (a) Sliding
   (b) Contraction
   (c) Double
   (d) Expansion
   (e) Shear

9. Cracks are frequently found in brick masonry walls due to
   (a) Introduction of new type of construction
   (b) Moisture absorption
   (c) Temperature variation
   (d) Deflection and shrinkage of concrete slabs resting on wall
   (e) All the above reasons

CHAPTER 18

Framed Structures

1. Structures which are formed by connecting a series of horizontal and vertical members in suitable positions is called
   (a) Pannelled structures
   (b) Framed structures
   (c) Composite structures
   (d) Rigid structures
   (e) None of the above

2. Choose the best advantage of a timber framed structure from the following.
   (a) Durability
   (b) Beauty
   (c) Shock absorbing
   (d) Electrical insulation
   (e) High strength

3. In order to resist wind and earthquake forces, steel framed structures need ________
   (a) Welding
   (b) Long length members
   (c) Bracing
   (d) High strength
   (e) Bolting

4. Buildings with more than ________ stories are called as multi-storey buildings.
   (a) 12
   (b) 10
   (c) 8
   (d) 7
   (e) 5

5. Separation of structure into two or more elements and join the entire structural elements such that it functions as a monolithic structure is called.
   (a) Free structure
   (b) Combined structure
   (c) Articulated structure
   (d) Framed structure
   (e) Balanced structure

6. In fully prefabricated construction it is the practice to use __________ elements.
   (a) Smaller
   (b) Shorter
   (c) Heavier
   (d) Larger
   (e) Longer

7. Multi-storey buildings may be built of standardised construction with column spaces based on some __________
   (a) Specific spacing
   (b) Modular grids
   (c) Small grids
   (d) Large unequal spanned grids
   (e) Equally spanned grids

8. Based on dimensions multi-storey buildings are erected by means of __________ placed one or both sides of the frame.
   (a) Hoists
   (b) Pillar cranes
   (c) Over head cranes
   (d) Tower cranes
   (e) All of the above

CHAPTER 19

Arches and Lintels

1. Mechanical arrangement of wedge-shaped blocks of stones or bricks mutually supporting each other and in turn supported at the ends by piers or abutments or walls is called
   (a) Bridge structure
   (b) Lintel
   (c) Rib structure
   (d) Arch
   (e) Open-space structure

2. Stability of an arch is disturbed by
   (a) Crushing of arch material
   (b) Sliding of wedge-shaped blocks
   (c) Rotation about an edge
   (d) Differential settlement of supports
   (e) Any one of the above

3. The inner curve of an arch is called
   (a) Extrados
   (b) Intrados
   (c) Abutment
   (d) Arch ring
   (e) Springing line

4. Wedge-shaped blocks of stone or brick masonry of which an arch ring is made is called
   (a) Soffit
   (b) Haunch
   (c) Voussoirs
   (d) Skew backs
   (e) Key stones

5. Two-centered arches are
   (a) Blunt arch
   (b) Gothic arch
   (c) Acute arch
   (d) All the above
   (e) None of the above

6. Brick arches are made out of the following
   (a) Gaughed arches
   (b) Axed or rough cut arches
   (c) Rough brick arches
   (d) Brick flat arches
   (e) All the above

7. Variety of bricks used for brick arches are
   (a) Ordinary standard bricks
   (b) Ordinary bricks cut to a wedge shape
   (c) Special bricks of different sizes
   (d) Soft bricks
   (e) All the above

8. As a general rule the depth of lintel can be adopted as 1/12th of the span or __________ whichever is greater.
   (a) 30 cm
   (b) 25 cm
   (c) 20 cm
   (d) 15 cm
   (e) 10 cm

9. Minimum depth of plinth beams __________ and should be provided to the full width of the wall.

(a) 10 to 15 cm
(b) 15 to 20 cm
(c) 20 to 25 cm
(d) 25 to 30 cm
(e) > 30 cm

10. Grade beams are provided in between __________ footings.
(a) Combined
(b) Continuous
(c) Strip
(d) Isolated
(e) Ring

CHAPTER 20

Doors and Windows

1. In order to keep the frame of door or window in position this member is used.
(a) Horn
(b) Holdfast
(c) Style
(d) Mullion
(e) Jamb

2. Doors are to be located so that there is
(a) Freedom of movement
(b) Maximum use of accommodation
(c) Privacy of the occupants
(d) (a) and (b)
(e) (a), (b) and (c)

3. A door comprising of a frame work of rails and styles and covered with plywood or hard board is called
(a) Sash door
(b) Ledge door
(c) Pannel door
(d) Flush door
(e) Framed and panelled door

4. A depression cut in the frame to receive the door is known as
(a) Recess
(b) Rebate
(c) Louver
(d) Tension
(e) Groove

5. A window provided on a flat roof of a room is known as
(a) Lantern window
(b) Skylight window
(c) Dormer window
(d) Gable window
(e) Louvered window

6. The window which projects outside the room of a building is called
(a) Dormer window
(b) Bay Window
(c) Clearstorey window
(d) Gable window
(e) Louvered window

7. The window which is provided on the sloping roofs is called
(a) Bay window
(b) Gable window
(c) Glazed window
(d) Dormer window
(e) Corner window

8. The window in which the shutters open like doors is called
(a) Casement window
(b) Gable window
(c) Dormer window
(d) Bay window
(e) Louvered window

9. Pivoted windows fixed near the top of the main roof is called
(a) Dormer window
(b) Gable window
(c) Bay window
(d) Casement window
(e) Clearstorey window

10. A narrow window of small height which is generally fixed at the top of door or window is called
(a) Ventilator
(b) Dormer window
(c) Sky light
(d) Lantern
(e) Auxiliary window

CHAPTER 21

Stairs and Elevators

1. The vertical member between treads is known as
(a) Flight
(b) Drop
(c) Rise
(d) Ascend
(e) Riser

2. It is an upper horizontal portion of a step used to keep the foot while ascending or descending.
(a) Tread
(b) Flight
(c) Riser
(d) Drop
(e) Ascend

3. It is a series of steps without any break such as a platform, landing, etc.
(a) Tread
(b) Flight
(c) Drop
(d) Ascend
(e) Riser

4. In order to make the ascend comfortable the stair-way should not have steps more than
(a) 20
(b) 18
(c) 15
(d) 12
(e) 10

5. In order to prevent undue exortion to the user the pitch or slope of a stair should not exceed
(a) 60°
(b) 50°
(c) 40°
(d) 30°
(e) 20°

6. A flat platform at the head of a series of steps is known as
(a) Stopper
(b) Rest space
(c) Relief
(d) Platform
(e) Landing

7. The posts set at the top and bottom of a stair supporting the hand rail, are called
(a) Baluster
(b) End posts
(c) Arm posts
(d) Pillars
(e) Newels

8. In this type of stairs the direction of flight is changed through 180° by introduction of landings and or window is called
(a) Quarter-turn stairs
(b) Straight stairs
(c) Dog-legged stairs
(d) Bifurcated staircase
(e) Geometric stairs

9. In this stair the bottom flight is wide which is divided into narrow flights from either side of the landing, is called
(a) Geometric stairs
(b) Bifurcated stairs
(c) Dog-legged stairs
(d) Straight stairs
(e) Half-turn stairs

10. Radiating or angular steps which give a change in directions to the stairs, are called
(a) Round steps
(b) Skew steps
(c) Radial steps
(d) Winders
(e) Circular steps

11. In case of stairs (Rise × Tread) should be between
(a) 200 and 250
(b) 250 and 300
(c) 300 and 400
(d) 400 and 500
(e) 500 and 600

12. For residential buildings, the common size of step is ________.
(a) 10 cm × 20 cm
(b) 15 cm × 30 cm
(c) 15 cm × 28 cm
(d) 12 cm × 24 cm
(e) 12 cm × 20 cm

13. Ramps are sloping surfaces which should be provided with a ________ surface.
(a) Flat
(b) Rough
(c) Smooth
(d) Non-slippery
(e) None of the above

14. Stairs which are moving stairs or ever-moving stairs is called
(a) Elevator
(b) Ramp
(c) Fast moving Elevator
(d) Hoist
(e) Escalator

CHAPTER 22

Temporary Supporting Structures

1. Formwork should have
   (a) Adequate strength and rigidity
   (b) Smooth inner surface
   (c) Less leakage
   (d) Easy removal
   (e) All the above

2. Temporary structures are needed when the height of construction exceeds about
   (a) 1.5 m
   (b) 2.5 m
   (c) 3.0 m
   (d) 3.2 m
   (e) 3.5 m

3. Vertical members of a scaffold which are either supported on the ground or embedded into the ground or rested on sand filled bags are called.
   (a) Ledgers
   (b) Standards
   (c) Put logs
   (d) Rakers
   (e) Braces

4. This type of scaffolding is commonly used and particularly in the construction of brickwork, which is called
   (a) Double scaffolding
   (b) Mason's scaffolding
   (c) Single scaffolding
   (d) Suspended scaffolding
   (e) Trestle scaffolding

5. In this type of shoring inclined members are adopted to provide temporary support to the external walls from the ground. This type of shoring is called __________ shore
   (a) Raking
   (b) Flying
   (c) Vertical
   (d) Horizontal
   (e) Dead

6. Method of supporting structures while providing new foundations or strengthening the foundation without affecting the stability of the existing structures.
   (a) Under pinning
   (b) Shoring
   (c) Scaffolding
   (d) Flying shores
   (e) Raking shores

7. Methods of under pinning are
   (a) Pit method
   (b) Pile method
   (c) Pier method
   (d) Chemical method
   (e) All the above

8. Pit method of under pinning is impracticable or uneconomical in the case of
   (a) Water-logged areas
   (b) Heavy loads are existing structure
   (c) Loads to be transferred to a deeper depth
   (d) All the above
   (e) None of (a), (b) and (c)

CHAPTER 23

Ground and Upper Floors

1. This is a part of a floor which is required to impart strength and stability to support a floor covering and all other super imposed loads including live loads. This is called
   (a) Plinth
   (b) Levelling course
   (c) Sub-floor
   (d) Sub-base course
   (e) None of the above

2. Functional requirements of a floor in general are
   (a) Strengths and stability
   (b) Heat and sound insulation
   (c) Fire resistance sand durability
   (d) Damp and moisture prevention
   (e) All the above

3. Selection of basement floor depends on
   (a) Availability of ventilation
   (b) Drainage of water from the floor
   (c) Adequate safety against fire
   (d) Ground water level
   (e) All the above

4. Special type of concrete flooring containing marble chips as aggregates is called
   (a) Granolithic floor
   (b) Terrazzo floor
   (c) Mosaic floor
   (d) Marble floor
   (e) None of the above
5. Floors made from materials such as PVC, linoleum, rubbles, etc are called
   (a) Plastic floor
   (b) Resilient Floor
   (c) Terrazzo Floor
   (d) Mosaic Floor
   (e) None of the above
6. In flat slab floor construction, if the reinforcement is provided between the columns and perpendicular to the columns is called __________.
   (a) Four-way system of reinforcement
   (b) Uniform system of reinforcement
   (c) Two-way system of reinforcement
   (d) Non-uniform system of reinforcement
   (e) None of the above
7. Rubber floor is composed of __________ tuber with various filling compounds.
   (a) Artificial
   (b) Processed
   (c) Synthetic
   (d) Natural
   (e) None of the above
8. The weight per metre length of structural hollow clay tiles should not exceed
   (a) 200 M
   (b) 160 M
   (c) 150 M
   (d) 140 M
   (e) 120 M

CHAPTER 24

Structural Steel Works

1. For construction works, steel should be available in certain forms which are known as __________.
   (a) Steel forms
   (b) Commercial forms
   (c) Steel section
   (d) Moulded section
   (e) Deformed section
2. Channel section has two equal flanges and a __________.
   (a) Leg
   (b) Web
   (c) Base
   (d) Angle
   (e) None of the above
3. Steel members are fastened together by means of
   (a) Rivets
   (b) Bolts
   (c) Welding
   (d) (a) and (b) only
   (e) (a), (b) and (c)
4. Rivets used in building construction are of soft steel with a tensile strength of the order of about
   (a) 6000 kg/cm^2
   (b) 5000 kg/cm^2
   (c) 4000 kg/cm^2
   (d) 3000 kg/cm^2
   (e) 2000 kg/cm^2
5. In order to make use of steel structures the standard sections are combined, such a combined section is called __________.
   (a) Made-up section
   (b) Arranged section
   (c) Built-up section
   (d) Formed-section
   (e) None of the above
6. Steel roof trusses are designed in such a way that the members are either in __________.
   (a) Tension or shear
   (b) Compression or tension
   (c) Compression or bond
   (d) Bond or tension
   (e) Compression or shear
7. Most suitable section for principal rafter is __________.
   (a) I-section
   (b) Channel section

(c) Angle section
(d) Tee-section
(e) None of the above

8. Steel trusses for spans more than 15 m are give below. Identify the wrong one.
(a) North-light vertical face truss
(b) Bow-string truss
(c) Arch truss
(d) North-light inclined face truss
(e) King-post truss

9. Types of steel lacings are given below. Identify the wrong one
(a) Single lacing
(b) Batten lacing
(c) X-type lacing
(d) Double lacing
(e) Z-type lacing

CHAPTER 25

Roof and Roof Coverings

1. The slope of a roof which may be expressed as degrees or inclination to the horizontal or the rise to the span is called
(a) Hip
(b) Ridge
(c) Eaves
(d) Pitch
(e) Gable

2. The lowest course of the roof is __________
(a) Hip
(b) Ridge
(c) Valley
(d) Eaves
(e) Gable

3. The apex line of a sloping roof is called __________
(a) Ridge
(b) Valley
(c) Hip
(d) Gable
(e) Eaves

4. This is a roof which has two rafters with a central ridge piece at the top. The rafters are fixed to the wall plates. Name the type of roof
(a) North-light roof
(b) Lean to roof
(c) Couple roof
(d) Couple close roof
(e) Collar tie roof

5. A type of roof used in factories is called
(a) Shed roof
(b) North light roof
(c) South light roof
(d) Saw tooth roof
(e) Either of (b) and (d)

6. Couple-close roofs are preferred for spans up to
(a) 3.0 m
(b) 3.5 m
(c) < 4.0 m
(d) 4.5 m
(e) > 5 m

7. A king-post truss is used for spans
(a) 3 to 5 m
(b) 6 to 9 m
(c) 10 to 15 m
(d) 15 to 18 m
(e) > 18 m

8. Water proofing of flat roof is done by the methods given below. Identify the wrong one
(a) Cement mortar finishing
(b) Bedding concrete and finishing
(c) Mastic asphalt finish
(d) Use of water-proof compounds
(e) Rough finishing

9. A ring constructed on the perimeter of a dome is to resist __________
(a) Shear
(b) Thrust
(c) Tension
(d) Bond
(e) Torque

10. This type of shell is developed by moving a straight line so that its ends lie on two fixed vertical curves.
(a) Cylindrical shell
(b) Spherical shell
(c) Rectangular shell
(d) Ruled surface shell
(e) Barrel vault

CHAPTER 26

Building Finishes

1. Building finishes contribute
   (a) For the protection of exposed surface
   (b) Provides a pleasing look
   (c) Adds to the aesthetic view
   (d) (a) and (b)
   (e) (a), (b) and (c)

2. The method of covering rough and un-even surfaces of the various components of a building with a material is called
   (a) Pointing
   (b) Painting
   (c) Colour washing
   (d) Plastering
   (e) Varnishing

3. For suitable application of plaster the background should be
   (a) Solid
   (b) Lathing
   (c) Boards
   (d) Slabs
   (e) All the above

4. The surface to which the first coat of plaster is applied is called
   (a) Base surface
   (b) Preliminary surface
   (c) Ground surface
   (d) First surface
   (e) Background surface

5. The development of one or more load swellings on the finished plastered surfaces, in called
   (a) Crazing
   (b) Blistering
   (c) Swelling
   (d) Boiling
   (e) Foaming

6. The roughening of solid backgrounds to provide suitable key for plastering is
   (a) Backing
   (b) Grounding
   (c) Hacking
   (d) Facing
   (e) Preparing

7. The process of working of the float for finishing coat is called
   (a) Blistering
   (b) Knetting
   (c) Peeling
   (d) Floating
   (e) Dubbing out

8. Finishing of mortar joints to have a good appearance is called
   (a) Pointing
   (b) Pressing
   (c) Roughening
   (d) Gap rendering
   (e) Refined plastering

9. In this type of pointing a V-shaped projection is made.
   (a) Vee-Pointing
   (b) Weathered pointing
   (c) Tuck pointing
   (d) Struck pointing
   (e) Beaded pointing

10. In wood and metal works, in order to give a hard and opaque covering, a basic coat is made which is called
   (a) Primary coating
   (b) Basic coating
   (c) Surface coating
   (d) Weathering coating
   (e) Resisting coating

CHAPTER 27

Essential Services in Buildings

1. Plumbing services comprises of
   (a) System of Piping
   (b) Fixtures
   (c) Providing water supply
   (d) Drainage arrangement
   (e) All the above

2. Pipe branching out of the water main which is used for feeding water from the main pipe to the building is called
   (a) Delivery pipe
   (b) Subsidiary pipe
   (c) Service pipe
   (d) Minor pipe
   (e) Branch pipe

3. Back flow of used or polluted water from a plumbing fixture or vessel into a water supply pipe due to negative pressure is called
   (a) Reverse flow
   (b) Back-Siphonage
   (c) Reverse-Siphonage
   (d) Negative flow
   (e) Controlled flow
4. The used water from bathrooms, kitchen, wash basins, sinks, etc is called
   (a) Sullage
   (b) Sewage
   (c) Waste water
   (d) Rubbish water
   (e) Polluted water
5. The capacity of an overhead tank in residences is
   (a) 100 to 250 litres
   (b) 200 to 300 litres
   (c) 300 to 350 litres
   (d) 300 to 500 litres
   (e) > 1000 litres
6. The requirement of water per head per day, as per Indian Standards is
   (a) 100 litres
   (b) 120 litres
   (c) 135 litres
   (d) 150 litres
   (e) 180 litres
7. Tap which operates as soon as the hand is placed below the tap and closes automatically as soon as the hand is removed is called
   (a) Magnetic tap
   (b) Self closing tap
   (c) Pillar tap
   (d) Bib tap
   (e) Stop Tab
8. This trap is intended to receive sullage water from baths, sinks and wash basins. Identify the trap
   (a) Flow trap
   (b) Gully trap
   (c) Intercepting trap
   (d) S-trap
   (e) P-trap
9. Disposal of sludge can be done by adopting any one of the method. Identify the correct method.
   (a) Dumping into waste bodies
   (b) Drying in beds
   (c) Lagooning
   (d) Shallow Burrial
   (e) All the above
10. A cover pit through which the effluent is allowed to be soaked or absorbed into the surrounding soil is called
    (a) Sludge pit
    (b) Waste pit
    (c) Soak pit
    (d) Septic pit
    (e) Sullage pit
11. A pool whose top portion acts as an absorption field and the bottom as a septic tank is known as
    (a) Waste pool
    (b) Soak pool
    (c) Drain pool
    (d) Cess pool
    (e) Absorption pool
12. Wiring of buildings are performed by
    (a) Sheathed wiring by surface fixing
    (b) Conduit installation as walls and ceilings
    (c) Concealed conduct wiring
    (d) (a) and (b) only
    (e) (a), (b) and (c)

CHAPTER 28

Special Services in Buildings

1. The branch of science dealing with the study of mixture of dry air and water vapour and the study of the behavior of moist air is called
   (a) Atmospherics
   (b) Chrometrics
   (c) Psychrometrics
   (d) Barometrics
   (e) None of the above
2. Air conditioning involves control of
   (a) Surrounding temperature
   (b) Humidity

(c) Air circulation
(d) (a) and (b) only
(e) (a), (b) and (c)

3. Air conditioning classified under Functional Point of view is
(a) Comfort-air-conditioning
(b) Industrial air-conditioning
(c) Summer air-conditioning
(d) Winter air-conditioning
(e) (a) and (b) only

4. Comfort air-conditioning is achieved by
(a) Temperature control
(b) Air motion control
(c) Humidity control
(d) (a) and (b) only
(e) (a), (b) and (c)

5. System of air condition is
(a) Central system
(b) Self contained system
(c) Semi-contained system
(d) Combined system
(e) All the above

6. The heat of air-conditioning system is
(a) Cooling
(b) Humidification
(c) Air circulation
(d) Dehumidification
(e) (a) and (c) only

7. Main purpose of thermal insulation is
(a) To conserve a constant heat or temperature
(b) To prevent condensation
(c) To reduce the risk of water freezing in case of pipes
(d) To reduce the heat loss in case of hot water system
(e) To provide comfortable living and working

8. Points of low sound intensity causing unsatisfactory heating are called
(a) Echoes
(b) Dead spots
(c) Fringes
(d) Foci
(e) All the above

9. The highest audible sound has a frequency of
(a) 5000 cps
(b) 7500 cps
(c) 15000 cps
(d) 20000 cps
(e) 50000 cps

10. Repeated reflection of sound is called
(a) Echoes
(b) Reverberation
(c) Repeated noise
(d) Acoustical reflection
(e) None of the above

CHAPTER 29

Protection of Buildings

1. White ants (Termites) are very fast in eating wood and other __________ materials as food.
(a) Plastic
(b) Asbestos
(c) Cellulose
(d) (b) and (c) only
(e) (a), (b) and (c)

2. Termites can not enter through
(a) Hardwood
(b) Heavy cardboard
(c) Mud mortar
(d) Dense concrete
(e) None of the above

3. The emulsifiable chemical recommended by code of soil treatment against termites is
(a) Chloropyrifos concentrate
(b) Heptachlor concentrate
(c) Chloride concentrate
(d) (b) and (c) only
(e) Any of (a), (b) and (c),

4. In order to prevent the termites entry through walls, impenetratable physical structural barriers may be provided continuously at __________ level.
(a) Foundations
(b) Top of base concrete
(c) 0.5 m above bottom of foundation
(d) Plinth level
(e) Sill level

5. Natural causes for dampness is
   (a) Penetration of rain
   (b) Rise of moisture from ground and moisture condensation
   (c) Drainage conditions of the site
   (d) Orientation of the site
   (e) All the above

6. Fire load is defined as the ratio of the weight of all combustible materials to the floor area. Fire load is classified as medium when the limit is
   (a) 1.15×10^6 kJ/m^2
   (b) 1.15×10^6 to 2.30×10^6 kJ/m^2
   (c) 2.30×10^6 to 4.6×10^6 kJ/m^2
   (d) 5×10^6 to 7.5×10^6 kJ/m^2
   (e) $> 7.5 \times 10^6$ kJ/m^2

7. Depending on the population and importance of the region, one hydrant for an area of ________ is provided
   (a) 1000 to 2000 m^2
   (b) 2000 to 3000 m^2
   (c) 3000 to 4000 m^2
   (d) 4000 to 10000 m^2
   (e) > 10000 m^2

8. The lightning protection system consists of an unbroken chain of conductors from the roof of a building to the
   (a) Top window level
   (b) Sill level
   (c) Plinth level
   (d) Ground
   (e) None of the above

9. Earthquakes are classified as
   (a) Tectonic earthquakes
   (b) Volcanic earthquakes
   (c) Reservoir-associated earthquakes
   (d) (a) and (b) only
   (e) (a), (b) and (c)

10. Richter's magnitude scale is in __________ scale.
    (a) Arithmetic scale
    (b) Semi-logarithmic scale
    (c) Logarithmic scale
    (d) (b) and (c) only
    (e) All the above

11. The zone map of India sub-divides India into
    (a) Two zones
    (b) Three zones
    (c) Four zones
    (d) Five zones
    (e) Regional zones

12. Intensity of earthquake is said to be moderate when the acceleration (cm/s/s) is
    (a) Less than 1
    (b) Over 1
    (c) Over 2.5
    (d) Over 5.0
    (e) Over 10.0

CHAPTER 30

Preventive Measures and Maintenance of Buildings

1. Maintenance of drainage system is
   (a) Fittings inside the building
   (b) Waste and soil pipes
   (c) Gulleys and chambers
   (d) Roof-drainage system
   (e) All the above

2. Maintenance of plastered brick work involves in
   (a) Removal of efflorescence
   (b) Repairing of cracks
   (c) Repointing old brick points
   (d) Repainting brick work
   (e) (a), (b) and (d)

3. Maintenance of stonework involves in
   (a) Removal of efflorescence
   (b) Removal of stains
   (c) Repair cracks
   (d) Water proofing
   (e) All the above

4. This type of joint is provided at the location where the construction is stopped at the end of day's work so as to bridge the old work and new work by a proper bond, which is called as
   (a) Expansion joint
   (b) Contraction joint
   (c) Construction joint
   (d) (a) or (b)
   (e) None of the above

5. Expansion and contraction joints are provided in all concrete structures when the length exceeds
   (a) 5 m
   (b) 8 m
   (c) 10 m
   (d) 12 m
   (e) 20 m

6. The process of injecting mortar with low water-cement ratio at a high pressure through a nozzle to repair cracks in concrete is called
   (a) Grouting
   (b) Shortcrete
   (c) Guniting
   (d) Injecting
   (e) None of the above

7. Leakage of flat roof can be attended by
   (a) Providing adequate slope
   (b) Junction of horizontal surface and vertical faces is to be provided with fillets
   (c) Rain water entry to be checked
   (d) Cracks in weathering courses are to be checked
   (e) All the above

8. Maintenance of electrical system is attended by
   (a) Providing ECCB trip switch
   (b) High amperage gadgets are to be plugged tightly
   (c) Phase-change device should be provided
   (d) Performance of regulators, condensers, etc should be checked
   (e) All the above

CHAPTER 31

Construction Planning and Scheduling

1. The process of devising of a scheme for doing, making or arranging a project before the commencement of a work is called
   (a) Planning
   (b) Scheduling
   (c) Arranging
   (d) Streamlining
   (e) None of the above

2. Consideration of the main construction activities and all the supporting elements such as labour, material, equipment and expenditure is called
   (a) Streamlining
   (b) Scheduling
   (c) Planning
   (d) Arranging
   (e) None of the above

3. Technical planning is carried out
   (a) To finalise design and estimate
   (b) To plan resources
   (c) To decide executing authority
   (d) To foresee obstacle
   (e) All the above

4. Limitations of planning depends on
   (a) Correct assumptions needed
   (b) Planning is expensive
   (c) Planning encourages false sense of activity
   (d) (a) and (b)
   (e) (a), (b) and (c)

5. Construction Schedule involves
   (a) Number of activities and the stages
   (b) Labour contribution
   (c) Equipment required
   (d) Quantity
   (e) All the above

6. The project scheduling techniques are concerned with resource
   (a) Materials
   (b) Equipment
   (c) Time
   (d) Labor
   (e) None of the above

7. Technical scheduling can be done effectively by the methods
   (a) Bar chart
   (b) Milestone chart
   (c) Network analysis
   (d) (a) and (b)
   (e) None of the above

8. Overall economy of a project can be achieved by a thorough coordination in planning of services and the construction plant. This is done by

(a) Job Layout
(b) Materials and Labour
(c) Equipment
(d) (b) and (c)
(e) None of the above

CHAPTER 32

Construction Equipment

1. Procurement cost of an equipment involves in
   (a) Cost price of the equipment
   (b) Interest on money invested
   (c) Taxes paid
   (d) Insurance cost
   (e) All the above

2. Operating cost depends mainly on
   (a) Cost of investment and depreciation cost
   (b) Cost of fuel and repair
   (c) Cost of labour
   (d) (a) and (b)
   (e) (a), (b) and (c)

3. Hydraulic excavators are more advantages since
   (a) Additional attachments can be fitted
   (b) Efficient hydraulic-unit
   (c) Efficient angle of dipping
   (d) (a) and (b)
   (e) (a) and (c)

4. This is a most popular shovel which consists of a boom, dipper stick, the bucket and the mechanism of operation. This shovel is called
   (a) Dipper showel
   (b) Drag showel
   (c) Clamshell
   (d) Hoe
   (e) Drag line

5. Bulldozers can be used to haul materials up to
   (a) 10 m
   (b) 20 m
   (c) 40 m
   (d) 60 m
   (e) 100 m

6. This is primarily a device for loading or finishing earth work. This is called
   (a) Grader
   (b) Tractor
   (c) Shovel
   (d) Hoe
   (e) None of the above

7. Trenchers can be used to make trenches of width up to
   (a) 18 m
   (b) 16 m
   (c) 14 m
   (d) 12 m
   (e) 10 m

8. This dewatering system can be installed outside the zone of construction operation and drainage is effected to the depth of excavation. This method of dewatering is called
   (a) Single stage well-point
   (b) Deep-well drainage
   (c) Vacuum-dewatering
   (d) Artesian dewatering
   (e) Multiple-stage dewatering

9. When a direct electric current is passed through a saturated soil stratum, water moves towards the cathode and the water is removed at the cathode well. This method is called
   (a) Vacuum-dewatering
   (b) Deep-well drainage
   (c) Electro-osmosis
   (d) Current drainage method
   (e) None of the above

10. Compaction is attained by applying the mechanical force by
    (a) Rolling
    (b) Kneading
    (c) Vibrations
    (d) Ramming
    (e) Any one of above

11. This is the simplest form of hammer, which does not use any external power.
    (a) Single acting hammer
    (b) Double acting hammer
    (c) Hydraulic hammer
    (d) Drop hammer
    (e) Vibratory hammer

12. Hauling equipment are given below. Identify the wrong one
    (a) Trucks
    (b) Dumpers
    (c) Trailers
    (d) Dump Trucks
    (e) Lift Trucks

ANSWERS TO OBJECTIVE TYPE QUESTIONS

Chapter 1

1. (c) 2. (a) 3. (b) 4. (c) 5. (e)
6. (b) 7. (d) 8. (c) 9. (e)

Chapter 2

1. (e) 2. (d) 3. (d) 4. (c) 5. (b) 6. (a) 7. (c)
8. (b) 9. (d) 10. (a) 11. (d) 12. (a) 13. (b) 14. (e)

Chapter 3

1. (c) 2. (d) 3. (c) 4. (b) 5. (d) 6. (a)
7. (e) 8. (a) 9. (b) 10. (c) 11. (d) 12. (a)

Chapter 4

1. (c) 2. (e) 3. (b) 4. (e) 5. (c) 6. (a) 7. (d)

Chapter 5

1. (e) 2. (a) 3. (c) 4. (e) 5. (a) 6. (e)
7. (a) 8. (d) 9. (e) 10. (a) 11. (b) 12. (e)

Chapter 6

1. (c) 2. (a) 3. (c) 4. (e) 5. (a) 6. (e) 7. (b)
8. (d) 9. (b) 10. (c) 11. (a) 12. (b) 13. (c) 14. (a)
15. (d) 16. (d) 17. (b)

Chapter 7

1. (a) 2. (b) 3. (c) 4. (d) 5. (e) 6. (c) 7. (b)
8. (d) 9. (e) 10. (b) 11. (d) 12. (e)

Chapter 8

1. (e) 2. (c) 3. (a) 4. (e) 5. (b) 6. (e) 7. (e)
8. (a) 9. (d) 10. (c) 11. (d) 12. (c) 13. (c) 14. (b)
15. (a) 16. (d) 17. (d) 18. (b)

Chapter 9

1. (c) 2. (d) 3. (c) 4. (e) 5. (b)

Chapter 10

1. (b) 2. (d) 3. (e) 4. (e) 5. (e) 6. (b) 7. (b)
8. (d) 9. (b) 10. (a) 11. (c) 12. (d) 13. (e) 14. (d)
15. (b) 16. (b)

Chapter 11

1. (e) 2. (d) 3. (a) 4. (e) 5. (e)
6. (a) 7. (e) 8. (d) 9. (b) 10. (e)

Chapter 12

1. (e) 2. (c) 3. (a) 4. (b) 5. (b)
6. (e) 7. (c) 8. (e) 9. (c) 10. (d)

Chapter 13

1. (a) 2. (d) 3. (e) 4. (e) 5. (a) 6. (e)
7. (c) 8. (a) 9. (c) 10. (d) 11. (b) 12. (d)

Chapter 14

1. (d) 2. (c) 3. (e) 4. (d) 5. (e)
6. (e) 7. (d) 8. (e) 9. (c)

Chapter 15

1. (c) 2. (e) 3. (d) 4. (e) 5. (e) 6. (d) 7. (c)
8. (d) 9. (e) 10. (c) 11. (d) 12. (c) 13. (d) 14. (a)
15. (a) 16. (b) 17. (e) 18. (e)

Chapter 16

1. (c) 2. (d) 3. (b) 4. (c) 5. (c)
6. (c) 7. (a) 8. (c)

Chapter 17

1. (c) 2. (e) 3. (c) 4. (d) 5. (c)
6. (e) 7. (e) 8. (d) 9. (e)

Chapter 18

1. (b) 2. (d) 3. (c) 4. (e) 5. (c)
6. (d) 7. (b) 8. (d)

Chapter 19

1. (d) 2. (e) 3. (b) 4. (c) 5. (d)
6. (e) 7. (e) 8. (d) 9. (a) 10. (d)

Chapter 20

1. (b) 2. (e) 3. (d) 4. (c) 5. (a)
6. (b) 7. (d) 8. (a) 9. (e) 10. (a)

Chapter 21

1. (e) 2. (a) 3. (b) 4. (d) 5. (c) 6. (e) 7. (a)
8. (c) 9. (b) 10. (d) 11. (d) 12. (c) 13. (d) 14. (e)

Chapter 22

1. (e) 2. (a) 3. (b) 4. (c) 5. (a)
6. (a) 7. (e) 8. (d)

Chapter 23

1. (c) 2. (e) 3. (e) 4. (b) 5. (b)
6. (c) 7. (d) 8. (d)

Chapter 24

1. (c) 2. (b) 3. (e) 4. (c) 5. (c)
6. (b) 7. (d) 8. (e) 9. (c)

Chapter 25

1. (d) 2. (d) 3. (a) 4. (c) 5. (e)
6. (d) 7. (b) 8. (e) 9. (b) 10. (d)

Chapter 26

1. (e) 2. (d) 3. (e) 4. (e) 5. (b)
6. (c) 7. (d) 8. (a) 9. (b) 10. (a)

Chapter 27

1. (e) 2. (c) 3. (a) 4. (a) 5. (d) 6. (c)
7. (a) 8. (b) 9. (d) 10. (c) 11. (d) 12. (e)

Chapter 28

1. (c) 2. (e) 3. (e) 4. (e) 5. (e)
6. (b) 7. (a) 8. (b) 9. (d) 10. (b)

Chapter 29

1. (c)	2. (d)	3. (e)	4. (d)	5. (e)	6. (b)
7. (d)	8. (d)	9. (e)	10. (c)	11. (d)	12. (d)

Chapter 30

1. (e)	2. (e)	3. (e)	4. (c)	5. (d)
6. (c)	7. (e)	8. (e)		

Chapter 31

1. (a)	2. (b)	3. (e)	4. (e)	5. (e)
6. (c)	7. (c)	8. (a)		

Chapter 32

1. (e)	2. (e)	3. (d)	4. (a)	5. (e)	6. (a)
7. (d)	8. (b)	9. (c)	10. (e)	11. (d)	12. (e)

BIBLIOGRAPHY

ACI, ACI Committee 211.1–91, Standard Practice for Selecting Proportions for Normal Heavy Weight and Mass Concrete, Part I, ACI Manual of Cement Practice, New Delhi, 1994.

Anbuvelan, K., Management Concepts for Civil Engineers, Laxmi Publications, New Delhi, 2007.

Arora, S.P. and S.P. Bhindra, A Text-Book of Building Construction, Dhampath Rai and Sons, Delhi, 2005.

Ataev, et al., S.S., Construction Technology, MIR Publishers, Moscow, 1985.

Bangar, K. M., Principles of Engineering Geology, Standard Publishers Distributors Pvt. Ltd., New Delhi, 2009.

Chellis, R. D., Pill Foundations, In: Foundation Engineering, Edited by G. A. Leonards, McGraw Hill, New York, 1962.

Dayaratnam, P., Brick and Reinforced Brick Structures, Oxford & IBH Publishing House, New Delhi, 1997.

Duggal, K. N., Elements of Public Health Engineering, Chand and Company Ltd., New Delhi, 1988.

Gambhir, M.L., Concrete Technology, Tata McGraw Hill Ltd., 2010.

Handa, S. and Rangaswamy, Civil Engineering, Objective Type, 'Sixth Edition', Satya Prakasham, New Delhi, 1995.

IS: 269, Specification for Ordinary Portland Cement, 33 grade, BIS, 1989.

IS: 303, Specification Plywood for General Purpose, 1989.

IS: 383, Specification for Coarse and Fine Aggregates from Natural Sources for Concrete, 1997.

IS: 401, Code of Practice for Preservation of Timber, 2001.

IS: 455, Specification for Portland Slag Cement (Fourth revision) 1990.

IS: 456, Code of Practice for Plain and Reinforced Concrete (third revision), BIS, 2000.

IS: 650, Specification for Standard Sand for Testing of Cement, BIS, 1991.

IS: 654, Mangalore Pattern Clay Roofing Tiles.

IS: 659 – Safety Code for Air Conditioning, BIS,1964.

IS: 712, Specification for Building Lime, BIS, 1984.

IS: 762 – Part I, Recommendations for Orientation of Buildings, BIS, 1974.

IS: 1077, Specification for Common Burnt Clay Building Bricks, BIS, 1976.

IS: 1121 – Parts 1 to 3, Methods of Test for Determination of Strength and Properties of Natural Building Stones, BIS, 1974.

IS: 1123, Methods of Identification of Natural Building Stones, BIS, 1975.

IS: 1124, Method of Test for Determination of Absorption, Apparent Specific Gravity, Porosity of Natural Building Stones, 1998.

IS: 1172, Code of Basic Requirements for Water Supply Drainage and Sanitation, 1983.

IS: 1196, Code of Practice for Laying Bitumen Mastic Flooring, BIS, 1978.

IS: 1346, Code of Practice for Waterproofing of Roofs with Bitumen Felts, BIS, 1991.

IS: 1443, Code of Practice for Laying and Finishing Cement Concrete Flooring Tiles, BIS, 1972.

IS: 1477, Parts 1 and 2, Code of Practice for Painting of Terrors Metals in Buildings, BIS, 1971.

IS: 1489, Specification for Portland Pozzolana Cement, Part I, Fly Ash Based, BIS 1981.

IS: 1489, Specification for Portland Pozzolana Cement, Part II, Calcived Clay Based, BIS, 1991.

IS: 1597, Code of Practice for Stone work. Part 1: Rubble Stone Masonry; Part 2: Ashlar Stone Masonry, BIS, 1992.

IS: 1609, Code of Practice for Laying Damp-proofing Treatment Using Bitumen Felts, BIS, 1991.

IS: 1624, Methods of Field Testing of Building Lime, BIS, 1986.

IS: 1634, Code of Practice for Design and Construction of Wood Stain in Houses, BIS, 1973.

IS: 1641, Code of Practice for Fire Safety of Buildings, BIS, 1988.

IS: 1649, Code of Practice for Fire Safety of Buildings (General), Materials and Details of Construction, BIS, 1960.

IS: 1658, Specification for Fibre Hard-board, BIS, 1966.

IS: 1661, Code of Practice for Application of Cement and Concrete Lime Plaster Finishes, BIS, 1972.

IS: 1742, Code of Practice for Building Drainage, BIS, 1972.

IS: 1860, Code of Practice for Installation Operation and Maintenance of Electric Passenger and Good Lifts, BIS, 1980.

IS: 1893, Indian Standard Criteria for Earthquake Resistant Design of Structures, BIS, New Delhi, 1984.

IS: 1893 (Part I): 2002 – Criteria for Earthquake Design of Structures – Part I: General Provisions and Buildings (Fifth revision), BIS, 2005.

IS: 1904, Design and Construction of Foundation in Soils, General Requirements (revised), BIS, 1986.

IS: 1905, Code of Practice for Structural Use of Unreinforced Masonry, BIS, 1987.

IS: 1911, Schedule of Unit Weights of Building Materials, BIS, 1967.

IS: 2065, Code of Practice for Water Supply in Buildings, BIS, 1983.

IS: 2114, Code of Practice for Laying In-situ Terrazzo Floor Finish, BIS, 1984.

IS: 2250, Code of Practice for Preparation and Use of Masonry Mortars, BIS, 1981.

IS: 2338, Parts 1 and 2, Code of Practice for Finishing of Wood and Wood – Based Materials, BIS.

IS: 2470, Code of Practice for Design and Construction of Septic Tanks, Part I, Small Installations, BIS, 1968.

IS: 2526, Code of Practice for Acoustical Design of Auditorium and Conference Halls, BIS, 1963.

IS: 2527, Code of Practice for Fixing Rainwater Gutters and Downpipes for Roof Drainage, BIS, 1984.

IS: 2571, Specifications for Laying In-situ Concrete Flooring, BIS, 1970.

IS: 2690 – Part 1, Specification for Burnt Clay Flat Terrace Tile, BIS, 1972.

IS: 2911 – Part 3, Code of Practice for Design and Construction of Pile 'Foundations – Under Reamed Piles, BIS, 1980.

IS: 2911 – Part 2, Code of Practice for Design and Construction of Pile Foundations, Timber Piles, BIS, 1980.

IS: 3007, Part 1: Code of Practice for Laying of Asbestos Cement Sheets, BIS, 1982.

IS: 3036, Code of Practice for Laying of Lime Concrete for Waterproofed Finish, BIS, 1992.

IS: 3037, Part 2: Specification for Bitumen Mastic for Use in Waterproofing of Roofs, BIS, 1986.

IS: 3102, Classification of Clay Solid Bricks, BIS, 1971.

IS: 3362, Code of Practice for Natural Ventilation of Residential Buildings, BIS, 1977.

IS: 3495 – Parts 1 to 4, Methods of Texts for Burnt Clay Building Bricks, BIS, 1985.

IS: 3792, Code of Practice for Guide for Heat Insulation of Non-industrial Buildings BIS, 1978.

IS: 3812, Specification of Fly Ash for Use as Pozzolana and Admixture, BIS, 1981.

IS: 4021, Specification for Timber Door, Window and Ventilator Frames, BIS, 1967.

IS: 4031, PT 1 to 15, Methods of Physical Test for Hydraulic Cement, BIS, 1989.

IS: 4326: 1993 – Code of Practice for Earthquake Resistant Design and Construction of Buildings, BIS, 1998.

IS: 4591, Code of Practice for Installation and Maintenance of Escalators, BIS, 1968.

IS: 5766, Code of Practice for Laying Burnt Clay Brick Flooring, BIS, 1970.

IS: 6276, Code of Practice for Whitewashing and Colour-washing, BIS, 1971.

IS: 6278, Parts 1 – 3, Code of Practice for Anti-termite Treatment of Buildings, BIS, 1978.

IS: 6452, Specification for High Alumina Cement for Structural Use, BIS, 1989.

IS: 6461, Code of Practice for Formwork for Concrete, BIS, 1972.

IS: 6909, Specification for Supersulphated Cement, BIS, 1990.

IS: 6932 – Parts 1 to 11, Methods of Tests for Building Lime, BIS, 1973.

IS: 8041, Specification for Rapid Hardening Portland Cement (Second revision), BIS, 1990.

IS: 8043, Specification for Hydrophobic Portland Cement, BIS, 1991.

IS: 8112, Specification for 43 Grade Ordinary Portland Cement, BIS, 1989.

IS: 9012, Recommended Practice for Short-creating, BIS, 1978.

IS: 9759, Guidelines for Dewatering Duration Construction, BIS, 1981.

IS: 10262, Indian Standard Recommended Method of Concrete Mix Design, BIS, 1982.

IS: 12269, Specification for 53 Grade Ordinary Portland Cement, BIS, 1987.

IS: 12330, Specification for Sulphate Resisting Portland Cement, BIS, 1988.

IS: 12600, Specification for Low Heat Portland Cement, BIS, 1989.

IS: 13182, Waterproofing and Damp-proofing Wet Areas in Buildings – Recommendations, BIS, 1991.

IS: 13920: Ductile Detailing of Reinforced Concrete Structures Subjected to Seismic Forces – Code of Practice, BIS, New Delhi, 1993.

IS: 13920: 1993 – Ductile Detailing of Reinforced Concrete Structures Subjected to Seismic Forces – Code of Practice, BIS, 1993.

Iyer, T.S.R., Marine Deposits, Proc. 5th Asian Regional Conference, Vol. 1, pp. 35–69, 1975.

John, C.G.S and Johnson, Steel Structures, Design and Behaviour, Hamper and Row Publishers, 2nd Edition, 1980.

Komar, A., Building Materials and Components, MIR Publishers, Moscow, 1994.

Leonards, G. A., Engineering Properties of Soils, In: Foundation Engineering, edited by G. A. Leonards, McGraw Hill, New York, pp. 66–240, 1962.

Leslie, J. R. and Cheesman, W. J., An Ultrasonic Method of Studying Deterioration and Cracking in Concrete Structures. ACI Journal, 1949.

Mahesh Varma, Construction Equipment and Its planning and Application, Metropolitan Book Co.(P) Ltd. New Delhi, 1979.

Mitchell, J. K., Inplace Treatment of Foundation Soils, ASCE, JSMFD, Vol. 96, pp. 73–110, 1970.

National Building Code of India, Indian Standard Institution, New Delhi, 1983.

Palanichamy, M. S., Basic Civil Engineering, 3rd Edn., Tata McGraw Hill Publishing Co. Ltd., New Delhi, 2007.

Parbin Singh, A Text Book of Engineering and General Geology, Katsan Publishing House, Ludhiana, 2009.

Pillai, S. V. and Menon, D., Reinforced Concrete Design, Third Edition, Tata McGraw Hill Education Private Ltd, New Delhi, 2012.

Popovin, S., Concrete Materials, Properties, Specifications and Testing, Standard Publishers Distributors, Delhi, 2002.

Purushothama Raj, P., Construction Techniques, Equipment and Practices, Sri Krishna Hitech Publishing Company Pvt. Ltd., Chennai, 2015.

Purushothama Raj, P, Basic Civil Engineering, 3rd Edn., Dhanam Publications, Chennai, 2001.

Purushothama Raj, P, Soil Mechanics and Foundation Engineering, Pearson Education, New Delhi, 2nd Edn, 2013.

Purushothama Raj, P, Ground Improvement Techniques, Laxmi Publications (P) Ltd., New Delhi, 2000.

Rangwala, S. C. et. al., Engineering Materials (Material Science), Charitor Publishing House Pub. Ltd., Anand, 2012.

Rangawala, K. S., P. S. Rangalwala, and K. S. Rangwala, Building Construction, Charotar Publishing House, Anand, 1996.

Rajput, R. K., Engineering Materials, S. Chand & Company Ltd., New Delhi, 2009.

SEAOC, Recommended Lateral Force Requirements and Commentary, Seismology Committee, Structural Engineers' Association of California, San Francisco, 1980.

Sharma, S. K., A Text Book of Building Construction, S. Chand and Company (Pvt) Ltd., New Delhi, 1988.

Sharma, S. K., and B. K. Kaul, Building Construction, S. Chand and Company (Pvt) Ltd., 1990.

Shetty, M.S., Concrete Technology – Theory and Practice, S. Chand and Co, New Delhi, 2008.

SP 20, Handbook for Masonry Design and Construction, BIS, 1991.

SP 23, Handbook for Design of Concrete Mixer, BIS, 11982.

SP 35, Handbook for Water Supply and Drainage, BIS, 1987.

SP 41, Handbook on Functional Requirements of Buildings (Other than Industrial Buildings), BIS, 1987.

SP 57, Handbook on Pipes and Fittings for Drinking Water Supply, BIS, 1993.

SP 62, Handbook for Building Construction Practices (Excluding Electrical Works).

Sushil Kumar, Building Construction, Standard Publishers Distributors, 10th Edn., Delhi, 1984.

Syal, I.C. and Goel, A.K., Reinforced Concrete Structures, A.H. Wheelers & Co. Pvt. Ltd., New Delhi, 1994.

Tomlinsan, M. J., Foundation Design and Construction, 5th ed., English Language Book Society, Essex, 842 pp. 2009.

Varghese, P. C., Building Materials, PHI Learning Private Ltd, New Delhi, 2010.

Varghese, P. C., Building Construction, PHI Learning Private Ltd., New Delhi, 2012.

Vazirani V. N., and S. P. Chandola, Civil Engineering Handbook, Vol.1, Khanna Publishers, New Delhi, 1989.

Vidal, H., The Principles of Reinforced Earth, Highway Research Record, No. 282, pp. 1–16, 1969.

INDEX

A

B

C

D

E

H

I

J

K

L

M

N

O

P

T

U

V

W